Approximate Methods for Weapon Aerodynamics

Approximate Methods for Weapon Aerodynamics

Frank G. Moore

ISBN 1-56347-399-2

Table of Contents

Preface

During my career, I have had the fortunate opportunity to spend over 30 years in weapon aerodynamics. A large portion of this time was spent in developing approximate analytical methods to compute aerodynamics with reasonable accuracy, cost, and time. These methods, many of which were new or extensions of existing methods, were integrated to a computer code that is known as the aeroprediction code. The first version of this code was developed in 1972 and was limited to body-alone aerodynamics at low angle of attack (AOA) and Mach numbers up to 3. It was the first total force and moment code, to the author's knowledge, at that time. Its ability to accurately predict axial force on unguided projectiles was its most outstanding feature.

As a result of the success of the 1972 version of the aeroprediction code (AP72) additional requests from various agencies were made for extensions to the AP72, and my sponsors were supportive of funding these requests. Hence, fins were added to the code capability in 1974 (AP74). Dynamic derivatives were added in the 1977 version (AP77) and the Mach number capability was extended from 3 to 8 in 1981 (AP81). Real gas effects were added, along with aeroheating information in 1993 (AP93). The Mach number range was extended from 8 to 20 and, for the first time, nonlinear aerodynamics were included empirically to analytical low AOA methods. Angles of attack to 30 deg were allowed. The AOA limit was extended to 90 deg in 1995 (AP95) and, for the first time, the code was downloaded to a personal computer with an interactive pre- and postprocessing software package. The most recent version of the code released to the public is the AP98. It adds significant new capability to the AP95. The new technology developed includes improvements to the axial force at AOA; nonaxisymmetric body capability; aerodynamics in the roll orientation of 45 deg (fins in cross or "x" orientation) in addition to roll attitude of 0 deg (fins in plus or "+" orientation); and, finally, all nonlinear loads are distributed over the body and lifting surfaces for use by structural engineers. Even as this book is being written, new technology is being developed to keep the code current with new and emerging requirements from the tactical weapons community.

While a history of the aeroprediction code is useful to set the stage for why this book is being written, it is not the purpose of the book to in any way "espouse" or "market" this code. Rather, the purposes of the book are severalfold. First, it is my desire to bring together in a single document innovative aerodynamic methods I have developed or found the most useful during my career. These methods are spread out in approximately 75 references over a 30-year career and, in some cases, are not easily accessible by beginning engineers. I have had significant encouragement during the last three years by many of my peers to bring these methods together in a single document. Second, I hope to provide a book that is useful to practicing engineers in terms of useful formulas, tables, and figures to allow rapid computation of aerodynamics. In so doing, I hope to always indicate the assumptions leading up to approximations. Third, I hope to bridge the gap between the academic textbook and practical application. This was one of my major difficulties when I started as a practicing engineer. I had all the fluid mechanics and computational fluid dynamics (CFD) coursework offered at my Alma Mata, Virginia Polytechnic Institute and State University (Blacksburg, Virginia). However, all of the examples that we considered were generally simple configurations (cones, wedges, etc.) that fit nicely into theoretical methods. When I went to work, there were practical shapes, not clean configurations, to which I had to apply theories. This is where many of my engineering approximations were made.

I hope this book will prove useful to the practicing engineer as well as the academic community. It could prove useful to the practicing engineer by comparison of CFD and approximate methods and as a quick reference guide for approximate aerodynamic methods in general. It could prove useful in the classroom in a missile design course at the undergraduate or graduate level. At the graduate level, the book could be used as a practical guide and students could be asked to derive many of the formulas from first principles, which are given in the book.

I have many acknowledgments to make. First, I would like to thank the good Lord for giving me the health and talents that have allowed me to do the work discussed in this book. Next, I am grateful to my deceased parents, who instilled in me a strong work ethic. Being one of ten children was also a blessing, particularly now when we get together at our family reunions. I would also like to thank my two best friends: my wife, Linda, who has been a constant source of encouragement, and my daughter, Wendy, who has traveled extensively with me. I am also thankful for my many teachers and coaches throughout my career. I would like to single out for special recognition Professor Fred DeJarnette, who was my advisor for both my M.S. and Ph.D. degrees. He has continued to be a source of technical support and help throughout my career. In fact, Professor DeJarnette was the person I asked to do the technical editing of this book.

While many of the ideas and methods given in the book were the author's, there have been many others who worked with me during my career. Without their assistance, I could not have been nearly as productive. These individuals include Dr. Roy McInville, Mr. Tom Hymer, Dr. David Robinson, Mr. Mike Armistead, Mr. Steve Rowles, Dr. Leroy Devan, Dr. John Sun, Mr. Larry Mason, Dr. Charlie Swanson, Mr. Gil Graff, and Mr. Bill McKerley. Special recognition is due Dr. Roy McInville and Mr. Tom Hymer, who have worked with me for 10 years on the last three versions of the aeroprediction code. In addition to these individuals, there were several contractors involved during the various stages of code development. These included Professor DeJarnette and several graduate students at North Carolina State University, Nielsen Engineering and Research, and Lockheed Martin. Appreciation to these individuals and contractors is also given.

I have many sponsors also to thank, because without the funding they provided and their confidence in me to deliver products, many of the methods discussed in this book would not be available. The first sponsor who took a risk to fund a new Ph.D. in 1971 to develop the AP72 was Mr. Lionel Pasiuk at the Naval Sea Systems Command (NAVSEA). He was joined by the late Mr. Bill Voltz in 1975 of the Naval Air Systems Command (NAVAIR). When technology responsibility was shifted from NAVSEA and NAVAIR to the Office of Naval Technology (now Office of Naval Research), Dave Siegel was the initial sponsor and, more recently, Mr. James Chew. At the Navy Laboratories where these technology funds are managed, Mr. Danny Brunson, Mr. Robin Staton, Mr. Gil Graff, Mr. Roger Horman, and Mr. John Fraysse have supported the effort at Naval Surface Warfare Center, Dahlgren Division (NSWCDD) and Dr. Craig Porter and Mr. Tom Loftus at Naval Air Warfare Center, China Lake (NAWCCL). In addition, within NSWCDD, Mr. Bob Steigler, Mr. Craig Melton, Mr. George Long, and Mr. Tim Ryan, among others, have provided supporting funds. Also, Mr. Ray Deep and Mr. Dave Washington at Army Missile Command in Huntsville provided a small level of support, as did Dr. Don Daniels of Eglin Air Force Base. I am grateful to all these sponsors for their support and confidence in me to do a good job for them.

Frank G. Moore
April 2000

Chapter 1

Introduction

Nomenclature

AP	= aeroprediction code (AP with a number behind it indicates the year of release of a given version, i.e., AP98)
3DOF, 5DOF, 6DOF	= 3, 5, or 6 degrees of freedom; 3 degrees of freedom normally refers to x, y, and θ; 5 degrees of freedom normally refers to x, y, z, θ, and ψ; and 6 degrees of freedom refers to x, y, z, φ, θ, and ψ.
a, b	= semimajor and semiminor axis, respectively, of an ellipse
h	= altitude
k	= parameter used to define corner radius for a square or triangle ($k = r_n/W_m$)
M_∞	= freestream Mach number
r	= local body radius
r_n	= radius of a corner of a triangle or square
x, y, z	= body fixed axis system with x along body axis, y out the right wing, and z up
W_m	= maximum diameter of a triangle or square as measured normal to the velocity vector
α	= angle of attack (degrees)
δ	= control deflection with positive being leading edge up
φ,θ,ψ	= roll, pitch, and yaw angles, respectively (degrees)

THE purpose of this chapter is to set the stage for the rest of the book. My career has been primarily devoted to developing simplified methods to compute aerodynamics for the range of conditions for which weapons fly. Simplified methods here are meant to define those theoretical and empirical techniques required to define a set of aerodynamics for a given configuration that are reasonably accurate, yet can be obtained in a matter of seconds on a personal computer. Aerodynamics are required throughout the design process of any flight vehicle. These aerodynamics are used for flight performance estimates, including range, maneuverability, miss distance, and stability analysis. In addition, they are used for structural analysis, including material requirements and selection; to determine structural member thicknesses required to withstand the loads; and as inputs for heat transfer or ablation analysis. Table 1.1 gives a summary of some of the uses for aerodynamics. Generally, an interactive design process occurs between the aerodynamicist, the structural designer, and the flight dynamicist to arrive at a configuration that meets some set of desired launcher constraints and performance requirements given a warhead and possibly a guidance system. Hence, the first part of this chapter deals with weapon system requirements from an aerodynamic standpoint. Next, the approaches to compute aerodynamics and aeroheating over the configuration and flight regimes of interest will be discussed. This will be followed by a tradeoff of the various aerodynamic methods in terms of cost, time, and educational requirements. Finally, an outline of the remainder of the book will be discussed, highlighting the various theories and empirical techniques that will be outlined in the remaining chapters of the book.

I. Weapon System Aerodynamic Requirements

Weapons include unguided projectiles, which are usually spin-stabilized; unguided rockets; guided projectiles, and missiles. In most cases, stability is provided by aerodynamic surfaces (fins). However, in some cases, side jets (thrust) are used in addition to, or in lieu of, fins to reduce the airframe time constant (the time it takes for the airframe to attain about 63% of its required maneuver level). At very high altitudes, where the air density is very low, stability and control are completely provided by side jets. This book will focus on aerodynamic requirements for spin- or fin-stabilized weapons. However, a discussion of stabilization by thrust will be given at the end of the book in terms of future technology needs.

Weapon system aerodynamic requirements are discussed in terms of Mach number, angle of attack (AOA), control deflection, altitude, roll orientation, and configuration geometry. These variables are the parameters that drive the requirements for aerodynamic methods. Angle of attack requirements are driven primarily by the launcher and the end game maneuver requirement to successfully engage a target. Many surface-launched

Table 1.1 What aerodynamics are used for

Flight dynamics	Structures
Range computation	Loads (pressure)
Engagement of target and miss distance	Aeroheating (inputs to heat transfer codes)
Maneuverability estimates	
Any trajectory analysis (3, 5, and 6 degrees-of-freedom)	Ablation analysis inputs

missiles are launched vertically and can have high crosswinds at launch. This means that AOA prediction capability approaching 90 deg is required at low Mach numbers. At transonic Mach numbers ($0.5 \leq M_\infty \leq 1.5$), the missile can undergo fairly high AOAs in both the surface- or air-launched mode due to pitchover requirements for minimum engagement range. Finally, in the terminal phase of engagement, AOAs approaching 40 deg at moderate supersonic Mach numbers can be expected. Figure 1.1 is a qualitative representation of AOA boundaries that aerodynamic methods need to operate. It should be pointed out that, when the missile is at very high AOA, it is still under thrust control, even though fins may be extended. On the other hand, when the missile is in the end game (last few seconds before target intercept), it may or may not have thrust available to assist in the control. Mach numbers can be as low as zero when a missile is launched from a ship launcher or as high as 12 to 15 if the missile is engaging a tactical ballistic missile. These higher Mach numbers are generally attained only at higher altitudes where the aerodynamic heating is the lowest. However, aerodynamic heating analysis is considered in this text, focused primarily on hypersonic Mach numbers. Figure 1.2 is a qualitative operational boundary of altitude versus Mach number for most weapons.

Another flight parameter of interest is roll orientation. Some weapons are flown in a roll controlled orientation of $\varphi = 0$ (fins in plus or "+" roll orientation) and some weapons are flown in the $\varphi = 45$-deg roll mode (fins in cross or "×" roll orientation). Some weapons are flown with an intentional low roll rate. In a general maneuver toward a target, one can have combined AOA and sideslip. This type of maneuver generates out-of-the-pitch-plane aerodynam-

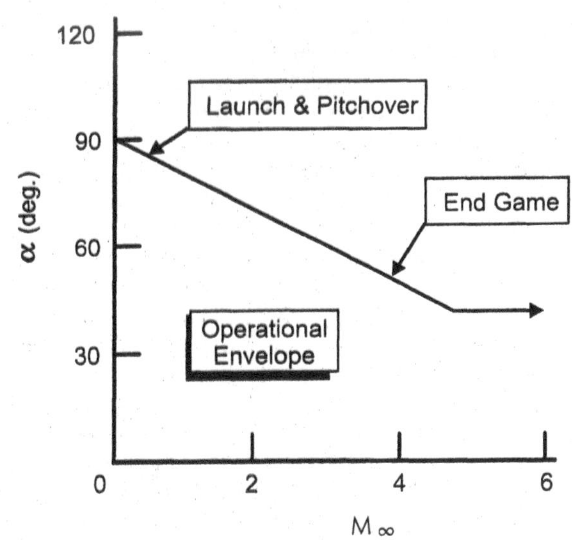

Fig. 1.1 Desired operational α–M envelope for aerodynamic methods.

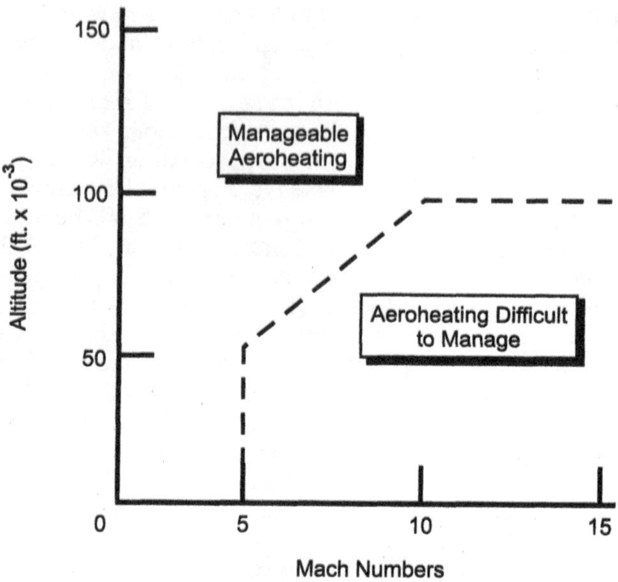

Fig. 1.2 Qualitative operational *h–M* envelope for most weapons.

ics. On the other hand, if one assumes maneuvers in the pitch plane only, then planar aerodynamics can be used and can be computed with acceptable accuracy. Acceptable accuracy is here being defined for engineering analysis in the preliminary or conceptual design phase of a weapon system. During this phase, average accuracy levels of $\pm 10\%$ for normal and axial force coefficients and $\pm 4\%$ of the body length for center of pressure have proved adequate. Accurate out-of-the-pitch-plane [or six-degree-of-freedom (6DOF)] aerodynamics will require more sophisticated aerodynamic methods than accurate planar aerodynamics.

The last flight parameter of interest is control deflection δ. Most fins are deflected to a maximum of ± 20 to ± 30 deg due to hinge moment constraints but some can be deflected as high as ± 50 deg at supersonic speeds. These control deflections can be made by canards, wings, or tail surfaces. In reality, the largest deflections are generally made by smaller control surfaces due to the required size of the control hardware within the weapon.

The last weapon system requirement that affects the aerodynamics and hence the aerodynamic methods selected is the configuration geometry. Typical geometry requirements are somewhat dependent on the weapon. Unguided projectiles typically have a truncated nose, due to the fuze design, with a slope discontinuity along the nose where the fuze joins the ogive. Many of the projectiles have boattails and rotating bands to provide spinup. Guided projectiles typically have either sharp or spherically blunt noses, depending on the type of guidance used. Most guided projectiles have two sets of lifting surfaces. The forward set is used for control and the rear set is used for stability. These lifting surfaces are typically fairly thick to withstand high acceleration forces at launch and typically have blunt leading and trailing edges. Rockets tend to be lower drag designs with tails designed for stability. Guided missiles are quite similar to guided projectiles, except the fins tend to have lower drag and, in most cases, lower aspect ratio. Figure 1.3 is representative of the axisymmetric body weapon cross sections prevalent in the world today.

In the last several years, increased interest has been shown in weapons that have nonaxisymmetric bodies. This increased interest has come primarily from a desire to increase range or improve volumetric efficiency for shipboard or aircraft launch systems. As a result, body cross-sectional shape tradeoff studies are performed to examine the improvements afforded by various nonaxisymmetric body alternatives. Figure 1.4 shows a variety of cross-sectional shapes of interest to weapon designers. These shapes can be

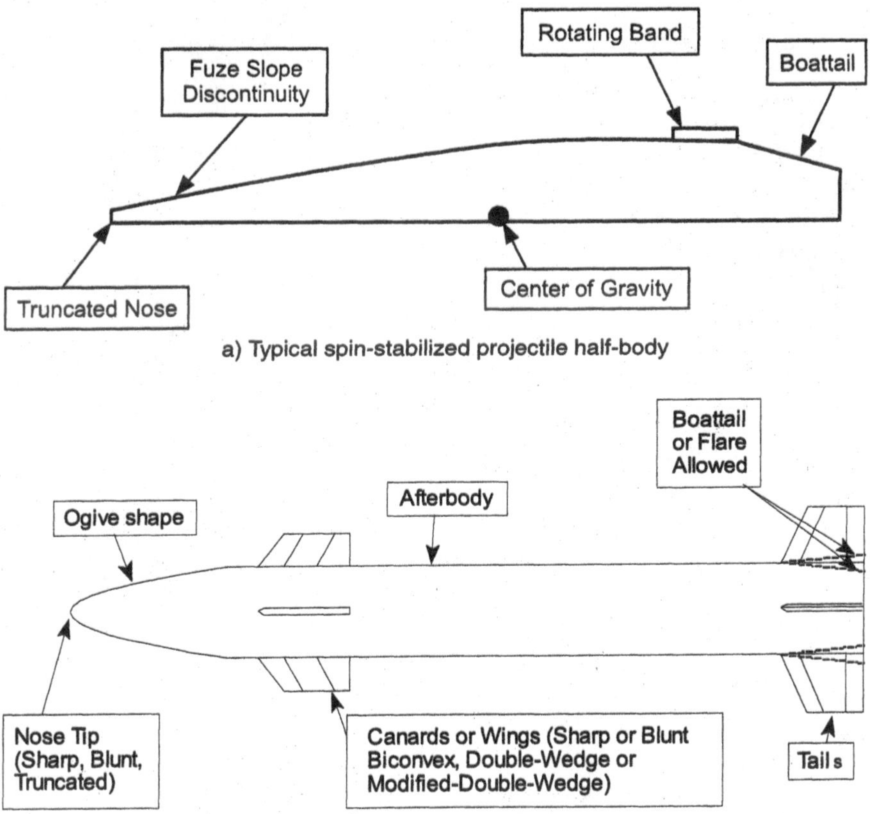

a) Typical spin-stabilized projectile half-body

b) Typical guided projectile or missile

Fig. 1.3 Typical weapon geometries.

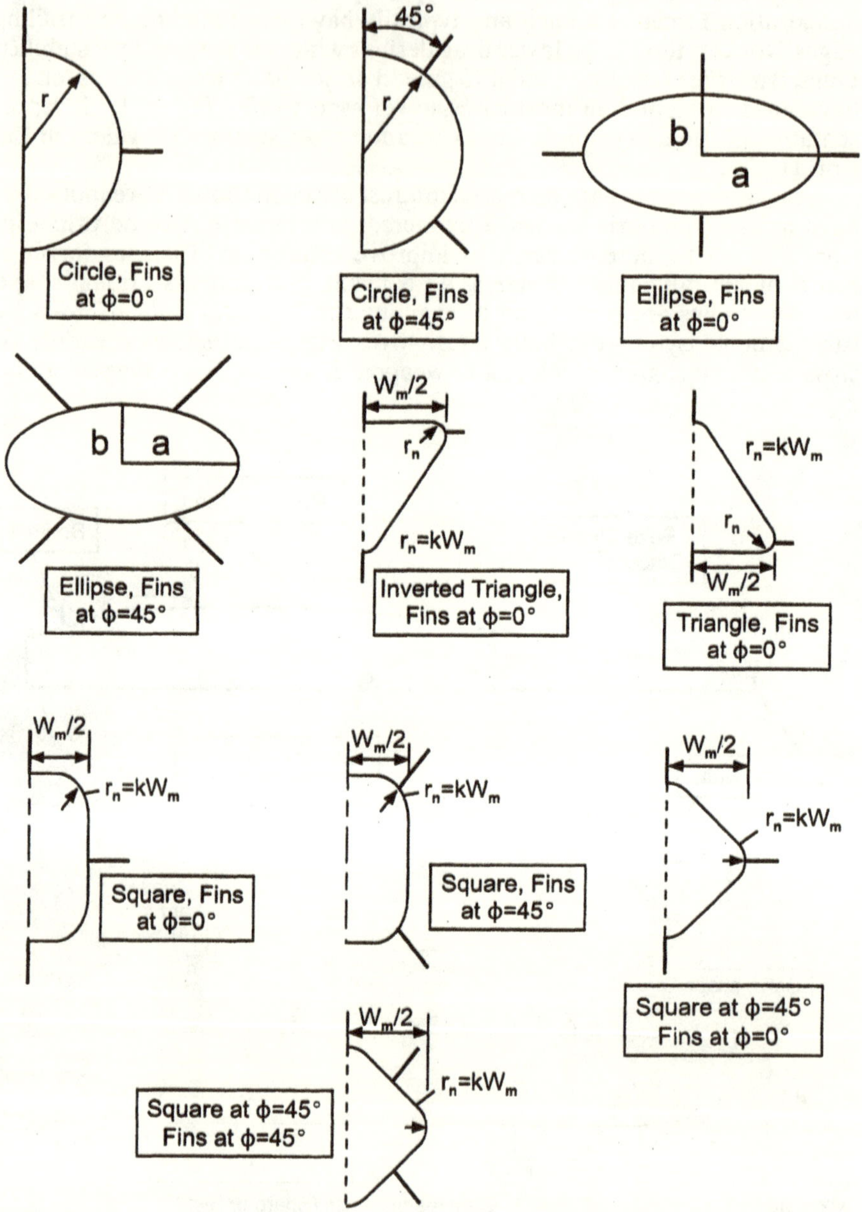

Fig. 1.4 Noncircular cross-sectional, wing–body configurations of interest for weapon trade studies.

Table 1.2 Summary of weapon flight and geometry requirements

Mach number	0–15
Angle of attack	0–90 deg
Control deflection	±30 deg
Roll orientation	0 deg, 45 deg for 3DOF aerodynamics; any roll for 6DOF aerodynamics
Sets of fins	zero, one, two required; three desired
Number of fins in a set	two, four, six, or eight are typical
Body geometry	Axisymmetric required; nonaxisymmetric treatment desired

made quite general by varying the corner radius r_n for the square and triangular cross sections and the ellipticity ratio a/b for elliptical cross sections.

Another configuration geometry consideration has to do with the number of fins used for stability and control. This is because many weapon designs are constrained by their launcher. Launchers such as guns or shoulder-launched configurations tend to be circular in shape. This circular shape puts a constraint on the weapon when it is fin-stabilized versus spin-stabilized. The typical results of these constraints are that, to get adequate stability, a four-fin configuration may need large spans to get adequate lifting surface area on the tail fins. This situation exists because the maximum chord length of a single fin is equal to one-half the body diameter in many cases of practical interest. The large span fins can have an adverse effect on the rest of the weapon design in terms of reducing either the rocket motor length, the warhead size, or both because the fins are either folded forward and into the projectile or rocket or folded rearward. In either case, the fins pop up or out after exit from the launcher to provide the static and dynamic stability required for successful flight.

In addition to effects on the weapon design, it could be desirable from an aircraft or trainable surface launcher standpoint to have a weapon designed with six or eight shorter span fins versus cruciform fins with larger spans. One option to reduce the span of a four-fin projectile or missile, and hence to reduce the effect on the rocket motor, warhead, or launch vehicle, is to increase the number of tail fins from four to six or eight. This design alternative could prove attractive when the fins are folded rearward and pop up after launch. In such a case, one could obtain the same level of static stability with a shorter length round, or the extra length could be used for additional rocket motor or warhead or other equipment.

Table 1.2 gives a top-level summary of the flight and geometrical requirements that must be considered when choosing theoretical or empirical methods to compute aerodynamics.

II. Uses of and Methods to Obtain Aerodynamics

Before discussing the uses and various general approaches to obtain aerodynamics based on the requirements of Table 1.2, it is believed prudent to discuss the physics involved in the Table 1.2 requirements. Figure 1.5 attempts to describe weapon aerodynamics in a generic way as a function of the AOA and Mach number regions desired from an operational standpoint shown in Fig. 1.1. At low AOA, the aerodynamics are mostly linear (except for transonic flow), and linearized and slender-body aerodynamic methods prove very effective in providing acceptable accuracy for approximate aerodynamic prediction codes.[1,2] As AOA increases above 10 deg, the nonlinearities associated with AOA become increasingly important at all Mach numbers, and the nonlinear compressibility effects become important above $M_\infty \approx 0.6$. Linearized codes can still be used in the range 10 deg $< a <$ 20 deg, but with degraded accuracy. As AOA increases above 20 deg, the flow becomes strongly nonlinear but still reasonably well behaved (in most cases) in terms of static aerodynamics as a function of AOA. At these AOAs nonlinear aerodynamic methods are required for reasonable accuracy. For approximate aerodynamic codes, these nonlinear methods are typically empirical or semi-empirical.[3,4,5] For numerical methods,[6,7] the nonlinearities

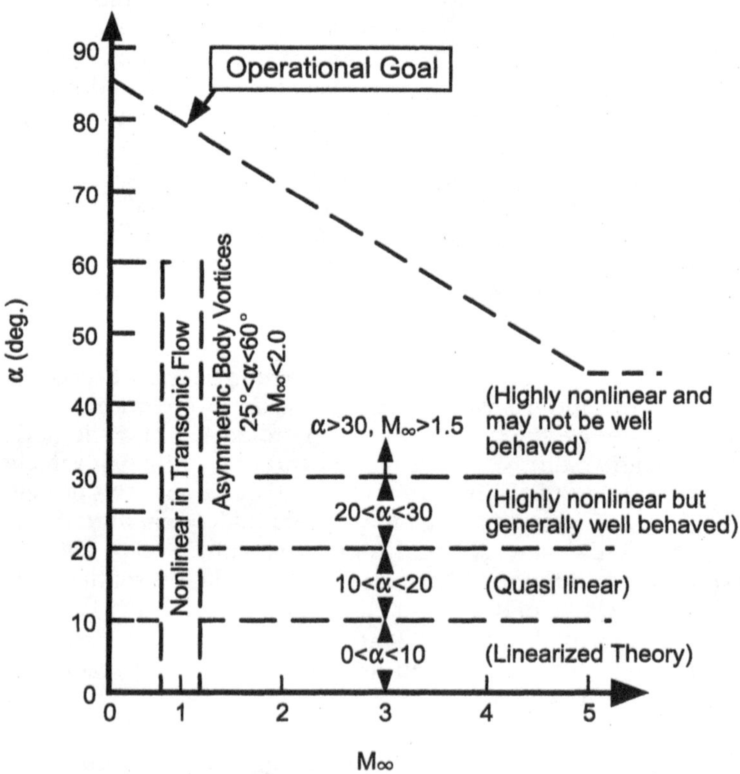

Fig. 1.5 Qualitative classification of weapon aerodynamics as a function of Mach number and AOA.

are accounted for naturally as a part of the overall solution. Of course, wind tunnel or flight tests also measure the nonlinearities and are generally viewed as the truth model by which all theoretical predictions are assessed.

Above α of about 25 deg and below M_∞ of about 2, another nonlinear phenomenon occurs on many configurations because of the asymmetric shedding of body vortices. This phenomenon is the strongest in the subsonic Mach number range and starts to dissipate at crossflow Mach numbers of about 0.5.[8] While the phenomenon is reasonably well understood, it is predictable in the general sense only in terms of upper bounds on side forces created by the shedding. Forward placed fins or control surfaces tend to minimize the effect.

Above α of about 30 deg and $M_\infty \approx 2.0$, strong shock interactions can cause very strong effects on the aerodynamics. These effects exhibit themselves in the form of loss of static stability and normal force. Typical shocks of interest are the bow shock intersecting a control surface or a forward lifting surface shock intersecting a tail surface. The latter problem appears to be more critical because the moment arm to the tail from the center of gravity is generally longer than to the wing or canard. Aerodynamics in this region may not be well behaved in terms of either monotonically increasing or decreasing.

Before 1971, the tactical weapons aerodynamicist could do one of three things to obtain aerodynamics. The individual could perform flight tests of a full-scale configuration; design, build, and test a wind tunnel model over the flight range of interest; or finally, utilize existing handbooks, wind tunnel data reports, and theoretical analysis to estimate empirically the aerodynamics of a given configuration.

The first two approaches were often more costly, time consuming, and accurate than needed in the preliminary design stages, whereas the latter approach was more time consuming than desired but also had no general accuracy assessment.

A fourth alternative (which did not exist before 1971), to compute aerodynamics on a complete configuration over the Mach number and AOA range of interest, is to have a general computer program to perform such a task. There are three alternative theoretical approaches to develop such a code (see Table 1.3). The first of these is solution of the full Navier–Stokes

Table 1.3 How we get aerodynamic data

1. Wind tunnel, free flight data, ballistic range
2. Empirical estimates: wind tunnel reports, handbooks, experience
3. Aeroprediction codes
   - A. Navier–Stokes: continuum flow
   - B. Euler equations + boundary-layer: inviscid outer layer + thin viscous layer near surface + some empirical techniques
   - C. Approximations to Euler and boundary-layer equations + empirical techniques

equations. The only assumptions associated with this set of equations is continuum flow (that is, the flowfield region is not sparsely populated with air molecules such as at altitudes greater than about 200,000 to 250,000 ft) and the transition and turbulence models selected. A second theoretical alternative is to assume the viscous flow region lies in a thin layer near the body and thus solution of the equations can be reduced to that of an inviscid flowfield plus a thin boundary layer near the surface. This, combined with empirical estimates of base drag and other protuberance aerodynamics, gives a complete set of aerodynamics for the configuration of interest. A third theoretical alternative is to assume that the body perturbs the flowfield only slightly and then to make appropriate approximations to the Euler and boundary-layer equations. These approximate theories are then combined with other theoretical approaches and empirical data for the complete aerodynamics code.

The three types of aeroprediction codes of Table 1.3 actually involve many subclassifications. For example, there are several levels of Navier–Stokes solutions [full Navier–Stokes (FNS), thin-layer Navier–Stokes (TLNS), parabolized Navier–Stokes (PNS)], Euler solutions with and without viscous effects, and several levels of approximate solutions. Table 1.4 expands on Table 1.3 and gives the assumptions associated with each theoretical approach. Most of the remainder of the book will cover these methods in more detail, with emphasis on numbers 7, 8, and 9 of Table 1.4.

III. Tradeoffs in Methods Selection

The approach or method selected to obtain aerodynamics from Table 1.3 is dependent on several things. First and foremost is how long and how much funding the engineer is given by a program or project manager to derive a set of aerodynamics. Second, for what are the aerodynamics to be used (Table 1.2)? Third, what is the stage of the design process? Table 1.5 illustrates this last question in terms of three design stages: missile synthesis, preliminary design, and detail design. Also shown in Table 1.5 are the typical aerocode design requirements, typical trade studies, and aerodynamic uses. The important point to gather from the table is that, as one gets farther and farther along in the design process, more accurate aerodynamics are needed.

Another tradeoff in the methods selection is the educational level of the user of the code. Table 1.6 lists the codes in current or recent use at Naval Surface Warfare Center, Dahlgren Division (NSWCDD) and the educational level of the user. Also shown in the table are typical learning times, setup times, computational times, and type of computer required.

One way to try to compare the level of sophistication versus accuracy and the cost of various codes is through the examination of the total cost to obtain a set of aerodynamics. To do this, Table 1.6, which compares educational, computer, and computational time requirements of various aeroprediction codes in use at NSWCDD, is quite useful. Referring to Table 1.6, the level of sophistication increases in going from top to bottom of the table. For example, the Missile III, Aeroprediction versions 95 and 98, HABP, and Missile DATCOM are all semiempirical codes. SWINT/ZEUS, ZEUS^{++},

Table 1.4 Assumptions of flowfield equations

1. Full Navier–Stokes (high AOA)
   A. Continuum flow
   B. Turbulence model and location of transition
2. Thin-layer Navier–Stokes (moderate separation)
   A. Neglect streamwise and circumferential gradients of stress terms
   B. Turbulence model and location of transition
   C. Continuum flow
3. Parabolized Navier–Stokes (small separation)
   A. Steady state
   B. Neglects streamwise viscous gradient
   C. Approximate streamwise pressure gradient in subsonic portion of flow near surface
   D. Turbulence model and location of transition
   E. Continuum flow
4. Euler equations + boundary-layer (small separation)
   A. Viscous region confined to thin region near body surface
   B. Large Reynold's number
   C. Neglect streamwise gradients of stress terms
   D. Neglect normal pressure gradient
   E. Turbulence model and location of transition
   F. Continuum flow
5. Euler equations
   A. Neglect all viscous terms
   B. Continuum flow
6. Full potential equations
   A. Neglect all viscous terms
   B. Flow is isentropic (no shock waves)
   C. Continuum Flow
7. Linearized Potential Equations
   A. Neglect all Viscous Terms
   B. Flow is isentropic (no shock waves)
   C. Body creates small disturbances in flowfield
   D. Continuum flow
8. Theoretical approximations
   A. Certain other simplifications to Euler, potential equations, or boundary-layer equations
   B. Continuum flow
9. Empirical data base
   A. Data base covers vehicles and flight regime of interest
   B. Enough data are available for good interpolations

Table 1.5 Aerodynamic code requirements and uses in various missile design stages

Design stage	Aerocode design requirements	Trade studies (typical)	Aerodynamics uses
Missile synthesis	Robustness Ease of use Minimal input parameters Extremely fast computationally 25% accuracy	Engine types Warhead types Material requirements Typical weights Guidance types Airframe control type	Range Maneuverability Response time
Missile preliminary design	Blend of robustness, ease of use, and accuracy Fast computationally 10% accuracy	Structural layout (material, thickness, etc.) Aeroshape vs engineering design, guidance, size, etc. Hot vs cold structure	Range Maneuverability Miss distance (3DOF) Structural design
Detailed design and problem solving (or analysis codes)	Accuracy (<5%) Computationally affordable User friendliness and robustness still important	Detailed structural design, including material selection Investigating critical problem areas	Range Maneuverability Miss distance (6DOF) Structural design

CFL3DE, and GASP are all numerical codes. The Aeroprediction 95/98, SWINT/ZEUS, and ZEUS++ were all developed at NSWCDD. The Missile III was developed by Nielsen Engineering and Research (NEAR);[9] HABP[10] and Missile DATCOM[11] by Wright Patterson Air Force Base and McDonnell–Douglas of St. Louis; and the Navier–Stokes codes were developed jointly by NASA Langley Research Center and Virginia Tech.

Included in Table 1.6 is the time required to learn how to use the code, the setup time for a typical geometry, and the computer time for one case referenced to the same computer (CRAY YMP). Also shown are other criteria, including typical educational level of the user and size of the computer required. To get the total cost of using a code, it is necessary to add the staff setup time to the computer cost and prorate the training time over some nominal expected usage. Experience has shown that most project and program managers are willing to pay the costs of SWINT/ZEUS type codes and any above that in Table 1.6. However, the cost and requirements of the full Navier–Stokes codes must come down substantially before they will be

Table 1.6 Educational and time requirements for aeroprediction codes in recent use at NSWCDD

Code	Typical user level of education	Typical time required to learn to use code	Setup time (one case)	Time to compute one case, same computer	Computer required
Aeroprediction 95[4] and 98[5]	Coop B.S. M.S. Ph.D.	≈ 1 wk	<1 h	<1 s	Personal computer
Missile III[9]	Coop B.S. M.S. Ph.D.	≈ 1 wk	<1 day	<1 s	Personal computer
HABP[10]	B.S. M.S. Ph.D.	≈ 2 wk	<1 wk	<1 s	Micro Vax
Missile[11] DATCOM	B.S. M.S. Ph.D.	≈ 2 wk	<1 wk	<1 s	Micro Vax
SWINT[12]/ZEUS[6]	M.S. Ph.D.	≈ 1 mo	<1 mo	1–3 min	Vax CDC Super Mini
ZEUS++[13]	B.S. M.S. Ph.D.	≈ 1 wk	<1 day	3–15 min	PC
Navier–Stokes (CFL3DE,[14] and GASP[7])	Ph.D. Some M.S.	≈ mo-yr	≈ day-mo	≈ h-mo	Cray or Super Mini

used on a routine basis for design. This means much additional research and advancements in computer speed are still needed in this area.

To illustrate this point, a particular example was chosen for cost comparison. The example is to develop a set of trim aerodynamics on a typical missile configuration to be used as an input to a 3DOF flight simulation model. This example is quite typical of what an empirical or semi-empirical code would be used for. By definition, trim is that combination of AOAs (α) and control deflections (δ) that give zero pitching moment about the vehicle center of gravity.[15] To determine the (α, δ) map as a function of Mach number, one must compute the static aerodynamics over enough α, δ, M_∞ conditions so that the flight envelope will be covered. Also, it will be assumed that the missile is a surface-launched, tail-controlled, cruciform fin configuration that has a Mach range of 0 to 4, AOA range 0 to 30 deg, control deflection of 0 to 20 deg, and attitude 0 to 80,000 ft. These conditions are reasonable for many of the world's missiles. To cover the flight envelope, 7

Mach numbers, 5 AOAs, and 5 control deflections are assumed. This gives a total of $7 \times 5 \times 5 = 175$ cases. Furthermore, skin friction varies with attitude so 5 altitudes will be chosen to define the zero lift change in skin friction with altitude, giving a total of 180 cases for which aerodynamics are to be computed on a single configuration.

Before costs of each computer code can be made for this particular example, some assumptions must be made. These assumptions are given in Table 1.7 and are based on NSWCDD experience in using various aeroprediction codes. The cost to perform the set of trim aerodynamics calculations using these codes is shown in Fig. 1.6. It should be noted that for the PNS and Euler solutions, one-half the cases are assumed to have supersonic flow throughout and one-half are assumed to have subsonic flow so an iterative solution is required.

Several points are worthy of note in analyzing Fig. 1.6. First, all methods of computation are below the cost of experiment. This is in contrast to the same calculation that was performed about five years ago. At that time, both FNS and TLNS were above the cost of wind tunnel tests. Since that time, computer costs have come down by a factor of about 25, while wind tunnel costs have stayed the same or increased. Second, many people have access to free computer time. This has a major cost reduction for FNS and TLNS computations. The FNS costs drop from $84,000 to $30,000, for example, if free computer time is available to the user.

Another point is that in both setup and postprocessing time, an experienced user is assumed, and no problems occur with Computational fluid dynamics (CFD) methods. Past experience at NSWCDD has shown this assumption to be quite optimistic. Some of the problems that occur are in

Table 1.7 Assumptions in cost estimates to compute set of trim aerodynamics with various levels of aeroprediction codes (180 cases)

Code	Setup time	Postprocessing time	Computer time
FNS	4 wk	6 wk	$15 \times 180 = 2700$ h
TLNS	4 wk	6 wk	$12 \times 180 = 2160$ h
PNS	3 wk	6 wk	$0.15 \times 90 + 1.5 \times 90 = 148.5$ h
Euler + BL + BD*	1 wk	2 wk	$0.05 \times 90 + 0.5 \times 90 = 49.5$ h
AP98	0.3 h	1 day	0.01 h

Estimated cost: Cray computer at $20/h for Navier–Stokes
Engineer time = 130,000/work year
Engineer assumed to be experienced on use of codes so no training time is included
Need enough resolution in grid size to predict skin friction drag
Wind tunnel includes model and test cost

*Euler + boundary layer + base drag.

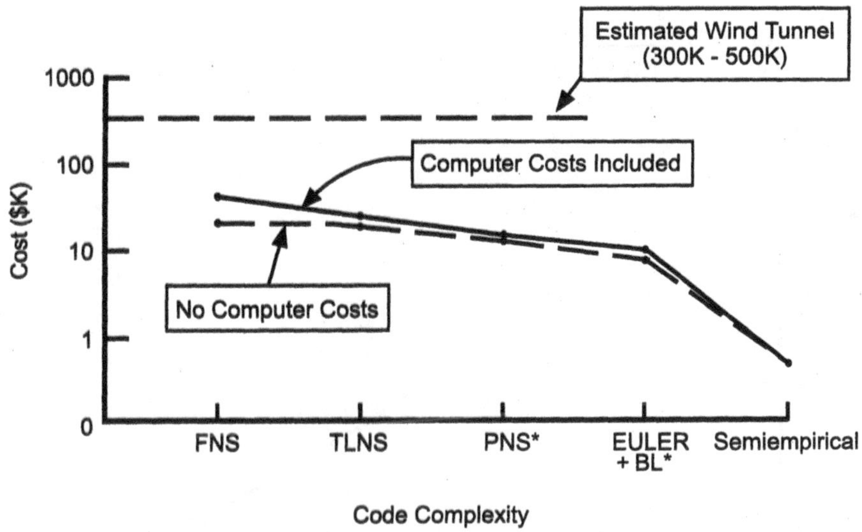

Fig. 1.6 Estimated cost to obtain set of trim aerodynamics (180 cases) (cost on log scale) (BL = boundary layer).

generating the approximate grid to get accurate results, accessing the computer on a priority basis, and numerical difficulties with a code, just to name a few. A factor of two or three can easily be used for the cost and time of the Navier–Stokes methods, over the most optimistic assumptions used in Table 1.7.

A second way of comparing aerodynamic computations is the total time it takes to get the complete set of computations performed. These results are estimated, again based on NSWCDD experience, as shown in Fig. 1.7. For most development programs, semi-empirical codes obviously have the most desirable turnaround time and cost. However, full Euler computational costs and turnaround time have now come down to the point where they can be used for many design issues. This is, in part, due to increased computer speed and, in part, the preprocessor developed for the ZEUS code (ZEUS++[14]). If the configuration is complex, then setup time will be longer than allotted in Fig. 1.6 and Table 1.7, however. The Navier–Stokes, TLNS, and experimental costs are higher than many programs can afford. However, major programs can afford these costs. The combination of cost, accuracy, and complexity of the various means of computing aerodynamics has led most agencies to a mix of approaches. The most used codes still remain the semi-empirical codes, with Euler plus boundary-layer becoming more and more prevalent as the robustness and ease of use improves. Navier–Stokes and TLNS are used for specialized problems or a few validation cases of other codes; much work is still needed to improve user friendliness for this class of codes. Wind tunnel data still remain the most reliable but time-consuming method to obtain aerodynamic data.

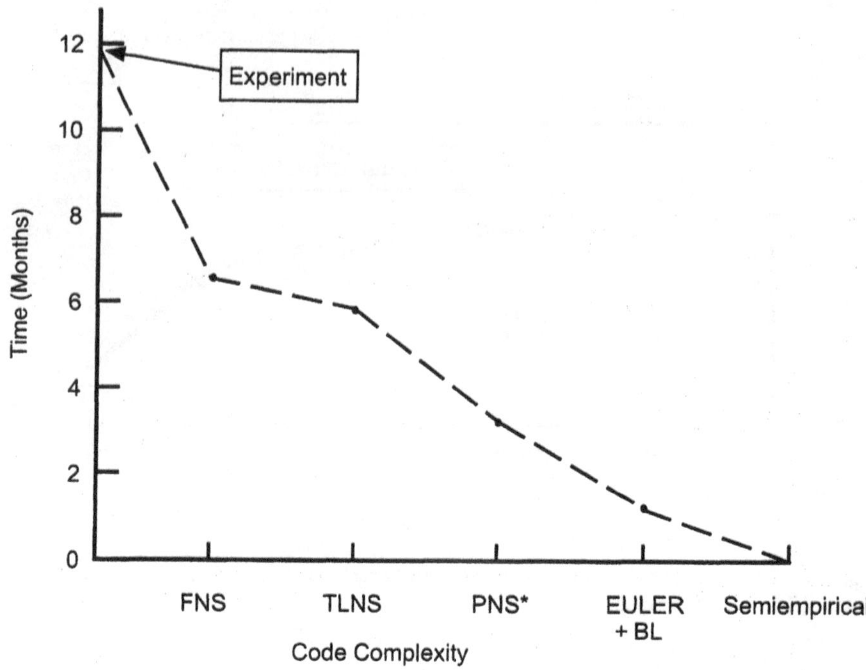

Fig. 1.7 Estimated time to generate set of trim aerodynamics (BL = boundary layer).

In concluding the section on methods selection, it may be useful to give a personal perspective on various approaches to obtaining aerodynamics of weapons. As mentioned already, when I first began my professional career in the late 1960s, the only alternatives available to obtain total weapon aerodynamics were either testing or the use of handbooks and other test data (empirically). This changed in the early 1970s, with the first version of the aeroprediction code. Today many semi-empirical codes are available in the literature, with the aeroprediction series,[1–5] Missile DATCOM[11], and NEAR[9] series being the most popular. To the author's knowledge, the SWINT[12] code developed in the late 1970s and early 1980s, was the first full Euler solver of a wing–body–tail configuration. Today many Euler solvers are available and, with the increase in computer power and improvements in pre- and postprocessing software, these codes can now be used readily in design (20 years after the first code was available). Today, there are many production codes available for full Navier–Stokes calculations (Refs. 7 and 14 are two examples). However, significant increases in computer power and improvements in pre- and postprocessing software are still needed before these codes can be readily used in the various phases of design. One can only say that their usage will surely increase with time. However, as Figs. 1.6 and 1.7 illustrate, the combination of reasonable accuracy, ease of use, and fast turnaround will keep semi-empirical codes as a part of the aerodynamicist's toolset for many years in the future.

IV. Book Outline

While this book is focused on approximate methods to compute aerodynamics, one always needs to keep in mind the assumptions that are being made to get to the physical model being used to calculate configuration aerodynamics. As such, Chapter 2 will briefly discuss the Navier–Stokes and Euler equations, with the objective of showing the assumptions made at each step of Table 1.4. Chapter 3 will then discuss those portions of perturbation and slender body theory that the author has found useful in applying to the low AOA linear region of Fig. 1.5. Chapter 4 will continue the approximate theory methods by focusing on local slope and a combination of local slope and perturbation methods. Chapter 5 will then address semi-empirical and empirical methods used to estimate nonlinearities that occur at moderate to high AOA, as shown in Fig. 1.5. Chapter 6 will address the aerodynamics of nonaxisymmetric bodies and Chapter 7 will discuss real gas effects that occur for hypersonic Mach numbers. Chapter 8 will include some important design considerations. Various example cases are included throughout Chapters 3 and 8 to illustrate the methods. Finally, Chapter 9 will discuss some possible future directions or aerodynamic prediction methods.

References

[1]Moore, F. G., "Body Alone Aerodynamics of Guided and Unguided Projectiles at Subsonic, Transonic and Supersonic Mach Numbers," Naval Weapons Lab., now Naval Surface Warfare Center, Dahlgren Div., Rept. NWL TR-2976, Dahlgren, VA, Nov. 1972.

[2]Moore, F. G., "Aerodynamics of Guided and Unguided Weapons: Part 1—Theory and Application," Naval Weapons Lab., now Naval Surface Warfare Center, Dahlgren, Div., Rept. NWL TR-3018 Dahlgren, VA, Dec. 1973.

[3]Moore, F. G., Hymer, T. G., and McInville, R. M., "Improved Aeroprediction Code: Part 1—Summary of New Methods and Comparison with Experiment," Naval Surface Warfare Center, Dahlgren Div., Rept. NSWCDD TR-93/91, Dahlgren, VA, May 1993.

[4]Moore, F. G., McInville, R. M., and Hymer, T. G., "The 1995 Version of the NSWC Aeroprediction Code: Part 1—Summary of New Theoretical Methodology," Naval Surface Warfare Center, Dahlgren Div., NSWCDD TR-94/379, Dahlgren, VA, Feb. 1995.

[5]Moore, F. G., McInville, R. M., and Hymer, T. G., "The 1998 Version of the NSWC Aeroprediction Code: Part 1—Summary of New Theoretical Methodology," Naval Surface Warfare Center, Dahlgren Div., Rept. NSWCDD TR-98/1, Dahlgren, VA, April 1998.

[6]Wardlaw, A. B., and Davis, S., "A Second-Order Gudonov Method for Supersonic Tactical Missiles," Naval Surface Warfare Center, now Naval Surface Warfare Center, Dahlgren Div., Rept. NSWC TR 86-506, Dahlgren, VA, 1986.

[7]Walters, R. W., Slack, D. C., Cimmella, P., Applebaum, M. P., and Frost, C., "A

Users Guide to GASP," Virginia Polytechnic Inst. State Univ., Dep. Aerospace Ocean Engineering, Blacksburg, VA, Nov. 1990.

[8]Ericsson, L. E., and Reding, J. P., "Asymmetric Vortex Shedding from Bodies of Revolution," *Tactical Missile Aerodynamics,* edited by M. J. Hemsch and J. N. Nielson, Vol. 104, AIAA Progress in Astronautics and Aeronautics, AIAA, New York, 1986, pp. 243–296.

[9]Lesieutre, D. J., Mendenhall, M. R., and Nazario, S. M., "Aerodynamics Characteristics of Cruciform Missiles at High Angles of Attack," AIAA Paper 87-0212, 1987.

[10]Gentry, A. E., Smyth, D. N., and Olivier, W. R., "The MARK IV Supersonic–Hypersonic Arbitrary Body Program," Air Force Flight Dynamics Lab., AFFDL TR-73-159, Wright Patterson Air Force Base, OH, 1973.

[11]Vukelich, S. R., and Jenkins, J. E., "Missile DATCOM: Aerodynamic Prediction on Conventional Missiles Using Component Build-up Techniques," AIAA Paper 84-0388, 1984.

[12]Wardlaw, A. B., Baltakis, F. P., Solomon, J. M., and Hackerman, L. B., "An Inviscid Computational Method for Tactical Missile Configurations," Naval Surface Warfare Center, now Naval Surface Warfare Center, Dahlgren Div., Rept. NSWC TR 81-457, Dahlgren, VA, 1981.

[13]Robinson, D. F., "ZEUS++—A Graphical User Interface Flowfield Analysis Tool," Naval Surface Warfare Center, Dahlgren Div., Rept. NSWCDD/TR-98/147, Dahlgren, VA, (submitted for publication).

[14]Thomas, J. L., Taylor, S. L., and Anderson, W. K., "Navier–Stokes Computations of Vortical Flows Over Low Aspect Ratio Wings," AIAA Paper 87-0207, Jan. 1987.

[15]Etkin, B., *Dynamics of Flight,* Wiley, New York, 1967, Chap. 2.

Errata for Approximate Methods for Weapon Aerodynamics

p. 19 Nomenclature: p_x, q_y= should be p_x, p_y
q_x, p_y= should be q_x, q_y

p. 47 Line beginning "The perturbation": the second Φ_x should be Φ_r.

p. 56 Eq. (55): U^2 should be u^2

p. 64 Second paragraph, third line: The revised sentence should be, "The reasons for the applicability of perturbation theory on wings at high Mach numbers are severalfold."

p. 68 First paragraph, last sentence: "IJL" should be "IJK"

p. 70 Second paragraph, second sentence: A period should be placed after 0.5

p. 119 Eq. (8): $C_N = C_{P_l} \cos(\theta_W) - C_{P_u} \sin(\theta_W)$ should be $C_N = (C_{P_l} - C_{P_u}) \cos(\theta_W)$

p. 119 Last paragraph starting with, "Figure 4.1 illustrated...": Replace "TOTWT" with "TDTWT"

p. 156 Fig. 4.23:

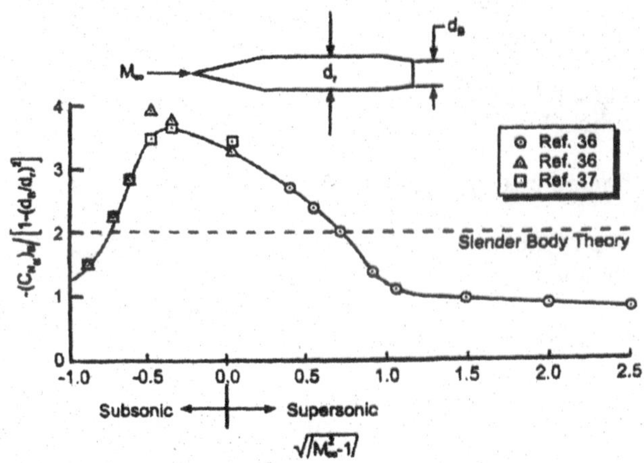

p. 158 Fig. 4.24: The label on the vertical axis should be $(x_{cp})_a / l_a$

p. 169 Eq. (73): The subscript "on" should be italicized.

p. 190 Fig. 5.3: The first bullet in the first box "$C_{N\alpha}$ Increases as α Increases" should be "$C_{N\alpha\alpha}$ Increases as α Increases"

p. 190 Fig. 5.3: The first bullet in the second box "$C_{N\alpha\alpha}$ Can Increase or Decrease Slightly as " Increases" should be "$C_{N\alpha\alpha}$ Can Increase or Decrease Slightly as α Increases"

p. 192 First sentence of first paragraph: "...> 0.5" should be "...< 0.5."

p. 197 Eq. (9): The following equation should be inserted after the first equation: $C_{NL} = (C_{NL})_{\alpha = 30^\circ}$; $30^\circ < \alpha \leq 45^\circ$

p. 199 Line 3: "However, data suggested that flow ..." should be "However, data suggested that the flow..."

p. 205 Line 8: "Moore and McInville[5]..." should be "Baker[5]..."

p. 205 Line 12: "Finally, the Ref. 9 database was made primarily from a combination of Refs. 8 and 9 data, with other available data. Moore and McInville[10] data are important in comparing the Refs. 8 and 9 data and in making a decision between the Refs. 8 and 9 databases on occasion." should be "Finally, the Ref. 9 database was made primarily from a combination of Refs. 5 and 8 data, with other available data. Nielsen et al.[9] data are important in comparing the Refs. 5 and 8 data and in making a decision between the Refs. 5 and 8 databases on occasion."

p. 249 Eq. (40): "$D\alpha^2$" should be "$D\alpha^3$"

p. 263 Eq. (53a): "$C_{N\delta w}$" should be "$C_{A\delta w}$"

p. 299 Second paragraph, line 11: "Programs" should be "Program"

p. 311 Third paragraph, first sentence: "To compute..." should be "To compute the..."

p. 322 Example computation for SBTSF: The "+1.64" should be "=1.64"

p. 326 Top of the page: The first "$k_{W(B)}$" should be "$K_{W(B)}$" and the second "$K_{W(B)}$" should be "$k_{W(B)}$"

p. 328 Fifth line from the bottom of the page: "Fig. 6.13" should be "Fig. 6.14"

p. 329 Fig. 6.14 caption: "Effect square..." should be "Effect of square..."

p. 385 Second paragraph, line 1: "As Mach number increases, shock..." should be "As Mach number increases, the shock..."

p. 453 Eq. (40): "1/2" should be set off by parentheses "(1/2)"

p. 454 Fig. 8.30, heading for 4th column: "Moderate/high supersonic $1.8 \leq M_\infty \leq 1.8$" should be "Moderate/high supersonic $1.8 \leq M_\infty \leq 6.0$"

Chapter 2

Navier–Stokes and Euler Equations

Nomenclature

a = speed of sound
C_{D_f} = skin friction drag coefficient
$\bar{c}_p, \bar{c}_v$ = specific heats at constant pressure and volume, respectively
d = diameter of body
$\hat{e}_t$ = internal energy
h = altitude
$\hat{\imath}, \hat{\jmath}, \hat{k}$ = unit vectors in x, y, and z directions, respectively
k = coefficient of heat conductivity
k_T = eddy coefficient of heat conductivity
ℓ = mean free path (distance air molecules travel between collisions)
M = Mach number
p = pressure
p_x, q_y = normal stress $p = -(p_x + p_y)/2$
q_x, p_y = heat transfer in x and y directions, respectively
R = gas constant [for air R = 1716 ft-lb/(slug − °R)]
R_N = Reynolds number
T = temperature
u, v, w = velocities in x, y, and z directions, respectively
x, y, z = orthogonal coordinate axis system with x fixed along body centerline
$-$ = indicates mean or average quantity
$'$ = indicates variation of flow quantity from mean value of flow variable
τ_{xy} = shear stress in y direction in plane normal to x axis
τ_{yx} = shear stress in x direction in plane normal to y axis
ρ = fluid density
μ = coefficient of viscosity

THE introductory chapter noted that the primary rationale for including Chapter 2 was to show the assumptions made in the full equations of motion that are needed to obtain more simplified methods used in semi-empirical codes. Thus, when an assumption is made, such as transitional Reynolds number, one can refer back to the full equations of motion to see the effect of this assumption. A secondary reason is to give some balance to this book, because two of the theoretical approaches to calculate aerodynamics of weapons are versions of the Navier–Stokes or full Euler equations in conjunction with some approximation of the viscous regions not accounted for by the Euler equations. The first part of the discussion on the Navier–Stokes and Euler equations addresses the continuum flow assumption. This assumption is inherent in this book, as seen in all of the theoretical options listed in Table 1.4. The second part of Chapter 2 will briefly show the various physical models associated with the full and reduced versions of the Navier–Stokes equations. An attempt will be made to summarize the important points, without resorting to details. Many references exist to provide details. The third part of the chapter will discuss the approach to compute aerodynamics using the full Euler equations plus approximations to viscous effects. Finally, a few comments will be offered with respect to numerical solutions of the Navier–Stokes and Euler equations. References used for this chapter include several excellent textbooks (Refs. 1–5), along with the author's many unpublished notes.

I. Continuum Flow Assumption

Continuum flow implies that the air is fairly dense. The parameter used to describe this denseness from a continuum or rarefied standpoint is the mean free path ℓ of the gas. The mean free path is defined as the distance that air molecules travel between collisions. Air is simply a gas composed primarily of nitrogen (N_2) and oxygen (O_2), with minute traces of carbon dioxide (CO_2), argon (Ar), nitrogen oxide (NO), and other trace elements. These air molecules are in constant motion and are in constant collision with one another. At low altitudes, where the air is very dense, the mean free path is very small. On the other hand, at very high altitudes, where the air molecules are very sparse, the mean free path is large and the flow is no longer continuum. If one defines the ratio of the mean free path to some characteristic body dimension, say d, as ℓ', then we have continuum flow if

$$\ell' \equiv \frac{\ell}{d} << 1 \tag{1}$$

If

$$\ell' >> 1 \tag{2}$$

we have free molecular flow. In between these two regions is what is referred to as the transitional region between the free molecular and continuum regions. Truitt[2] uses the kinetic theory of gases to relate the mean free path to kinematic viscosity, speed of sound, and ratio of specific heats of air as follows:

$$\ell = 1.255 \frac{\nu}{a} \sqrt{\gamma} \tag{3}$$

Figure 2.1 plots the variation of ℓ as a function of altitude, where an average seasonal temperature is assumed at each altitude.

As shown in Fig. 2.1, as long as one considers altitudes below about 250,000 ft, the mean free path is much less than the characteristic body dimension of any practical vehicle and, from Eq. (1), the flow can be considered continuum. For altitudes on the order of 300,000 ft to about 450,000 to 500,000 ft, the flow is in the transitional region, where ℓ is of the order of the characteristic dimension of the body. For altitudes greater than about 500,000 ft, flow should be treated as free molecular because Eq. (2) is obeyed.

In the transitional flow region, the Navier–Stokes equations can still be used if the wall boundary condition is changed from no slip (i.e., $u = v = 0$ at the body surface) to a slip boundary condition ($u = f(\partial u/\partial y), v = 0$). As a result, the Navier–Stokes equations and the various approximations discussed in this book can be used for most practical aerodynamics problems of interest. To validate this last statement, a practical example is given to illustrate the point. Consider a vehicle that flies to high altitude. To get to high altitude, it must have significant thrust and hence, most probably, a reasonable diameter and weight. Assume the first-stage booster has dropped off in the lower atmosphere, leaving a second stage that weighs 1000 lb, is 1.5 ft in diameter, and is traveling at Mach 5 to 15 after its rocket engine has burned out. Figure 2.2 plots a typical drag for this vehicle as a function of altitude, beginning at 100,000-ft altitude. Note that at h = 275,000 ft, which is about the upper boundary for the Navier–Stokes equations without the slip flow boundary-layer condition, the drag is just greater than 1 lb at M_∞ = 5 and less than 10 lb at M_∞ = 15. This is for a vehicle that is 1.5 ft in diameter. When viewed in terms of drag-to-weight ratio, these values are less than 1%. Therefore, for these high altitudes where Navier–Stokes and gas dynamics equations considered in this book are not valid, the forces and moments from external aerodynamics are quite small and can be neglected

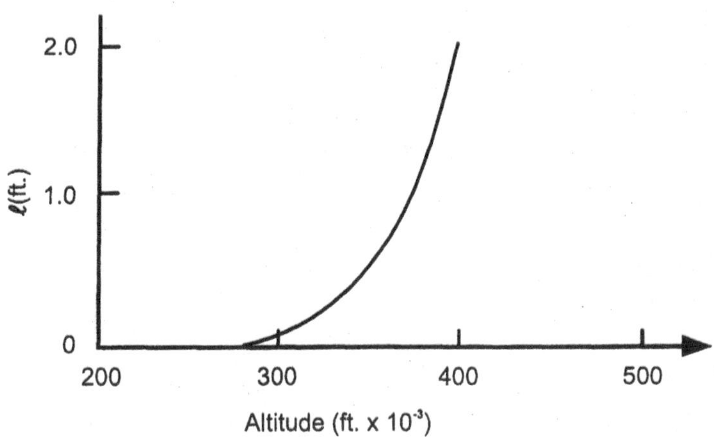

Fig. 2.1 Approximate mean free path of air as a function of altitude.

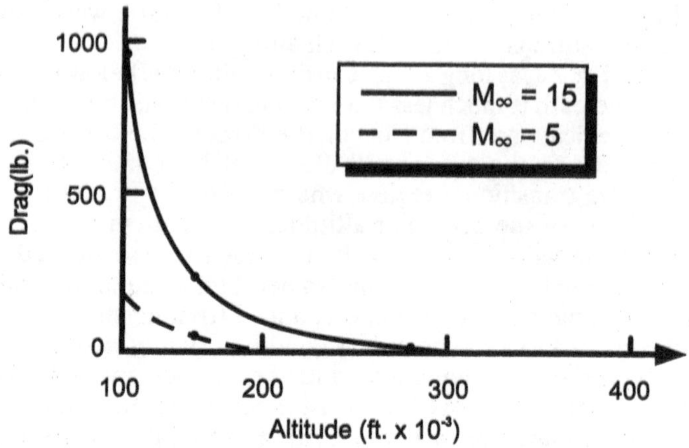

Fig. 2.2 Approximate drag versus altitude for 18-in.-diam missile.

for many applications. However, for some applications, such as ballistic coefficient of reentering ballistic missiles or the body dynamics at high altitude, even small forces and moments may still be important. For these types of applications, the continuum flow assumption is probably unacceptable.

II. Navier–Stokes Equations

One goal of this book is to show the reader the options available for computing aerodynamics of weapons, with a focus on approximate methods given in Chapter 3 to 7. Thus, to try to keep the length of equations down and focus on assumptions made in mathematical models, only two-dimensional flow for the Euler and Navier–Stokes equations will be considered. Warsi[4] can be referred to for the full three-dimensional equations.

To derive the full Navier–Stokes equations, one examines a small element of fluid in a flowfield. This element of fluid, in a two-dimensional sense, is shown in Fig. 2.3.

Figure 2.3 illustrates both the normal stresses along the x and y axes as well as the shear stresses on the surface of the element of fluid. The first subscript of the shear stress indicates the axis to which the stress is normal and the second subscript indicates the axis to which the shear stress is parallel. The subscript of the normal stresses indicates the axis to which the stress is parallel.

The next step in derivation of the Navier–Stokes equations is to apply the conservation of mass, momentum, and energy to this element of fluid shown in Fig. 2.3. Then Stokes law of friction for a Newtonian fluid, which relates the rate of strain of the fluid element in Fig. 2.3 to the normal and shear stresses, is utilized. To obtain the so-called Reynolds-averaged Navier–Stokes equations for two-dimensional flow, each of the dependent variables in the flow

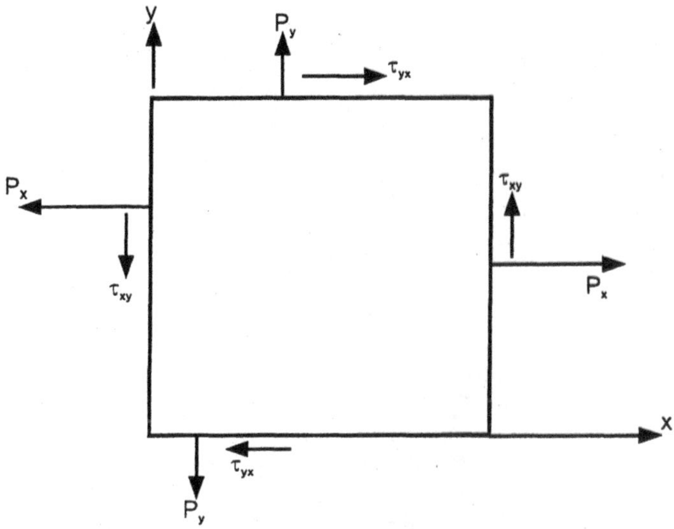

Fig. 2.3 Element of fluid showing normal and shear stresses.

equations is assumed to consist of a mean and a fluctuation component that arise as a result of the turbulence in the flow. For example, the u component of velocity is

$$u = \bar{u} + u'$$

Then, triple products of fluctuations are neglected in comparison to second-order terms.

Going through these steps, the Reynolds-averaged Navier–Stokes equations for two-dimensional turbulent flow written in flux vector format are (see Ref. 3):

$$\frac{\partial Q}{\partial t} + \frac{\partial Q'}{\partial t} + \frac{\partial E}{\partial x} + \frac{\partial F}{\partial y} = \frac{\partial E_v}{\partial x} + \frac{\partial F_v}{\partial y} + \frac{\partial E'}{\partial x} + \frac{\partial F'}{\partial y} \tag{4}$$

where

$$Q = \begin{bmatrix} \bar{\rho} \\ \bar{\rho}\bar{u} \\ \bar{\rho}\bar{v} \\ \bar{\rho}\bar{e}_t \end{bmatrix} \tag{5a}$$

$$Q' = \begin{bmatrix} 0 \\ 0 \\ 0 \\ \frac{1}{2}\bar{\rho}\left(\overline{u'^2} + \overline{v'^2}\right) \end{bmatrix} \tag{5b}$$

$$E = \begin{bmatrix} \bar{\rho}\bar{u} \\ \bar{\rho}\bar{u}^2 + \bar{p} \\ \bar{\rho}\bar{u}\bar{v} \\ (\bar{\rho}\bar{e}_t + \bar{p})\bar{u} \end{bmatrix} \tag{5c}$$

$$F = \begin{bmatrix} \bar{\rho}\bar{v} \\ \bar{\rho}\bar{u}\bar{v} \\ \bar{\rho}\bar{v}^2 + \bar{p} \\ (\bar{\rho}\bar{e}_t + \bar{p})\bar{v} \end{bmatrix} \tag{5d}$$

$$E_v = \begin{bmatrix} 0 \\ \bar{p}_x \\ \bar{\tau}_{xy} \\ \bar{u}\bar{p}_x + \bar{v}\bar{\tau}_{xy} - \bar{q}_x \end{bmatrix} \tag{5e}$$

$$F_v = \begin{bmatrix} 0 \\ \bar{\tau}_{xy} \\ \bar{p}_y \\ \bar{u}\bar{\tau}_{xy} + \bar{v}\bar{p}_y - \bar{q}_y \end{bmatrix} \tag{5f}$$

$$E' = \begin{bmatrix} 0 \\ -\bar{\rho}\overline{u'^2} \\ -\bar{\rho}\overline{u'v'} \\ -\frac{1}{2}\bar{\rho}\bar{u}\left(\overline{u'^2} + \overline{v'^2}\right) - \bar{\rho}\left(\overline{e'u'}\right) - \bar{\rho}\bar{u}\left(\overline{u'^2}\right) - \bar{\rho}\bar{v}\left(\overline{u'v'}\right) - \overline{p'u'} \\ \quad + \mu\left(\frac{4}{3}\overline{u'\frac{\partial u'}{\partial x}} - \frac{2}{3}\overline{u'\frac{\partial v'}{\partial y}}\right) + \mu\left(\overline{v'\frac{\partial u'}{\partial y}} + \overline{v'\frac{\partial v'}{\partial x}}\right) \end{bmatrix} \tag{6a}$$

$$F' = \begin{bmatrix} 0 \\ -\bar{\rho}\overline{u'v'} \\ -\bar{\rho}\overline{v'^2} \\ -\frac{1}{2}\bar{\rho}\bar{v}\left(\overline{u'^2} + \overline{v'^2}\right) - \bar{\rho}\left(\overline{e'v'}\right) - \bar{\rho}\bar{u}\left(\overline{u'v'}\right) - \bar{\rho}\bar{v}\left(\overline{v'^2}\right) - \overline{p'v'} \\ \quad + \mu\left(\overline{u'\frac{\partial u'}{\partial y}} + \overline{u'\frac{\partial v'}{\partial x}}\right) + \mu\left(\frac{4}{3}\overline{v'\frac{\partial v'}{\partial y}} - \frac{2}{3}\overline{v'\frac{\partial u'}{\partial x}}\right) \end{bmatrix} \tag{6b}$$

$\bar{p}$ is related to the normal stresses $\bar{p}_x$ and $\bar{p}_y$ by

$$\bar{p} = -\frac{1}{2}\left(\bar{p}_x + \bar{p}_y\right)$$

The normal sign convention for pressure is positive along a normal toward the surface, whereas normal stress is positive along a normal away from a surface. Equation (4) is actually a set of four equations, as seen from the flux vectors defined by (5) and (6). However, there are more dependent variables than four. These include $\overline{p}$, $\overline{\rho}$, $\overline{u}$, $\overline{v}$, $\overline{e}_t$, $\overline{\tau}_{xy}$, $\overline{q}_x$, $\overline{q}_y$ as well as all the fluctuation quantities denoted by primes. Additional equations required for a complete mathematical model include the equation of state,

$$\overline{p} = \overline{p}\,(\overline{\rho}, \overline{e}_t) \tag{7a}$$

which, for a perfect gas, is

$$\overline{p} = \overline{\rho}\,\overline{e}_t\,(\gamma - 1) \tag{7b}$$

Also, τ_{xy} is related to strain through Stokes law as

$$\overline{\tau}_{xy} = \mu\left(\frac{\partial \overline{u}}{\partial y} + \frac{\partial \overline{v}}{\partial x}\right) \tag{8}$$

where μ is the coefficient of viscosity. The heat transfer in the x and y directions is given by

$$\overline{q}_x = k\frac{\partial \overline{T}}{\partial x} \text{ and } \overline{q}_y = k\frac{\partial \overline{T}}{\partial y} \tag{9}$$

where k is the coefficient of heat conductivity. Both μ and k of Eqs. (8) and (9) are generally known functions of temperature. Also, $\overline{T}$ of Eq. (9) is related to the internal energy of the system through the equation of state. That is,

$$\overline{e}_t = \overline{e}_t\,(\overline{\rho}, \overline{T}) \tag{10a}$$

For a perfect gas, this can be written as

$$\overline{e}_t = \overline{c}_v\,\overline{T} \tag{10b}$$

where $\overline{c}_v$ is the specific heat at constant volume and is known for a given gas. The primed quantities, which arise due to turbulence in the flow, are defined by some turbulence models. These turbulence models are all semi-empirical in nature and relate fluctuating variables to the mean correlation variables. These turbulence models are composed of either algebraic ordinary differential or partial differential equations. There is no one turbulence model for all classes of problems. Hence, a person experienced in computational fluid dynamics will generally pick a turbulence model that has proven most effective for a class of configurations or problems based on past history.

Equations (4) through (10), with an appropriate turbulence model, a set of boundary conditions at the body surface and far from the body surface, and a definition of transition, thus gives a set of sufficient equations to solve for all of the dependent flow variables at any point in space. This set of equations is quite general (except for the restriction to two-dimensional flow to reduce the size of the mathematical model), with the only assumptions being continuum flow and the turbulence and transition model. This model for computing aerodynamics is the first option listed in Table 1.4.

The next level of mathematical model shown on Table 1.4 is the set of thin-layer Navier–Stokes (TLNS) equations. These equations can be readily obtained from Eqs. (4) through (6) by assuming that streamwise and circumferential gradients of stress terms are small compared to normal gradients. Usually, mixed partial derivatives of flux vectors are also dropped. These assumptions have the effect of reducing the number of terms in Eqs. (4) through (10) somewhat and reducing the number of gridpoints. However, solution of the flowfield is still a time marching process due to the elliptic nature of the equations, and thus the computational time is reduced only by about 15 to 25% over the full Navier–Stokes equations as illustrated in Fig. 1.7. The full Reynolds-averaged Navier–Stokes equations are applicable at any angle of attack (given an appropriate turbulence model and transition location), whereas TLNS equations are limited to moderate AOAs. We have actually used thin-layer equations to 40-deg AOA at Mach numbers 1.5 to 4.6 with good success. It is suspected that subsonic Mach numbers would present difficulties at the higher AOA of 30 to 40 deg for TLNS equations, however.

A much more significant reduction in computational time can be achieved from the so-called parabolized Navier–Stokes assumption. Here the flowfield is assumed to be steady and streamwise stress gradients are assumed small compared to normal gradients. Also, the streamwise pressure gradient in the subsonic portion of the boundary layer near the surface must be approximated. These assumptions allow Eqs. (4) through (10) to be marched in space rather than time. In effect, it means that the flowfield is solved only once as opposed to hundreds or even thousands of times as with the full Navier–Stokes or TLNS equations. Because the equations are parabolic in nature, subsonic conditions in the flowfield are not permitted. This limits one to lower AOAs at lower supersonic Mach numbers. However, at hypersonic Mach numbers, AOAs can be increased. The parabolized Navier–Stokes method has proved to be very useful at high Mach numbers where the shock layer is thin and the viscous layer occupies a large portion of it. Figures 1.6 and 1.7 show a significant reduction in both cost and time to obtain a set of trim aerodynamics using the parabolized Navier–Stokes versus the full Navier–Stokes or TLNS approaches.

Before completing this section on the Reynolds-averaged Navier–Stokes equations, it should be pointed out that, for laminar flow, E' and F' of Eqs. (6) are zero and the bars over all of the flow variables in Eqs. (4) and (5) can be removed as there is no turbulence. This set of equations is then the original set of Navier–Stokes equations as shown in Ref. 1. Also, in examining Eqs. (6a) and (6b), it is seen that three additional stresses in the momentum equations arise as a result of turbulence in the flowfield. These additional stresses are

$$-\bar{\rho}\overline{u'^2};\ -\bar{\rho}\overline{u'v'};\ -\bar{\rho}\overline{v'^2} \tag{11}$$

They are termed "apparent" or virtual or Reynolds stresses of turbulent flow. These stresses are added to normal and shear stresses in laminar flow when turbulence is present. In most regions of turbulent flow, they are the dominant viscous terms.

In addition to the transition and turbulence model, boundary conditions are necessary to solve the Navier–Stokes equations. For no slip, the velocity at the body surface is zero. Also, at the outer edge of the computational boundary, flow conditions are assumed to go to the freestream conditions without effect by the presence of the body, assuming a shock-capturing numerical method is used for the computations.

III. Euler Plus Boundary Layer Plus Base Drag

Another alternative to computing weapon aerodynamics at high AOA is to assume that the flowfield can be calculated by an inviscid region plus a thin layer of viscous flow near the body surface known as a boundary layer. In addition, if accurate axial forces are desired, then empirical predictions of base drag are needed. This approach, compared to solution of Navier–Stokes equations, has several advantages. First, it gives a significant reduction in computational time, cost, and grid generation. Second, most of the nonlinearities in missile aerodynamics can be accounted for by this approach. Exceptions to this include high AOA at any Mach number and moderate AOA at low Mach number ($M_\infty < 1.5$). However, for many applications, these exceptions do not occur.

The fully conservative Euler equations can be obtained by neglecting the viscous terms E_v and F_v of Eq. (4). The fully conservative form of the Euler equations refers to the equations written in their divergent form. In vector form, they are (neglecting external body forces):

Continuity equation

$$\frac{\partial \rho}{\partial t} + \nabla \cdot \left(\rho \vec{V}\right) = 0 \tag{12}$$

Momentum equations

$$\frac{\partial\left(\rho \vec{V}\right)}{\partial t} + \nabla \cdot \left(\rho \vec{V}\vec{V}\right) = -\nabla p \tag{13}$$

Energy

$$\frac{\partial\left(\rho e_t\right)}{\partial t} + \nabla \cdot \left[(\rho e_t + p)\vec{V}\right] = 0 \tag{14}$$

Equation of state

$$e_t = e_t(T, \rho) \tag{15a}$$

or, for a perfect gas,

$$e_t = c_V T \tag{15b}$$

and

$$T = \frac{p}{\rho R} \tag{15c}$$

For two-dimensional flow,

$$\vec{V} = u\vec{i} + v\vec{j} \tag{16a}$$

Equations (12) through (15) consist of a set of six equations for the six unknowns ρ, p, u, v, e_t, and T. If the flow is three-dimensional,

$$\vec{V} = u\vec{i} + v\vec{j} + w\vec{k} \tag{16b}$$

and Eqs. (12) through (15) consist of a set of seven equations and the seven unknowns ρ, p, u, v, w, e_t, and T.

For inviscid flow, the no-slip boundary condition at the wall (that is used for the Navier–Stokes equations) is replaced with the condition that the velocity is parallel to the surface. That is, the velocity component normal to the surface is required to be zero and, in addition, the conditions at a large distance from the body are required to go to conditions of the freestream for subsonic flow and conditions just behind the shock for supersonic flow. If the flow is supersonic, the inviscid equations are hyperbolic in nature and the solution is marched downstream in a computational sense and fairly fast flowfield solutions can be obtained. The example shown in Figs. 1.6 and 1.7 required 3 to 15 min per supersonic case on a personal computer with a 200-megahertz chip. The subsonic Euler solution in some cases required several hours per data point on a workstation.

If one is interested only in aerodynamics associated with pressure distributions (normal force, pitching moment, side force, side moment, and rolling moments), the inviscid solution is generally quite adequate to give an acceptable solution. The viscous solution has only second-order effects on these aerodynamics for most cases. However, if one is interested in accurate axial force information, the viscous effects must be estimated. The Euler solution will give quite accurate wave drag, but skin friction and base drag are both viscous components of the axial force and must be predicted by a viscous solution of boundary-layer equations or by an empirical method.

There are several different approaches to estimate the skin friction drag of the weapon. The first and most rigorous approach is to reduce the Navier–Stokes equations to the three-dimensional, boundary-layer equations by assuming that the viscous region occurs in a thin layer near the body surface. This thin layer typically begins as a laminar boundary layer and then transitions to a turbulent boundary layer somewhere on the body surface. In the most sophisticated analysis approach, one then adds the boundary-layer displacement thickness to the body profile and then solves the inviscid equations of motion. This approach is known as a viscous–inviscid interaction approach. While either the viscous–inviscid interaction approach or the inviscid plus three-dimensional laminar/turbulent boundary-layer approaches are fairly rigorous, the author does not favor either of these approaches for estimating the viscous skin-friction drag. The reason is that while a three-dimensional boundary-layer solution on a body alone is reasonably straightforward, when wings and tail fins are added to the body, with configuration discontinuities, the numerical problems in a boundary-layer solution can be substantial. A Navier–Stokes solution where the entire configuration is considered at one time in a single grid is probably not much more difficult. In addition, the payoff of the sophisticated boundary-layer approach is minimal. That is because there is generally very small, if any, effect on any force and moment other than axial force. Also, experience has shown that acceptable accuracy on skin friction drag can be obtained by simpler boundary-layer methods than the full three-dimensional equations.

The second approach to estimating the skin friction component of axial force is to assume two-dimensional flow, reduce the Navier–Stokes equations to the two-dimensional laminar and turbulent boundary-layer equations, and then apply the cone rule of Van Driest[6] to the body-alone portion of skin friction drag. There are several advantages to this approach, which was first introduced by the author in 1972[7] and later adopted by other approximated codes.[8] First, the body and wings can be considered independently for skin friction drag computations and their results added together. Second, computational times and numerical difficulties are minimal. Third, the accuracy of skin friction drag is almost as good as the more sophisticated approach. Fourth, because skin friction drag is a smaller portion of total drag than wave or base drag for many applications, this loss in accuracy between the more sophisticated and practical approaches can hardly be seen in the total drag predictions. This approach is the one the author favors and is, therefore, the one that will be discussed here in more detail.

To obtain the two-dimensional laminar boundary-layer equations from the two-dimensional, Reynolds-averaged, Navier–Stokes equations given by Eqs. (4) through (6), the following assumptions are necessary:

1) The flow is laminar and hence the turbulence terms of Eq. (6) can be neglected.
2) The viscous effects are confined to a very thin layer near the body surface. This assumption is really valid only where Reynolds number

$$R_N \equiv \frac{\rho V \ell}{\mu} >> 1 \tag{17}$$

These assumptions allow the y momentum equation to be reduced to

$$\frac{\partial p}{\partial y} = 0 \tag{18}$$

which means that the pressure distribution in the boundary layer is a function of x only. Equation (18) also implies that boundary-layer equations are invalid near sharp corners. They are valid for surfaces that have large radii of curvature and hence the velocity and temperature gradients in the x direction are smaller than those in the y direction. That is

$$\frac{\partial u}{\partial x} << \frac{\partial u}{\partial y}$$

$$\frac{\partial T}{\partial x} << \frac{\partial T}{\partial y} \tag{19}$$

With the assumptions given by 1 and 2 and Eqs. (17) to (19), the laminar boundary-layer equations for two-dimensional steady flow become

Continuity
$$\frac{\partial(\rho u)}{\partial x} + \frac{\partial(\rho v)}{\partial y} = 0 \tag{20}$$

x Momentum
$$\rho u \frac{\partial u}{\partial x} + \rho v \frac{\partial u}{\partial y} = -\frac{\partial p}{\partial x} + \frac{\partial}{\partial y}\left(\mu \frac{\partial u}{\partial y}\right) \tag{21}$$

y Momentum
$$\frac{\partial p}{\partial y} = 0 \tag{18}$$

Energy
$$\rho u \bar{c}_p \frac{\partial T}{\partial x} + \rho v \bar{c}_p \frac{\partial T}{\partial y} = u \frac{\partial p}{\partial x} + \frac{\partial}{\partial y}\left(k \frac{\partial T}{\partial y}\right) + \mu\left(\frac{\partial u}{\partial y}\right)^2 \tag{22}$$

Equations (18), (20), (21), and (22) contain the dependent variables ρ, u, v, p, T, $\bar{c}_p$, k, and μ, which exceed the number of equations by four. However, $\bar{c}_p$, k, and μ are all defined as functions of temperature and the equation of state

$$p = \rho R T \tag{23}$$

provides the remaining equation to make the system complete. The boundary conditions are no-slip at the wall ($u_w = v_w = 0$), and conditions far from the surface approach inviscid surface conditions.

Equations (20) through (23) can be solved numerically or additional assumptions can be made to find approximate closed-form solutions for more limited cases. These approximate solutions will be discussed in Chapter 4. Suffice it to say that if the two-dimensional solution is used for wings or fins, the values of heat transfer and skin friction drag apply directly. If the solutions are used for body-alone skin friction and heat transfer, the cone rule,[2] which states

$$\left(C_{D_f}\right)_{\text{cone}} = \frac{2}{\sqrt{3}}\left(C_{D_f}\right)_{\text{two dimensional}} \tag{24}$$

should be used to increase the skin friction drag of the axisymmetric body by about 14% over that due to the two-dimensional solution.

For steady, turbulent, two-dimensional compressible boundary layers, Eqs. (4) through (6b) reduce to[6]

Continuity
$$\frac{\partial}{\partial x}\overline{(\rho u)} + \frac{\partial}{\partial y}\overline{(\rho v)} = 0 \tag{25}$$

x Momentum
$$\overline{\rho u}\,\frac{\partial \bar{u}}{\partial x} + \overline{\rho v}\,\frac{\partial \bar{u}}{\partial y} = -\frac{\partial \bar{p}}{\partial x} + \frac{\partial}{\partial y}\left(\bar{\mu}\,\frac{\partial \bar{u}}{\partial y}\right) + \frac{\partial}{\partial y}\left[-\overline{(\rho v)'u'}\right] \tag{26}$$

y Momentum
$$\frac{\partial \bar{p}}{\partial y} = 0 \tag{27}$$

Energy
$$\overline{\rho u}\,\bar{c}_p\,\frac{\partial \bar{T}}{\partial x} + \overline{\rho v}\,\bar{c}_p\,\frac{\partial \bar{T}}{\partial y} = \bar{u}\,\frac{\partial \bar{p}}{\partial x} + \frac{\partial}{\partial x}\left(\bar{k}\,\frac{\partial \bar{T}}{\partial y}\right) + \bar{\mu}\left(\frac{\partial \bar{u}}{\partial y}\right)^2 + \frac{\partial}{\partial y}\left[-\bar{c}_p\overline{(\rho v)'T'}\right] - \overline{(\rho v)'u'}\,\frac{\partial \bar{u}}{\partial y} \tag{28}$$

In comparing Eqs. (25) through (28) with the laminar counterparts, Eqs. (18), (20), (21), and (22), three additional terms are present with the turbulent boundary layer. The additional term in the x Momentum and energy equations, $[-\overline{(\rho v)'u'}]$, is the well-known Reynolds stress or apparent stress of turbulent flow. The additional term, $\bar{c}_p\,\overline{(\rho v)'T'}$, in the energy equation is the apparent heat flux term. In many turbulent flow problems, these additional terms can dominate the other viscous stress terms.

Equations (25) through (28), with appropriate boundary conditions, can be solved numerically for the stress and skin friction drag on a flat plate. They can be used to approximate the viscous aerodynamic properties on other body shapes as well. The cone rule mentioned earlier can then be applied to convert the two-dimensional results to approximate those on a body. Additional approximations to Eqs. (25) through (28) discussed later in Chapter 4 will allow estimates of skin friction drag to be made much faster than numerical solution of the partial differential equations given by Eqs. (25) through (28).

The question arises as to where the laminar boundary layer ends and where the turbulent boundary layer begins. There is no simple answer to this question. The transition from laminar to turbulent flow is a function of many variables, including Reynolds number, configuration roughness, and configuration shape, just to name a few. In general, the flow in the nose region or leading edge region of a configuration is laminar and, at some distance downstream, this laminar flow transitions to turbulent flow. Experience has shown that, for most weapons in development, the surface roughness on bodies is such that a transition Reynolds number of 5×10^5 is adequate for the start of transition and 1×10^6 for the flow to be fully turbulent. On flat surfaces, such as wings, a lower Reynolds number for transition to start is typical. A value of 5×10^5 appears reasonable for the fully turbulent flow. Wind tunnel models in general are much smoother than development weapons, and values of transition Reynolds numbers two to four times higher than those for development weapons are typical. However, some older wind tunnel models, which may have been used numerous times for wind tunnel testing, can become fairly rough because of sandblasting from the air, plastered screw holes, or metal tape. For rough wind tunnel models, the transition Reynolds number may be closer to that of a weapon in development, than a new wind tunnel model. To obtain the total skin friction drag, the area of all laminar flow is computed and the laminar skin friction coefficient used for this area. Likewise, the area for all turbulent flow is computed and the turbulent skin friction coefficient used for this area. For the transition region, an average of the laminar and turbulent skin friction coefficients is acceptable. The mathematics of this process will be given in Chapter 4.

In addition to the skin friction component of viscous drag, two- and three-dimensional components of base pressure drag must be computed as well. Base drag can only be found using full Navier–Stokes solutions or empirically. Empirical estimates are in general as good as Navier–Stokes solutions, and because they can be done at lower cost, they are the preferable approach. Empirical estimates of base drag are discussed in Chapter 4. Finally, most Euler solutions assume leading edges of fins are sharp for ease of computations. Thus, if fin leading edges are truncated or blunt, additional axial force drag needs to be added in to account for this approximation as well. A method to approximate this component of axial force will be discussed in Chapter 3.

To summarize the Euler plus boundary layer plus empirical base drag approach to compute weapon aerodynamics, it should be stated that this approach is gaining widespread use in a production sense. The reasons are that most weapon nonlinearities can be accounted for; aerodynamics is accurate enough for most applications; aerodynamics can be obtained for a reasonable cost and in a reasonable time frame; and, finally, both planar and nonplanar aerodynamics can be calculated. It is true that empiricism is needed for accurate axial force information, but the positive aspects of this approach far outweigh this minor problem.

IV. Numerical Flowfield Solutions

The Navier–Stokes equations (4) through (6), Euler equations (12) through (14), boundary-layer equations (20) through (22), and Eqs. (25) through (28) were all written in physical body coordinates for illustration purposes. Here, body coordinates refer to the x axis being along the centerline of the body, positive nose to rear, the y axis out the right wing of a wing–body configuration, and z normal to the x–y plane. For boundary-layer computations, physical coordinates are many times defined along and normal to the body surface. To numerically solve the flowfield, the most convenient approach is to transform the physical domain to a computational domain through a set of coordinate transformations.[3,5] These coordinate transformations change the computational space from the Cartesian x–y–z system to general curvilinear coordinates where two of the coordinates in a three-dimensional case lie along the body surface. This type of computational domain requires additional computations to transform the results back to the Cartesian x–y–z system. However, this downside is more than offset by the ease of implementation of the body boundary conditions to obtain accurate derivatives and hence an accurate numerical solution. This is particularly important in the case of an elliptic problem solution that is time marched where several hundred or even a few thousand iterations are required for convergence. Any numerical errors propagating as a result of inaccurate values of the dependent variables and their derivatives at the boundary can thus cause the entire flowfield solution to lose accuracy or become unstable. Both Refs. 3 and 5 give details of the various computational approaches to the solution of the Navier–Stokes, Euler, and boundary-layer equations for interested readers.

Before leaving the chapter discussing more complex ways to compute aerodynamics, the author believes the engineer should give careful consideration to the problem at hand before deciding whether an approximate or a more exact method should be used. A major portion of this decision process goes back to Table 1.5, which discusses aerodynamic code requirements in the various stages of weapon design. The author thinks that, while an agency should maintain the capability to predict aerodynamics numerically, using this approach if it is not warranted is excessive. The methods that will be discussed in the remainder of the book are preferred approaches if the accuracy of these methods is adequate to meet the agency's needs.

References

[1]Schlichting, H., *Boundary Layer Theory,* 4th ed., McGraw–Hill, New York, 1960.

[2]Truitt, R. W., *Fundamentals of Aerodynamic Heating,* Ronald, New York, 1960.

[3]Hoffman, K. A., *Computational Fluid Dynamics for Engineers,* Engineering Education System, Austin, TX, 1989.

[4]Warsi, Z. U. A., *Fluid Dynamics, Theoretical and Computational Approaches,* CRC Press, Boca Raton, FL, 1993.

[5]Roache, P. J., *Fundamentals of Computational Fluid Dynamics,* Hermosa Publishers, Albuquerque, NM, 1998.

[6]Van Driest, E. R., "Turbulent Boundary Layer in Compressible Fluids," *Journal of Aeronautical Sciences,* Vol. 18, No. 3, March 1951, pp. 145–160.

[7]Moore, F. G., "Body Alone Aerodynamics of Guided and Unguided Projectiles at Subsonic, Transonic and Supersonic Mach Numbers," Naval Warfare Lab., now Naval Surface Warfare Center, Dahlgren Div., NWL TR-2796, Nov. 1972

[8]Vukelich, S. R., and Jenkins, J. E., "Missile DATCOM: Aerodynamic Prediction on Conventional Missiles Using Component Buildup Techniques," AIAA Paper 84-0388, 1984.

Chapter 3

Perturbation Methods

Nomenclature

A	= area of the spherical surface
A_F	= frontal area of wing or tail
AR	= aspect ratio = b^2/A_W
a	= speed of sound (ft/s)
A_{ref}	= reference area (maximum cross-sectional area of body, if a body is present, or planform area of wing, if wing along)(ft^2)
b	= wing span (not including body)(ft)
C_A	= axial force coefficient
C_{A_W}	= component of axial force coefficient
C_D, C_d	= total and local drag coefficient, respectively
C_ℓ	= rolling moment coefficient of wing planform
c_ℓ	= local rolling moment coefficient
C_{ℓ_P}	= roll damping moment coefficient, $\partial(C_\ell)/\partial(p\ell_{ref}/2V_\infty)$
C_M, c_m	= total and local pitching moment coefficient, respectively (based on reference area and body diameter, if body present, or mean aerodynamic chord, if wing alone)
C_{M_B}	= pitching moment coefficient of body alone
$\Delta C_{M_{B(T)}}, \Delta C_{M_{B(W)}}$	= additional pitching moment on body due to presence of tails or wings, respectively
$C_{M_q} + C_{M_{\dot{\alpha}}}$	= total pitch damping moment coefficient
C_{M_q}	= pitch damping term due to a constant pitch rate = $\partial(C_M)/\partial(q\ell_{ref}/2V_\infty)$
$C_{M_{\dot{\alpha}}}$	= pitch damping term due to a constant vertical acceleration = $\partial(C_M)/\partial(\dot{\alpha}\,\ell_{ref}/2V_\infty)$
$C_{M_{T(B)}}, C_{M_{W(B)}}$	= pitching moment of tail or wing, respectively, in presence of body
$C_{M_{T(V)}}$	= additional pitching moment on tail surfaces caused by wing-shed vortices
C_N, c_n	= total and local normal-force coefficient, respectively
C_{N_B}	= normal-force coefficient of body alone
$C_{N_{B(V)}}$	= negative afterbody normal-force coefficient due to canard or wing-shed vortices

$C_{N_{B(W)}}, C_{N_{B(T)}}$ = normal-force coefficient on body in presence of wing or tail
$\Delta C_{N_{B(T)}}, \Delta C_{N_{B(W)}}$ = additional normal-force coefficient on body due to presence of the tail or wing, respectively
C_{N_L} = linear component of normal-force coefficient
$C_{N_{NL}}$ = nonlinear component of normal-force coefficient
c_{n_p} = sectional normal-force coefficient derivative due to rolling motion
c_{n_q} = sectional normal-force coefficient derivative due to pitching motion
$C_{N_{T(V)}}$ = negative normal-force coefficient component on tail due to wing or canard-shed vortex
$C_{N_{W(B)}}, C_{N_{T(B)}}$ = normal-force coefficient of wing or tail, respectively, in presence of body
C_{N_a} = normal-force coefficient derivative
$c_{n_{\dot{a}}}$ = sectional normal-force coefficient derivative due to constant vertical acceleration
C_P = pressure coefficient $\left(\dfrac{p - p_\infty}{1/2\, \rho_\infty V_\infty^2}\right)$
(ΔC_P) = difference in pressure coefficients of upper and lower surfaces of wing planform
$(\Delta C_P)_P$ = difference in pressure coefficients of upper and lower surfaces of wing planform due to a rolling velocity p
$(\Delta C_P)_q$ = difference in pressure coefficients of upper and lower surfaces of wing planform due to a pitch rate q
$(\Delta C_P)_{\dot{a}}$ = difference in pressure coefficients of upper and lower surfaces of wing planform due to a constant vertical acceleration $\dot{a}$
c = chord (ft)
c_r = root chord (ft)
c_t = tip chord (ft)
cal = caliber(s) (one body diameter)
d = body diameter (ft)
e_i = span efficiency factor
d_{ref}, r_r = reference body diameter (ft) or radius, respectively
f, h = lateral and vertical position of wing vortex
f_W, f_T = lateral location of wing or tail vortex (measured in feet from body center line)
i = tail interference factor
$K_{B(W)}, K_{B(T)}$ = ratio of additional body normal-force coefficient in presence of wing, or tail to wing, or tail alone normal-force coefficient at $\delta = 0$ deg
$k_{B(W)}, k_{B(T)}$ = ratio of additional body normal-force coefficient due to presence of wing or tail at a control deflection to that of wing or tail alone at $a = 0$ deg
$K_{W(B)}, K_{T(B)}$ = ratio of normal-force coefficient of wing or tail in

presence of body to that of wing or tail alone at $\delta = 0$ deg

$k_{W(B)}, k_{T(B)}$ = ratio of wing or tail normal-force coefficient in presence of body due to a control deflection to that of wing or tail alone at $\alpha = 0$ deg

ℓ = length (ft)

ℓ_n = nose length (can be in calibers or feet)

m = $\beta \cot \Lambda_{LE}$

M = Mach number = V/a

p = pressure (lb/ft²) when used in pressure coefficient and roll rate (rad/s) when used in roll damping

q = pitch rate (rad/s)

Q = mass rate of flow

Q_∞ = dynamic pressure ($1/2\ \rho_\infty V_\infty^2$), (lb/ft²)

r_b = radius of body (ft)

R = spherical radius

S = entropy (ft-lb/slug-°R)

s = wing or tail semispan plus the body radius in wing–body lift methodology (ft)

S_{ref} = surface area of wing planform (ft²)

t = thickness except when used as a partial derivative which means time (ft)

t/c = thickness to chord ratio of wing or tail

u, v, w = perturbation velocities in x, y, z directions (ft/s)

V = magnitude of velocity vector, $\overline{V}$ (ft/s)

Vol_n = volume of the nose

x_a, y_a = coordinates of wing tip (ft)

x_{CP} = center of pressure (in feet or calibers from some reference point that can be specified) in x direction (ft)

x, r, φ = polar or cylindrical coordinate system with x along body axis, r along radius from centerline to point on body or flowfield, and φ around body

x, y, z = axis system fixed with x along centerline of body (ft)

y_{CP} = center of pressure of wing lift in y direction (ft)

α = AOA (deg)

β = $\sqrt{M^2 - 1}$, or $\sqrt{1 - M^2}$ if $M < 1$

Γ = vortex circulation, positive counterclockwise facing upstream (ft²/s)

γ = specific heat ratio

δ = control deflection (deg), positive leading edge up

δ_{eq} = angle between velocity vector and tangent to body surface

δ_W, δ_T = deflection of wing or tail surfaces (deg), positive leading edge up

θ = local surface slope of body with respect to body axis

$\Lambda_{LE}, \Lambda_{TE}$ = leading or trailing edge sweep angle, respectively, of wing or tail (deg)

λ	= taper ratio of a lifting surface = c_t/c_r
μ	= Mach angle = $\sin^1 (1/M)$
ρ	= density of air (slug/ft^3)
σ	= $(y/x) \tan \Lambda_{LE}$
Φ	= perturbation velocity potential
$\overline{\Phi}$	= full velocity potential
φ	= circumferential position around body where $\varphi = 0$ is leeward plane (deg)
Ψ_1, ζ_1	= first-order axial and crossflow solutions of velocity potential equation
Ψ_2	= second-order solution to full potential equation

Subscripts

b	= on body surface
fb	= force break Mach number
u, ℓ	= upper, lower surface, respectively
L, NL	= linear, nonlinear, respectively
∞	= free-stream conditions

I. Introduction

CHAPTER 2 discussed two approaches to obtaining a complete set of aerodynamics for weapon systems. The first method, full or thin-layer Navier–Stokes solution, is still too costly and time consuming for more than a few validation and example cases, whereas the second approach, Euler plus boundary-layer plus empirical estimates of base drag, is becoming more accepted as a production option for aerodynamics. The third approach, which has been the author's area of concentration for most of his career and which is the subject of most of this book, is to use perturbation or local slope methods, with generic empirical databases to form a complete set of aerodynamics. This third approach to computing aerodynamics has several basic steps. These steps are

1) Use slender body, linear theory, or other theoretical methods at low angle of attack (AOA) as much as practical.
2) Use the slender body theory and linear theory methods in conjunction with generic wind tunnel databases to approximate nonlinear aerodynamic terms as functions of key flight and geometric parameters.
3) Develop analytical approximations or tables of data that allow nonlinear aerodynamics to be estimated on configurations other than those in databases.
4) Compare new nonlinear methods to a broad class of configurations outside the generic database.
5) Refine empirical tables or approximations of nonlinear terms based on comparisons to cases outside the generic wind tunnel database to minimize wind tunnel and other errors.

There are several advantages to this approach. First, it is based on primarily theoretical methods at low AOA, so a good foundation exists for the initial estimate of aerodynamics as a function of flight conditions and configuration geometry. Second, there have been several large generic wind tunnel databases developed over the last 20 years that allow predictions of nonlinear aerodynamic terms to be made with reasonable robustness in geometry and flight conditions. Third, as seen in Figs. 1.6 and 1.7, an approach like this is very cost-effective and provides very rapid turnaround. Fourth, it has been shown that average accuracies in static planar aerodynamics of ±10% for normal and axial force and ±4% of body length for center of pressure can be consistently achieved with this approach.

The major disadvantages of an approach like this are

1) Out-of-the-pitch-plane aerodynamics are difficult to predict with the same accuracy as planar aerodynamics. Induced roll is a particularly difficult term to accurately predict. Euler codes have been shown to be quite successful in predicting induced roll, however, and that is the approach recommended by the author for out-of-the-pitch-plane aerodynamic computations.
2) For issues outside of databases, such as side jets for control, the method is not applicable and, once again, numerical solutions of Navier–Stokes or Euler equations are required.

As a result of the advantages and disadvantages of various approaches to computing aerodynamics, the author recommends that each organization involved in aerodynamics try to maintain a capability in all three approaches discussed, that is, Navier–Stokes, Euler plus boundary layer plus empirical, and semi-empirical. In addition, wind tunnel data will still be needed during the design process. Many organizations find it is more cost-effective to contract out to small specialty firms for their computational fluid dynamics calculations. Contracting for computational fluid dynamics to specialist firms is primarily driven by the educational and experience requirements of the user of a computational fluid dynamics code, as illustrated in Table 1.6.

The remainder of this book will be primarily devoted to the semi-empirical approach to computing a reasonably accurate and cost-effective set of planar aerodynamics for use in the earlier design stages of weapons. Chapters 3 and 4 will be devoted to the low AOA theoretical approaches primarily. Chapter 5 will discuss nonlinear empirical methods used to compute the higher AOA aerodynamic terms. Chapter 6 will then discuss approximate methods to predict aerodynamics of noncircular cross-sectional bodies. Chapter 7 will then discuss modifications necessary to compute convective heat transfer information at hypersonic Mach numbers. Finally, Chapter 8 will use the aerodynamic methods of Chapters 3 to 7 to highlight some design considerations and Chapter 9 will discuss future directions of aerodynamic codes.

II. Component Buildup of Aerodynamics

Component buildup of aerodynamics is commonly used in semi-empirical aerodynamic codes. It gets its theoretical basis from linearized theory where linear solutions of the equations of motion can be added together in a linear sense. This approach is not allowed for the Navier–Stokes and Euler solution methods, where entire flowfields are solved simultaneously for all configuration components as opposed to solving for configuration components separately and then combining results. Component buildup of aerodynamics applies to components of aerodynamics associated with individual missile components as well as aerodynamic terms within a component. To illustrate these two types of component buildup of aerodynamics, consider the missile configuration of Fig. 3.1.

The configuration of Fig. 3.1 has a body, cruciform wings, and cruciform tails. There are thus several separate aerodynamic force and moment terms that must be considered separately in the component buildup approach. These include the body alone; the wing alone; the tail alone; the additional force on the body due to the presence of the wings and tails; the additional force on the wings and tails due to the presence of the body; and, finally, the additional force on the tail as a result of being in the downwash field of the wings. Experience has shown that the axial force interference effects of the body and lifting surfaces can be neglected in most cases, except for the tail's effect on the base drag. However, the normal force and pitching moments of the total configuration are affected by all these terms. In equation form, these aerodynamic components can be summed together to obtain the total aerodynamics. That is,

$$C_A = C_{A_{\mathrm{BODY}}} + C_{A_{\mathrm{WING}}} + C_{A_{\mathrm{TAIL}}} \tag{1}$$

$$C_N = C_{N_B} + C_{N_{W(B)}} + \Delta C_{N_{B(W)}} + C_{N_{T(B)}} + \Delta C_{N_{B(T)}} + C_{N_{T(V)}} \tag{2}$$

$$C_M = C_{M_B} + C_{M_{W(B)}} + \Delta C_{M_{B(W)}} + C_{M_{T(B)}} + \Delta C_{M_{B(T)}} + C_{M_{T(V)}} \tag{3}$$

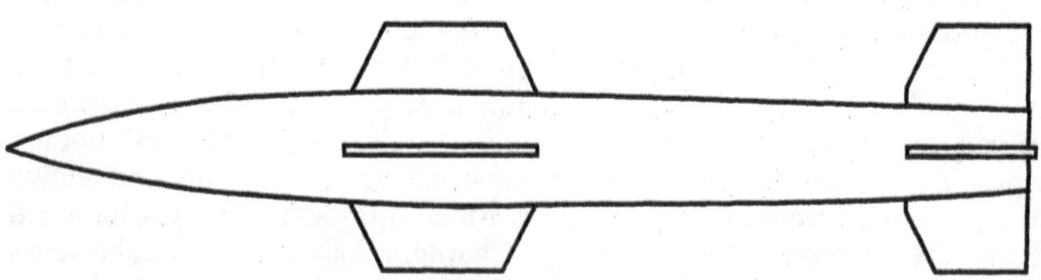

Fig. 3.1 Typical missile configuration.

$$x_{CP} = -\frac{C_M}{C_N} \tag{4}$$

$$\begin{aligned} C_{M_q} + C_{M_{\dot{\alpha}}} &= \left(C_{M_q} + C_{M_{\dot{\alpha}}}\right)_{\text{BODY}} + \left(C_{M_q} + C_{M_{\dot{\alpha}}}\right)_{\text{WING}} \\ &\quad + \left(C_{M_q} + C_{M_{\dot{\alpha}}}\right)_{\text{TAIL}} \end{aligned} \tag{5}$$

$$C_{\ell_P} = \left(C_{\ell_P}\right)_{\text{BODY}} + \left(C_{\ell_P}\right)_{\text{WING}} + \left(C_{\ell_P}\right)_{\text{TAIL}} \tag{6}$$

Each aerodynamic component in Eqs. (1) to (6) is computed separately and then added together linearly in the component buildup approach in analogy to solutions of linear equations. It is assumed here that the body is rolling slowly enough so that magnus moment is zero.

The second type of component buildup used in semi-empirical codes is a breakdown within a force or moment. For example, consider the axial force. It is common practice to compute wave drag, base drag, and skin friction drag components separately for all of the configuration components (body, wing, tail) and then add all of these components together to get the total axial force. This practice is in contrast to solution of the full Navier–Stokes equations, where all of the individual axial force components are computed simultaneously as part of the solution. Also, consider the body-alone normal force coefficient. In semi-empirical codes, each of the aerodynamic terms in Eqs. (1) to (6) can be broken down into a linear and a nonlinear component. For example, normal force coefficient of the body alone can be written as

$$C_{N_B} = \left(C_{N_L}\right)_B + \left(C_{N_{NL}}\right)_B \tag{7}$$

The linear terms of each of these forces or moments is generally computed by slender body or linear theories. The nonlinear terms are computed empirically or semi-empirically, again, in contrast to Navier–Stokes or Euler equation solutions, where all of the nonlinearities are a part of the solution.

The remaining parts of this chapter and Chapter 4 will discuss theoretical methods used to compute the low AOA solutions for the linear terms of each of the aerodynamic force and moments of Eqs. (1) to (6). References 1 to 3 are primary references used in this part of the discussion. Chapters 5 and 6 will then discuss empirical and semi-empirical methods used for the nonlinear components of Eqs. (1) to (6).

III. Linearized Flow and Slender Body Assumptions

The linearized flow and slender body theory follows from use of full inviscid or Euler equations [Eqs. (12) to (15) in Chapter 2], with certain simplifying assumptions. The assumptions generally made for linearized flow or slender body theory are that the flow is irrotational, homentropic, and steady and the body is thin or slender. It should be emphasized here that assumptions mentioned are those generally made for linearized and slender body theory and will be the ones made in this text. For example, it is not necessary to assume irrotational flow for slender body theory and it is not necessary to consider it only for low AOA approximations. In mathematical form, these assumptions imply

Irrotational flow: $\nabla \times \vec{V} = 0$
Homentropic: Shocks are weak, and flow can be assumed isentropic throughout so that $S = \not{c}$ everywhere
Steady: $\dfrac{\partial}{\partial t} = 0$
Thin bodies: Changes in flow velocities due to presence of body are small

From the first assumption, the implication is that

$$\vec{V} = \nabla\overline{\Phi} \tag{8}$$

where $\overline{\Phi}$ is a scalar potential function. The bar is used to distinguish the full velocity potential from the perturbation velocity potential Φ, which is used throughout this chapter. Equation (8) is arrived at based on the fact that $\nabla \times \nabla\,(\;) = 0$ and, because for irrotational flow, $\nabla \times \vec{V} = 0$, Eq. (8) naturally follows. Substituting Eq. (8) into Eqs. (12) to (15) of Chapter 2 using the first

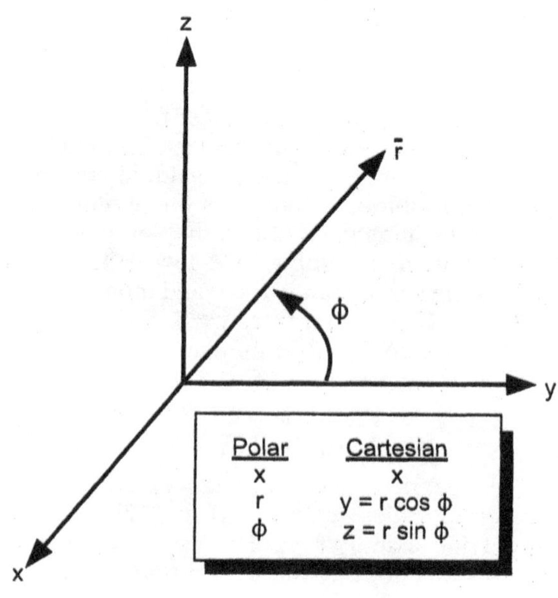

Fig. 3.2 Cartesian and polar coordinates.

three assumptions, the full potential equation of motion in Cartesian coordinates can be shown to be[1]

$$\begin{aligned}&\left(\overline{\Phi}_x^2 - a^2\right)\overline{\Phi}_{xx} + \left(\overline{\Phi}_y^2 - a^2\right)\overline{\Phi}_{yy} + \left(\overline{\Phi}_z^2 - a^2\right)\overline{\Phi}_{zz} + 2\overline{\Phi}_x\overline{\Phi}_y\overline{\Phi}_z \\ &\quad + 2\overline{\Phi}_z\overline{\Phi}_y\overline{\Phi}_{yz} + 2\overline{\Phi}_x\overline{\Phi}_z\overline{\Phi}_{xz} = 0\end{aligned} \tag{9}$$

Equation (9) is most convenient for use on two- or three-dimensional lifting surfaces. Equation (9), expressed in polar or cylindrical coordinates (see Fig. 3.2) for use on bodies of revolution, is

$$\begin{aligned}&\left(\overline{\Phi}_x^2 - a^2\right)\overline{\Phi}_{xx} + \left(\overline{\Phi}_r^2 - a^2\right)\overline{\Phi}_{rr} + \left(\frac{\overline{\Phi}_\varphi^2/r^2 - a^2}{r^2}\right)\overline{\Phi}_{\varphi\varphi} \\ &\quad - \frac{\overline{\Phi}_r}{r}\left(\frac{\overline{\Phi}_\varphi^2}{r_2} + a^2\right) + 2\left[\overline{\Phi}_x\overline{\Phi}_r\overline{\Phi}_{xr} + \frac{\overline{\Phi}_x\overline{\Phi}_\varphi\overline{\Phi}_{x\varphi}}{r^2} + \frac{\overline{\Phi}_r\overline{\Phi}_\varphi\overline{\Phi}_{r\varphi}}{r^2}\right] = 0\end{aligned} \tag{10}$$

The velocity vector can be written in terms of perturbation velocities as

$$\vec{V} = \left(V_\infty + u\right)\vec{i} + v\vec{j} + w\vec{k} = \left(V_\infty + \Phi_x\right)\vec{i} + \left(\Phi_y\right)\vec{j} + \left(\Phi_z\right)\vec{k} \tag{11}$$

in Cartesian coordinates. Referring to Fig. 3.3, Eq. (11) can be written in polar coordinates as

$$\begin{aligned}\vec{V} &= \left(V_\infty \cos a + u\right)\hat{e}_x + \left(V_\infty \sin a \cos\varphi + v\right)\hat{e}_r + \left(-V_\infty \sin a \sin\varphi + w\right)\hat{e}_\varphi \\ &\cong \left(V_\infty + \Phi_x\right)\hat{e}_x + \left(V_\infty\, a \cos\varphi + \Phi_r\right)\hat{e}_r + \left(-V_\infty a \sin\varphi + \frac{\Phi_\varphi}{r}\right)\hat{e}_\varphi\end{aligned} \tag{12}$$

The perturbation velocity components, u, v, w of Eqs. (11) and (12), are assumed to be much smaller than the freestream velocity V_∞. Substituting Eq. (11) into Eq. (9) and Eq. (12) into (10), carrying out the algebra, and neglecting products of perturbation quantities compared to first order terms, one arrives at the small perturbation equations in Cartesian and polar coordinates as

$$\left(M_\infty^2 - 1\right)\Phi_{xx} - \Phi_{yy} - \Phi_{zz} = 0 \tag{13}$$

and

$$\left(M_\infty^2 - 1\right)\Phi_{xx} - \Phi_{rr} - \frac{1}{r}\Phi_r - \frac{1}{r^2}\Phi_{\varphi\varphi} = 0 \tag{14}$$

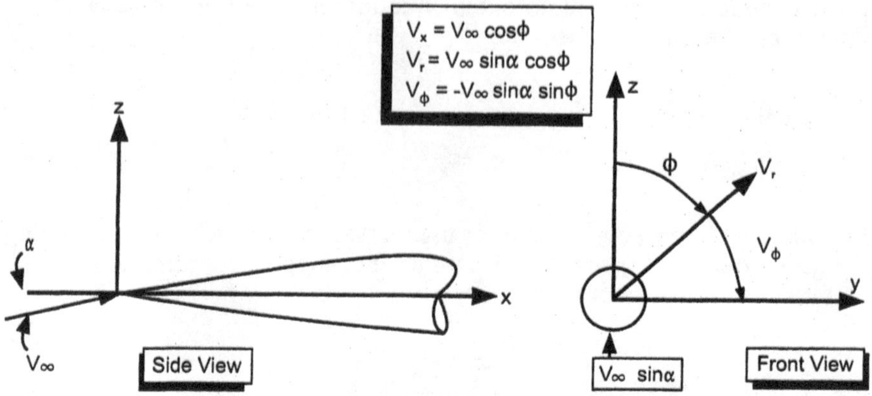

Fig. 3.3 Velocity components in polar coordinates.

Here we will call Eq. (13) the linearized flow equation and Eq. (14) the slender body flow equation, although Eq. (14) is also linear. Both Eqs. (13) and (14) are applicable at subsonic and supersonic Mach numbers. However, around $M_\infty = 1.0$, a nonlinear term enters Eqs. (13) and (14) that is of the same order of magnitude as the first term of these equations. Hence, for transonic flow, no linearized solutions are possible. In this region, a modified form of Eqs. (13) and (14) is required.[2]

The boundary conditions on Eqs. (13) and (14) are that the flow must be tangent to the body surface and that perturbation velocities go to zero far from the body. If $f(x,y,z) = 0$ describes the planar surface, then the flow tangency condition can be applied by writing

$$\vec{V} \cdot \nabla f = 0 \tag{15}$$

This is because the gradient of a function is by definition normal. Thus, requiring the dot product of the velocity vector to the gradient of the surface to be zero is the same thing as requiring the flow to be parallel to the surface $f(x,y,z)$. Substituting Eq. (11) into (15), neglecting products of small terms, and evaluating near $y = 0$, one obtains

$$v_b(x,0) = V_\infty \frac{\mathrm{d}y}{\mathrm{d}x} = V_\infty y' \tag{16}$$

Equation (16) says that the velocity in the y direction at the body surface is related to the local slope of that surface relative to the freestream direction. Thus, y' actually consists of an AOA term, a camber term, and a thickness distribution term. If the wing is undergoing rotational or plunging motion such as occurs in a dynamic sense, the local AOA at any point on the wing can change as a function of the rotation or plunge velocity. However, all of these terms can be considered independently due to the linearity of Eq. (13).

In analogy to thin-wing theory where the boundary condition [Eq. (16)] was determined for Eq. (13) by assuming thin planar surfaces, slender body theory assumes slender body surface slopes. We then evaluate the surface boundary condition as r approaches zero. For bodies of revolution, Eq. (16) is replaced by

$$v_b(x, r) = V_\infty \frac{dr_b}{dx} \tag{17}$$

for use in conjuction with Eq. (14).

Equations (13) and (14) are both second-order, linear, partial differential equations with variable coefficients. The axial solution to both Eqs. (13) and (14) is found by placing a distribution of "sources" and "sinks" along the x axis with the strength of these sources and sinks chosen to match boundary conditions at the wing or body surface. For lifting surfaces, one also needs a distribution of vortices or, for lifting bodies, a distribution of doublets. Solutions are also different for subsonic and supersonic flow. To illustrate this process, consider Eqs. (14) and (17) for the body case. Figure 3.4 illustrates a "source": a source is a flow radially outward, emanating in all directions.

Applying the conservation of mass to this spherical source at the origin, one obtains

$$\rho A V = ¢ \tag{18}$$

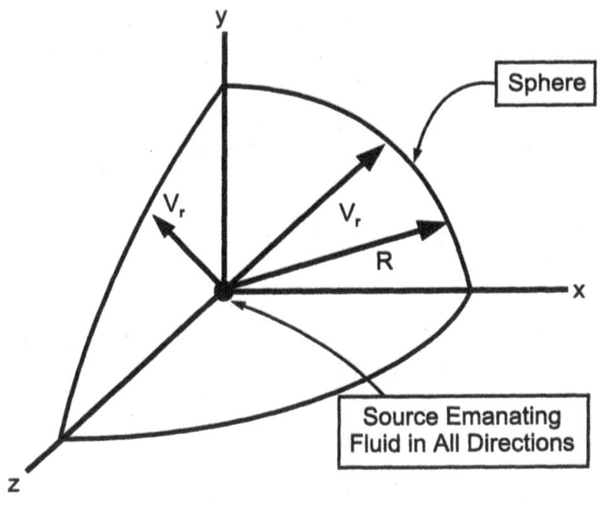

Fig. 3.4 Illustration of a spherical source.

For illustration purposes, assume incompressible flow so that $\rho = \not{c}$ and Eq. (18) can be written as

$$V_r = \frac{C_1}{A} \tag{19}$$

where A is the area of the spherical surface, that is, $4\pi R^2$, and R is spherical radius. Also, defining Q as the mass rate of flow or source strength, where

$$Q = C_1/2$$

then Eq. (19) can be written as

$$V_r = \overline{\Phi}_r = \frac{Q}{2\pi R^2} \tag{20}$$

Integration of Eq. (20) yields

$$\overline{\Phi} = -\frac{Q}{2\pi R} \tag{21}$$

This is the velocity potential for a single source at the origin. It can be shown that this is a solution to Eq. (14) if R is taken to be $\sqrt{x^2 + \beta^2 r^2}$. The source strength Q must be determined in the general sense by the boundary condition at the body surface, Eq. (17). Because the slope of the body surface changes, a distribution of "sources" and "sinks" (a sink is a source with a minus sign or a flow that goes radially inward versus outward) is required along the body axis to match the boundary condition. Figure 3.5 illustrates a distribution of sources and sinks along the x axis.

From Eq. (21), the velocity potential for the element of length $d\xi$ of Fig. 3.5, which has a local source strength per unit length of $q(\xi)$, is

$$d\Phi = -\frac{q(\xi)d\xi}{2\pi\sqrt{(x-\xi)^2 + r'^2}} \tag{22a}$$

where $r' = \beta r$.

The total potential is then the integral of Eq. (22a), or

$$\Phi = -\frac{1}{2\pi}\int_0^{\ell} \frac{q(\xi)d\xi}{\sqrt{(x-\xi)^2 + r'^2}} \tag{22b}$$

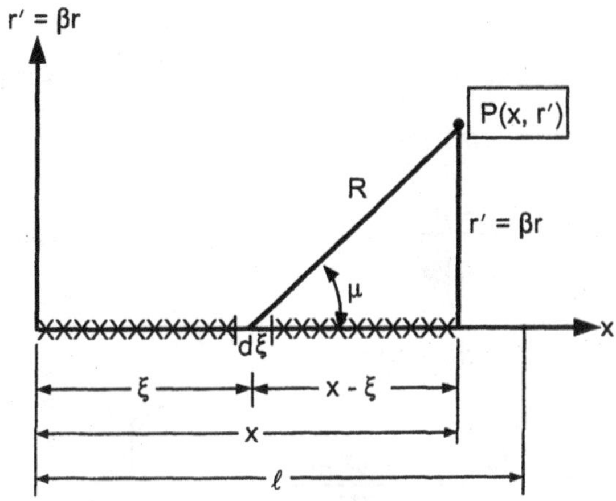

Fig. 3.5 Distribution of sources and sinks along x axis.

The perturbation velocity components are Φ_x and $\Phi_{x'}$, which are obtained by differentiating Eq. (22), that is,

$$u = \Phi_x = \frac{1}{2\pi}\int_0^{\ell} \frac{(x-\xi)\,q(\xi)\mathrm{d}\xi}{\left[(x-\xi)^2 + r'^2\right]^{3/2}} \tag{23}$$

$$\frac{v}{\beta} = \Phi_{r'} = \frac{r'}{2\pi}\int_0^{\ell} \frac{q(\xi)\mathrm{d}\xi}{\left[(x-\xi)^2 + r'^2\right]^{3/2}} \tag{24}$$

Applying the boundary condition, Eq. (17), to Eq. (24), one obtains

$$\frac{1}{r_b}\frac{\mathrm{d}r_b}{\mathrm{d}x}\frac{V_\infty}{\beta^2} = \frac{1}{2\pi}\int_0^{\ell} \frac{q(\xi)\mathrm{d}\xi}{\left[(x-\xi)^2 + r_b'^2\right]^{3/2}} \tag{25}$$

To solve Eqs. (23) and (24), one solves for the distribution of sources and sinks $q(\xi)$ that satisfies the boundary condition Eq. (25). This is done numerically by assuming a polynomial for the source distribution and revising the assumed polynomial until Eq. (25) is satisfied along the body surface. Using this value of $q(\xi)$, Eqs. (23) and (24) are then solved numerically for u and v. Using the compressible Bernoulli's equation,

$$\frac{V^2}{2} + \frac{\gamma}{\gamma-1}\frac{p}{\rho} = \not{C} \tag{26}$$

and the isentropic pressure relation,

$$p/\rho^{\gamma} = \mathcal{C}_1 \tag{27}$$

and the definition of pressure coefficient,

$$C_P = \frac{p - p_\infty}{\frac{1}{2}\rho_\infty V_\infty^2} \tag{28}$$

the exact pressure coefficient can be arrived at as

$$C_P = \frac{2}{\gamma M_\infty^2}\left\{\left[1 - \frac{\gamma - 1}{2} M_\infty^2\left(\frac{2u}{V_\infty} + \frac{u^2 + v^2 + w^2}{V_\infty^2}\right)\right]^{\frac{\gamma}{\gamma-1}} - 1\right\} \tag{29}$$

Using the perturbation process of neglecting higher-order products of small quantities, Eq. (29) reduces to

$$C_P = -\frac{2u}{V_\infty} \tag{30}$$

for two-dimensional and planar flows and to

$$C_P = -\frac{2u}{V_\infty} - \frac{v^2}{V_\infty^2} \tag{31}$$

for axisymmetric bodies. Eqs. (23), (24), and (31) give the complete value of pressure coefficient all along the body surface for zero AOA. Knowing the pressure coefficients, forces, and moments can be determined by numerical integration of these pressure coefficients along and around the body. If the body is at an AOA, the perturbation theory allows one to separate the axial and AOA solutions and add the results together linearly because the equation of motion is linear. It should be emphasized here that, because of the isentropic flow assumption, solutions to Eq. (31) in particular are generally limited to about $M_\infty = 3$ and below.

This section has attempted to show how linearized and slender body theory is derived from the full inviscid equations of motion. It has also attempted to generally show how the perturbation equations are solved using an example of a body of revolution at zero AOA in subsonic flow. Section 3.4 of this chapter will treat the body of revolution in more detail, Section 3.5 and 3.6 will treat the wing-alone solutions at subsonic and supersonic speeds, and Section 3.7 will treat the interference effects.

Before leaving the linearized theory and slender body theory, it is believed worthwhile to give some of the more useful closed-form results of linearized and slender body theory that can be used by engineers to estimate "ballpark" numbers for aerodynamics. These numbers can be estimated with a hand calculator, as opposed to the numerical solutions to the linear theory and slender body theory that will be discussed in the following sections of this chapter.

For two-dimensional flat wings in incompressible flow, the normal force coefficient is

$$C_{N_O} = 2\pi a \tag{32}$$

(where a is in radians) and the center of pressure is at a quarter of the mean geometric chord. In many references, C_{N_O} of Eq. (32) is equal to the lift coefficient C_{L_O}. For small AOAs, where Eq. (32) is primarily applicable, C_{N_O} and C_{L_O} are approximately equal. Compressibility can be accounted for by use of the Prandtl–Glauert rule as

$$C_N = \frac{C_{N_O}}{\sqrt{1 - M_\infty^2}} \tag{33}$$

The term C_{N_O} of Eq. (33) comes from Eq. (32).

The center of pressure remains at the quarter chord. Equation (33) is valid up to $M_\infty \approx 0.7$ before linear theory starts becoming invalid due to neglect of the transonic nonlinear term. At supersonic speeds, $M_\infty \geq 1.2$, Eq. (33) is replaced by

$$C_N = \frac{4a}{\sqrt{M_\infty^2 - 1}} \tag{34}$$

where a is in radians. The theoretical center of pressure at supersonic speeds is at the half-chord versus the quarter-chord in subsonic flow. In reality, the center of pressure transitions in a smooth fashion from the forward point in subsonic flow to the more aft point supersonically. This transition typically occurs between $M_\infty = 0.7$ and about 2.0. Equations (32) to (34) are all based on wing area. Axial force of the flat, zero-thickness wings is due entirely to skin friction.

For flat wings of finite span in incompressible flow, an approximate relationship for the normal force is[2]

$$C_{N_O} \cong \frac{2\pi}{1 + 2/(e_1 \mathrm{AR})} a \tag{35}$$

Equation (35) can be improved upon by using the actual local C_{N_a} of a given airfoil section, as opposed to the two-dimensional value of 2π if the local value of C_{N_α} is known. In many cases, the local value is not known, so Eq. (35) is the more generally used formula. The term e_1 of Eq. (35) is known as the span efficiency factor and is typically about 0.85 to 0.95. If an average value of 0.9 is used, Eq. (35) can be reduced to

$$C_{N_O} \approxeq \left(\frac{2\pi}{1 + \dfrac{2.22}{\mathrm{AR}}} \right) \alpha \tag{36}$$

For infinite aspect ratio, Eq. (36) reduces to Eq. (32). To account for compressibility effects subsonically, one must use the Gothert extension to the Prandtl–Glaurt rule,[2] which states that, for zero-thickness wings,

$$\left(C_N\right)_{M_\infty, \alpha, \mathrm{AR}} = \left(\frac{1}{1 - M_\infty^2} \right) C_{N_O, \beta\alpha, \beta\mathrm{AR}} \tag{37}$$

Equation (37) states that, to compute the compressible normal force of a wing of aspect ratio AR at an AOA α and Mach number M_∞, we must use Eq. (36) where AR and α of Eq. (36) are replaced by βAR and $\beta\alpha$, respectively. At subsonic speeds, the wing center of pressure theoretically is at the quarter-chord.

Unfortunately for finite wings in supersonic flow, no simple expression exists similar to Eqs. (36) and (37) in subsonic flow. One must use the methods described in Section 3.5. This is because of the tip effects at the wing leading edge and root chord caused by the perturbations of the Mach waves emanating from these points and affecting the wing. If the flow is all two-dimensional, Eq. (34) can be utilized. Also, Eq. (34) could be used as an upper bound for supersonic wing normal force at small AOAs.

Another useful result for very low aspect ratio wings that was derived from slender body theory is[1]

$$C_N = \frac{\pi}{2} \mathrm{AR}\, \alpha \tag{38}$$

This result is independent of Mach number and has proved useful for a first approximation to wing normal force coefficient for low AOAs and aspect ratios of 0.5 and less. Strictly speaking, it was derived based on slender triangular wings from slender body theory. However, the author has found it useful for any low aspect ratio wing.

Turning now to the body, slender body theory gives the quite useful result that[1,3]

$$C_{N_B} = 2\left(\frac{A_B}{A_{\mathrm{REF}}} \right) \alpha = 2\left(\frac{R_b}{R_r} \right)^2 \alpha \tag{39}$$

and

$$\frac{x_{CP}}{\ell_n} = 1 - \frac{\text{Vol}_n}{\pi R_n^2 \ell_n} \tag{40}$$

where Vol_n is the volume of the nose.

Equation (39) is quite reasonable for configurations that are primarily a nose shape, as slender body theory gives no additional normal force due to a cylindrical afterbody. Experience has shown for most practical missile shapes, Eq. (39) gives a low value of C_N by as much as a factor of two around $M_\infty = 3$ to 4. However, at lower Mach numbers, the slender body theory results are closer to reality. Likewise Eq. (40) gives values of x_{CP} that vary from about 45 to 67% of the nose length, depending on the nose shape. For configurations with no afterbody, these results are reasonable. Experience has shown that, for practical missile shapes with no boattail, the low AOA center of pressure is around the point where the nose joins the afterbody.

Finally, for slender wing–body configurations[1,3]

$$C_N = 2\left(\frac{R_B}{R_r}\right)^2 a\left[\left(\frac{b}{2R_B}\right)^2 + \frac{b}{R_B} + \left(\frac{R_B}{R_B + b/2}\right)^2\right] \tag{41}$$

Equation (41) reduces to Eq. (39) when the wing span is zero. It also shows a positive contribution to normal force when the wings are present. To estimate the center of pressure of the entire configuration, one can divide the lift of the wing and body by subtracting Eq. (39) from (41) to obtain the wing and interference terms. The configuration center of pressure at low AOA can be approximated by (relative to nose tip)

$$\frac{x_{CP}}{\ell} = \frac{(x_{CP}/\ell)_B C_{N_B} + (x_{CP}/\ell)_{W(B)} C_{N_{W(B)}}}{C_N} \tag{42}$$

The $(x_{CP}/\ell)_B$ term is assumed at the body nose and afterbody juncture as a first approximation; C_{N_B} comes from Eq. (39); $(x_{CP}/\ell)_{W(B)}$ is at a point on the body corresponding to the 1/4 to 1/2 chord point; and $C_{N_{W(B)}}$ comes from subtracting Eq. (39) from Eq. (41). Equation (42) is simply a weighted average of the center of pressure of the body normal force and wing normal force contributions. The body center of pressure is typically near the nose–cylinder juncture and the wing–body normal force center of pressure is typically around the part on the body where the root chord of the wing meets the body. This can vary, depending on Mach number and wing sweep angle, from about 10% of the root chord length to about 75% of the root chord length.

Linear theory and slender body theory are not nearly as useful for rough estimates of axial force because both theories assume inviscid flow and two of the drag components, skin friction and base drag, are viscous components. Hence, the axial force will be dealt with in more detail in the remainder of this book.

IV. Hybrid Theory of Van Dyke

The first of the linearized theory methods to be discussed is the hybrid theory of Van Dyke (HTVD).[4] The HTVD combines a second-order axial solution to the potential Eq. (14) with a first-order crossflow solution first espoused by Tsien.[5] The advantage of this method is that it gives second-order accuracy in the axial direction where first-order accuracy is generally unacceptable for drag computations. On the other hand, first-order accuracy in the crossflow plane is typically acceptable for normal force and center of pressure computations. The fundamental reason for the second-order requirement in the axial direction is that perturbations in the flow, due to the presence of a body, have more effect in the axial compared to the normal force direction. The main reason for this larger perturbation in the axial direction has to do primarily with the nose shape. If the nose is long and slender (say greater than 3 calibers), perturbations are smaller and one may be able to use a first-order accurate solution for axial force coefficients. However, few practical configurations have nose lengths that long with accompanying low values of body slope along the nose. Hence, to get axial force accuracy compatible with a goal of $\pm 10\%$ requires second-order methods, whereas $\pm 10\%$ accuracy on C_N can be obtained with first-order methods in many cases.

As already mentioned, the hybrid theory comes from the potential equation of fluid dynamics. It is limited to supersonic flow (we have used this method down to $M_\infty = 1.2$) where the assumption of isentropic flow (shock waves are weak) can be made. Thus, the upper Mach number range is limited to about $M_\infty = 2.0$ to 3.0, depending on the body shape. Also, the slope of the body surface must be less than the Mach angle. The Tsien solution, or crossflow part of the solution, comes from the linearized perturbation equation. On the other hand, the second-order solution to the axial flow is found by obtaining a particular solution to a reduced version of the full potential equation. This is the key to the accuracy improvement afforded by Van Dyke's solution, in that some of the nonlinearity inherent in the axial flow problem is brought into the solution by this process. The beauty of the Van Dyke method is that this particular second-order solution is given entirely in terms of the first-order solution. That is, one simply solves the first-order perturbation solution for the axial flow and then solves a nonlinear algebraic equation for the second-order solution where the boundary condition at the body is satisfied.

In equation form, the general first-order perturbation problem is Eq. (14) repeated here,[4]

$$\Phi_{rr} + \Phi_r/r + \Phi_{\varphi\varphi}/r^2 - (M_\infty^2 - 1)\Phi_{xx} = 0 \tag{43}$$

with boundary conditions that do not allow any upstream disturbances:

$$\Phi_r(0,r,\varphi) = \Phi_x(0,r,\varphi) = 0 \tag{43a}$$

and that require the flow to be tangent to the body surface:

$$\Phi_r(x, r_b, \varphi) + \sin a \cos \varphi = \frac{dr}{dx}\left[\cos a + \Phi_x(x, r_b, \varphi)\right] \tag{43b}$$

The subscripts in Eq. (43) indicate partial derivatives. The solution to Eq. (43) is satisfied identically by

$$\Phi(x, r, \varphi) = \psi_1(x, r)\cos a + \zeta_1(x, r)\sin a \cos \varphi \tag{44}$$

The first term of Eq. (44) is the first-order axial solution, and the second term is the first-order crossflow solution. Because the equation is linear, these two solutions can be found independently and then added together. The axial solution, $\psi_1(x,r)$, for a general body is found by placing a distribution of sources and sinks along the x axis and satisfying the boundary conditions at each point, as discussed in Section 3.3. The crossflow solution, $\zeta_1(x,y)$, is found by placing a distribution of doublets along the axis, again satisfying the boundary conditions. A doublet is simply the limit of a source and sink of equal strength and, as the distance between them approaches zero, the doublet strength tends toward a finite value.[1]

Equation (43), which is linear, separates out naturally into two independent solutions for the axial and crossflow. These two solutions can then be linearly combined for total velocities and pressures. The first-order axial flow problem is

$$\psi_{1rr} + \psi_{1r}/r - \beta^2\psi_{1xx} = 0 \tag{45}$$

with boundary conditions

$$\psi_{1r}(0,r) = \psi_{1x}(0,r) = 0$$

$$\psi_{1r}(x,r_b) = r'_b[1 + \psi_{1x}(x,r_b)] \tag{46}$$

The first-order crossflow problem is

$$\zeta_{1rr} + \zeta_{1r}/r - \zeta_1/r^2 - \beta^2\zeta_{1xx} = 0 \tag{47}$$

with boundary conditions:

$$\zeta_{1r}(0,r) = \zeta_{1x}(0,r) = 0$$

$$1 + \zeta_{1r}(x,r_b) = r_b'\,\zeta_{1x}(x,r_b) \tag{48}$$

The general solution of Eqs. (45) and (47) is found by a distribution of sources or sinks along the x axis for Eq. (45) and doublets for Eq. (47). In equation form, these solutions are

$$\psi_1(x,r) = -\int_0^{x-\beta r} \frac{f(\xi)\,d\xi}{\sqrt{(x-\xi)^2-\beta^2 r^2}} \tag{49}$$

$$\zeta_1(x,r) = \frac{1}{\beta r}\int_0^{x-\beta r} \frac{(x-\xi)g(\xi)d\xi}{\sqrt{(x-\xi)^2-\beta^2 r^2}} \tag{50}$$

The functions $f(\xi)$ and $g(\xi)$ are defined for linear, quadratic, or square root strength sources; sinks; or doublets, depending on the shape of the body profile. If a first-order-only solution were required, only linear elements would suffice. However, the second-order solution will require second derivatives of the first-order solution, which means the strength of the source/sink and doublet functions must be higher order to preserve meaning. Van Dyke then formed a second-order accurate axial perturbation equation. This equation is given by

$$\begin{aligned}\psi_{2rr} + \psi_{2r}/r - \beta^2\psi_{2xx} = {} & \mathrm{M}_\infty^2\{2\psi_{1r}\psi_{1xr} + [2+(\gamma-1)M_\infty^2]\,\psi_{1x}\psi_{1xx} \\ & + \psi_{1r}^2\psi_{1rr} + [(\gamma-1)/2]M_\infty^2\psi_{1r}^2\psi_{1xx}\end{aligned} \tag{51a}$$

with boundary conditions

$$\begin{aligned}\psi_{2r}(0,r) &= \psi_{2x}(0,r) = 0 \\ \psi_{2r}(x,r_b) &= r_b'\big[1+\psi_{2x}(x,r_b)\big]\end{aligned} \tag{51b}$$

Note that the left-hand side of Eq. (51a) involves the second-order axial solution, which is defined in terms of the first-order axial solution on the right-hand side. The solution of Eq. (51a) was simplified considerably by Van Dyke when he found a particular solution defined by

$$(\psi_P)_2 = M_\infty^2\big[\psi_{1x}(\psi_1 + Nr\psi_{1r}) - (1/4)r\psi_{1r}^3\big] \tag{52a}$$

where $N = \big[(\gamma+1)/2\big]\,M_\infty^2/\beta^2)$

The importance of Eq. (52a) is the fact that a second-order particular solution can be computed entirely in algebraic terms from the definition of the first-order solution. Velocity components can be found by differentiating Eq. (52a). These velocity components are

$$\psi_{2x} = M_\infty^2 \left[\psi_{1xx}(\psi_1 + Nr\,\psi_{1r}) + \psi_{1x}\,(\psi_{1x} + Nr\,\psi_{1xr}) - 3/4r\,\psi_{1xr}\,\psi_{1r}^2\right] \tag{52b}$$

$$\begin{aligned}\psi_{2r} = M_\infty^2 \{&\psi_{1xr}(\psi_1 + Nr\,\psi_{1r}) + \psi_{1x}\left[(N+1)\psi_{1r} + Nr\,\psi_{1rr}\right] \\ &- 1/4\,\psi_{1r}^2\,(\psi_{1r} + 3r\,\psi_{1rr})\}\end{aligned} \tag{52c}$$

Note that the velocity components are also defined totally in terms of the first-order solution.

The complete solution to Eq. (51) involves the particular plus a complementary function. That is,

$$\psi_2 = (\psi_P)_2 + (\psi_C)_2 \tag{53a}$$

where $(\psi_P)_2$ and $(\psi_C)_2$ denote the second-order particular and complementary solutions, respectively. The complementary solution is found by a solution of the first-order equation but with the boundary condition, Eq. (51b), replaced by Eq. (53b)

$$(\psi_P)_{2r} + (\psi_C)_{2r} = r'_b[1 + (\psi_P)_{2x} + (\psi_C)_{2x}] \tag{53b}$$

The new constants in the distribution function for the sources and sinks are defined to satisfy Eq. (53b).

The hybrid theory then combines the second-order accurate axial solution with a first-order accurate crossflow solution to get more accurate normal forces as well.

Once the second-order axial perturbation velocity components ψ_{2x}, ψ_{2r} are computed with the first-order crossflow components ζ_{1x} and ζ_{1r}, the total perturbation velocities are then

$$\frac{u}{V_\infty} = (\cos a)\,(1 + \psi_{2x}) + (\sin a \cos\varphi)\,\zeta_{1x} \tag{54a}$$

$$\frac{v}{V_\infty} = \cos a\,(\psi_{2r}) + (\sin a \cos\varphi)\,(1 + \zeta_{1r}) \tag{54b}$$

$$\frac{w}{V_\infty} = -(\sin a \sin\varphi)\left(1 + \frac{\zeta_1}{r}\right) \tag{54c}$$

where w is the circumferential velocity component.

The pressure coefficient at each body station is then

$$C_P(x,\varphi) = \frac{2}{\gamma M_\infty^2}\left\{\left[1+\frac{\gamma-1}{2}M_\infty^2\left(1-\frac{U^2+v^2+w^2}{V_\infty^2}\right)\right]^{\frac{\gamma}{\gamma-1}}-1\right\} \tag{55}$$

Finally, the force coefficients are

$$C_A = \frac{2}{\pi r_r^2}\int_0^\ell\int_0^\pi C_P(x,\varphi)\frac{r\mathrm{d}r}{\mathrm{d}x}\,\mathrm{d}\varphi\,\mathrm{d}x \tag{56}$$

$$C_N = -\frac{2}{\pi r_r^2}\int_0^\ell\int_0^\pi C_P(x,\varphi)\cos(\varphi) r\,\mathrm{d}\varphi\,\mathrm{d}x \tag{57}$$

$$C_M = \frac{1}{\pi r_r^3}\int_0^\ell\int_0^\pi C_P(x,\varphi)\cos(\varphi) xr\,\mathrm{d}\varphi\,\mathrm{d}x \tag{58}$$

and the center of pressure in calibers from the nose is

$$\frac{x_{CP}}{d} = -C_M/C_N \tag{59}$$

It should be pointed out that, in the actual numerical integration of Eqs. (56), (57), and (58), the integration must be carried out in segments of the body between each discontinuity due to the discontinuous pressure distribution.

Van Dyke[6,7] defined quadratic solutions for the distribution of sources and sinks to allow Eq. (49) to be solved on other than smooth bodies of revolution. Therefore, the flexibility of the HTVD is greatly increased, as bodies with curvature discontinuities can be treated in closed form. Also, the HTVD is limited to pointed bodies of revolution. Bluntness will be considered later.

Figures 3.6 and 3.7 illustrate the pressure distribution along a 10-deg cone–cylinder at zero- and 12-deg AOAs, respectively, at $M_\infty = 2.07$ and $M_\infty = 2.0$. Good agreement of the HTVD and experiments is obtained. Figure 3.8 compares the pressure distribution on a 3.5-caliber tangent ogive nose–cylinder computed by the HTVD and the method of characteristics. Excellent agreement between the second-order theory and the exact numerical solution is obtained for this $M_\infty = 3.24$ case. Finally, Fig. 3.9 presents the pressure distribution on a short 2.0-caliber tangent ogive nose at $M_\infty = 1.2$ and 2.0. The HTVD is compared to computational fluid dynamics solutions of full Euler equations. Again, very good agreement is obtained. Figures 3.6 to 3.9 illustrate the effectiveness of the second-order perturbation solution of Van Dyke in terms of its accuracy at fairly low supersonic Mach numbers and over a range of nose lengths. Integration of pressure coefficients allows reasonably accu-

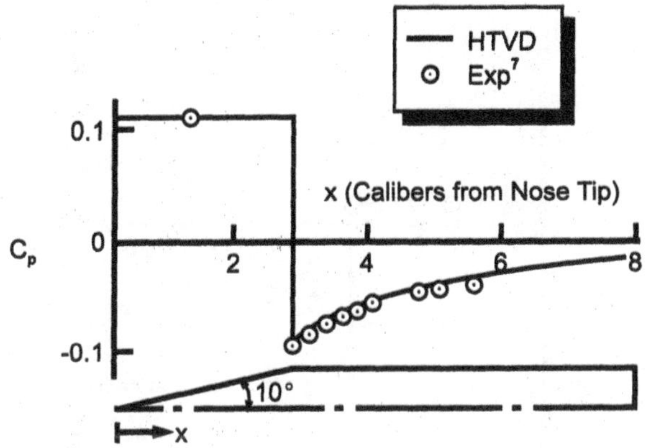

Fig. 3.6 Pressure coefficient on cone–cylinder ($M_\infty = 2.07$, $a = 0°$).

rate axial and normal forces to be computed. The accuracies are generally within the desired goals of ±10% of experiment.

V. Lifting Surface Theory

Lifting surface theory (LST) refers to the solution of the flow over a thin three-dimensional wing where the distribution of pressure is allowed to vary in both the spanwise and chordwise directions. The fundamental equation is the three-dimensional perturbation equation, here written in rectangular coordinates as

$$\left(1 - M_\infty^2\right)\Phi_{xx} + \Phi_{yy} + \Phi_{zz} = 0 \tag{60}$$

The flow tangency boundary condition requires

$$\left.\begin{aligned} \Phi_z &= \frac{\partial z_u}{\partial x} \text{ at } z = 0^+ \\ \Phi_z &= \frac{\partial z_\ell}{\partial x} \text{ at } z = 0^- \end{aligned}\right\} \text{ for } (x, y) \text{ on } S \tag{60a}$$

The assumptions involved in this LST, as applied to most missile configurations, are therefore small perturbations in the flow due to the presence of the wing, and the thickness and camber effects are zero or small compared to AOA effects. If the wing thickness is neglected and we limit ourselves to missiles, then wing camber can also be neglected. Then the boundary conditions in Eq. (60a) become

$$\Phi_z = -a \tag{60b}$$

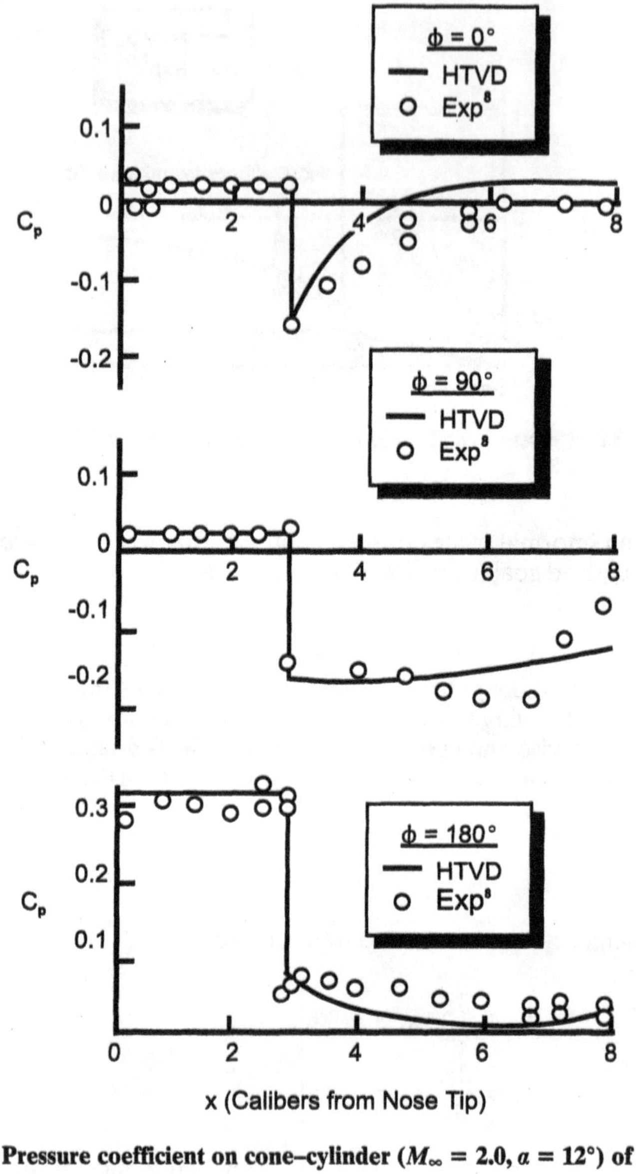

Fig. 3.7 Pressure coefficient on cone–cylinder ($M_\infty = 2.0$, $\alpha = 12°$) of Fig. 3.6.

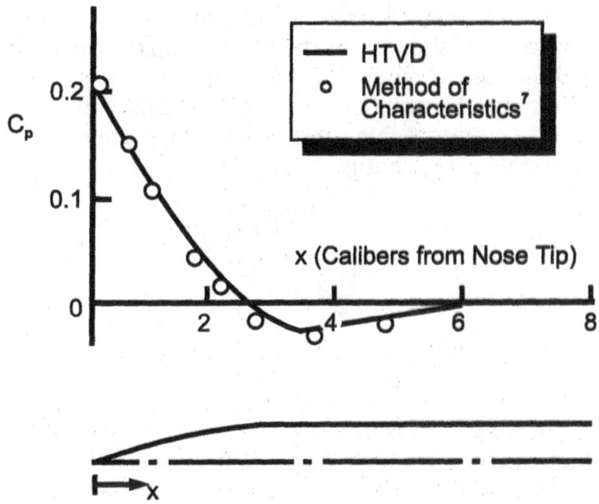

Fig. 3.8 Pressure coefficient on tangent ogive–cylinder (M_∞ = 3.24, α = 0°)

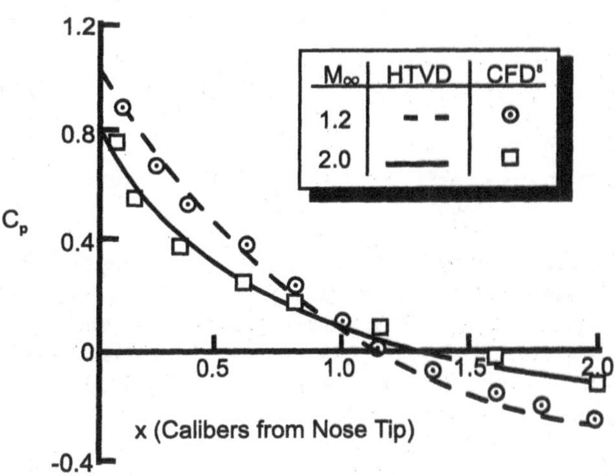

Fig. 3.9 Pressure coefficient on 2-caliber tangent ogive (α = 0). (CFD = computational fluid dynamics).

for both the upper and lower surfaces. In addition to this boundary condition, the Kutta condition (which requires the velocity on the upper and lower surfaces at the trailing edge to be equal) is also imposed for subsonic flow.

Equation (60) may be simplified somewhat by using the Gothert rule (see Ref. 2) to relate the compressible subsonic normal force or pitching moment to the incompressible case. That is,

$$(C_N)_{M_\infty, \mathrm{AR}, a} = \frac{(C_N)_{0, \beta\mathrm{AR}, \beta a}}{1 - M_\infty^2} \tag{61a}$$

$$(C_M)_{M_\infty, \mathrm{AR}, a} = \frac{(C_M)_{0, \beta\mathrm{AR}, \beta a}}{1 - M_\infty^2} \tag{61b}$$

The way this rule is applied is as follows: given a wing of aspect ratio AR, freestream Mach number M_∞, and AOA a, the normal force and pitching moment can be obtained by calculating the normal force and pitching moment on another wing with Mach number zero, aspect ratio βAR, and AOA βa.

Using the above relations, the normal force and pitching moment on a given wing at any subsonic Mach number may be found by calculating the aerodynamics of a similar wing at zero Mach number.

For $M_\infty = 0$, Eq. (60) reduces to La Place's equation

$$\nabla^2 \Phi = 0 \tag{62}$$

with boundary condition, eq. (60b).

There are many methods to solve Eq. (62). The one used here is that of Chadwick et al.,[9] which closely follows Ashley et al.[1] The velocity potential Φ is given by

$$\Phi(x, y, z) = -\frac{1}{8\pi} \iint_S \frac{\Delta C_P(x_1, y_1)}{(y - y_1)^2 + z^2} z\left[1 + \frac{x - x_1}{\sqrt{(x - x_1)^2 + (y - y_1)^2 + z^2}}\right] \mathrm{d}x_1 \mathrm{d}y_1 \tag{63}$$

Here, x_1, y_1 are coordinates of an element of the lifting surface that has a differential pressure coefficient of ΔC_P between the lower and upper surfaces at the point (x_1, y_1). It is required to determine the pressure loading over the entire surface. Following Chadwick,[9] Eq. (63) is first differentiated with respect to z and the limit as $z \to 0$ taken. The result is then equated to the boundary condition, Eq. (60b) to obtain

$$a(x_1, y_1) = \frac{1}{8\pi} \iint \frac{\Delta C_P(x_1, y_1)}{(y - y_1)^2} \left[1 + \frac{x - x_1}{\sqrt{(x - x_1)^2 + (y - y_1)^2}}\right] \mathrm{d}x_1 \mathrm{d}y_1 \tag{64}$$

The cross on the y_1 integral indicates a singularity at $y = y_1$, in which case Mangler's principal-value technique[1] can be applied. Details of the solution of the integral Eq. (64) for $\Delta C_P\,(x,y)$ will not be repeated here as they are given in detail in many references (see, for example, Chadwick[9]). Worthy of note, however, is the fact that Eq. (64) is an integral equation for which the wing loading ΔC_P is to be found as a linear function of AOA. This wing loading is first approximated by a series expansion with a set of unknown coefficients of number equal to the number of surface elements on the wing planform. That allows each ΔC_P to be influenced by all other elements of the wing. The unknown coefficients in each ΔC_P series are found by solution of an inverse matrix. The value of $\Delta C_P\,(x,y)$ is then calculated.

Once the span loading $\Delta C_P\,(x,y)$ is known over the entire wing surface, the local normal force coefficient at a given spanwise location is

$$c_n = \frac{1}{c}\int_{x_{LE}}^{x_{TE}} \Delta C_P \mathrm{d}x \tag{65}$$

The total normal force coefficient for the entire wing is

$$C_N = \frac{2}{S_{\mathrm{ref}}}\int_0^{b/2} c c_n \mathrm{d}y \tag{66}$$

The pitching moment coefficient of a given airfoil section, about the point where the wing leading edge intersects the body, is then (positive leading edge up)

$$c_m = -\frac{1}{c\ell_{\mathrm{ref}}}\int_{x_{LE}}^{x_{TE}} x\Delta C_P \mathrm{d}x \tag{67}$$

The total pitching moment coefficient becomes

$$C_M = \frac{2}{S_{\mathrm{ref}}}\int_0^{b/2} c c_m \mathrm{d}y \tag{68}$$

If it is desired to calculate the pitching moment about some other reference point, then

$$C_{M_O} = C_M + C_N\frac{x_0}{\ell_{\mathrm{ref}}} \tag{69}$$

where x_0 is the distance from the reference point to the juncture of the wing leading edge with the body. The center of pressure of an airfoil section is

$$(x_{CP})_i = -\frac{c_m}{c_n} \tag{70}$$

or, for the entire wing,

$$x_{CP} = -\frac{C_M}{C_N} \tag{71}$$

Finally, the spanwise center of pressure of a wing semispan is

$$y_{CP} = \frac{\int_0^{b/2} c c_n y dy}{\int_0^{b/2} c c_n dy} \tag{72}$$

Equations (65), (66), (67), (68), and (72) can be solved by numerical quadrature, such as Simpson's rule, with special attention given to the leading edge singularity.

It should also be mentioned that, if one is interested in dynamic derivatives,[10] these aerodynamic parameters can be obtained by a modification to the boundary condition, Eq. (60a). That is, for rolling and pitching motions, the AOA in Eq. (60a) is replaced by

$$a(x, y) = a_o + \frac{py}{V_\infty} + \frac{q(x - x_{\text{ref}})}{V_\infty} + \frac{\dot{a}t}{V_\infty} \tag{73}$$

Equation (62) is a linear partial differential equation, so that solutions can be combined together in a linear fashion. For roll damping, then, simply set $a_o = q = \dot{a} = 0$, and the boundary condition is

$$a(x, y) = \frac{py}{V_\infty} \tag{73a}$$

Likewise, for constant pitching velocity q, $a_0 = p = \dot{a} = 0$, and

$$a(x, y) = \frac{q(x - x_{\text{ref}})}{V_\infty} \tag{73b}$$

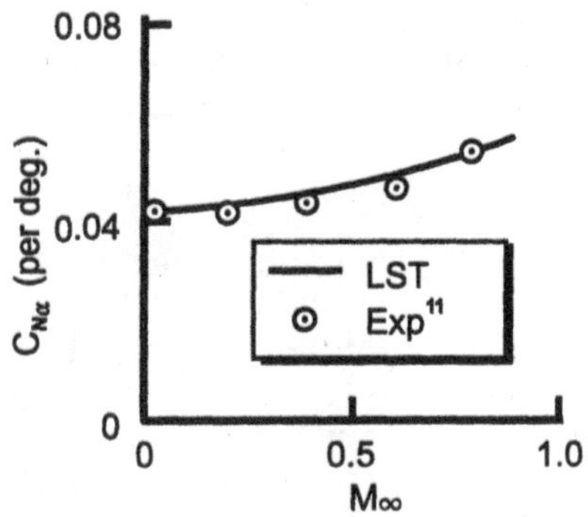

Fig. 3.10 Normal force coefficient slope for a thin wing at subsonic speeds (AR = 2.31, λ = 0.27, and Λ_{LE} = 56.5°).

Finally, for constant vertical acceleration, simply set $a_0 = p_0 = q = 0$, and the boundary condition is then

$$a(x, y) = \frac{\dot{a}t}{V_\infty} \tag{73c}$$

Pitch damping moment is then the sum of the results from the linearized Eq. (60) with separate boundary conditions of Eqs. (73b) and (73c). The dynamic derivatives will be discussed in Sections 3.7 and 3.8.

Figures 3.10 and 3.11 illustrate the application of LST to two wing planforms. The first case is an aspect ratio 2.31 case with a taper ratio of 0.27 and a leading edge sweep angle of 56.5 deg. Figure 3.10 plots the normal force coefficient slope as a function of Mach number. As seen in the figure, the LST agrees quite well with the data. The second example is for an aspect ratio 1.0 rectangular wing at a low Mach number of 0.2. Here, lift coefficient is shown as a function of AOA. Note the nonlinearity that occurs above about a 5-deg AOA. For this reason, LST by itself is limited to low AOA applications. Chapter 5 will discuss the treatment of nonlinear aerodynam-

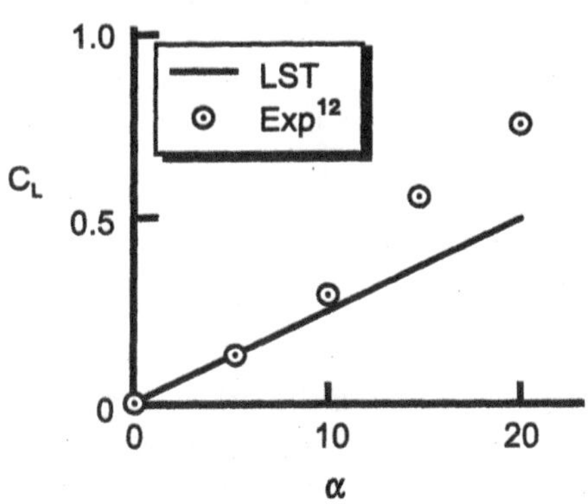

Fig. 3.11 Lift coefficient for a rectangular wing (M_∞ = 0.2, AR = 1.0, λ = 1.0, and Λ_{LE} = 0°).

ics. It is worthy to note that the LST gives a good estimate of the wing aerodynamics at low AOA. This fact is one of the fundamental underpinnings upon which semi-empirical methods discussed in this book are based.

VI. Three-Dimensional Thin Wing Theory

Three-dimensional thin wing theory (TDTWT)[1] is quite similar to the LST in the sense that the same perturbation equation (60) is used. The only difference is that TDTWT is normally used to represent the supersonic flow solutions of Eq. (60) versus LST for the subsonic solutions. Because, for supersonic flow, Eq. (60) is hyperbolic, versus elliptic for the subsonic case, solutions are generally easier to obtain because no upstream influence is felt by a disturbance at a given point on the wing surface. In contrast, subsonic solutions required a matrix inversion at each wing element to determine the unknown coefficients used to determine the pressure differential from lower to upper surfaces. On the other hand, the assumptions of TDTWT are the same as for LST in that they both assume small perturbations in an isentropic flow. The isentropic flow assumption means that only weak shock waves are allowed.

In contrast to the body solutions generated by Van Dyke, adequate wing solutions for axial force can be obtained at higher Mach numbers using perturbation theory. Reasons for the applicability of perturbation theory on wings at high Mach numbers is severalfold. First, low slopes are present on most wing planforms (thickness is generally very small); second, wing frontal area is generally less than 10% of the body frontal area; and, finally, in the region of leading edge bluntness, where perturbation theory is invalid, modified Newtonian theory can be used for wave drag calculation.

The most general boundary conditions for Eq.(60) in isentropic flow are the flow tangency condition specified by

$$\frac{w(x, y,0\pm)}{V_\infty} = \Phi_z = \frac{\partial F}{\partial x} = \left(\frac{\partial z}{\partial x}\right)_y + a + \frac{py}{V_\infty} + \frac{q(x - x_{\mathrm{ref}})}{V_\infty} + \dot{a}t \qquad (74)$$

and the perturbation velocities must vanish upstream from the point where the disturbance originates. Here, $F(x,y,z)$ describes the wing surface. Mathematically, this can be stated in the form

$$u(o^-,y,z) = v(o^-,y,z) = w(o^-,y,z) = 0 \qquad (75)$$

Because Eq. (60) is linear, individual solutions can be added together. This allows individual treatment of Eq. (74) boundary condition for drag, lift, roll, and pitch damping computations. For wave drag calculations, only the first term of Eq. (74) is retained and the other terms are set to zero. For lift calculations, the AOA a is retained and the other terms are set to zero. For roll damping, the third term of Eq. (74) is retained and the other terms are set to zero. For pitching rate, the q term of Eq. (74) is retained and the other terms are set to zero. Finally, for a constant vertical acceleration, the last term is retained and the other four terms are set to zero. Pitch damping moment, $C_{M_q} + C_{M_{\dot{\alpha}}}$, normally refers to the sum of the terms due to a constant pitch rate and constant vertical acceleration.

A. Axial Force Wave Drag

The solution to Eq. (60), using the first term of Eq. (74) as the boundary condition, will give the axial force coefficient of a sharp wing. If the leading edge is blunt, modified Newtonian theory can be used in conjunction with perturbation theory. The general solution to Eq. (60) is[1] a distribution of sources and sinks defined by

$$\Phi(x,y,0) = -\frac{w(xy)}{\pi} \iint_R \frac{dx_1 dy_1}{\sqrt{(x-x_1)^2 - \beta^2(y-y_1)^2}} \tag{76}$$

The pressure coefficient at any point on the wing surface is

$$C_P = -2\Phi_x\,(x,y,0) \tag{77}$$

The perturbation velocity Φ_x, at a given point p, is dependent on the location of the point with respect to the line of sources and sinks that generates the wing leading edge or other discontinuity and whether this point is in a subsonic or supersonic flow region. For example, referring to Fig. 3.12a, if point P is at P_1, and the wing generator is a subsonic source or sink line (SOSL), then[13]

$$\Phi_x = -\frac{2w(x_{p1}, y_{p1})}{\pi\beta\sqrt{\eta^2 - 1}} \cosh^{-1}\sqrt{\frac{\eta^2 - 1}{\sigma^2 - 1}} \tag{78}$$

where w is determined from the boundary condition and is (for the airfoil section at $y = y_{p1}$):

$$w(x_{p1}, y_{p1}) = \left.\frac{\partial z}{\partial x}\right|_{x=x_{P_1}}$$

In Eq. (78), the definitions

$$\eta = \frac{k}{\beta}$$

$$k = \tan\Lambda \tag{78a}$$

$$\sigma = \frac{ky_p}{x_p}$$

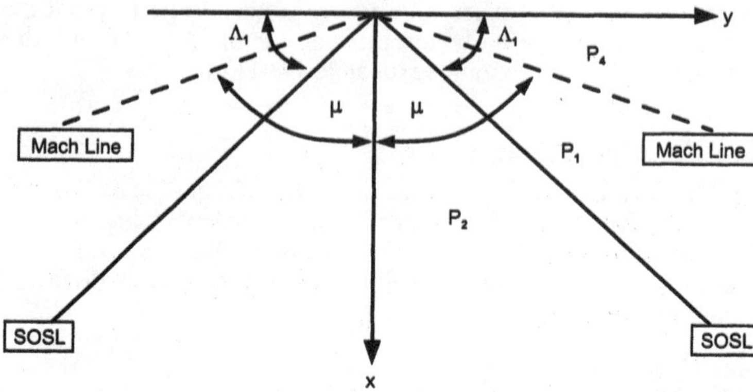

Fig. 3.12a Triangular surface symmetric about x axis for subsonic SOSL.

have been used. If $P = P_2$, the induced velocity at P_2 due to a given SOSL is

$$\Phi_x = -\frac{2w(x_{p2}, y_{p2})}{\pi\beta\sqrt{\eta^2 - 1}} \cosh^{-1}\sqrt{\frac{\eta^2 - \sigma^2}{1 - \sigma^2}} \tag{79}$$

At the wing tip, there is an additional disturbance within the Mach line emanating from the tip leading edge (Fig. 3.12b). The induced velocity in this region, $P = P_3$, is

$$\Phi_x = -\frac{w(x_{p3}, y_{p3})}{\pi\beta\sqrt{\eta^2 - 1}} \cosh^{-1}\left[\frac{\eta^2 + |\sigma|}{\eta(|\sigma| + 1)}\right] \tag{80}$$

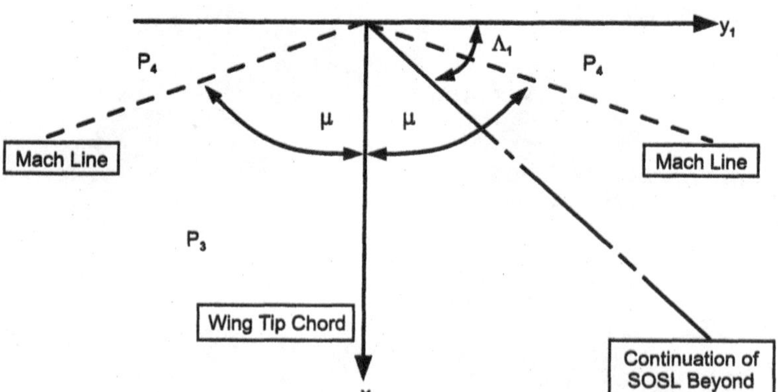

Fig. 3.12b Wing tip effects for subsonic SOSL.

The absolute value of σ is taken because σ is actually negative for the point P_3. The induced velocity at any point, say $P = P_4$, outside of the Mach lines emanating from the beginning of the SOSL is zero because this point is out of the zone of influence.

If the wing generator is supersonic, the Mach lines from point 0 in Fig. 3.13a lie behind the SOSL. If in Fig. 3.13a, $P = P_1$, then the induced velocity at P_1 due to the disturbance caused by the SOSL is[13]

$$\Phi_x = -\frac{w(x_{p1}, y_{p1})}{\beta\sqrt{1-\eta^2}} \tag{81}$$

If $P = P_2$, the induced velocity is

$$\Phi_x = -\frac{w(x_{p2}, y_{p2})}{\pi\beta\sqrt{1-\sigma^2}}\left[\pi - 2\sin^{-1}\sqrt{\frac{\eta^2-\sigma^2}{1-\sigma^2}}\right] \tag{82}$$

Referring to Fig. 3.13b, the additional induced velocity inside the area bounded by the tip and the Mach line emanating from the tip ($P = P_3$) is:

$$\Phi_x = -\frac{w(x_{p3}, y_{p3})}{\pi\beta\sqrt{1-\eta^2}}\cos^{-1}\left[\frac{|\sigma|+\eta^2}{\eta(1+|\sigma|)}\right] \tag{83}$$

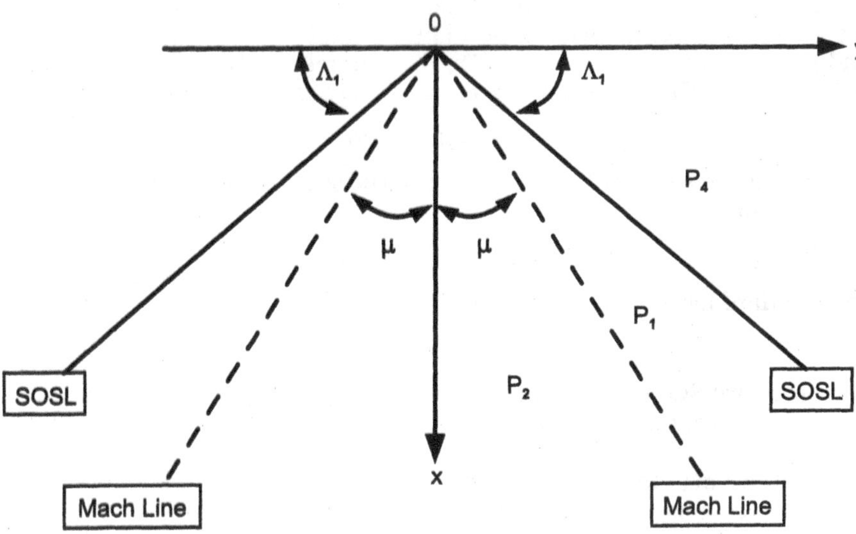

Fig. 3.13a Triangular surface symmetric about x axis for supersonic SOSL.

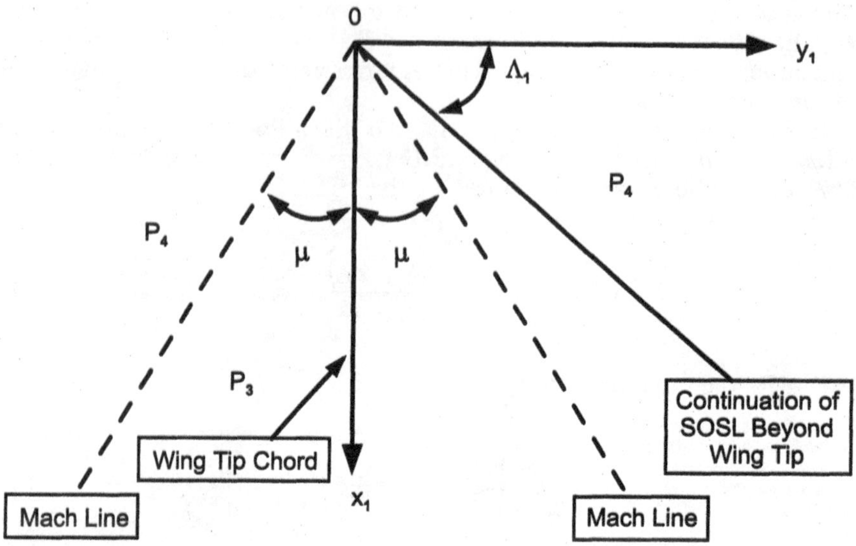

Fig. 3.13b Wing tip effects for supersonic SOSL.

Again, if $P = P_4$, the point is out of the zone of influence of the SOSL, and thus the induced velocity is zero.

Because of the linear nature of the governing flowfield equation (60), the induced velocity at a given point on any wing geometry can now be computed by the proper superposition of the triangular SOSL shown in Figs. 3.12a and b and 3.13a and b. As an example of how the superposition principle works, consider the wing shown in Fig. 3.14. For simplicity, the slopes χ_1 and χ_2 are constant. The wing AHJD can be represented by the superposition of five SOSLs. The first has the planform AEH and source intensity

$$w(x_p, y_p) = V_\infty \chi_1 \tag{84a}$$

where χ_1 is the slope of the segment AB. The second has the planform BIF and intensity

$$w(x_p, y_p) = (\chi_2 - \chi_1)V_\infty \tag{84b}$$

and the third has the planform DJG and intensity

$$w(x_p, y_p) = -\chi_2 V_\infty \tag{84c}$$

The other two SOSLs represent tip effects. They are the planforms HJL and IJL and have source intensities of opposite signs than those representing the wing.

The above procedure can be applied to a wing of general planform. The only difference is that, for each point in question, the slope may not be constant, as was the case in the simplified example. Then for some general point located on the wing surface, the total induced velocity due to all

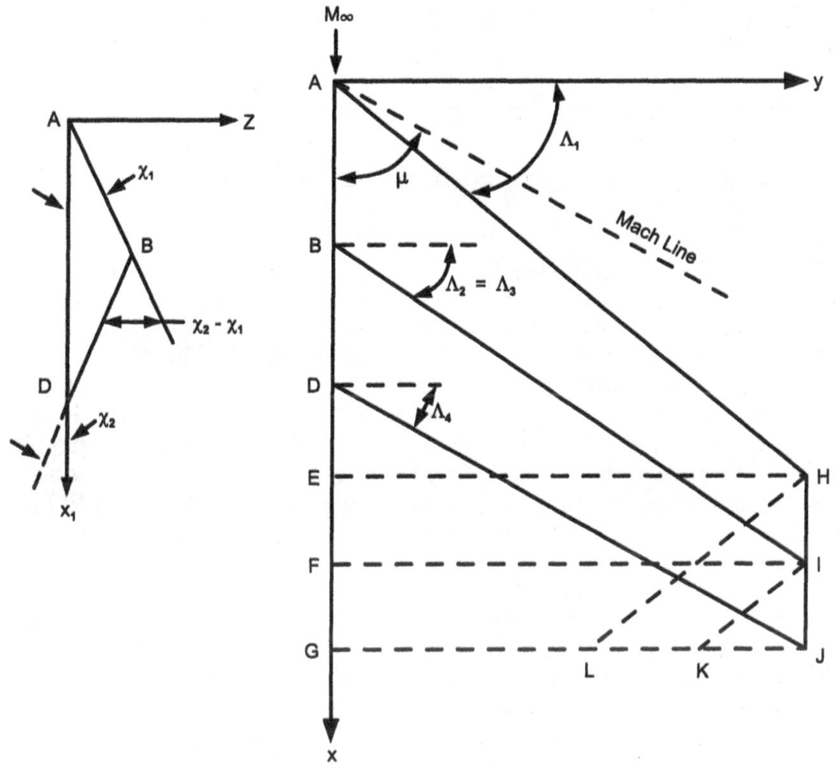

Fig. 3.14 Linear superposition of triangular source and sink distributions.

sources and sinks is found by applying one of Eqs. (3.78) through (3.83) for each SOSL. The particular equation applied depends on the location of the point relative to the SOSL and the Mach line as discussed earlier. These individual contributions are then summed to get the total induced velocity. Knowing the total induced velocity at a point allows one to calculate the pressure coefficient at the given point by Eq. (77).

The pressure coefficient can be calculated at a given number of spanwise and chordwise locations. The drag coefficient of a given airfoil section at the spanwise station $y = y_A$ is then

$$c_d = \frac{2}{c(y_A)} \int_0^{c(y_A)} C_P(x, y_A) w(x, y_A) \mathrm{d}x \tag{85}$$

The total drag coefficient for one fin of semispan $b/2$ is then

$$C_D = \frac{1}{S_w} \int_0^{b/2} c_d c(y) \mathrm{d}y \tag{86}$$

where $S_w = b/2(c_r + c_t)$. For cruciform fins, the total drag coefficient is

$$C_D = \frac{4}{S_w} \int_0^{b/2} c_d c(y) \mathrm{d}y \tag{87}$$

If it is desired to base the drag coefficient on the body cross-sectional area, then Eq. (87) must be multiplied by the factor S_w/S_{ref}.

Equations (85) and (87) can be integrated by numerical quadrature if the generators of the wing surface are supersonic. If the generators are subsonic, linear theory indicates that the pressure coefficients go to infinity at the wing generators. Physically, this cannot be true, which means that, for a subsonic SOSL, linear theory is not valid at the SOSL. The reason for this leading edge singularity is that velocity perturbations in the vicinity of discontinuities are no longer small, violating one of the assumptions in linear theory. However, velocity perturbations are small a slight distance from the SOSL so that linear theory can be applied. Numerical experiments indicated that a distance of five thousandths of the chord length from the SOSL is sufficient, and the value of pressure calculated at this point can be assumed to exist up to the SOSL.

As an example application of the linear theory approach to compute wave drag of wings, a simple rectangular planform is taken from Ref. 14. Figure 3.15a shows an aspect ratio of 0.5 Note that the wing was made fairly thick to mount pressure measurement taps as shown on the figure. If one includes the front and rear wedge segments, this wing has an effective t/c of 0.267. The wedge half-angle on both the leading and trailing edges is 15 deg. Figure 3.15b compares the pressure coefficient predictions of linear theory with the experimental data of Ref. 14 at the point where $y = 2.7$ and as a function of x/c for $M_\infty = 2.16$. Note that the linear theory gives lower values than the experiment and exact wedge solution on the leading edge and slightly more negative values on the trailing edge than the experiment or the value computed by a Prandtl–Meyer expansion. These differences are not unexpected due to the large wedge half-angles of the case considered.

Moore et al.[15] then computed the axial force coefficients by integrating the pressures at several $2y/b$ planes along the span and at several points along the chord. The actual front view of the configuration tested[14] is shown on the left part of Fig. 3.15c. The AP98[15] will not allow this type of frontal area distribution to be input to the code, as only a root and tip thickness are allowed. Experience has shown that a reasonable representation of the front view can be obtained by keeping the frontal area that the flowfield sees as constant. Then the wing thickness varies linearly from its maximum at the root chord to the minimum at the tip as shown on the right side of Fig. 3.15c. Figure 3.15d compares the wave portion of the wing axial force coefficient of the linear theory with the experimental data as a function of Mach number and for two aspect ratio rectangular wings. Linear theory values are low by as much as 20% for these cases. However, when this value for C_{A_W} is compared to the total C_A of the overall missile configuration, it is generally only a fraction of the total drag coefficient. For this reason, linear theory

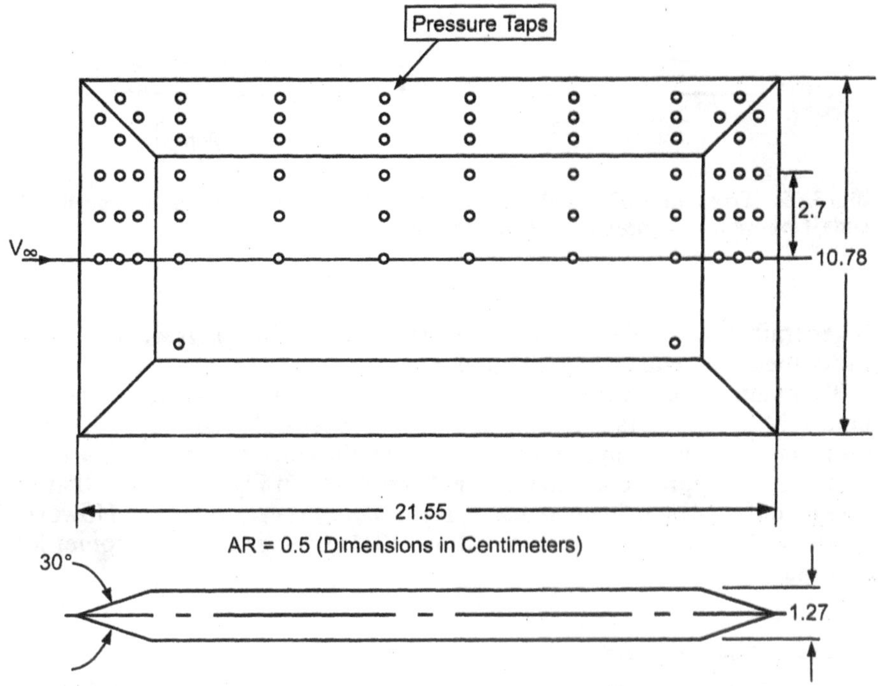

Fig. 3.15a Planform and airfoil cross section of wind tunnel model[14] showing location of pressure taps.

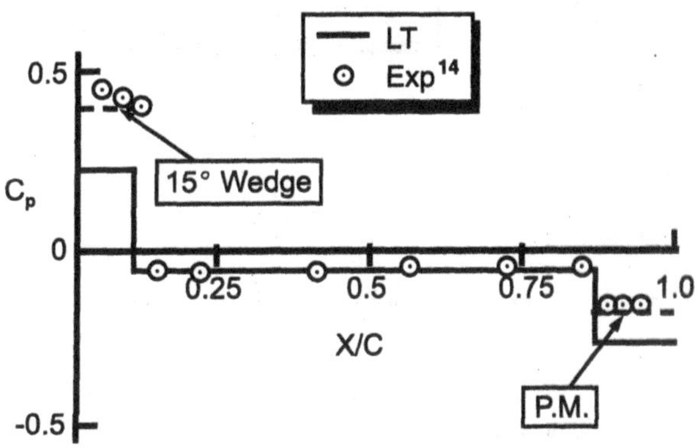

Fig. 3.15b Pressure coefficients at y = 2.7 cm as a function of x/c (M_∞ = 2.16 and α = 0).

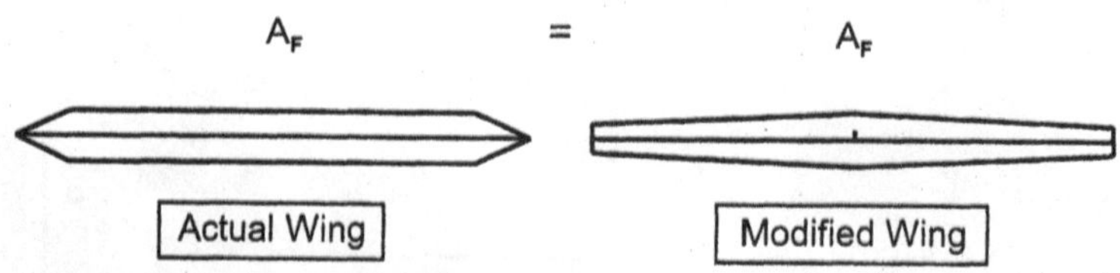

Fig. 3.15c Front view of actual wing of Fig. 3.15a and modified for aeroprediction code[15] where frontal area A_F is held constant.

is generally acceptable for wave drag estimation of the wings where second-order theory is required for the body.

To quantify the accuracy of linear theory, a simple wedge case is considered at Mach numbers of 2 and 5. Here Ames Research Staff[16] give the exact pressure coefficients and Moore et al.[15] give those for linear theory. Results of exact and approximate theories are compared in Fig. 3.16. Note that for wedge half-angles of 5 deg and less, linear theory is very accurate. However, as the wedge half-angle increases above 10 deg, linear theory becomes less accurate.

B. Wing Normal Force and Center of Pressure

The TDTWT, Eq. (60) with boundary conditions, Eqs. (60a) and (60b), is the appropriate mathematical model. As with axial force calculations, the solution to Eq. (60) is dependent on whether the flow normal to the leading edge is subsonic or supersonic. Consider the supersonic leading edge case first.

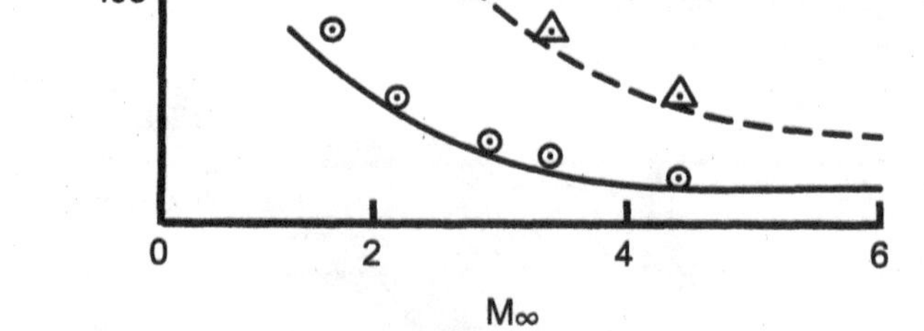

Fig. 3.15d Wave drag coefficient of rectangular wings (λ = 1.0).

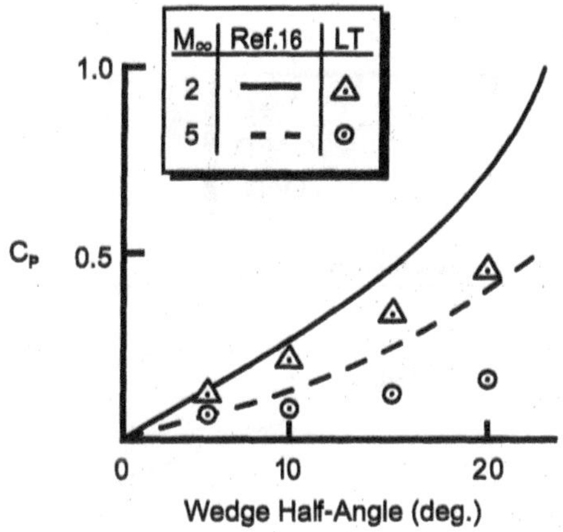

Fig. 3.16 Pressure coefficient comparison of exact solution and linear theory for wedge flow.

1. Supersonic Leading Edge

Ahead of the Mach line, in region 1 of Fig. 3.17a, the flow is two-dimensional, so the lifting pressure is constant at[17]

$$\Delta C_P = \Delta C_{P_1} = \frac{4a}{\beta\sqrt{1-\eta^2}} \tag{88}$$

Because the flow is conical, flow properties are constant along rays emanating from the vertex point and lying behind the Mach line. Referring therefore to Fig. 3.17a, where the leading edge Mach line intersects the wing trailing edge, if the flow properties at one point on the ray are known, they are known all along the ray. It is then a matter of computing the induced velocities and hence lifting pressures on several rays emanating from point 0. The lifting pressure on any one of these rays in region 2 is[17]

$$\Delta C_P = \Delta C_{P_2} = \frac{4a}{\beta\sqrt{1-\eta^2}}\left(1 - \frac{2}{\pi}\sin^{-1}\sqrt{\frac{\eta^2-\sigma^2}{1-\sigma^2}}\right) \tag{89}$$

where η and σ are defined by

$$\eta = \frac{k}{\beta}$$

$$k = \tan\Lambda$$

$$\sigma = \frac{ky_p}{x_p}$$

Note that ΔC_{P_2} includes ΔC_{P_1} as the first term.

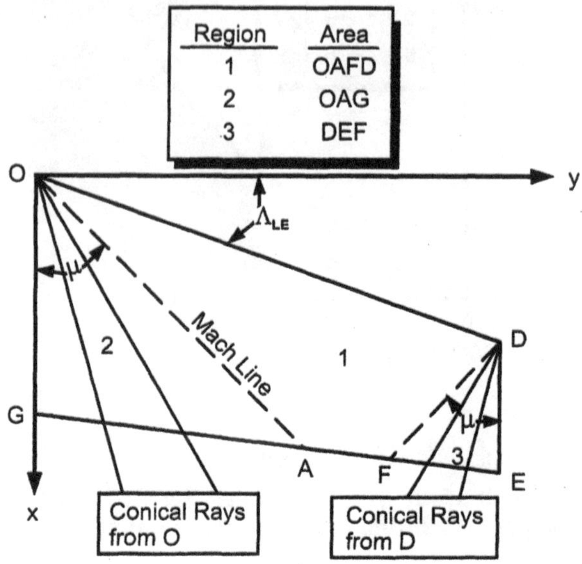

Fig. 3.17a Flat plate wing planform with supersonic leading and trailing edges; Mach line intersects wing trailing edge.

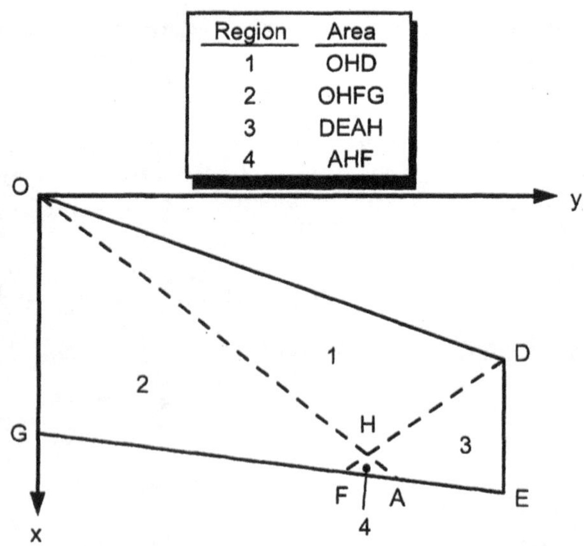

If the wing were tapered to a point, the above two relationships would allow complete determination of lifting pressures over the wing. For most practical cases, the wing tip is not pointed so tip effects must be accounted for. The tip affects the pressure within the Mach line from the tip leading edge (region 3 of Fig. 3.17a). Within this region, the flow is again conical along rays from point D so the lifting pressure caused by the tip is[1]

$$\Delta C_{P_3} = \frac{-8a}{\pi\beta}\sqrt{1-\eta^2}\tan^{-1}\frac{\sqrt{\eta+\sigma_D}}{-\sigma_D(1+\eta)} \tag{90}$$

The original equation of motion, Eq. (60), is linear so that superposition of solutions is allowed. Thus, the total lifting pressure in region 3 is

$$\Delta C_P = \Delta C_{P_1} + \Delta C_{P_3}$$

If the situation arises such that Mach lines intersect as shown at the bottom of Fig. 3.17a, then an additional pressure of ΔC_{P_3} is created in region 4. Hence, the total lifting pressure in region 4 is

$$\Delta C_P = \Delta C_{P_2} + \Delta C_{P_3}$$

The second case to consider, when the leading edge is supersonic, is when the leading edge Mach line intersects the tip as illustrated in Fig. 3.17b. The lifting pressures in regions 1, 2, 3, and 4 are calculated in the same manner as when the Mach line intersects the trailing edge. However, in region 5, the lift to be canceled along the tip is variable [see Eq. (89)] compared to the constant value canceled in regions 3 and 4 [see Eq. (88)]. Therefore, the problem is complicated somewhat in that a summation (or integration) must now be performed to calculate the cancellation lift in region 5. Cancellation lift here is used to denote the fact that the additional lifting pressure behind the Mach line is actually negative and, therefore, cancels some of the two-dimensional lifting pressure ahead of the Mach line. Referring to Fig. 3.17b, region 5 is broken down into a finite number of intervals in the following manner. Conical rays from the vertex at 0 are projected behind the Mach line OA and intersect the wing tip at equal intervals along the tip (for example, rays OJ and OK). Lines are then drawn from the points of intersection on the wing tip parallel to the Mach line AI until the point $P(x,y)$ is enclosed. The difference in ΔC_P across one of these intervals is what must be canceled throughout region 5. For any given interval then, say JK, this difference in pressure is found by applying Eq. (89) to each of the rays from O passing through J and K. That is,

$$\Delta C_{P_{\mathrm{JK}}}(x, b/2) = \Delta C_{P_J}(x, b/2) - \Delta C_{P_K}(x, b/2)$$

or

$$\Delta C_{P_{\mathrm{JK}}}(x, b/2) = \frac{8a}{\pi\beta\sqrt{1-\eta^2}}\left[\sin^{-1}\sqrt{\frac{\eta^2-\sigma_{0\mathrm{K}}^2}{1-\sigma_{0\mathrm{K}}^2}} - \sin^{-1}\sqrt{\frac{\eta^2-\sigma_{0\mathrm{J}}^2}{1-\sigma_{0\mathrm{J}}^2}}\right]$$

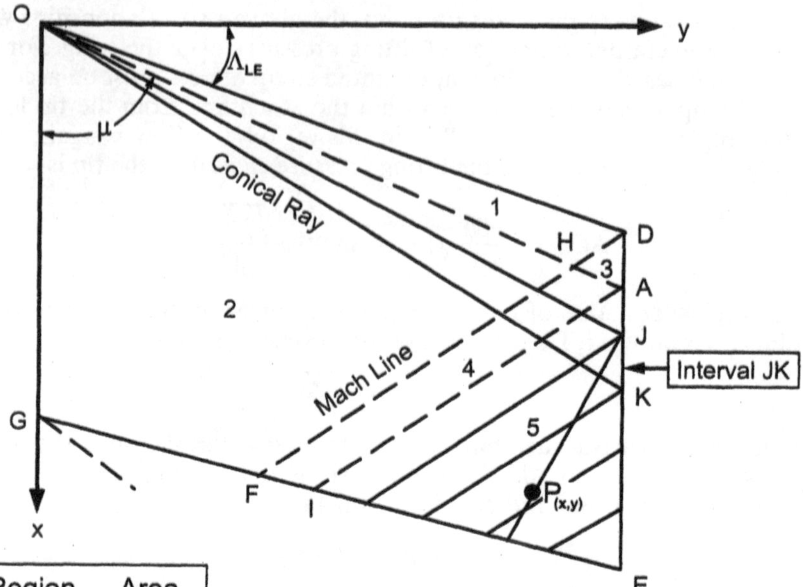

Region	Area
1	ODH
2	OHFG
3	AHD
4	FHAI
5	AEI

Fig. 3.17b Flat plate wing planform with supersonic leading and trailing edges; Mach line intersects wing tip.

The effect of this cancellation pressure at any point $P(x,y)$ is then

$$\Delta C_{P_{JP}}(x, y) = \Delta C_{P_{JK}}(x, b/2)\,\frac{2}{\pi}\tan^{-1}\sqrt{\frac{\eta + \sigma_{JP}}{-\sigma_{JP}(1+\eta)}} \tag{91}$$

Now the interval JK is any interval upstream of the Mach line from P passing through the wing tip. If m is the total number of intervals upstream of P, then the total lifting pressure coefficient at point $P(x,y)$ within region 5 is

$$\Delta C_P = \Delta C_{P_2} + \Delta C_{P_4} + \sum_{i=n}^{i=m} \Delta C_{P_{iP}}$$

where $\Delta C_{P_{iP}}$ is given at each interval by Eq. (91).

The force and moment coefficients can now be found by substituting the expressions for ΔC_P in each region into Eqs. (65) through (72) and performing the indicated operations. Integrations could be carried out in closed form for regions 1 through 4, but the formulas are very lengthy for even the simplest cases.[18] A more straightforward approach is to simply numerically integrate the integrals, Eqs. (65) through (68) over the entire surface, particularly because pressures in region 5 have to be numerically integrated.

2. Subsonic Leading Edge

For subsonic leading edges, the velocity and lifting pressures approach infinite values near the leading edge. The solution for ΔC_P is complicated somewhat by this singularity, but due to the conical nature of the flow from the vertex points 0 and D (Fig. 3.18), a closed-form solution can be obtained. For region 1 of Fig. 3.18, the closed-form solution is[1]

$$\Delta C_P = \Delta C_{P_1} = \frac{4a}{E(k)\tan\Lambda_1\sqrt{1-\sigma_0^2}} \tag{92}$$

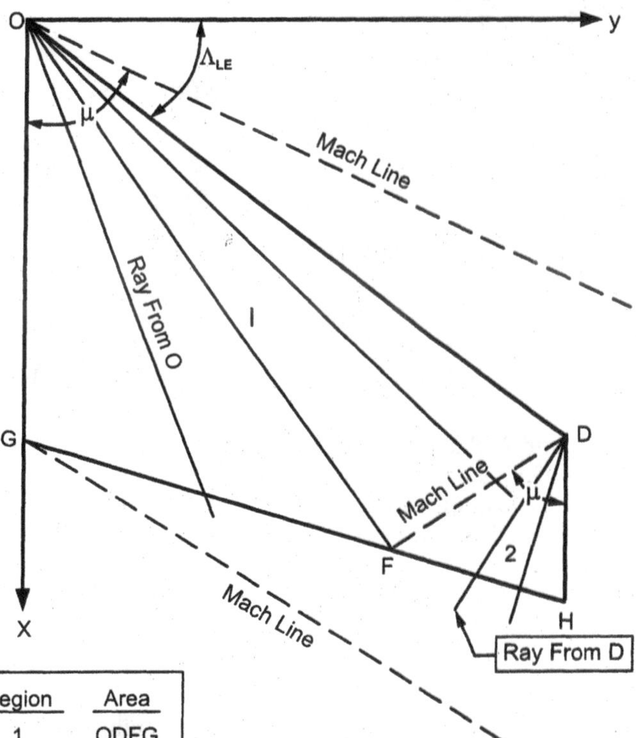

Region	Area
1	ODFG
2	DFH

Fig. 3.18 Flat plate wing planform with subsonic leading edge and supersonic trailing edge.

where $E(k)$ is a constant dependent only on the leading edge sweep angle and Mach number. That is,

$$E(k) = \int_0^{\pi/2} \sqrt{1 - k^2 \sin^2 \theta} d\theta \tag{93}$$

and $k = \sqrt{1 - \beta^2/\tan^2 \Lambda_{LE}}$ The values of the complete elliptic integral $E(k)$ have been tabulated and appear in standard mathematical handbooks.

The lift to be canceled at the tip is variable, as was the case for supersonic leading edges when the Mach line intersected the tip. However, in this case, the integration can be carried out in closed form. Thus, the cancellation lifting pressure is[18]

$$\Delta C_{P_2} = \frac{4a}{\pi \beta \eta^{3/2} E\left(\frac{\eta^2 - 1}{\eta}\right)} \left[K(k) \sqrt{\frac{2\beta(b/2 - y)}{x + \beta y}} \right.$$
$$\left. -2x \sqrt{\frac{\eta}{x^2 - \beta^2 y^2 \eta^2}} \left\{ \frac{F(k', \psi)}{K(k')} \left[\frac{\pi}{2} - K(k)E(k') \right] + K(k)E(k', \psi) \right\} \right] \tag{94}$$

where

$$k = \sqrt{\frac{(1 - a_0\eta)(\eta - 1)}{2\eta(a_0 + 1)}}, \quad k' = \sqrt{1 - k^2}$$

$$\psi = \sin^{-1} \sqrt{\frac{a_0(x + \beta y \eta)}{\beta\, b/2\, (1 + \eta a_0)}}, \quad a_0 = \frac{\beta\, b/2}{x + \beta(y - b/2)}$$

Here $F(\theta_1,\theta_2)$ and $E(\theta_1,\theta_2)$ are elliptic integrals of the first and second kind, respectively, which again can be found in any standard mathematical handbook. The complete elliptic integral K is related to $F(\theta_1,\theta_2)$ by $K = F(\theta_1,\pi/2)$ and the complete elliptic integral E of Eq. (93) is related to $E(\theta_1,\theta_2)$ by $E = E(\theta_1,\pi/2)$.

The total lifting pressure coefficient at any point within region 2 is then

$$\Delta C_P = \Delta C_{P_1} + \Delta C_{P_2}$$

Integrations for chordwise and spanwise lift, pitching moment, and center of pressure can be obtained by integrating Eqs. (65) through (72) numerically. However, caution must be taken in these integrations because, for subsonic leading edges, the lifting pressure tends toward $1/\sqrt{x}$ near the leading edge. In this vicinity, more mesh points must be added to the flowfield to get an accurate integration.

C. Transonic Flow

As mentioned earlier, airfoil thickness has a second-order effect on lift in subsonic and supersonic flow. However, this is not true in transonic flow, so the assumption of a flat plate with zero thickness is no longer valid except for very thin wings. Furthermore, as M_∞ approaches unity, Eq. (60) cannot be applied because the term $(1 - M_\infty^2)\varphi_{xx}$ becomes of the same order as nonlinear terms that have been neglected in deriving this equation from small perturbation theory. Although progress in the field of transonic aerodynamics has greatly advanced the state of the art, practical methods for flowfield computation are still severely limited. For example, to the author's knowledge, solutions for three-dimensional swept and tapered wings with thickness do not exist, even in approximate form. In light of these considerations, an empirical approach to wing lift in transonic flow is suggested. This method will be discussed in the empirical methods section of Chapter 5.

As an example application of the methods presented for calculating wing normal force at subsonic and supersonic speeds, consider the simple rectangular wing planform in Fig. 3.15a. Consider the plane where $y = 2.7$ or $2y/b \approx 0.5$. We will consider an AOA of 5 deg so linear theory should give a reasonable solution for the values of ΔC_P. Figure 3.19 shows the results of the wing loading at $M_\infty = 1.6$ for the AR = 0.5 case of Fig. 3.15a. Here, experimental data at both $\alpha = 5$ deg and $\alpha = -5$ deg are shown on the figure, where the $\alpha = -5$ deg data have the sign reversed. Experimental data[14] differ significantly from the linear theory at any given point on Fig. 3.19, partly because experimental data at $\alpha = 5$ deg have some slight nonlinearities included, whereas linear theory does not. Secondly, the wing of Fig. 3.15a is fairly thick, giving additional loads near the leading edge that may be thickness effects. When the load is integrated over the entire wing, linear theory gives a value of $C_N = 0.083$ and experiment gives $C_N = 0.11$. C_{N_α} from experiment is 0.02/deg versus 0.017 for linear theory. Thus linear theory gives reasonable, but low, values of normal force coefficient and normal force coefficient slope at low AOAs. The wing center of pressure experimentally is at $x/c \approx 0.27$ versus the linear theory prediction of $x/c \approx 0.28$.

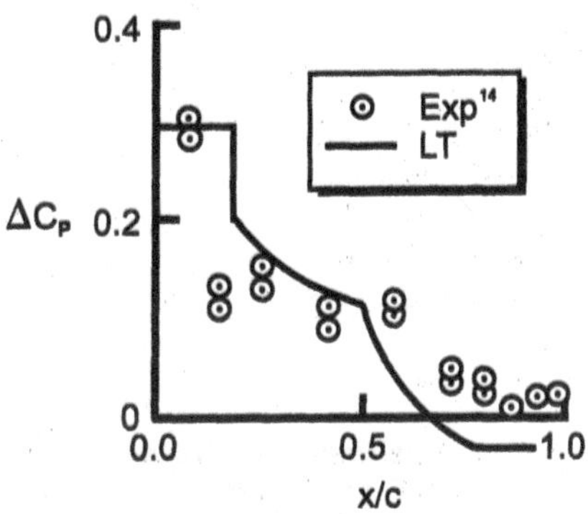

Fig. 3.19 Pressure coefficient loading on rectangular wing of AR = 0.5 [$y/(b/2)$ = 0.5, M_∞ = 1.6, and α = 5°].

Figure 3.19 presents only a single wing case for a fairly thick wing. The database of Ref. 14 consists of wings of aspect ratios 0.5 to 4.0; taper ratios of 0, 0.5; and 1.0; M_∞ = 1.6, 2.16, 2.86, 3.5, and 4.6; and AOAs to 60 deg. Baker[19] and Nielsen et al.[20] contain complementary wing-alone data at subsonic Mach numbers. Figure 3.20 shows the normal force coefficient slope near $a = 0$ for a variety of wing planforms and plotted as a function of Mach number. All wing planforms have no sweepback on the trailing edges. Figure 3.20a is for λ = 0, 3.20b is for λ = 0.5, and 3.20c is for λ = 1.0. Reasonably good agreement between theory and experiment is obtained for all planform shapes. This is even in light of the fact that the wing planforms were quite thick and the first value of AOA considered in the data was 5 deg, so some nonlinearity could be present. The linear theory results were computed at an AOA 1 deg.

VII. Roll Damping Moment

When a missile rolls about its own x axis at zero AOA, the lifting surfaces are subjected to a normal velocity vector along the wing span that is induced by the rolling velocity. The induced velocity results in an induced AOA on the wing surface that varies linearly with the radial distance from the rota-

a)

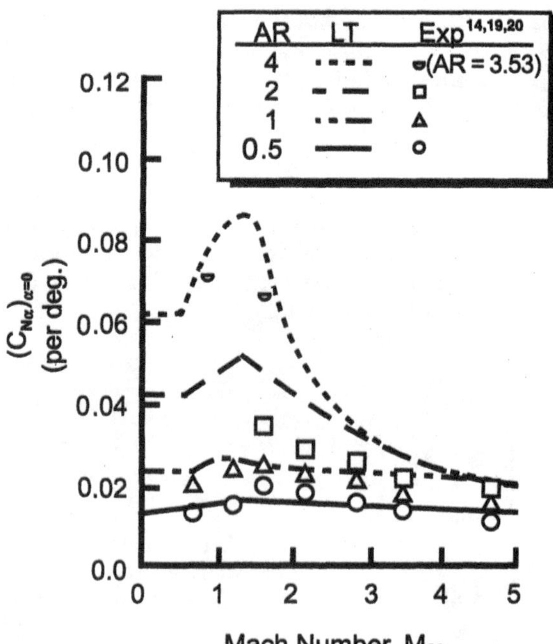

Fig. 3.20 Low AOA lift curve slope comparisons of experiment and linear theory (Λ_{TE} = 0) for various wing planforms. a) λ = 0, b) λ = 0.5, c) λ = 1.0

b)

c)

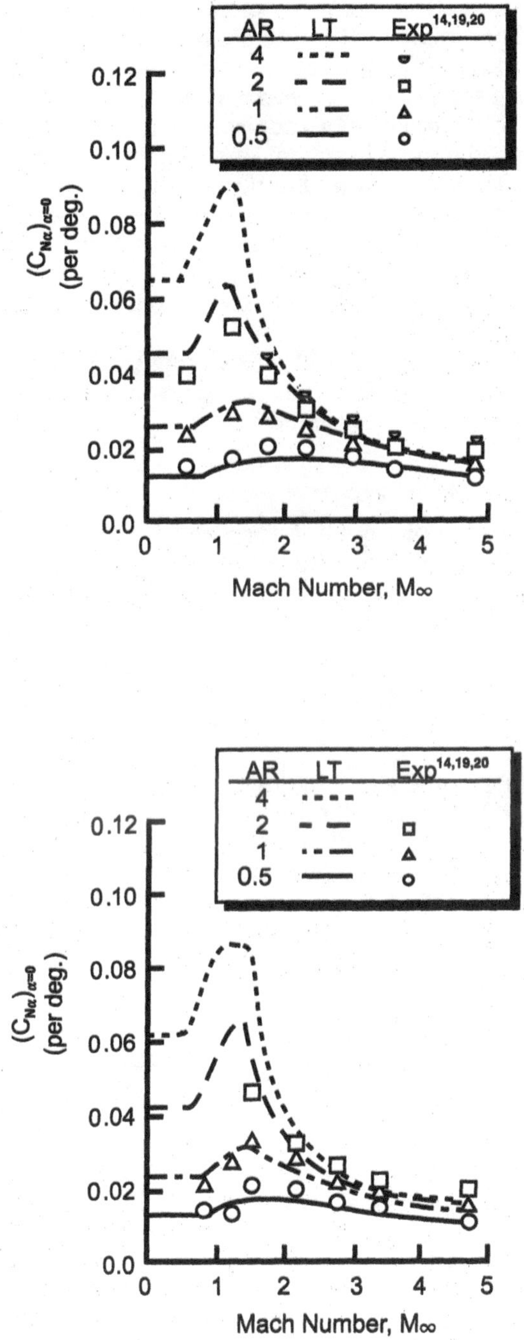

Fig. 3.20 *(Continued)*

tional x axis. The roll damping moment is defined as the roll moment induced by and opposing or damping the rolling velocity.

For a wing–body configuration, the wing completely dominates the roll damping moment so that it is of primary concern to have a good estimate of the wing-alone roll damping. For supersonic and subsonic flow calculations, the wing is assumed to be thin so that small perturbation theory can be used for flowfield calculations. In addition, the wing is assumed to have zero camber with aeroelastic effects being small. In transonic flow, empirical methods are used for roll damping calculations. Body interference is estimated using slender body theory. Tail fin to tail fin interference is zero unless the nondimensional spin $pd/2V_\infty$ is high enough so that the shed vortices impinge on an adjacent fin.

The individual methods used for calculating the wing roll damping moment coefficient derivatives in subsonic, supersonic, and transonic flow with the interference effects are discussed below.

A. Subsonic Flow ($M_\infty < M_{crit}$)

The small perturbation equation for three-dimensional steady flow, Eq. (60), applies to both subsonic and supersonic Mach number regimes. The boundary condition, Eq. (74), for the roll damping problem is

$$\frac{w(x,y)}{V_\infty} = \Phi_z = \frac{\partial F}{\partial x} = \frac{py}{V_\infty} \tag{95}$$

with the vanishing of the perturbation velocities a large distance from the surface. In Eq. (95), p is the roll rate. For subsonic flow, Gothert's extension to the Prandt–Glauert transformation can be used so that solution of the Laplace's equation (62), with boundary condition (95) allows solution for the velocity potential at a Mach number of zero.

The solution of Eq. (62) is the same as Eq. (63), except that ΔC_p is determined by applying the boundary condition, Eq. (95), to obtain

$$\frac{py}{V_\infty} = \frac{1}{8\pi} \int\!\!\!\!\times\!\!\int \frac{\Delta C_P(x_1, y_1)}{(y-y_1)^2}\left[1 + \frac{x-x_1}{\sqrt{(x-x_1)^2 + (y-y_1)^2}}\right] \mathrm{d}x_1 \mathrm{d}y_1 \tag{96}$$

The cross on the y_1 integral indicates a singularity at $y = y_1$, in which case Manglers principal-value technique[1] can be applied. Details of the solution of the integral equation (96) for $\Delta C_P(x,y)$ will not be repeated here as they are given in many references (see, for example, Ref. 1). Worthy of note, however, is the fact that Eq. (96) is an integral equation for which the wing loading ΔC_P is to be found as a linear function of the induced AOA, py/V_∞.

Once the wing loading $\Delta C_P(x,y)$, due to roll, is known over the entire wing, the local rolling moment coefficient at a given spanwise airfoil section y is

$$c_\ell = \frac{y}{c\ell_{\text{ref}}} \int_{x_{\text{LE}}}^{x_{\text{TE}}} (\Delta C_P)_p \mathrm{d}x \tag{97}$$

where the subscript p indicates the loading due to a roll rate p. In Eq. (97), ℓ_{ref} is the reference length one chooses by which to nondimensionalize the local rolling moment. It could be for example, the root chord, mean aerodynamic chord, body diameter, or wing semispan. The total rolling moment on the entire wing is then

$$C_\ell = \frac{2}{S_{\text{ref}}} \int_0^{b/2} c c_\ell \mathrm{d}y \tag{98}$$

Assuming the rolling moment is a linear function of roll rate, the roll damping moment coefficient derivative is

$$C_{\ell P} = \frac{C_\ell}{p\ell_{\text{ref}}/2V_\infty} \tag{99}$$

For Mach numbers other than zero, one applies the Gothert rule at each subsonic Mach number, similar to Eq. (61). That is,

$$(C_\ell)_{M_\infty, \text{AR}, \bar{p}} = \frac{(C_\ell)_{0, \beta\text{AR}, \beta\bar{p}}}{1 - M_\infty^2} \tag{100a}$$

$$\left(C_{\ell P}\right)_{M_\infty, \text{AR}, \bar{p}} = \frac{\left(C_{\ell P}\right)_{0, \beta\text{AR}, \beta\bar{p}}}{1 - M_\infty^2} \tag{100b}$$

where $\bar{p}$ of Eq. (100) is the nondimensional spin rate $\bar{p} = p\ell_{\text{ref}}/2V_\infty$.

B. Supersonic Flow ($M_\infty \geq 1.2$)

For supersonic flow past thin wings, Eq. (60) is still applicable along with the associated boundary condition, Eq. (95). Because the flow is supersonic, disturbances in the flowfield are not felt upstream of the point of disturbance. Thus the boundary condition that requires perturbations to vanish at a large distance from the wing planform can be modified to

$$u(0^-,y,z) = v(0^-,y,z) = w(0^-,y,z) = 0 \tag{101}$$

where it is assumed the disturbance occurs at $x = 0$. It will be assumed that the wing trailing edge is supersonic (Mach number normal to wing trailing edge is greater than one) so that the Kutta condition need not be applied. The solution to Eq. (60) is[18]

$$\Phi(x, y, o) = -\frac{1}{\pi} \iint_{S_1} \frac{py/V_\infty}{\sqrt{(x - x_1)^2 - \beta^2(y - y_1)^2}} \, dx_1 dy_1 \tag{102}$$

where S_1 is the wing planform. The lifting pressure coefficient, due to a constant roll rate p, is then related to the perturbation velocity potential Φ through the relation

$$(\Delta C_P)_p = -4\Phi_{x_0} \tag{103}$$

The limits of integration of Eq. (102) are dependent on whether the leading edge is subsonic or supersonic, just as for the TDTWT of Section 3.6. Each of these cases will be considered separately below.

For a supersonic leading edge, the Mach number normal to the leading edge is greater than one. Referring to Fig. 3.17a, this implies that the Mach line emanating from the root chord, OA, lies behind the wing leading edge. If the Mach line intersects the wing tip, as shown in Fig. 3.17b, there are five distinct disturbance regions present, and hence, five different perturbation solutions. If the Mach line intersects the wing trailing edge, as shown in Fig. 3.17a, the disturbance created by the Mach line OA impinging on the wing tip is eliminated so that only four perturbation solutions are needed.

Generalized formulas for the lifting pressure distributions on a wing in steady roll are derived in Ref. 21. For convenience, the final equations are repeated here. Referring to Figs. 3.17a and 3.17b for region 1, the flow is two-dimensional and the resulting lifting pressure is

$$(\Delta C_P)_P = \left[(\Delta C_P)_P\right]_1 = \frac{4pm^2x_0(m^2\sigma - 1)}{\beta^2 V_\infty (m^2 - 1)^{3/2}} \tag{104}$$

where $m = \beta \cot \Lambda_{LE}$.

In region 2, the total lifting pressure is

$$(\Delta C_P)_P = \left[(\Delta C_P)_P\right]_2 = \frac{4pm^2x_0}{\pi\beta^2 V_\infty (m^2 - 1)^{3/2}} \left\{(1 + m^2\sigma)\cos^{-1}\left[\frac{1 + m^2\sigma}{m(1 + \sigma)}\right] - (1 - m^2\sigma)\cos^{-1}\left[\frac{1 - m^2\sigma}{m(1 - \sigma)}\right]\right\} \tag{105}$$

The induced pressure caused by the tip Mach line DF is

$$[(\Delta C_P)_P]_3 = \frac{4pm}{\pi V_\infty \beta (m^2-1)^{3/2}} \left\{ \left[\frac{mx_a}{\beta} - m^2 y_a - \frac{b}{2}(m^2-1) \right] \cos^{-1} \left[\frac{\frac{mx_1}{\beta} + y_a(2m+1)}{y_a - \frac{mx_a}{\beta}} \right] - 2m\sqrt{-\frac{my_a}{\beta}(x_a + \beta y_a)(m+1)} \right\} \quad (106)$$

The total lifting pressure in region 3 is then

$$(\Delta C_P)_P = [(\Delta C_P)_P]_1 + [(\Delta C_P)_P]_3 \quad (107)$$

The lifting pressure in region 4 is a combination of that in regions 2 and 3. Thus, in region 4

$$(\Delta C_P)_P = [(\Delta C_P)_P]_2 + [(\Delta C_P)_P]_3 \quad (108)$$

Again, if the Mach line OA intersects the wing trailing edge, Eqs. (104) through (108) allow one to determine the complete lifting pressure distribution over the wing surface. If the Mach line OA intersects the wing tip, another perturbation is induced in the flow field. The total pressure differential in region 5 is then

$$(\Delta C_P)_P = \frac{4pm}{\pi \beta V_\infty (m^2-1)^{3/2}} \left\{ \left[\frac{mx_a}{\beta} + m^2 y_a + \frac{b}{2}(m^2+1) \right] \cos^{-1}\left(\frac{mx_a/\beta - y_a(1-2m) + b}{mx_a/\beta + y_a + b} \right) - 2m\sqrt{-y_a(m-1)(mx_a/\beta + my_a + b)} \right\} \quad (109)$$

If the Mach number normal to the leading edge is less than 1 (and the trailing edge Mach number is greater than 1), there are only two flow regions to consider. However, the solution is complicated somewhat by the leading edge singularity where the velocity is $1/x$.[22] Referring to Fig. 3.18, the pressure differential of upper and lower surfaces in region 1 is[22]

$$(\Delta C_P)_p = \frac{2I(m)pm^2 x_0 \sigma}{\beta^2 V_\infty \sqrt{1-\sigma^2}} \quad (110)$$

where

$$I(m) = \frac{2(1-m^2)}{(2-m^2)E(m) - m^2F(m)}$$

$$E(m) = \int_0^{\pi/2} \sqrt{1-k^2\sin^2\theta}\,d\theta$$

$$F(m) = \int_0^{\pi/2} \frac{d\theta}{\sqrt{1-k^2\sin^2\theta}}$$

$$k = \sqrt{1-m^2}$$

Values of the complete elliptic integrals of the first and second kind [$F(m)$ and $E(m)$, respectively] have been tabulated and appear in standard mathematical handbooks.

Region 2, which is affected by the wing tip Mach cone, has a lifting pressure differential given by

$$(\Delta C_P)_P = \frac{-8pm}{\pi\beta V_\infty}\sqrt{b/2 - y}\left[\frac{3mx_0/\beta + y(1-2m) - b/2\,(1+m)}{3(1+m)\sqrt{(1+m)(mx_0/\beta + y_0)}}\right] \tag{111}$$

For a flow with a subsonic leading edge and supersonic trailing edge, Eqs. (110) and (111) determine the complete pressure distribution due to a steady roll, p.

The local roll moment, total rolling moment, and roll damping moment for both the subsonic and supersonic leading edge cases can then be determined by Eqs. (97), (98), and (99), respectively.

C. Transonic Flow ($M_{fb} \leq M_\infty < 1.2$)

There are currently no simple, accurate analytical methods available for calculating transonic roll damping. With the increased emphasis on transonic aerodynamics in the last few years, it is hoped that a simple theoretical method will become available within a reasonable time frame. Until that time, one must resort to the full Navier–Stokes equations of motion or empirical techniques. Because the former approach is beyond the scope of this work, the empirical approach will be followed.

Without a theoretical model to calculate transonic roll damping, it has been current practice by some engineers[23] to estimate roll damping in direct proportion to the lift. That is,

$$\left(C_{\ell_P}\right)_M = \left(C_{\ell_P}\right)_{M=0.8} \frac{\left(C_{N_\alpha}\right)_M}{\left(C_{N_\alpha}\right)_{M=0.8}} \tag{112}$$

where the quantities C_{ℓ_p} and C_{N_α} are total configuration values. This means that if the roll damping is known at say $M_\infty = 0.8$ and C_{N_α} is known throughout the transonic speed regime, then the roll damping can be estimated according to Eq. (112). This is the procedure recommended for calculating roll damping.

One of the main problems inherent in calculating transonic aerodynamics is a need to account for thickness and, in some cases, aeroelastic effects. An empirical technique proposed by Edmondson[24] to correct for thickness and aeroelastic effects on rectangular wings is

$$C_{\ell_p} = \left(C_{\ell_p}\right)_{\text{theory}} (1 - t/c)^{2AR/3} \tag{113}$$

where t/c is the overall thickness-to-chord ratio of the wing planform. However, if values of C_{ℓ_p} in transonic flow are calculated based on the lift–curve slope, which has already accounted for thickness, then Eq. (113) should not be used. In addition, intuitively one would expect a Mach number effect to be present in Eq. (113), because thickness is most important in transonic flow. For these two reasons, Eq. (113) should be applied with caution.

As mentioned earlier, the wing dominates roll damping. However, the body does have some small value of C_{ℓ_p}. This value is best estimated empirically through use of an empirical program such as that of Ref. 25. The method of Whyte[25] is limited in length of the body. One approach to estimate roll damping moment for longer bodies is to assume that C_{ℓ_p} varies linearly with length. Thus,

$$C_{\ell_p} = \left(C_{\ell_p}\right)_1 \frac{\ell}{\ell_1} \tag{114}$$

where the subscript represents known data. This estimate should prove reasonable as long as data available are for a configuration of similar shape and dimensions to the configuration for which data are required.

Two example cases are considered to illustrate the roll damping predictive capability of the linear theory at subsonic and supersonic speeds in conjunction with empirical estimates at transonic speeds and the body-alone term. The two configurations are illustrated in Fig. 3.21. The first configuration is the Army–Navy Finner, which has an extensive database on static and dynamic derivatives. Experimental data used to compare to the theoretical roll damping derivative consist of wind tunnel data taken at Arnold Engineering Development Corporation[26] in Tullahoma, Tennessee, ballistic range data[27] taken at the range in Aberdeen, Maryland and wind tunnel data taken at the supersonic tunnel at the former Naval Surface Warfare Center, White Oak, Maryland[28]. Figure 3.22 compares linear theory predictions to experimental data at subsonic and supersonic Mach numbers. At transonic speeds ($0.8 < M_\infty < 1.2$), the empirical process described by Eq. (112) is used. Linear theory in general gives values lower than experiment by about 10 to 20% for this case. However, this is probably acceptable for most

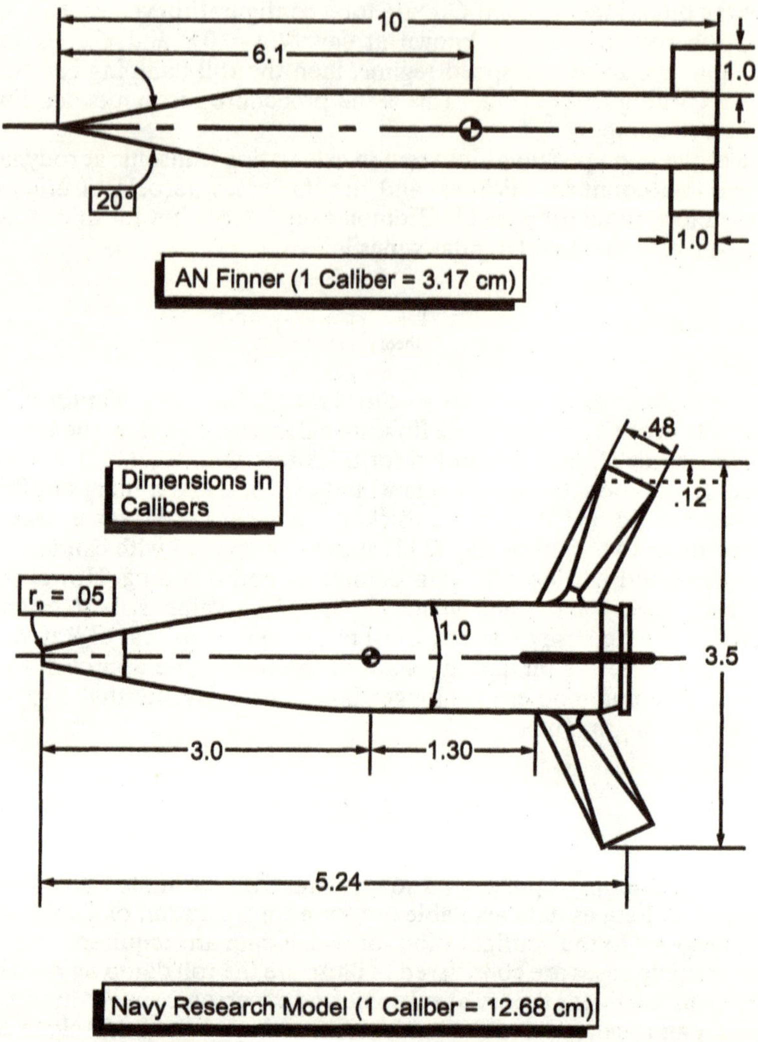

Fig. 3.21 Body–tail configurations used for dynamic derivative comparisons.

dynamic derivative predictions because, in general, they are not expected to be as accurate as static aerodynamics.

Figure 3.23 presents the comparison of the linear theory to experimental data[29] for the Navy research model of Fig. 3.21. Frierson[29] had roll damping data only for subsonic and transonic Mach numbers but pitch damping data at subsonic, transonic, and supersonic conditions. For the Mach numbers where data are available, linear theory predicts roll damping quite well. It should be noted that the Navy research model does not have streamwise

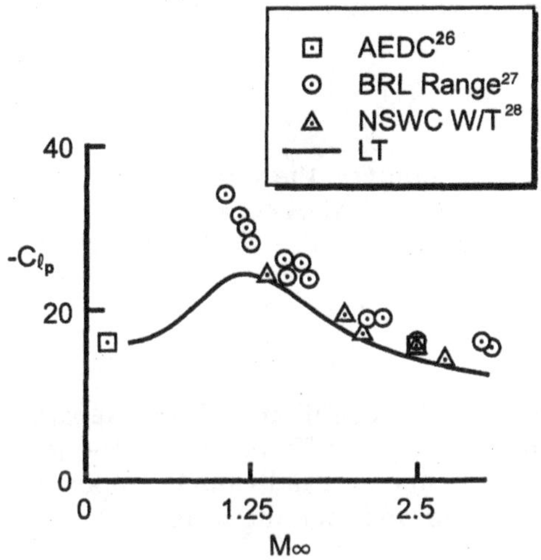

Fig. 3.22 Roll damping coefficient for AN Finner.

tips. As a result, the span has been modified as shown by the dotted line in Fig. 3.21, where wing area remains constant.

VIII. Pitch Damping Moment

There are two contributions to damping in pitch. These are the steady-state derivative C_{M_q} (the moment coefficient due to steady pitching velocity) and the un-steady-state or time-dependent derivative $C_{M_{\dot{\alpha}}}$ (the moment coefficient due to a constant vertical acceleration). The steady pitching velocity q induces a variable AOA as a function of the distance of the center of rotation,[2] similar to the case for a rolling velocity of p. However, in the case of the pitching velocity, the moment arm between the center of rotation and the wing, tail, or canard can be much larger than that due to roll, which

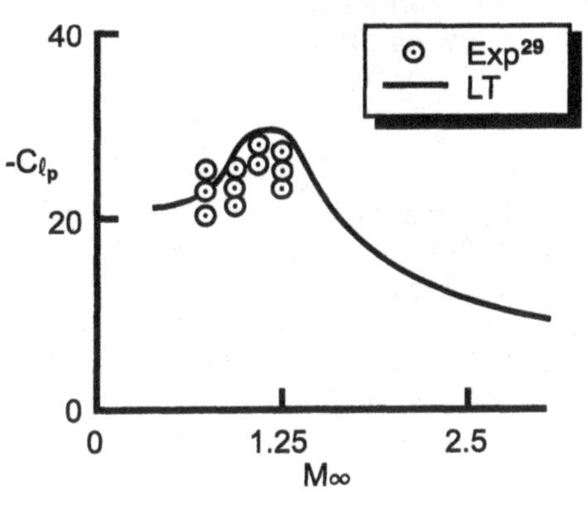

Fig. 3.23 Roll damping coefficient for Navy research model.

is generally the body radius plus wing semispan. For pitching velocity, the local AOA at the wing is

$$(\Delta a)_q = q\left(\frac{x_W - x_{CG}}{V_\infty}\right)$$

where $x_w - x_{CG}$ is the distance between a point on the wing and the center of rotation. The other component of the induced AOA considered as part of the pitch damping moment is that due to a constant vertical acceleration $\dot{a}$. Here, the local AOA is

$$a = \frac{\dot{w}}{V_\infty}$$

Because Eq. (60) is linear, we can solve this equation with two separate boundary conditions and add the results together. To determine the total damping of missile configurations, the damping of individual components (i.e., body, wings, and canards) is calculated, and then the wing–body interference is included.

The empirical procedure of the "Spinner" program[25] is recommended to estimate the body-alone pitch damping parameter. Thus, the quantity $C_{M_q} + C_{M_{\dot{a}}}$ for a particular body is related to the damping value for a dimensionally similar body by a function of the body length, boattail length, and center of gravity.

The lifting surfaces of missiles (canard–body–tail configurations) and fin-stabilized projectiles (body–tail configurations) generally represent the major contribution of the total damping in pitch. Therefore, it is important to have a good estimate of the wing-alone pitch damping.

For subsonic and supersonic flow calculations, once again a thin wing is assumed so that small perturbation theory is applicable. Also, as with the roll damping moment, the wing is assumed to have no camber or twist and negligible aeroelastic effects. In transonic flow, empirical methods are applied in longitudinal damping computations, similar to Eq. (114). The component interference effects are then approximated by assuming a slender body and using the apparent-mass method of Bryson.[30]

Individual procedures used for calculating the wing pitch damping moment coefficient derivatives in all flight regimes under consideration and the interference effects are presented in the following discussion.

A. Subsonic Flow ($M_\infty < 0.8$)

The stability derivatives C_{L_q} and C_{M_q} for a thin wing in subsonic flow are determined from the pressure distribution corresponding to a steady pitching rate q. In the case of incompressible flow, the pressure variation is obtained by solving Eq. (60), the small perturbation equation for three-dimensional steady flow, with the appropriate boundary condition. For steady-state pitch damping calculations, the necessary boundary condition is found by setting $a = p = dz/dx = 0$ in Eq. (74) so that

$$\Phi_z = q(x - x_{\text{ref}})/V_\infty \tag{115}$$

Moreover, the pressure loading due to a constant pitch rate is realized by solving the integral equation given by Eq. (96) with py/V_∞ replaced with $q(x - x_{ref})/V_\infty$. The resulting solution is the wing loading ΔC_p. Compressible subsonic calculations are based on the incompressible solution by applying Gothert's extension to the Prandtl–Glauert rule, analogous to the process for normal force, Eq. (61) and roll moment, Eq. (100).

From the chordwise pressure distribution, the sectional moment about the axis of rotation of the wing due to q is determined as

$$c_{m_q} = \frac{1}{c\ell_{ref}} \int_{x_{LE}}^{x_{LE}} x(\Delta C_P)_q \mathrm{d}x \tag{116}$$

where the quantity $(\Delta C_p)_q$ represents the loading due to a pitch rate q. The total pitching moment due to q on the entire wing is given by

$$(C_M)_q = \frac{2}{S_{ref}} \int_0^{b/2} c c_{M_q} \mathrm{d}y \tag{117}$$

Assuming the pitching moment is a linear function of pitch rate, the steady-state pitch damping moment coefficient derivative is

$$C_{M_q} = \frac{\partial(C_M)}{\partial\left(\dfrac{q\ell_{ref}}{2V_\infty}\right)} \tag{118}$$

The other part of the pitch damping moment, $C_{M_{\dot{\alpha}}}$, is not as easy to estimate at subsonic speeds. The author has actually been able to neglect this term for wings alone subsonically and achieve acceptable comparison to experimental data (within the ±20% goal for dynamic derivatives). If one does not want to neglect the wing alone $C_{M_{\dot{\alpha}}}$ at subsonic Mach numbers, a simple estimate is given by slender body theory,[3]

$$C_{M_{\dot{\alpha}}} = -\frac{\pi}{9\mathrm{AR}} \tag{119}$$

Equation (119) is based on wing span and wing area for reference length and reference area, respectively.

B. Supersonic Flow ($M_\infty \geq 1.2$)

While the component of pitch damping due to pitch rate q can be computed from the steady-state perturbation equation (60), the component of pitch damping due to a constant vertical acceleration $\dot{\alpha}$ requires consideration of the unsteady version of Eq. (60). However, Gardner[31] found that a proper solution of the unsteady wave equation for the $\dot{\alpha}$ term of pitch damping could be expressed as the sum of two steady-state potentials, one for a constant AOA and the other for steady pitching. This greatly simplifies the solution of the corresponding derivatives $C_{N_{\dot{\alpha}}}$ and $C_{M_{\dot{\alpha}}}$ as they can now be obtained by the solution of Eq. (60) to two steady-state problems and the results added together linearly.

For a supersonic leading edge (refer to Fig. 3.17) wing, Cole and Margolis[32] and Martin et al.[33] list the pressure loading required to define the load over the wing due to a constant pitch rate q and a constant vertical acceleration $\dot{\alpha}$. For region 1 of Fig. 3.17a, the pressure loading is

$$(\Delta C_P)_q = \left[(\Delta C_P)_q\right]_1 = \frac{4qmx}{\beta V_\infty (m^2-1)^{3/2}}[m^2 - 2 + \sigma] \tag{120}$$

$$(\Delta C_P)_{\dot{\alpha}} = \left[(\Delta C_P)_{\dot{\alpha}}\right]_1 = \frac{4\dot{\alpha}mx(1-\sigma)}{\beta V_\infty (m^2-1)^{3/2}}\left(\frac{m^2}{\beta^2}+1\right) \tag{121}$$

In region 2, the total lifting pressure is

$$(\Delta C_P)_q = \left[(\Delta C_P)_q\right]_2 = \frac{8qmx}{\pi\beta V_\infty}\left\{\frac{\sqrt{1-m^2\sigma^2}}{(m^2-1)} + \frac{(m^2-2-\sigma)}{2(m^2-1)^{3/2}}\cos^{-1}\left[\frac{1+m^2\sigma}{m(1+\sigma)}\right] + \frac{(m^2-2+\sigma)}{2(m^2-1)^{3/2}}\cos^{-1}\left[\frac{1-m^2\sigma}{m(1-\sigma)}\right]\right\} \tag{122}$$

$$(\Delta C_P)_{\dot{\alpha}} = \left[(\Delta C_P)_{\dot{\alpha}}\right]_2 = \frac{4\dot{\alpha}x}{V_\infty \beta^2 \pi (m^2-1)^{3/2}}\left\{\frac{2M^2m}{\beta}\sqrt{(1-m^2\sigma^2)(m^2-1)} - (m^2+\beta^2)\frac{m}{\beta}\left[(1+\sigma)\cos^{-1}\left(\frac{1+m^2\sigma}{m(1+\sigma)}\right) + (1-\sigma)\cos^{-1}\left(\frac{1-m^2\sigma}{m(1-\sigma)}\right)\right]\right\} \tag{123}$$

The induced pressure caused by the tip Mach line DF added to the contribution of region 1 yields the total lifting pressure in region 3:

$$\left[(\Delta C_P)_q\right]_3 = \frac{8q}{\pi V_\infty}\left[\frac{\frac{mx_a}{\beta}(m^2-2)+y_a+\frac{b}{2}(m^2-1)}{2(m^2-1)^{3/2}}\right.$$

$$\cdot\cos^{-1}\left(\frac{\frac{mx_a}{\beta}+(2m+1)y_a}{\frac{mx_a}{\beta}-y_a}\right)$$

$$\left.-\frac{(1+m-m^2)}{m^2-1}\sqrt{-my_a\left(\frac{x_a}{\beta}+y_a\right)}\right] \tag{124}$$

$$\left[(\Delta C_P)_{\dot{\alpha}}\right]_3 = \frac{4\dot{\alpha}}{V_\infty\beta\pi(m^2-1)^{3/2}}\left\{\left[\beta(m^2-m-1)-m\right]\right.$$

$$\sqrt{-4y_a(m+1)\left(\frac{mx_a}{\beta}+my_a\right)}-\beta\left(\frac{mx_a}{\beta}-y_a\right)$$

$$\left.(m^2+1)\cos^{-1}\left(\frac{\frac{mx_a}{\beta}+(2m+1)y_a}{\frac{mx_a}{\beta}-y_a}\right)\right\} \tag{125}$$

The lifting pressure in region 4 is a combination of that in regions 2, 3, and 1. Thus, in region 4

$$\left[(\Delta C_P)_q\right]_4 = \left[(\Delta C_P)_q\right]_2 + \left[(\Delta C_P)_q\right]_3 - \left[(\Delta C_P)_q\right]_1 \tag{126}$$

$$\left[(\Delta C_P)_{\dot{\alpha}}\right]_4 = \left[(\Delta C_P)_{\dot{\alpha}}\right]_2 + \left[(\Delta C_P)_{\dot{\alpha}}\right]_3 - \left[(\Delta C_P)_{\dot{\alpha}}\right]_1 \tag{127}$$

The total pressure differential in region 5 is

$$\left[(\Delta C_P)_q\right]_5 = \frac{4q}{\pi V_\infty (m^2-1)^{3/2}} \left\{ \left[(m^2-2)\frac{mx_a}{\beta} - y_a + (m^2-3)\frac{b}{2} \right] \cos^{-1} \left(\frac{mx_a/\beta - y_a(1-2m) + b}{mx_a/\beta + y_a + b} \right) + 2(m^2+m-1) \sqrt{-y_a(m-1)\left(\frac{mx_a}{\beta} + my_a + b \right)} \right\} \tag{128}$$

$$\left[(\Delta C_P)_{\dot{a}}\right]_5 = \left(\frac{4\dot{a}}{\beta^2} \right) \frac{M^2}{\pi V_\infty (m^2-1)^{3/2}} \left\{ -\left(\frac{mx_a}{\beta} + y_a + b \right)\left(1 + \frac{(m^2-1)}{M^2} \right) \cdot \cos^{-1} \left(\frac{mx_a/\beta - y_a(1-2m) + b}{mx_a/\beta + y_a + b} \right) + 2\left[(m^2-1)\left(1 - \frac{1}{M^2} \right) + m \right] \sqrt{-y_a(m-1)\left(\frac{mx_a}{\beta} + my_a + b \right)} \right\} \tag{129}$$

For a wing with a subsonic leading edge (see Fig. 3.18), the pressure differentials of the upper and lower surfaces in region 1 are[34]

$$(\Delta C_P)_q = 4q \frac{m}{\beta} \frac{G(m)}{V_\infty} \left[\frac{x(2-\sigma^2)}{\sqrt{1-\sigma^2}} \right] \tag{130}$$

$$(\Delta C_P)_{\dot{a}} = \frac{4\dot{a}}{V\beta^2} \left\{ M^2 \frac{m}{\beta} G(m) \frac{(2-\sigma^2)}{\sqrt{1-\sigma^2}} x - M^2 \frac{m\,xE'(m)}{\beta\sqrt{1-\sigma^2}} - \frac{m}{\beta} xE'(m)\sqrt{1-\sigma^2} \right\} \tag{131}$$

where

$$G(m) = \frac{1-m^2}{(1-2m^2)E(m) + m^2 F(m)},$$

$$E'(m) = \frac{1}{E(m)},$$

and the elliptic integrals $E(m)$ and $F(m)$ are defined after Eq. (110).

Region 2, which is affected by the wing tip Mach cone, has lifting pressures given by

$$(\Delta C_P)_q = \frac{4q}{3\pi V_\infty}\left\{\left[\frac{\frac{6mx}{\beta}(3+2m)+(12+8m+m^2)y-bm(1+m)}{(m+1)^2}\right] \cdot \sqrt{\frac{(m+1)(b-2y)}{2\left(\frac{mx}{\beta}+y\right)}}\right\} \tag{132}$$

$$(\Delta C_P)_{\dot\alpha} = \frac{4\dot\alpha}{V\beta^2}\left\{M^2\left[\frac{\frac{6mx}{\beta}(3+2m)+(12+8m+m^2)y-bm(1+m)}{3\pi(m+1)^{3/2}}\right] \cdot \sqrt{\frac{b/2-y}{\frac{mx}{\beta}+y}} - \frac{2M^2mx}{\pi\beta}\sqrt{\frac{b/2-y}{(m+1)\left(\frac{mx}{\beta}+y\right)}} - \frac{4}{\pi}\sqrt{\frac{\left(\frac{mx}{\beta}+y\right)\left(\frac{b}{2}-y\right)}{m+1}}\right\} \tag{133}$$

For a flow with a subsonic leading edge and supersonic trailing edge, Eqs. (130) through (133) determine the complete pressure distribution due to steady pitching rate q and constant vertical acceleration $\dot\alpha$.

The local pitching moment due to q and the total pitching moment due to q are calculated using Eqs. (116) through (118). The same quantities due to $\dot\alpha$ are obtained using Eqs. (116) through (118) by replacing q with $\dot\alpha$.

C. Transonic Flow ($0.8 \le M_\infty < 1.2$)

The problem of no simple, accurate analytical method to estimate dynamic aerodynamic coefficient derivatives is encountered again. As in previous cases, the only viable approach consistent with the present computational cost-effective objective is an empirical one.

The first step in the approximation of the transonic pitch damping is to determine the value at Mach number of 0.8 by applying the aforementioned subsonic calculation procedure. Then, using the empirical method discussed in Ref. 35, sweep, Mach number, aspect ratio, and thickness are taken into account. That is,

$$\left(C_{M_q} + C_{M_{\dot{\alpha}}}\right)_M = \left(C_{M_q} + C_{M_{\dot{\alpha}}}\right)_{M=0.8} \frac{\left(C_{N_\alpha}\right)_M}{\left(C_{N_\alpha}\right)_{M=M_{fb}}}, \tag{134}$$

where the quantities $C_{M_q} + C_{M_{\dot{\alpha}}}$ and C_{N_α} are the total configuration values. Moreover, if the damping in pitch is known at the nominal force break Mach number of 0.8, and C_{N_α} is known throughout the transonic speed regime, an estimate of pitch damping in transonic flow can be made with Eq. (134).

It should be pointed out that pitch damping derivative formulations are at the wing apex. For changes in the reference location about which aerodynamics are computed, the terms due to a vertical downward velocity follow the usual convention similar to normal force and pitching moment. However, the terms due to a constant pitch rate do not follow the usual transformation. This is because if the center of rotation is changed, the downwash or local spanwise loading changes on the wing planform due to the different point about which the wing is assumed to rotate. To illustrate this point, consider Fig. 3.24.

Suppose the original set of aerodynamics were calculated about the nose tip, and the center of rotation for the pitching rate term was about the wing apex. This original set of aerodynamics is defined by $(C_{N_{\dot{\alpha}}})_A$, $(C_{M_{\dot{\alpha}}})_A$ and $(C_{N_q})_A$ and $(C_{M_q})_A$. We wish to change these aerodynamics to the center of gravity of the body, which lies a distance x_2 behind the nose tip and a distance $-x_1$ ahead of the wing apex. Then according to Ref. 3, the new derivatives ($C_{N_{\dot{\alpha}}}$, $C_{M_{\dot{\alpha}}}$, C_{N_q}, and C_{M_q}) about the center of gravity, in terms of those calculated about the nose tip and wing apex, are

$$\begin{aligned}
C_{N_{\dot{\alpha}}} &= \left(C_{N_{\dot{\alpha}}}\right)_A \\
C_{M_{\dot{\alpha}}} &= \left(C_{M_{\dot{\alpha}}}\right)_A - (x_1/\ell_{\text{ref}})\left(C_{N_{\dot{\alpha}}}\right)_A \\
C_{N_q} &= \left(C_{N_q}\right)_A - 2C_{N_\alpha}(x_2/\ell_{\text{ref}}) \\
C_{M_q} &= (C_M)_A - 2\left(C_{M_\alpha}\right)_A(x_2/\ell_{\text{ref}}) - \left(C_{N_q}\right)_A(x_1/\ell_{\text{ref}}) + 2\,x_1x_2/\ell^2_{\text{ref}}\left(C_{N_\alpha}\right)_A
\end{aligned} \tag{135}$$

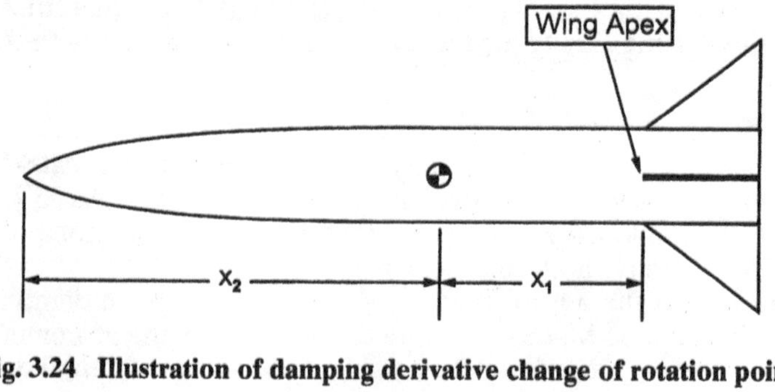

Fig. 3.24 Illustration of damping derivative change of rotation point.

The same two examples considered in Section 3.7 will be considered here for pitch damping. Figure 3.21 shows the configurations that are the Army–Navy Finner and the Navy research model, respectively. The first configuration is 10 calibers long with aspect ratio tail fin of 2, and the second configuration is 5.24 calibers long with effective aspect ratio fins of about 4.1. The centers of moment are at 6.1 and 3.0 calibers from the nose tip, respectively, for configurations 1 and 2 of Fig. 3.21. Fig. 3.25 presents comparison of the linear theory and empirical transonic and body-alone methods discussed here compared to experiment for the Army–Navy Finner.

Referring to Fig. 3.25, the wind tunnel and ballistic range data of Refs. 36 and 37, respectively, for the Army–Navy Finner compare rather well with the linear theory and transonic empirical methods. No data were available for subsonic Mach numbers. The one data point at Mach 2.0 of Ref. 26 is about 25 to 35% lower than the other data and linear theory. In Ref. 26, tests were conducted of dynamic derivatives as functions of AOA. Uselton and Jenke[26] pointed out that Ref. 36 data were taken for large oscillations, whereas Ref. 26 data were taken for small oscillations. Uselton and Jenke[26] also showed that for a 10-deg AOA, the value of $C_{M_q} + C_{M_{\dot{\alpha}}}$ was about 400 versus the value of 200 at 0-deg AOA. This value is shown on Fig. 3.25 with the symbol ơ. Uselton and Jenke[26] showed strong nonlinearity in dynamic derivatives as a function of AOA.

Figure 3.26 presents comparison of the linear theory to experiment for the Navy research model configuration of Fig. 3.21. Here, data were available at Mach number 0.6 up to 2.7 from Ref. 29. In general, the theory is higher than experimental data, particularly as Mach number increases. It is not known why the overprediction of the pitch damping moment occurs. However, as seen by the two wind tunnel data points at Mach 2 of Fig. 3.25 for $\alpha = 0$ and $\alpha = 10$ deg, where $C_{M_q} + C_{M_{\dot{\alpha}}}$ varies by a factor of two, it is not all that surprising to see theory and experiment to disagree by up to 30%. It is suspected that the short moment arm between the center of gravity and the wing apex also contributes to the larger percent error here as opposed to the previous case.

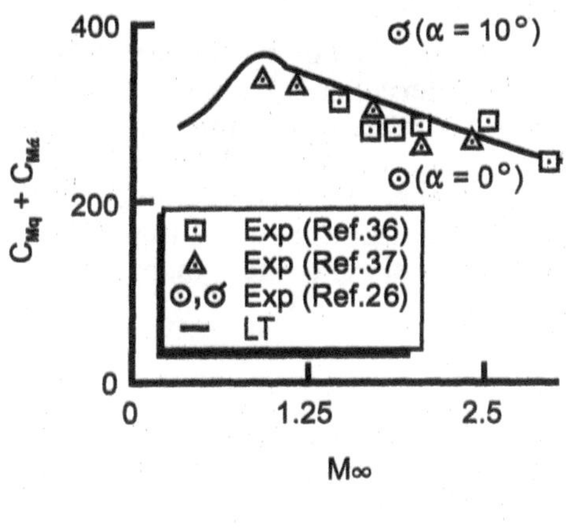

Fig. 3.25 Comparison of theory and experiment for pitch damping on Army–Navy Finner.

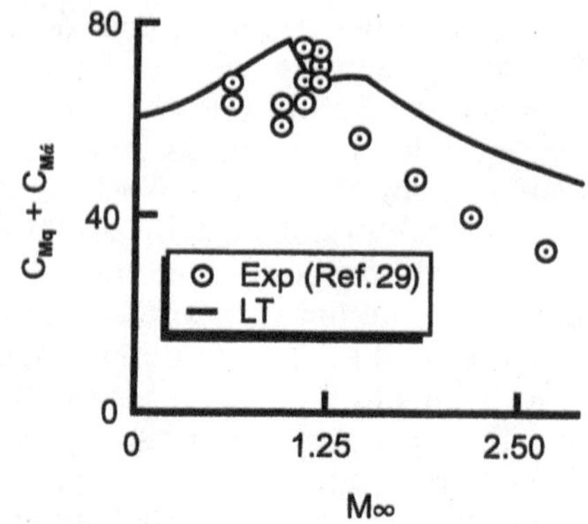

Fig. 3.26 Comparison of theory and experiment for pitch damping on Army–Navy research model.

IX. Interference Effects

There are several types of interference effects that occur in aerodynamics. First of all, when a wing–body configuration is present, there is an additional pressure load induced on the body as a result of the fins being present and, likewise, there is an additional pressure load induced on the fins as a result of being in the presence of the body. Both of these additional loads are created by the flowfield wrapping around the body (or wing) and, instead of continuing its path, hitting the adjacent structure, causing an increase in the pressure in the vicinity of the place where the wing is attached to the body. This is true whether the body is at a constant AOA, a constant pitching rate, or a constant vertical acceleration.

A second type of interference is that due to wing- and body-shed vortices as they affect the tail surfaces. The number of vortices shed is dependent on the roll orientation of the forward lifting surfaces. At roll orientation of zero (fins in plus "+" roll orientation), it is satisfactory to assume a single rolled up vortex is shed from each of the two forward lifting surfaces. If the fins are at roll orientation of 45 deg (fins in cross "×" roll orientation), then four vortices are shed if the forward lifting surfaces are cruciform. Each of these vortices induces a downwash on the tail fins, which in general decreases fin effectiveness and thus static stability.

Other types of interference effects, which will not be discussed here, include fin-to-fin interference effects and internal shock interactions. The first problem is generally small unless the number of fins present is large. The internal shock interaction problem generally occurs only at high AOA and moderate to high supersonic Mach numbers. Hence, the internal shock interaction problem will be discussed somewhat in Chapter 5. The remainder of this section will discuss slender body theory and linear theory results for wing–body and wing–tail interference.

A. Wing–Body Interference

The method almost universally used for including interference among various missile components into approximate aeroprediction codes is that due to Pitts et al.[38] There are two primary types of interference lift (note that lift and normal force are used interchangeably here) to be concerned with. These are the effects on the wing due to the presence of the body and the effect on the body due to the presence of a wing.

To better understand the interference lift components, it is instructive to examine the total normal force of a configuration as defined by Pitts et al.,[38] which is given by

$$C_N = C_{N_B} + \left[\left(K_{W(B)} + K_{B(W)}\right)\alpha + \left(k_{W(B)} + k_{B(W)}\right)\delta_W\right]\left(C_{N_\alpha}\right)_W + \left[\left(K_{T(B)} + K_{B(T)}\right)\alpha + \left(k_{T(B)} + k_{B(T)}\right)\delta_T\right]\left(C_{N_\alpha}\right)_T + C_{N_{T(V)}} + C_{N_{B(V)}} \tag{136}$$

The first term in Eq. (136) is the normal force of the body alone; the second term is the contribution of the wing (or canard), including interference effects and control deflection; the third term is the contribution of the tail, including interference effects and control deflection; and the last two terms are the negative downwash effect on the tail or body due to wing- or body-shed vortices. Each K represents the interference of the configuration with respect to AOA; and each k represents the interference with respect to control deflection. Each of these interference factors is estimated by slender body or linear theory.[38] As such, they are independent of AOA.

Various interference factors, as defined by slender body theory, are[38]

$$K_{W(B)} = 2/\pi \left\{ \frac{(1 + r^4/s^4\left[\frac{1}{2}\tan^{-1}\frac{1}{2}(s/r - r/s) + \pi/4\right]}{(1 - r/s)^2} - \frac{r^2/s^2[(s/r - r/s) + 2\tan^{-1}(r/s)]}{(1 - r/s)^2} \right\} \tag{137}$$

$$K_{B(W)} = (1 + r/s)^2 - K_{W(B)} \tag{138}$$

$$k_{W(B)} = \frac{1}{\pi^2}\left\{ \frac{\pi^2(s/r+1)^2}{4(s/r)^2} + \frac{\pi\left[(s/r)^2+1\right]^2}{(s/r)^2(s/r-1)^2}\sin^{-1}\left[\frac{(s/r)^2-1}{(s/r)^2+1}\right] - \frac{2\pi(s/r+1)}{s/r\,(s/r-1)} + \frac{\left[(s/r)^2+1\right]^2}{(s/r)^2(s/r-1)^2}\left(\sin^{-1}\left[\frac{(s/r)^2-1}{(s/r)^2+1}\right]\right)^2 - \frac{4(s/r+1)}{s/r\,(s/r-1)}\sin^{-1}\left[\frac{(s/r)^2-1}{(s/r)^2-1}\right] + \frac{8}{(s/r-1)^2}\log\left[\frac{(s/r)^2+1}{2\,s/r}\right] \right\} \tag{139}$$

$$k_{B(W)} = K_{W(B)} - k_{W(B)} \tag{140}$$

Figure 3.27 plots the interference lift factors given by Eqs. (137) through (140) as functions of the body radius to wing semispan plus body radius ratio (r/s).

As the Mach number increases supersonically, slender body theory gives values of $K_{B(W)}$ that are too high if the wing is near the missile rear. In this case, much of the carryover lift onto the body is actually lost to the wake of the vehicle. Figure 3.28 is an example of the no-afterbody, infinite-afterbody, and short-afterbody cases. Linear theory formulations are available for infinite-and no-afterbody cases to replace Eq. (138) if the parameter

$$\beta \text{AR}\,(1 + \lambda)\,[1/m + 1] > 4 \tag{141}$$

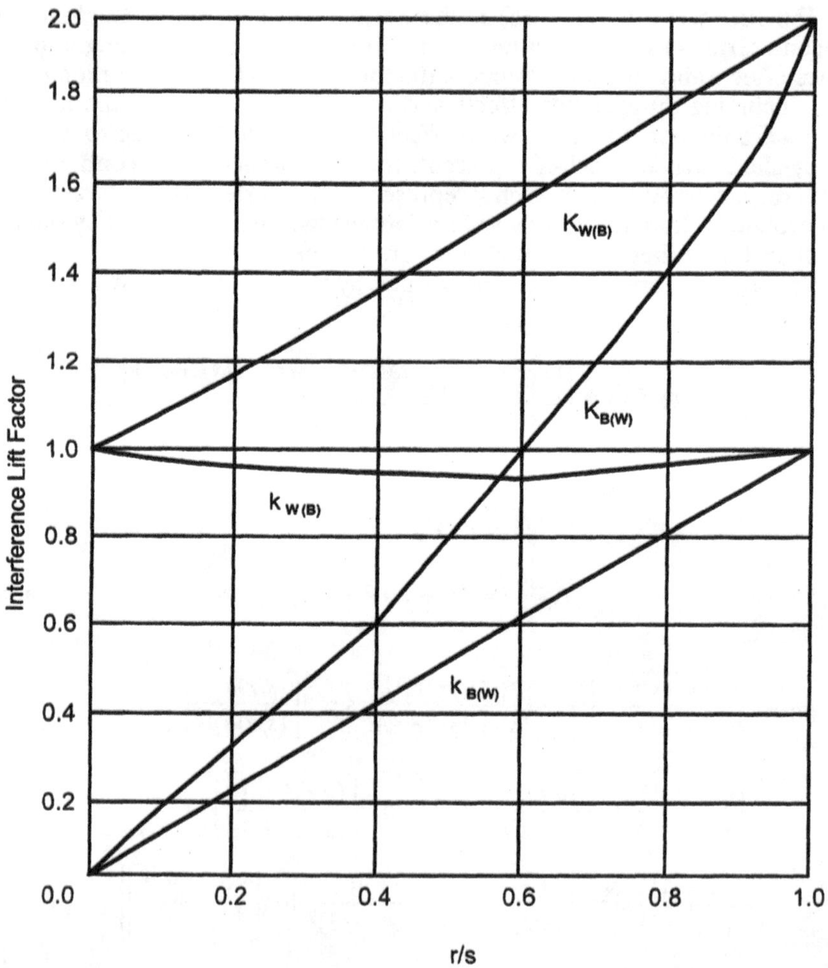

Fig. 3.27 Slender body theory interference lift factors.

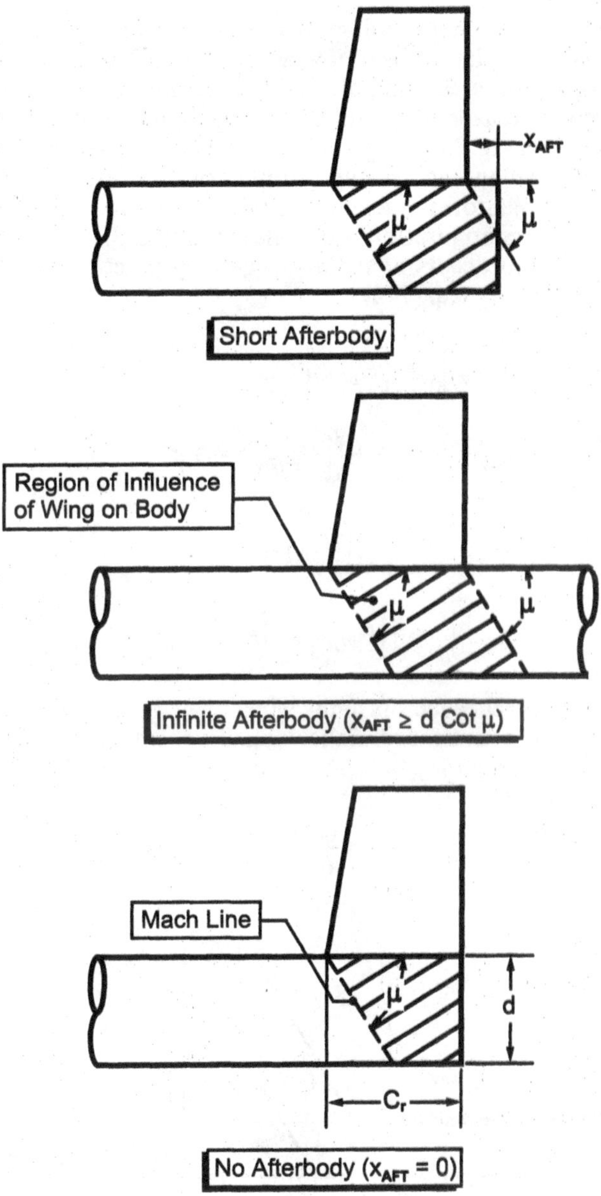

Fig. 3.28 Determination of $K_{B(W)}$ for high-aspect-ratio range at supersonic speeds.

Moore[39] then linearly interpolated between the infinite- and no-afterbody cases as functions of the area covered by the Mach lines to obtain $K_{B(W)}$ for the short-afterbody case.

Strictly speaking, the methodology discussed here is limited to slender bodies with triangular planforms of low aspect ratio. Experience has shown that if the correct value of wing-alone lift is computed, the interference factors can give very reasonable results for wings that do not have triangular planforms (such as in Fig. 3.29a) or even for wings that have low aspect ratio. Moore[39] showed how an engineering estimate of interference lift could be obtained, even for planforms such as that shown in Fig. 3.29b. The actual slender body theory configuration is that shown in Fig. 3.29a. Because most of the interference lift occurs near the wing–body juncture, Moore[39] used approximations given by Eq. (142)

$$\left[K_{B(W)}\right]_{\mathrm{II}} = \left[K_{B(W)}\right]_{\mathrm{I}} G$$

$$\left[K_{W(B)}\right]_{\mathrm{II}} = 1 + \left(\left[K_{W(B)}\right]_{\mathrm{I}} - 1\right) G$$

$$\left[k_{W(B)}\right]_{\mathrm{II}} = 1 + \left(\left[k_{W(B)}\right]_{\mathrm{I}} - 1\right) G$$

$$\left[k_{B(W)}\right]_{\mathrm{II}} = \left(\left[k_{W(B)}\right]_{\mathrm{I}} - \left[k_{W(B)}\right]_{\mathrm{I}}\right) G \tag{142}$$

to estimate the interference factors of the wing in Fig. 3.29a. The term G in Eq. (142) is the ratio of the root chord of the wing for which the interference factor is desired to that of the wing that slender body theory assumes. That is,

$$G = \frac{(c_r)_{\mathrm{II}}}{(c_r)_{\mathrm{I}}}$$

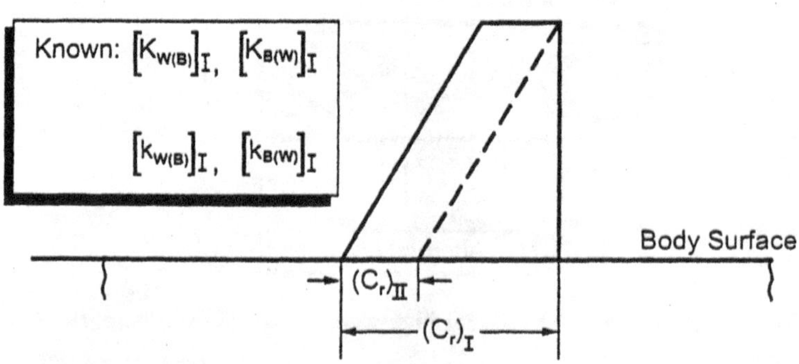

Fig. 3.29a Assumed slender body representation.

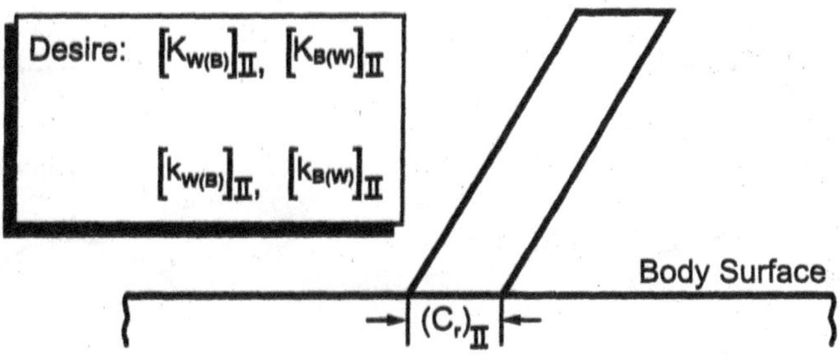

Fig. 3.29b Wing for which interference lift is desired.

The center of pressure of the wing–body interference is the same as that of the wing alone. The center of pressure of the body–wing interference term is at the centroid of the affected area on the body described by the cross-hatched portion of Fig. 3.28. If the flow is subsonic, of course the centroid of the body–wing interference is at $c_r/2$, or at the midchord point on the body.

B. Wing–Tail Interference

The last two terms of Eq. (136) are also interference terms. $C_{N_{T(V)}}$ is the lift on the tail caused by the vortices shed by the wing or canard upstream. $C_{N_{B(V)}}$ is the negative lift on the afterbody due to wing-shed vortices. The terms can be calculated analytically, but $C_{N_{T(V)}}$ is different for different roll orientations. This was not the case for cruciform wing–body configurations as shown in Ref. 3, so no mention of roll dependence was discussed in Section 3.8. The equation for $C_{N_{B(V)}}$, as defined in Ref. 38, is

$$C_{N_{B(V)}} = \frac{-4\Gamma}{A_W V_\infty}\left[\frac{f_W^2 - r_W^2}{f_W} - f_T + \frac{r_T^2}{\sqrt{f_T^2 + h_T^2}}\right] \tag{143}$$

While this term can be calculated analytically, the present author includes it as a part of the $K_{B(W)}$ term. The reason is that, in the nonlinear aerodynamics model, which will be discussed in Chapter 5, wind tunnel data have a portion of this term included, and it is hard to separate out the true $K_{B(W)}$ term from the $C_{N_{B(V)}}$ term.

A weapon can be rolling at any roll orientation during flight. However, most trim aerodynamic models assume a missile flying at a roll-stabilized mode of $\varphi = 0$ or 45 deg. Figure 3.30 illustrates the weapon flying in these two roll control orientations and the reasons these positions are chosen. We will consider the roll orientation position of 0 deg for illustration purposes and point out the process for determining the more complicated $\varphi = 45$ deg aerodynamics. The details of the $\varphi = 45$ deg aerodynamics can be found in Ref. 40.

The major new complexity that results from adding two sets of lifting surfaces to a missile configuration (versus one) is the vortices shed from the

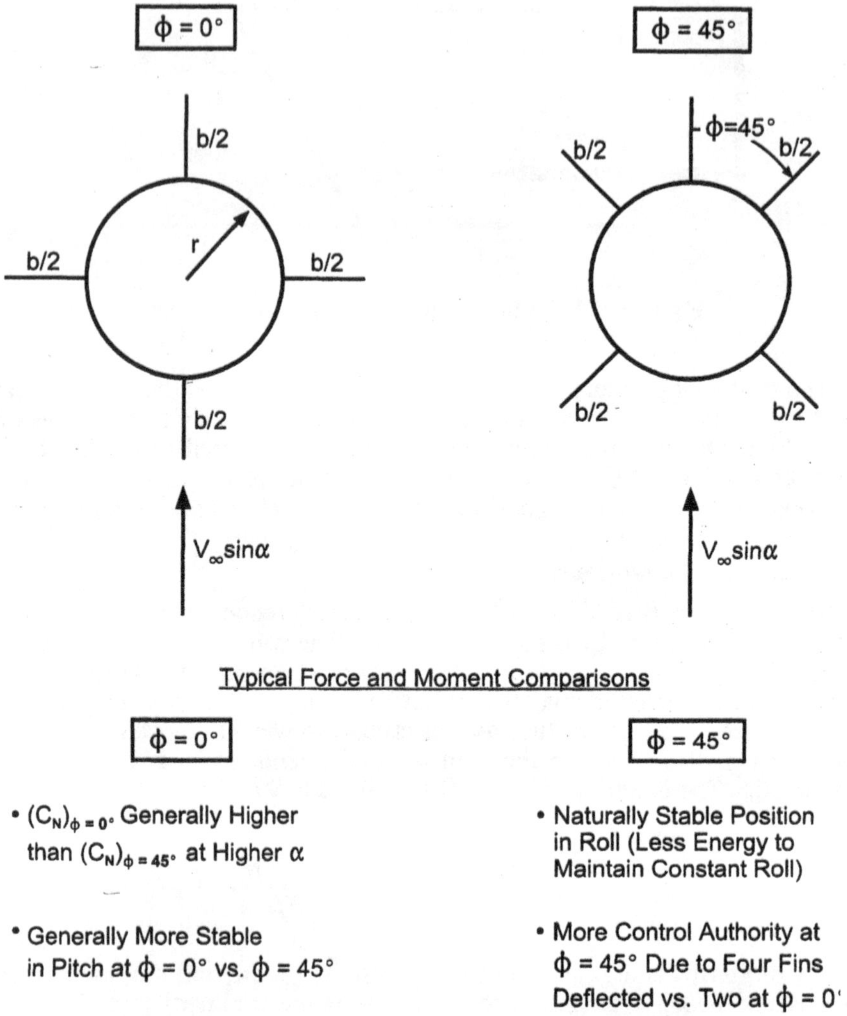

Fig. 3.30 Cruciform wing–body–tail missile configuration flying at roll of $\varphi = 0°$ and 45°.

forward surfaces affecting the rear surfaces. For this discussion, wings and canards will be used interchangeably as the forward surfaces, and tails will be the aft lifting surfaces. If the missile is at supersonic Mach numbers at an AOA greater than about 20 to 30 deg, shock wave interference between the forward and aft lifting surfaces or between the bow shock wave and the lifting surfaces can also occur. This problem will not be discussed in the linear theory and slender body theory discussion but is discussed in the nonlinear methods discussion of Chapter 5.

If the missile is in the $\varphi = 0$ deg roll orientation, only one wing planform will release a vortex at $a > 0$ deg for fairly thin wings of zero camber. However, at $\varphi = 45$ deg for a cruciform wing arrangement, both wings release vortices and these vortices follow different paths depending on whether they are released in the windward or leeward plane. It is instructive to examine the slender body theory results for the $\varphi = 0$ deg plane in terms of the loss of lift on the tail, or $C_{N_{T(V)}}$ term of Eq. (136). In reviewing the slender body theory results, Refs. 1 and 3 will be the primary sources of material.

Figure 3.31, taken from Ref. 3, is a schematic of the $\varphi = 0$ deg vortex model for a planar wing–body configuration. The wing-shed vortices are shown above the tail–body cross section. It is assumed here that one is looking upstream of a wing–body–tail configuration at a positive AOA. The strength of the wing-shed vortices, Γ, is assumed to be positive in a counter-clockwise direction looking upstream.

The term $C_{N_{T(V)}}$ of Eq. (136) is defined by[1]

$$C_{N_{T(V)}} = \frac{\left(C_{N_a}\right)_W \left(C_{N_a}\right)_T \left[K_{W(B)}a + k_{W(B)}\,\delta_W\right] i_T (s_T - r_T) A_W}{2\pi\, \mathrm{AR}_T (f_W - r_w) A_{\mathrm{REF}}} \tag{144}$$

Some of the parameters of Eq. (144) are shown in Fig. 3.31. The parameter f_W, the point on the wing where the single vortex separates, is approximated by the centroid of the approximate elliptic circulation distribution. That is,

$$f_W = r_W + (\pi/4)\,(b_W/2) \tag{145}$$

Utilizing Fig. 3.32, the height of the wing-shed vortex above the tail center of area is

$$h = \left\{(x_{\mathrm{LE}})_T - (x_{\mathrm{LE}})_W - \frac{\pi}{4}\left(\frac{b_w}{2}\right)\tan\left(\Lambda_{\mathrm{LE}}\right)_W - c_{r_W}\left[1 - \frac{\pi}{4}(1 - \lambda_W)\right] + \bar{x}_T\right\}\tan a$$
$$- \sin(\delta_W)\left\{(c_r)_W/2 + \left[\tan(\Lambda_{\mathrm{LE}})_w + \frac{(c_r)_W}{b_W/2}(\lambda_W - 1)\right]\left(\frac{\pi}{4}\right)\left(\frac{b_W}{2}\right)\right\} \tag{146}$$

Equation (146) assumes that the vortex leaves the wing at the trailing edge outboard location given by Eq. (145). The vortex then follows the velocity vector straight back and affects the tail center of pressure (here approximated along the root midchord). To satisfy the boundary condition on the body surface of zero normal velocity requires that a vortex of equal and

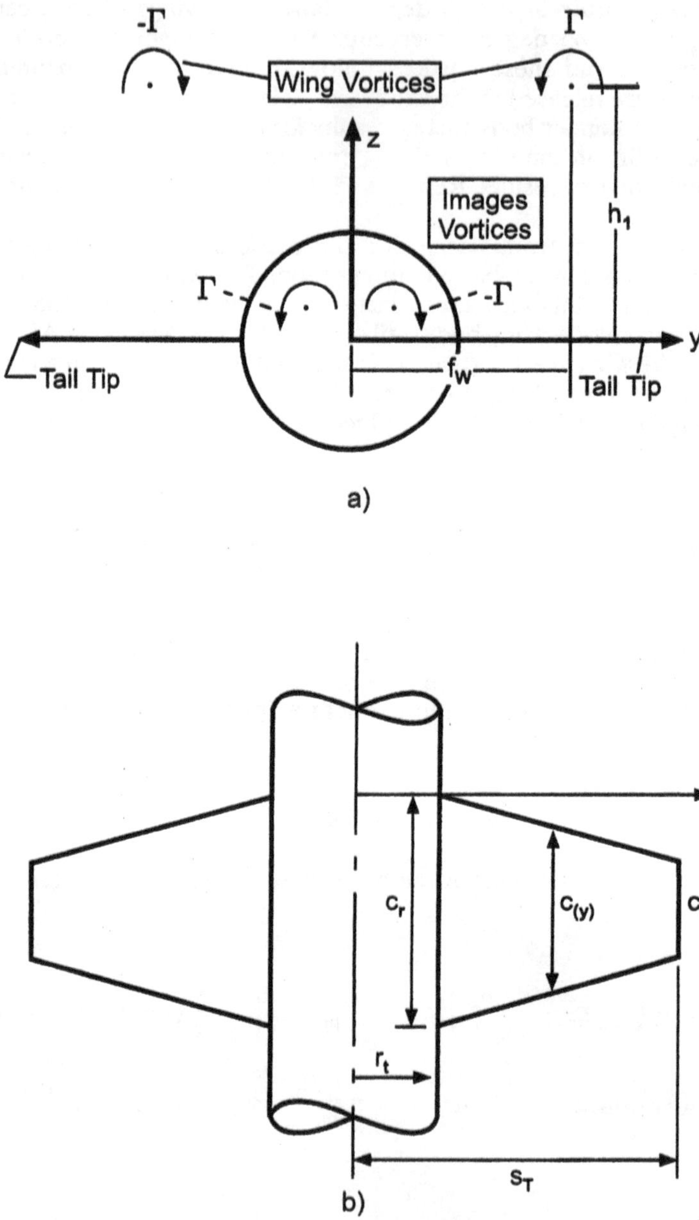

Fig. 3.31 Model and dimensions for determination of tail interference factor by strip theory for $\varphi = 0°$ (from Ref. 3): a) wing vortices in crossflow plane of tail and b) tail planform dimensions.

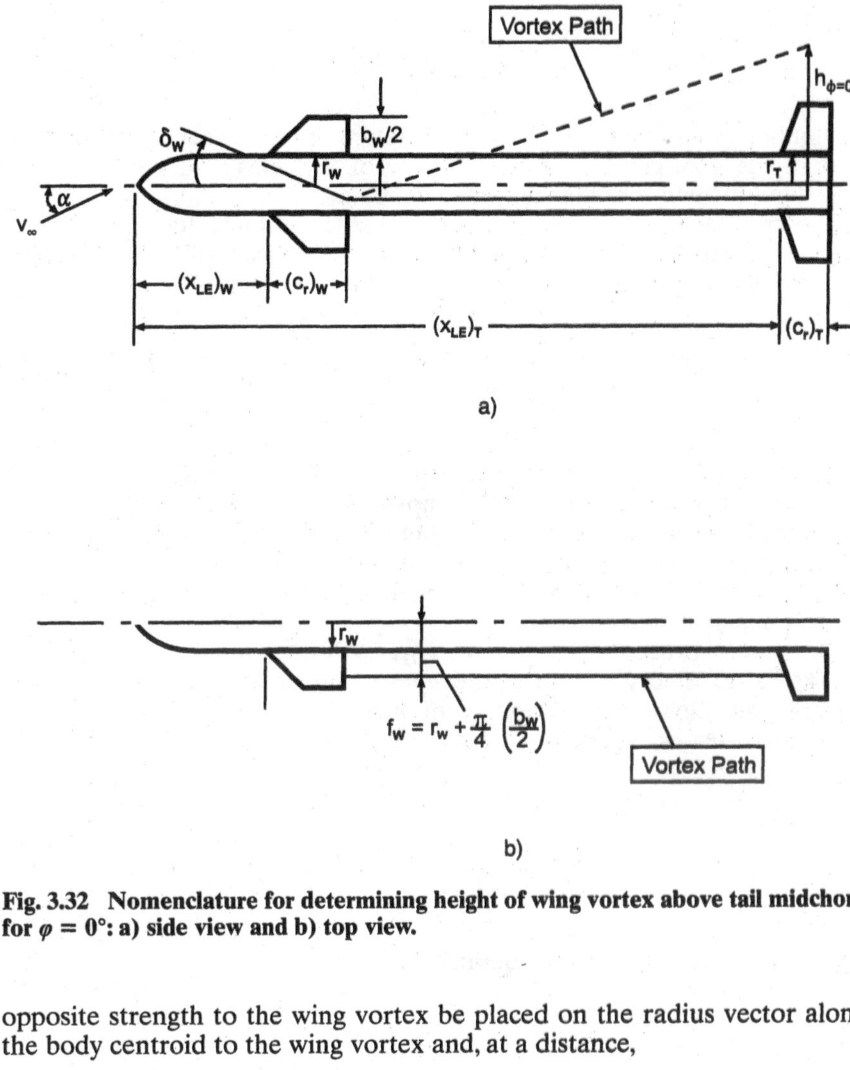

Fig. 3.32 Nomenclature for determining height of wing vortex above tail midchord for $\varphi = 0°$: a) side view and b) top view.

opposite strength to the wing vortex be placed on the radius vector along the body centroid to the wing vortex and, at a distance,

$$\frac{r^2}{\sqrt{f^2 + h^2}}$$

from the body centroid. Using proportionality of triangles, the coordinates of the image vortex are

$$f_i = f_W\left(\frac{r^2}{f_W^2 + h^2}\right) \tag{147}$$

$$h_i = h\left(\frac{r^2}{f_W^2 + h^2}\right) \tag{148}$$

It should be noted that image vortices are required to account for the body part of the configuration in the vicinity of the wing and are equal and opposite in strength to the wing-shed vortices.

The only term not defined in Eq. (144) is the tail interference factor i_T. The tail interference factor is a nondimensional measure of the interference of a vortex on a lifting surface. In mathematical terms, this definition is

$$i_T = \left(\frac{L_{T(V)}}{L_T}\right)\left(\frac{2\pi V_o(s_T - r_T)\, a}{\Gamma}\right) \tag{149}$$

In Eq. (149), $L_{T(V)}$ should be interpreted as a lift or normal force on the tail–body region due to a vortex shed upstream; L_T should be interpreted as a lift on the tail in isolation. Because the velocity of the vortices is counterclockwise in the tail region, a downward velocity, and hence the lift on the tail due to the wing-shed vortex, is negative. Hence, i_T is negative and Eq. (144) is negative.

The general procedure to compute $L_{T(V)}$ is through the use of strip theory because strip theory simplifies the integration of the lift over the wing surface and allows for a closed-form solution. The lift due to the right external vortex on the right external tail panel (looking upstream in Fig. 3.31) is

$$L_1 = \frac{4Q_\infty \Gamma (c_r)_T}{2\pi\left(\sqrt{M_\infty^2 - 1}\right)V_\infty} L\left(\lambda_T, \frac{r_T}{s_T}, \frac{f_W}{s_T}, \frac{h_1}{s_T}\right) \tag{150}$$

The function L of Eq. (150) is defined by

$$L\left(\lambda_T, \frac{r_T}{s_T}, \frac{f_W}{s_T}, \frac{h}{s_T}\right) = \left\{\frac{(s_T - r_T\lambda_T) - f_W(1-\lambda_T)}{2(s_T - r_T)}\, \ell n\left[\frac{h^2 + (f_W - s_T)^2}{h^2 + (f_W - r_T)^2}\right]\right.$$
$$\left. - \frac{(1-\lambda_T)}{(s_T - r_T)}\left[(s_T - r_T) + h\tan^{-1}\left(\frac{f_W - s_T}{h}\right) - h\tan^{-1}\left(\frac{f_W - r_T}{h}\right)\right]\right\} \tag{151}$$

The lift on the right panel due to the left vortex is then

$$L_2 = -\frac{4Q_\infty \Gamma (c_r)_T}{2\pi\left(\sqrt{M_\infty^2 - 1}\right)V_\infty} L\left(\lambda_T, \frac{r_T}{s_T}, -\frac{f_W}{s_T}, \frac{h_2}{s_T}\right) \tag{152}$$

The lifts of the right and left image vortices on the right wing panel are then given by

$$L_3 = -\frac{4Q_\infty \Gamma(c_r)_T}{2\pi\left(\sqrt{M_\infty^2 - 1}\right)V_\infty} L\left(\lambda_T, \frac{r_T}{s_T}, \frac{f_i}{s_T}, \frac{h_i}{s_T}\right) \tag{153}$$

$$L_4 = \frac{4Q_\infty \Gamma(c_r)_T}{2\pi\left(\sqrt{M_\infty^2 - 1}\right)V_\infty} L\left(\lambda_T, \frac{r_T}{s_T}, -\frac{f_i}{s_T}, \frac{h_i}{s_T}\right) \tag{154}$$

Because the left tail panel has the same lift as the right (because of symmetry at $\varphi = 0$ deg), the total lift on the tail due to the vortices shed in the wing–body region is

$$L_{T(V)} = \frac{8Q_\infty \Gamma(c_r)_W}{2\pi\left(\sqrt{M_\infty^2 - 1}\right)V_\infty} \left[L\left(\frac{f_W}{s_T}\right) - L\left(-\frac{f_W}{s_T}\right) - L\left(\frac{f_i}{s_T}\right) + L\left(-\frac{f_i}{s_T}\right)\right] \tag{155}$$

Equation (155) abbreviates the parameters in the L function equation (151) given by Eqs. (150), (152), (153), and (154) for simplicity.

The lift on the tail, again using strip theory is

$$L_T = \frac{4a\,Q_\infty(s_T - r_T)c_{r_T}(1 + \lambda_T)}{\sqrt{M_\infty^2 - 1}} \tag{156}$$

The nondimensional interference factor i_T then becomes

$$i_T = \left(\frac{2}{1 + \lambda_T}\right)\left[L\left(\frac{f_W}{s_T}\right) - L\left(-\frac{f_W}{s_T}\right) - L\left(\frac{f_i}{s_T}\right) + L\left(-\frac{f_i}{s_T}\right)\right] \tag{157}$$

It should be noted that while strip theory has been used for the integration of the lift on the tail in isolation and in the presence of the vortex, these errors in the numerator and denominator of Eq. (149) from using strip theory tend to cancel because they are nondimensionalized by one another. The same process for obtaining the interference factors $K_{W(B)}$, $K_{B(W)}$, $k_{W(B)}$ and $k_{B(W)}$ has been used with demonstrated success.

If the body is at $\varphi = 45$ deg roll, then four vortices are shed from the wings as shown in Fig. 3.33a. Again, to satisfy the boundary condition at the body surface of no velocity normal to the surface, four equal and opposite strength image vortices are placed on the radius vector along the body centroid to the wing vortex. In the plane of the tails (see Fig. 3.33b), dis-

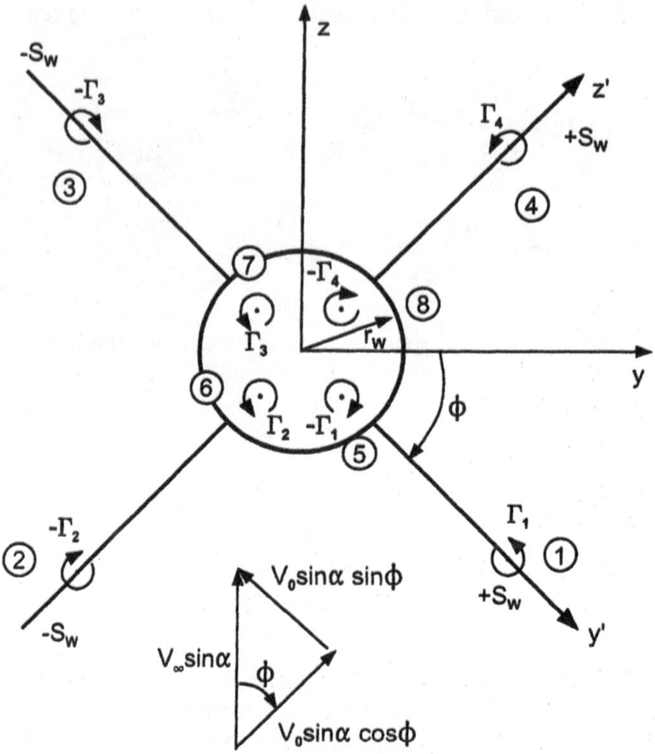

Fig. 3.33a Vortex model for cruciform wing–body arrangement.

tances y_i,Z_i can be determined for each of the eight vortices analogous to Eqs. (145) through (148) for the $\varphi = 0$ deg roll position. From symmetry, the strengths of opposite wing vortices are equal, that is

$$\begin{aligned} \Gamma_1 &= -\Gamma_3 \\ \Gamma_2 &= -\Gamma_4 \end{aligned} \tag{158}$$

As a result, only i_1 and i_4 need to be computed in analogy to Eq. (150). Finally, Eq. (144) is modified to the form

$$C_{N_{T(V)}} = \frac{\left(C_{N_\alpha}\right)_W \left(C_{N_\alpha}\right)_T [K_{W(B)}\alpha + k_{W(B)}\delta_W]}{2\pi \mathrm{AR}_T (f_W - r_W) A_{\mathrm{REF}}} (s_T - r_T)[i_1 \cos\varphi + i_4 \sin\varphi] A_W \tag{159}$$

It is interesting to note that while eight vortices, as opposed to four, must be considered for wing–tail interference computations at $\varphi = 45$ deg, the

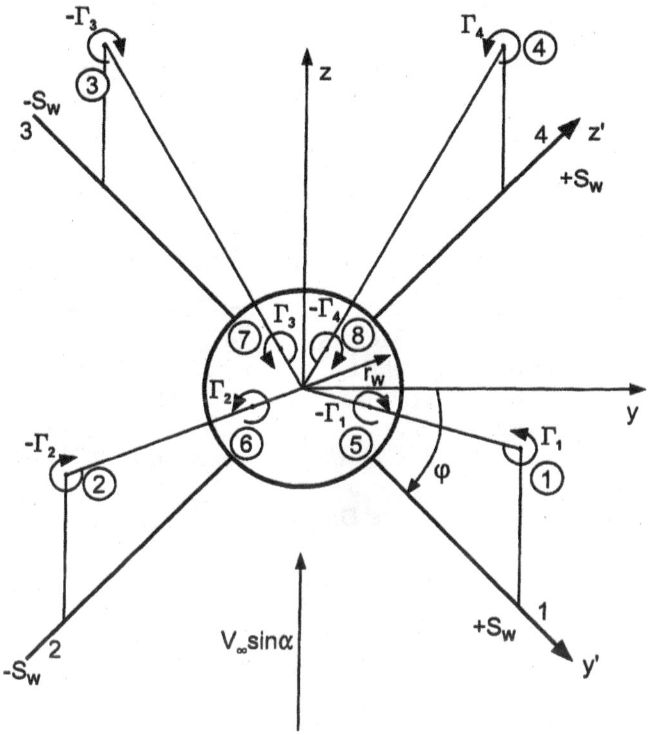

Fig. 3.33b Location of wing–body vortices in plane of tail.

results for $C_{N_{T(V)}}$ from Eqs. (159) and (144) are the same for small AOAs. This equality is similar to the results obtained for cruciform missiles at roll, where for low AOA the total normal force is independent of roll.

References

[1]Ashley, H., and Landahl, M., *Aerodynamics of Wings and Bodies,* Addison Wesley Longman, Reading, MA, 1965.

[2]Shapiro, A. H., *The Dynamics and Thermodynamics of Compressible Fluid Flow,* Vols. 1 and 2, Ronald, New York, 1954.

[3]Nielsen, J., *Missile Aerodynamics,* Engineering and Research, Mountain View, CA, 1988.

[4]Van Dyke, M. D., "First and Second-Order Theory of Supersonic Flow Past Bodies of Revolution," *Journal of Aeronautical Sciences,* Vol. 18, No. 3, March 1951, pp. 161–179.

[5]Tsien, H. S., "Supersonic Flow Over an Inclined Body of Revolution," *Journal of Aeronautical Sciences,* Vol. 5, No. 12, Oct. 1938, pp. 480–483.

[6]Van Dyke, M. D., "Practical Calculation of Second-Order Supersonic Flow Past Nonlifting Bodies of Revolution," NACA Technical Note TN 2744, July 1952.

[7]Van Dyke, M. D., "A Study of Second-Order Supersonic Flow Theory," NACA Rept. TR 1081, April 1952.

[8]Science Applications, "Theoretical Analysis of the Flow Field Over a Family of Ogive Bodies," Rept. TR TD-CR-77-5, U. S. Army Missile Research Development Command, Huntsville, AL, Aug. 1977.

[9]Chadwick, W. R., "The Application of Non-Planar Lifting Surface Theory to the Calculation of External Store Loads," *AIAA Journal,* Vol. 11, March 1974, pp. 181–188.

[10]Moore, F. G., and Swanson, R. C., "Aerodynamics of Tactical Weapons to Mach Number 3 and Angle of Attack 15 Degrees: Part 1 — Theory and Application," Naval Surface Warfare Center, Dahlgren Lab., (now Naval Surface Warfare Center, Dahlgren Div.), Rept. NSWCDL TR 3584, Feb. 1977.

[11]Murray, H. E., "Comparison with Experiment of Several Methods of Predicting Lift of Wings in Subsonic Compressible Flow," NACA TN 1739, 1948.

[12]Bradley, R. G., Smith, C. W., and Bhateley, I. C., "Vortex-Lift Prediction for Complex Wing Planforms," *Journal of Aircraft Engineering Note,* Vol. 10, No. 6, June 1973, pp. 379–381.

[13]Margolis, K., "Supersonic Wave Drag of Sweptback Tapered Wings at Zero Lift," NACA TN 1448, 1947.

[14]Stallings, R. L., Jr., and Lamb, M., "Wing-Alone Aerodynamic Characteristics for High Angles of Attack at Supersonic Speeds," NASA 1889, July 1981.

[15]Moore, F. G., McInville, R. M., and Hymer, T. C., "The 1998 Version of the NSWC Aeroprediction Code: Part 1 — Summary of New Theoretical Methodology," Naval Surface Warfare Center, Dahlgren Div., Rept. NSWCDD/TR-98/1, April 1998.

[16]Ames Research Staff, "Equations, Tables, and Charts for Compressible Flow," NACA Rep. 1135, 1953.

[17]Ferri, A., *Elements of Aerodynamics of Supersonic Flows,* MacMillan, New York, 1949, pp. 292–343.

[18]Jones, R. T., and Cohen, O., *High Speed Wing Theory,* Princeton Aeronautical Paperbacks, NO. 6, 1960.

[19]Baker, W. B., Jr., "Static Aerodynamic Characteristics of a Series of Generalized Slender Bodies With and Without Fins at Mach Numbers from 0.6 to 3.0 and Angles of Attack from 0 to 180°," Arnold Engineering Development, Rept. TR-75-124, Vols. 1 and 2, Tullahoma, TN, May 1976.

[20]Nielsen, J. N., Hemsch, M. H., and Smith, C. A., "A Preliminary Method for Calculating the Aerodynamic Characteristics of Cruciform Missiles to High Angles of Attack Including Effects of Roll Angle and Control Deflections," Rept. ONR-CR215-216, 4F, Office Naval, *Research,* Arlington, VA, Nov. 1977.

[21]Martin, J. C., and Jeffreys, I., "Span Load Distribution Resulting from Angle-of-Attack, Rolling, and Pitching for Tapered Sweptback Wings with Streamwise Tips," NACA TN 2643, 1952.

[22]Malvestuto, F. S., Jr., Margolis, K., and Retiner, H., "Theoretical Lift and Damping in Roll at Supersonic Speeds of Thin Sweptback Tapered Wings With Streamwise Tips, Subsonic Leading Edges and Supersonic Trailing Edges," NACA Rept. TR 970, 1950.

[23]Lockwood, V. E., "Effects of Sweep on the Damping-in-Roll Characteristics of

Three Sweptback Wings Having an Aspect Ratio of 4 at Transonic Speeds," NACA RM L 50J19, Dec. 1950.

[24]Edmonson, J. L., "Damping and Roll of Rectangular Wings of Several Aspect Ratios and NACA 65A Series Airfoil Sections of Several Thickness Ratios at Transonic and Supersonic Speeds as Determined with Rocket-Powered Models," NACA RM L50E25, 1950.

[25]Whyte, R. E., "Spinner—A Computer Program for Predicting the Aerodynamic Coefficients of Spin Stabilized Projectiles," General Electric Class 2 Repts., 1969.

[26]Uselton, B. L., and Jenke, L. M., "Experimental Missile Pitch and Roll-Damping Characteristics at Large Angles of Attack," *Journal of Spacecraft and Rockets*, Vol. 14, No. 4, April 1977, pp. 241–247.

[27]Nicolaides, J. D., and Bolz, R. E., "On the Pure Rolling Motion of Winged and/or Finned Missiles in Varying Supersonic Flights," *Journal of the Aeronautical Sciences*, Vol. 8, No. 3, March 1953, pp. 160–168.

[28]Regan, F. J., "Roll Damping Moment Measurements for the Basic Finner at Subsonic and Supersonic Speeds," Rept. 6652, Naval Ordnance Lab., White Oak, MD, now Naval Surface Warfare Center, Dahlgren Div., Dahlgren, VA, March 1964.

[29]Frierson, L., "Supersonic Aerodynamics of the 5-Inch Beamrider Guided Projectile, NSWCDL, Aerodynamics Memo No. DG 44-75-67, Dahlgren, VA, Oct. 1975.

[30]Bryson, A. E., "Stability Derivatives for a Slender Missile with Application to a Wing–Body–Vertical Tail Configuration," *Journal of the Aeronautical Sciences*, Vol. 20, No. 5, 1953.

[31]Gardner, C., "Time-Dependent Linearized Supersonic Flow Past Planar Wings," *Communications on Pure and Applied Math*, Vol. 3, No. 1, March 1950, pp. 33–38.

[32]Cole, I. J., and Margolis, K., "Lift and Pitching Moment at Supersonic Speeds Due to Constant Vertical Acceleration for Thin Sweptback Tapered Wings With Streamwise Tips, Supersonic Leading and Trailing Edges," NACA TN 3196, July 1954.

[33]Martin, J. C., Margolis, K., and Jeffreys, I., "Calculation of Lift and Pitching Moments Due to Angle of Attack and Steady Pitching Velocity at Supersonic Speeds for Thin Sweptback Tapered Wings with Streamwise Tips and Supersonic Leading and Trailing Edges, NACA TN 2699, June 1952.

[34]Malvestuto, F. S., and Hoover, D. M., "Lift and Pitching Derivatives of Thin Sweptback Tapered Wings with Streamwise Tips and Subsonic Leading Edges at Supersonic Speeds," NACA TN 2294, Feb. 1951.

[35]Douglas Aircraft Co., "USAF Stability and Control DATCOM," rev. by Wright Patterson Air Force Base, OH, July 1963, 2 Vols.

[36]Shantz, I., and Groves, R. T., "Dynamic and Static Stability Measurements of the Basic Finner at Supersonic Speeds," Naval Ordnance Lab., now Naval Surface Warfare Center, Dahlgren Div., Rept. 4516, Dahlgren, VA, Sept. 1960.

[37]MacAllister, L. C., "The Aerodynamic Properties of a Simple Non Rolling Finned Cone-Cylinder Configuration Between Mach Numbers 1.0 and 2.5," U.S. Army Ballistics Research Lab. (BRL) Rept. No. 934, Aberdeen, MD, May 1955.

[38]Pitts, W. C., Nielsen, J. N., and Kaattari, G. E., "Lift and Center of Pressure of Wing–Body–Tail Combinations at Subsonic, Transonic, and Supersonic Speeds," NACA TR 1307, 1957.

[39]Moore, F. G., "Aerodynamics of Guided and Unguided Weapons: Part 1—Theory and Application," Naval Weapons Lab., now Naval Surface Warfare Center, Dahlgren Div., Rept. NWL TR 3018, Dec. 1973.

[40]Moore, F. G., and McInville, R., "Extension of the NSWCDD Aeroprediction Code to the Roll Position of 45 Degrees," Naval Surface Warfare Center, Dahlgren Div., Rept. NSWCDD/TR-95/160, Dec. 1995.

Chapter 4

Local Slope and Empirical Methods

Nomenclature

AR	= aspect ratio
A_{ref}	= reference area used for force and moment calculations = $\pi d^2/4$ (ft²)
$C_{A_{\text{RB}}}$	= component of axial force due to a rotating band
$C_{A_{\text{vis}}}$	= component of axial force due to separation that occurs at low Mach number on a large cone angle in the nose region
C_{A_W}	= wave component of axial force coefficient
C_{f_∞}	= mean skin friction coefficient
C_{ℓ_P}	= roll damping moment coefficient, $C_\ell/(Pd/2V_\infty)$
C_M	= pitching moment coefficient
$C_{M_q} + C_{M_{\dot\alpha}}$	= pitch damping moment coefficient $(C_M)_q/(qd/2V_\infty)$ + $(C_M)_{\dot\alpha}/(\dot\alpha d/2V_\infty$
C_{M_α}	= pitching moment coefficient derivative
C_N	= normal-force coefficient
$C_{n_{P_\alpha}}$	= magnus moment coefficient, $C_n/(Pd/2V_\infty)/\alpha$
C_{N_α}	= normal-force coefficient derivative
C_P	= pressure coefficient
C_{P_B}	= base pressure coefficient
$(C_{P_B})_{NF,\alpha=0}$	= body-alone base pressure coefficient at $\alpha = 0$
$(C_{P_B})_{\alpha,\delta,t/c,x/c}$	= base pressure coefficient of wing–body configuration at AOA and wings with control deflection, thickness to chord ratio, and location upstream of base
C_{P_O}	= axial force coefficient at zero AOA
d	= diameter of body (ft), used as reference length in moment calculations
$\hat{e}_t,\hat{e}_n,\hat{e}_\varphi$	= unit vectors with $\hat{e}_t$ tangent to the body surface, $\hat{e}_n$ normal to the surface, and $\hat{e}_\varphi$ tangent to the body in the φ direction
H	= total enthalpy (ft²/s²)
h	= specific enthalpy (ft²/s²)
M	= Mach number
P	= roll rate (rad/s)
p	= pressure (lb/ft²)
r	= local body radius

r_n = nose radius
R_B = radius at base of body
RMF = jet momentum flux ratio
R_N = nose radius of spherical cap
s = streamline coordinate along body
T = temperature (°R)
t = wing or tail thickness
V = velocity (ft/s)
V_{max} = maximum velocity (ft/s)
V_r, V_θ = velocity components along and normal to a conical ray
x_{CP} = center of pressure measured from some reference location
x_{LE} = distance from nose tip to leading edge of wing
x, y, z = rectangular coordinate system
ℓ = body length
ℓ_n, ℓ_a, ℓ_B = nose, afterbody, and boattail length, respectively
α = angle of attack (deg)
β = $\sqrt{M^2 - 1}$ if $M > 1$; $\sqrt{1 - M^2}$ if $M < 1$
γ = ratio of specific heats
δ = angle between velocity vector and tangent to the body at a given point
δ^* = angle that the body nose makes with the afterbody at the nose–shoulder juncture
η, η_1 = exponent used in second-order shock expansion theory
θ = local body slope
θ_c, θ_w = cone or wedge angle, respectively
Λ = sweepback angle of wing or tail surface
λ = $\rho V^2/\beta$; also taper ratio = c_t/c_r
μ = Mach angle, $\mu = \sin^{-1}(1/M)$
ν = Prandtl–Meyer angle (deg)
ρ = density (slug/ft^3)
σ = shock wave angle (deg)
Φ = velocity potential
φ = roll orientation with $\varphi = 0$ being leeward plane

Subscripts

c = cone
fb = Mach number where wing stall begins to occur as a result of shock wave standing on surface
j = exit conditions of rocket motor
ℓ = lower surface
LE, TE = leading and trailing edge of wing, respectively
o = total conditions
r, t = root chord and tip chord
u = upper surface
∞ = freestream conditions
1,2 = upstream, downstream of a point, respectively; also ahead of and behind shock wave

LOCAL slope methods are approximate methods to compute pressure coefficients on a given segment of a body or wing surface based on a freestream Mach number and the local slope of a segment of the body or wing with respect to the freestream direction. Several of these methods are dependent on a wedge or cone solution of the inviscid equations of motion, Eqs. (12) to (15) in Chapter 2. Local slope methods in general are fairly simple and straightforward, yet yield reasonable results for force coefficients in many cases.

I. Tangent Wedge Method

Figure 4.1 illustrates the tangent wedge method. As seen in Fig. 4.1, the wedge is at an angle of attack (AOA) so the deflection angle on the lower surface is $\theta_w + \alpha$, whereas the deflection angle on the upper surface is $\theta_w - \alpha$. The shock wave angle on the lower surface has an angle of σ_L and the upper surface has a shock angle of σ_u. Mach numbers high enough to allow the shock to be attached to the wedge tip are required for the tangent wedge method to be valid. If this assumption is violated, Newtonian theory (to be discussed later in this chapter) can be used, or an empirical relation can be used for subsonic flow if the wedge angle is large enough to have a viscous separation bubble on the leading edge.

To solve for pressure coefficients on the upper or lower surfaces, inviscid equations of motion are applied to the two-dimensional configuration of Fig.

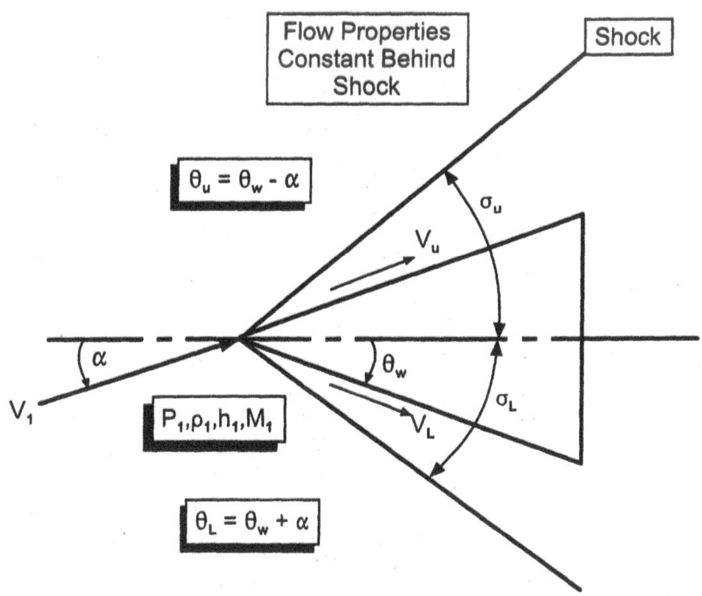

Fig. 4.1 Flow over a two-dimensional wedge at angle of attack.

4.1. These equations yield the standard oblique shock wave relations[1] for a perfect gas. The equation for the shock angle σ is

$$\sin^6\sigma + b\sin^4\sigma + c\sin^2\sigma + d = 0 \tag{1}$$

where

$$b = -\frac{M_1^2+2}{M_1^2} - \gamma\sin^2\theta$$

$$c = \frac{2M_1^2+1}{M_1^4} + \left[\frac{(\gamma+1)^2}{4} + \frac{\gamma-1}{M_1^2}\right]\sin^2\theta$$

$$d = -\frac{\cos^2\theta}{M_1^2}$$

Equation (1) can be solved numerically by guessing a value of σ and iterating until the value of σ changes little from the previous value. Once σ is known, other properties can be obtained from the following oblique shock relations:

$$\frac{p_2}{p_1} = \frac{2\gamma M_1^2\sin^2\sigma - (\gamma-1)}{\gamma+1} \tag{2}$$

$$\frac{\rho_2}{\rho_1} = \frac{(\gamma+1)M_1^2\sin^2\sigma}{(\gamma-1)M_1^2\sin^2\sigma + 2} \tag{3}$$

$$\frac{T_2}{T_1} = \frac{\left[2\gamma M_1^2\sin^2\sigma - (\gamma-1)\right]\left[(\gamma-1)M_1^2\sin^2\sigma + 2\right]}{(\gamma+1)^2M_1^2\sin^2\sigma} \tag{4}$$

$$M_2^2 = \frac{(\gamma+1)^2M_1^4\sin^2\sigma - 4(M_1^2\sin^2\sigma - 1)(\gamma M_1^2\sin^2\sigma + 1)}{\left[2\gamma M_1^2\sin^2\sigma - (\gamma-1)\right]\left[(\gamma-1)M_1^2\sin^2\sigma + 2\right]} \tag{5}$$

The term θ in Eq. (1) is θ_L corresponding to σ_L. This process is repeated for σ_u, where θ_L is replaced by θ_u. Note that flow properties between the upper shock and body are constant as are flow properties between the lower shock and body.

Knowing the pressure on the upper and lower surface, the pressure coefficient is then

$$C_{P_{u,\ell}} = \frac{p_{u,\ell} - p_1}{\frac{1}{2}\rho_1 V_1^2} \tag{6}$$

The wave component of axial force coefficient is

$$C_{A_W} = C_{P_\ell} \sin(\theta_W) + C_{p_u} \sin(\theta_W) \tag{7}$$

The normal force and pitching moment coefficients (about the leading edge) are

$$C_N = C_{P_\ell} \cos(\theta_W) - C_{P_u} \sin(\theta_W) \tag{8}$$

$$C_M = -C_N\left(\frac{\ell}{2}\right) \tag{9}$$

Note that equations 7 and 8 are for conditions where the wedge is the entire airfoil length (that is, $x/c = 1$).

In a case where θ_u is negative but the shock is still attached, one of two alternatives can be used. If the Mach number is high, the Newtonian flow assumption of zero-pressure coefficient in shadowed regions can be used. At lower supersonic Mach numbers, Prandtl–Meyer expansion can be used to compute the flow properties. Both of these methods will be discussed later in this chapter.

Figure 4.1 illustrated the tangent wedge method as the starting solution for a true two-dimensional case. For this initial condition, the starting solution is an exact solution of the flowfield. However, in principal, this process can be applied away from the leading edge of a circular shaped airfoil where the local slope is the angle between a tangent to the two-dimensional surface and the freestream velocity vector. Of course, aerodynamic coefficients are summed up over all of the piecewise linear segments of the airfoil or two-dimensional section of the configuration profile. The tangent wedge method is most useful on high aspect ratio wings or flat surfaces, where conditions are nearly two-dimensional. As the aspect ratio decreases to more practical conditions typical of weapons, significant regions of three-dimensional pressure effects near the root and tip chords are present. The three-dimensional thin wing theory (TOTWT) of Section 3.6 is more accurate for both axial and normal force computations for three-dimensional flow. However, in comparing Section 3.6 mathematics with the tangent wedge method of this section, it is clear that the tangent wedge method has an advantage in terms of simplicity over the TOTWT. Also, some aerodynamic codes have used the tangent wedge method on segments of a body as well, particularly on flat segments, because of its simplicity.

II. Tangent Cone Method

The tangent cone method is similar to the tangent wedge method, except here the configuration is a body of revolution as opposed to a two-dimensional surface. To get the local angle between the velocity vector and the tangent to a point of the surface is more complicated than for the wedge, however. Referring to Fig. 4.2, this angle δ is

$$\delta = \sin^{-1}[\sin\theta\cos a - \sin a\cos\varphi\cos\theta] \tag{10}$$

Equation (10) assumes $\varphi = 0$ is the leeward plane.

Also, the flow on a cone is more complicated than that for a wedge. Applying the conservation laws, Eqs. (12) to (15) in Chapter 2 to the flowfield on the cone of Fig. 4.3, one obtains the fact that flow properties are constant along rays emanating from the cone tip as opposed to being constant behind the shock as was the case for the wedge flow. The equation governing the flow over a cone at zero AOA is the well-known Taylor–McCall equation[2] defined by

$$\begin{aligned}\frac{\gamma-1}{2}\left[V_{\max}^2 - V_r^2 - \left(\frac{dV_r}{d\theta}\right)^2\right]\left[2V_r + \frac{dV_r}{d\theta}\cos^2\theta + \frac{d^2V_r}{d\theta^2}\right] \\ - \frac{dV_r}{d\theta}\left[V_r\frac{dV_r}{d\theta} + \frac{dV_r}{d\theta}\left(\frac{d^2V_r}{d\theta^2}\right)\right] = 0\end{aligned} \tag{11}$$

where

$$V_{\max}^2 = 2H = \text{const.}$$

To solve Eq. (11), we assume a shock angle σ for a given point on the body surface that has an angle δ between the velocity vector and the local slope of the body with respect to the x axis, θ. This angle δ replaces the angle θ in Eq. (11) for each point around the body. [If the body is anything other than a cone, Eq. (11) must be solved along and around the body at enough points that force coefficients can be computed.] Equation (11) is integrated between the shock and body where the boundary conditions at the shock are the shock relations and the boundary condition at the body surface that requires $V_\theta = 0$. If V_θ is not zero, then a new value of σ is guessed and an iteration occurs until the boundary condition of $V_\theta = 0$ at the body surface is satisfied. The Mach number at this surface point is then

$$M_2^2 = \left[\frac{(V_r/V_{\max})^2}{1-(V_r/V_{\max})^2}\right]\frac{1}{\gamma-1} \tag{12}$$

Because the flow is adiabatic, the total temperature throughout the flowfield is constant. That is,

$$T_{O_2} = T_{O_1} = T_1\left(1+\frac{\gamma-1}{2}M_1^2\right) = T_2\left(1+\frac{\gamma-1}{2}M_2^2\right) \tag{13}$$

Knowing M_2 and T_{O_2}, T_2 can thus be computed. The pressure and density immediately behind the shock where the angle σ has already been computed can be determined from the oblique shock wave relations, Eqs. (2) and (3).

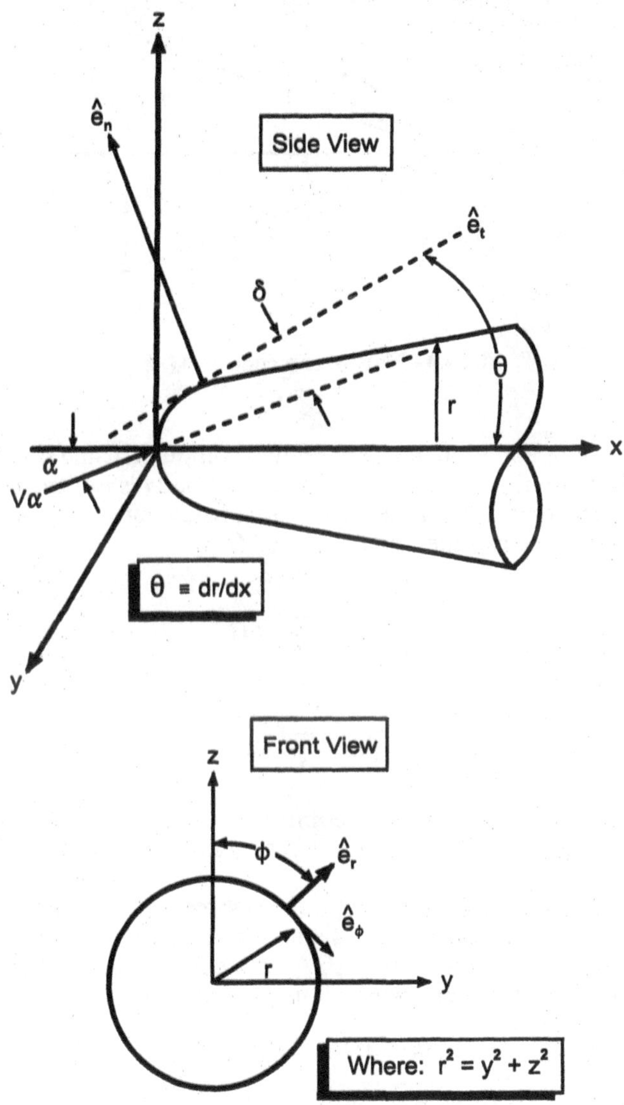

Fig. 4.2 Determination of angle δ for a body of revolution.

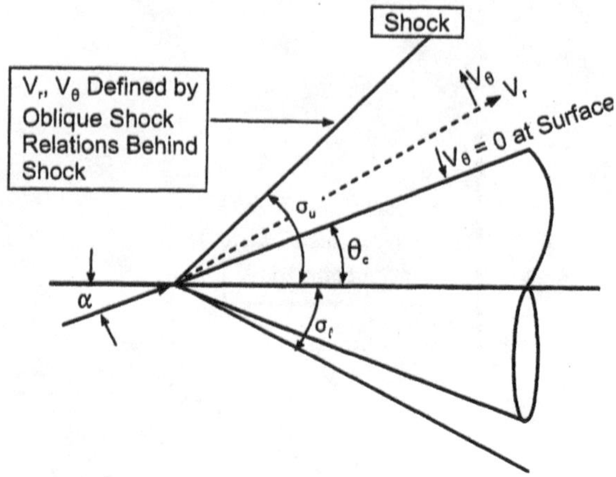

Fig. 4.3 Flow over a cone at AOA.

The total pressure and total density are then computed using these values of p_2 and ρ_2. Because flow between the shock and body is isentropic, then p_{O_2} and ρ_{O_2} are constant throughout this region. Knowing M_2 at the surface, p_2 and ρ_2 at the surface can then be computed from

$$\frac{p_{O_2}}{p_2} = \left(1 + \frac{\gamma - 1}{2} M_2^2\right)^{\frac{\gamma}{\gamma - 1}} \tag{14}$$

$$\frac{\rho_{O_2}}{\rho_2} = \left(1 + \frac{\gamma - 1}{2} M_2^2\right)^{\frac{\gamma}{\gamma - 1}} \tag{15}$$

Pressure coefficients at each body point are then determined from Eq. (6) and force and moment coefficients are determined from Eqs. (56) to (59) in Chapter 3.

Moore et al.[3] derived a simple method to estimate the pressure coefficient on a cone at AOA. In that method, slender body theory, Newtonian theory, an approximate expression for the AOA-zero cone pressure coefficient, and a Taylor series expansion in a were all used to obtain

$$C_P(a, \theta, \varphi, M) = C_{P_{a=0}} + \Delta C_P \tag{16a}$$

where

$$\Delta C_P = -2a \sin 2\theta \cos \varphi + a^2 \cos^2 \theta \left[\left(2 - \frac{1}{\beta}\right)(1 - \tan^2 \theta) - \left(2 + \frac{2}{\beta}\right) \sin 2\varphi\right] + \frac{4}{3} a^3 \sin 2\theta \cos \varphi \tag{16b}$$

$$C_{P_{a=0}} = \sin^2\theta_c\left[1 + \frac{(\gamma+1)K^2+2}{(\gamma-1)K^2+2}\,\ell n\left(\frac{\gamma+1}{2}+\frac{1}{K^2}\right)\right] \tag{16c}$$

and

$$K^2 = \left(M_1^2 - 1\right)\sin^2\theta_c$$

Equation (16) can thus be used to approximate the value of pressure coefficient at any point on a body surface in lieu of solving the second-order differential Eq. (11) at many points around and along the body surface. In applying Eq. (16), it was found that unacceptable performance occurred in the leeward plane of a blunt nosed configuration. Thus, in the leeward plane of a blunt nosed body, and behind the blunt nose region, an expression for C_P superior to Eq. (16) was found to be

$$C_P(a,\theta,\varphi,M) = C_{P_{a=0}} - \frac{2a\sin 2\theta\cos\varphi}{3} \tag{16d}$$

For blunt bodies, modified Newtonian theory is used to compute pressure coefficients in the nose region. Equation (16b) can be used in a tangent cone approach for the windward plane region (180 deg $\leq \varphi <$ 30 deg) and Eq. (16d) can be used in the leeward plane area ($\varphi \leq$ 30 deg).

Figures 4.4 and 4.5 compare the exact cone solution[4] results to those of Eq. (16b) and (13). Here, actual pressures versus pressure coefficients are plotted in Figs. 4.4 and 4.5 as well as local surface temperatures. Note that, in all four conditions in Figs. 4.4 and 4.5, good agreement on pressure and temperature is obtained compared to the exact solution. Most cases have given errors of less than 3% where Eqs. (16) have been used. Figure 4.5 shows errors as high as 15% for pressures. Better accuracy can be obtained at higher Mach numbers, as seen in Fig. 4.4. Also, if Eq. (11) is used to solve the cone flowfield numerically at $a = 0$, slight improvements in accuracy are achievable over the approximate formula given by Eq. (16c). It should be noted that the exact cone solution at AOA involves solution of a partial differential equation as opposed to the ordinary differential Eq. (11). Figures 4.4 and 4.5 illustrate that, if one were to use the Jones cone solution[4] for the tangent cone method, improved accuracy could be obtained, but at a significant increase in computational time.

III. Shock Expansion Theory

First-order expansion theory was first proposed by Eggers et al.[5], Eggers and Savin[6,7], and Savin[8] for bodies of revolution flying at high supersonic speeds. Basically, the Shock-expansion theory computes the flow parameters at the leading edge of a two dimensional surface with the oblique shock wave relations and with the solution for a cone at the tip of a three-dimensional

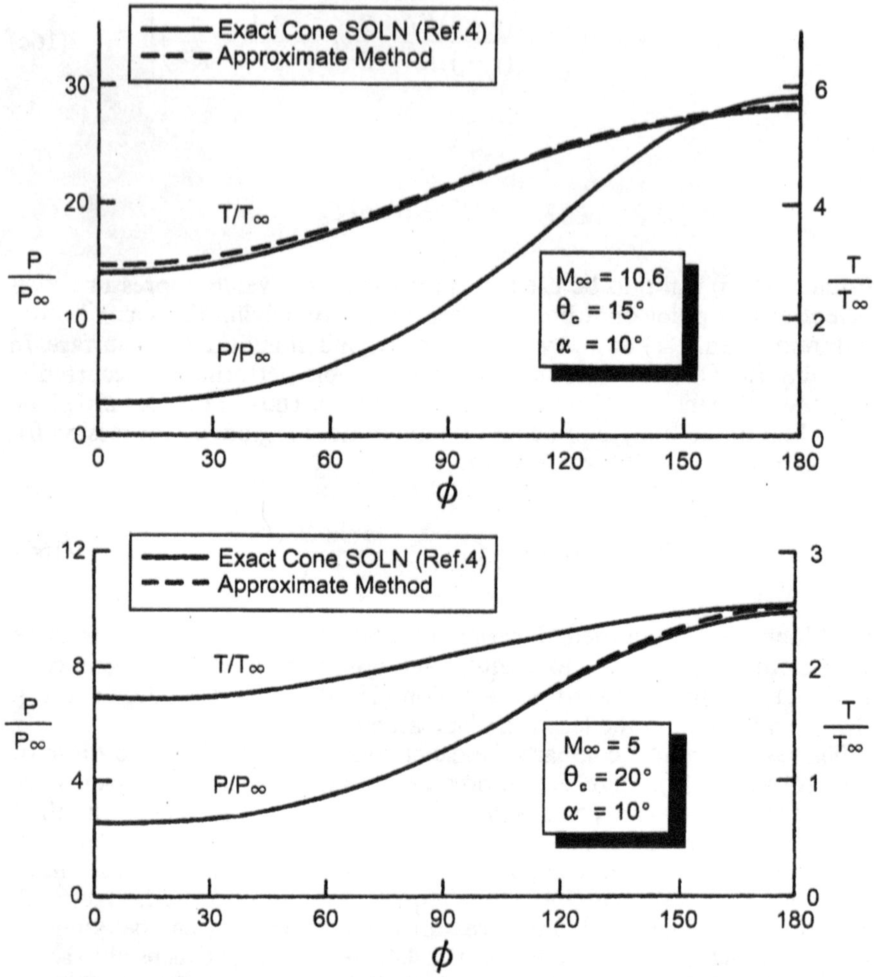

Fig. 4.4 Perfect gas comparison of exact and approximate cone solutions.

body. Both of these methods were described in Sections 4.1 and 4.2. Standard Prandtl–Meyer Expansion is then applied along the surface behind the leading edge or tip solution to get the complete pressure distribution over the body surface. Referring to Fig. 4.6, this shock expansion theory inherently assumes that the expansion waves created by the change in curvature around the body are entirely absorbed by the shock and do not reflect back to the body surface. Because the theory assumes constant pressure along one of the conical tangent elements of the surface, fairly slender surfaces must be assumed or many points along the surface assumed to obtain a fairly accurate pressure distribution. In other words, to minimize the strength of the distur-

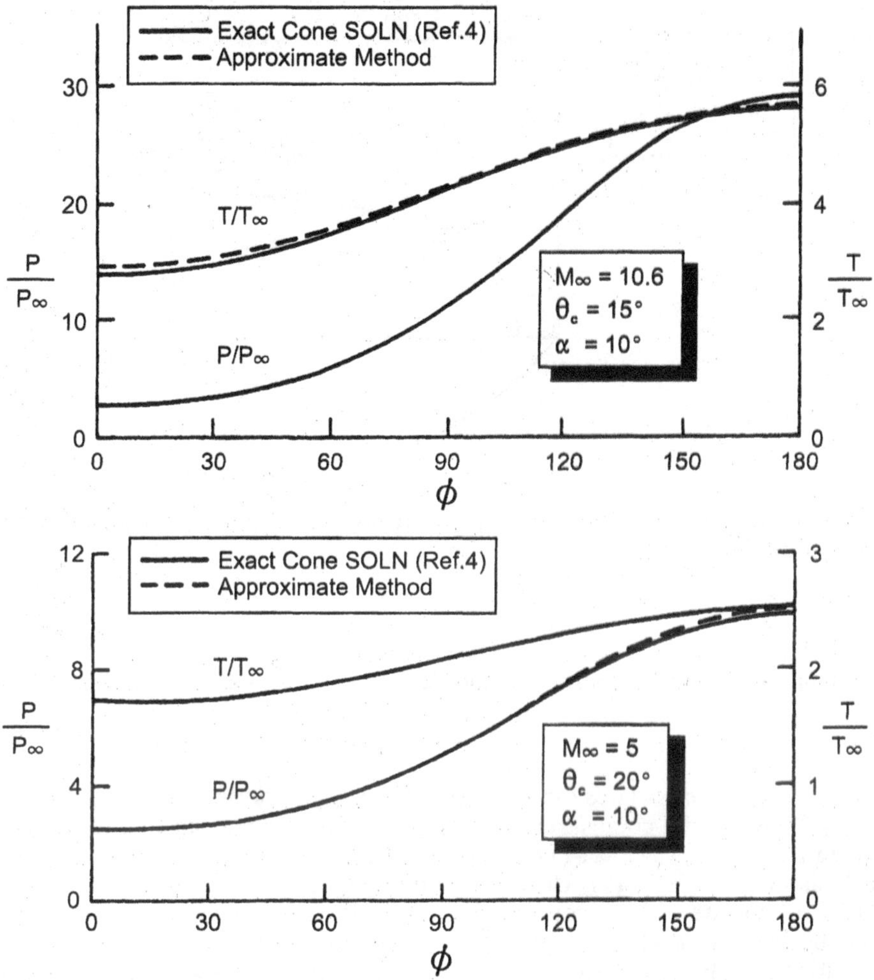

Fig. 4.5 Perfect gas comparison of exact and approximate cone solutions.

bance created by Mach waves emanating from the expansion corner and intersecting the shock, the degree of turn should be small.

Once the starting solution and the various flow properties are known at the initial surface of a two- or three-dimensional body, the Prandtl–Meyer expansion theory is applied for points downstream.

For a perfect gas, one can write[9]

$$\nu(M) = \sqrt{\frac{\gamma+1}{\gamma-1}}\tan^{-1}\sqrt{\frac{\gamma-1}{\gamma+1}(M^2-1)} - \tan^{-1}\sqrt{M^2-1} \tag{17}$$

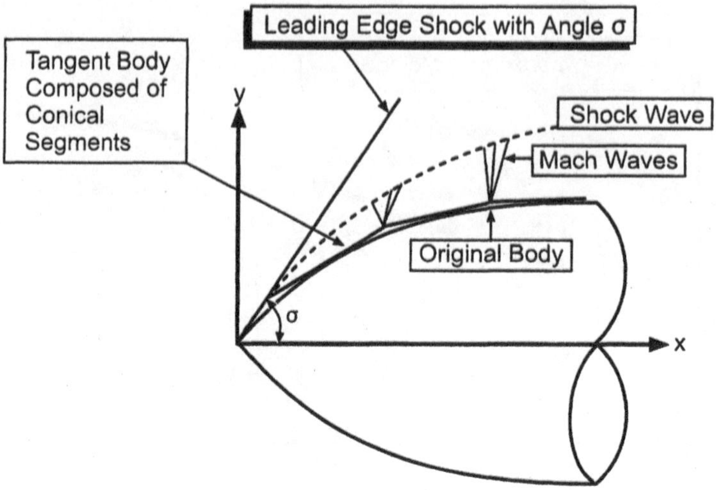

Fig. 4.6 Approximation of true body by one composed of straight line segments tangent to surface.

If $\delta\theta$ is the change in the local surface slope in going from one tangent segment of the body or airfoil to another tangent surface (see Fig. 4.6), then

$$\nu_2(M_2) = \delta\theta + \nu_1(M_1) \tag{18}$$

One, therefore, calculates the Prandtl–Meyer angle for the first surface with Eq. (17), where M_1 comes from two- or three-dimensional starting solutions given by either Eq. (5) or (12), respectively. A new value of $\nu_2(M_2)$ is then calculated from Eq. (18). This new value of ν_2 is then substituted into Eq. (17) and a value of M_2 is computed numerically. This process is repeated on the body at several other points downstream. Knowing the local Mach number and total pressure and density across the shock, one can then compute the local thermodynamic properties from Eqs. (13), (14), and (15). Force and moments can be determined from integration of pressure coefficients according to Eqs. (56) to (59) in Chapter 3.

It was mentioned earlier that, for many configurations, first-order theory is not adequate for axial force accuracy. Fortunately, Syvertson and Dennis extended the generalized Shock-expansion theory on pointed bodies and sharp airfoils to what they called a second-order theory.[10] They defined the pressure along a conical frustum by

$$p = p_C - (p_C - p_2)e^{-\eta} \tag{19}$$

instead of a constant on each segment, as was the case in the generalized theory. Here p_C is the pressure on a cone with the given cone half-angle equal to the slope of the conical segment with respect to the axis of symmetry. p_2 is the pressure just aft of a conical segment (see Fig. 4.7), which is calculated from the Prandtl–Meyer expansion, Eqs. (17) and (18).

Also,

$$\eta = \frac{\left(\frac{\partial p}{\partial s}\right)_2 (x - x_2)}{(p_C - p_2)\cos\theta_2} \tag{20}$$

Thus examining p from Eq. (19), it can be seen, for example, on the frustum element 2–3 of Fig. 4.7, that the pressure varies from the pressure of the generalized theory at point 2 to that of a cone of angle θ_2 with freestream Mach number M_∞ as s gets large. Syvertson and Dennis[10] approximated the pressure gradient as

$$\left(\frac{\partial p}{\partial s}\right)_2 = \frac{B_2}{r}\left(\frac{\Omega_1}{\Omega_2}\sin\theta_1 - \sin\theta_2\right) + \frac{B_2}{B_1}\frac{\Omega_1}{\Omega_2}\left(\frac{\partial p}{\partial s}\right)_1 \tag{21}$$

where

$$B_{1,2} = \frac{\gamma p_{1,2} M_{1,2}^2}{2\left(M_{1,2}^2 - 1\right)}$$

$$\Omega_{1,2} = \frac{1}{M_{1,2}}\left[\frac{1 + \frac{\gamma - 1}{2} M_{1,2}^2}{\frac{\gamma + 1}{2}}\right]^{\frac{\gamma+1}{2(\gamma-1)}}$$

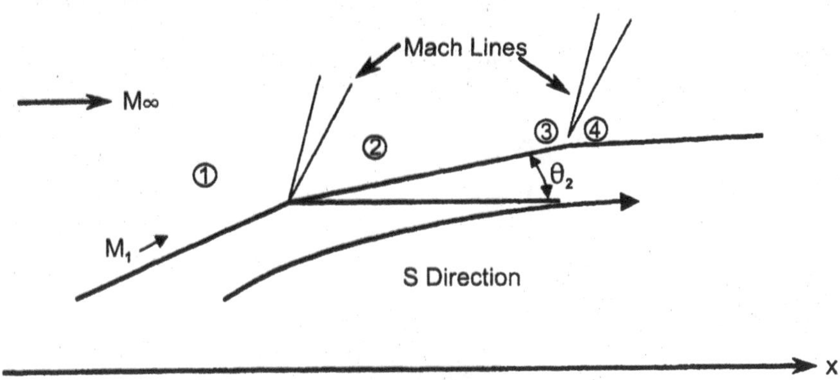

Fig. 4.7 Flow about a frustum element.

Here the subscript 1,2 refers to before and after a point of turn as seen in Fig. 4.7. Finally, for negative angles such as would occur on a boattailed configuration, p_C was replaced by p_∞. No discussion was given for blunt bodies. It should be noted that if η of Eq. (20) becomes negative, the second-order shock expansion theory (SOSET) reverts to the generalized or first-order shock-expansion theory discussed earlier. For negative values of η, Eq. (19) will not give the correct asymptotic cone solution.

Moore et al.[3] gave two other ways to compute the pressure gradient behind the corner, in lieu of Eq. (21). The first was an exact solution[11] computed through a numerical integration of the equation

$$-\left(\frac{\partial G}{\partial s}\right)\frac{\gamma+1}{2}\frac{\Omega M^4}{(M^2-1)} = \frac{\mathrm{d}}{\mathrm{d}\theta}\left[\Omega\left\{(\sqrt{M^2-1})2\left(\frac{\partial G}{\partial s}\right)+\frac{\sin\theta}{r}\right\}\right] \tag{22}$$

Integration of this equation gives $\partial G/\partial s$, and $\partial p/\partial s$ is then computed from

$$\frac{\partial p}{\partial s} = \left(\frac{\partial G}{\partial s}+\frac{\partial \theta}{\partial s}\right)\lambda \tag{23}$$

where

$$\lambda = \frac{2\gamma p}{\sin 2\mu}$$

and $\partial\theta/\partial s$ is the curvature of the body at a given point.

The second method was derived for real gas applications but is still applicable to perfect gases. It is defined by

$$\left(\frac{\partial p}{\partial s}\right)_2 = \frac{\left(\frac{1}{2\lambda}\frac{\partial p}{\partial s}\right)_1\left[\frac{s_5-s_4}{b_1}+4\cos(\mu_1)\right]+\frac{1}{r}\left[\sin(\mu_1)\sin(\theta_1)-\left(\frac{b_2}{b_1}\right)\sin(\mu_2)\sin(\theta_2)\right]}{\left(\frac{b_2}{b_1}\right)\frac{2\cos(\mu_2)}{\lambda_2}-\frac{s_5-s_4}{2\lambda_2 b_1}} \tag{24}$$

where

$$\frac{s_5-s_4}{b_1} = \left[\left(\frac{\rho_1 V_1\sin(\mu_1)}{\rho_2 V_2\sin(\mu_2)}-1\right)^2 + (\mu_1+\theta_1-\mu_2-\theta_2)^2\left(\frac{\rho_1 V_1\sin(\mu_1)}{\rho_2 V_2\sin(\mu_2)}\right)^2\right]^{\frac{1}{2}}$$

and

$$\frac{b_2}{b_1} = \frac{\rho_1 V_1 \sin \mu_1}{\rho_2 V_2 \sin \mu_2}$$

Subscript 1 refers to conditions before the Prandtl–Meyer turn and subscript 2 refers to conditions after the turn.

It is of interest to compare the three methods for predicting $\left(\frac{\partial p}{\partial s}\right)_2$ given by the Eqs. (21), (23), and (24).

For this comparison, a simple biconic nose shape was considered as indicated in Fig. 4.8. This allowed a simple starting solution for the first conical segment based on Section 4.2 methods and then a computation of the properties around a turn of a given $\Delta\theta = \theta_2 - \theta_1$ as shown in Fig. 4.7. Results were obtained for the parameter

$$\left(\frac{r}{P_\infty}\frac{\partial p}{\partial s}\right)_2$$

under various conditions using the approximate method of Syvertson and Dennis,[10] the exact prediction by DeJarnette et al.,[11] and the approximate technique given in Ref. 3. Because results by Syvertson and Dennis[10] and DeJarnette et al.[11] were for perfect gas only, the cases considered here for comparison of the three techniques mentioned were for perfect gas. The

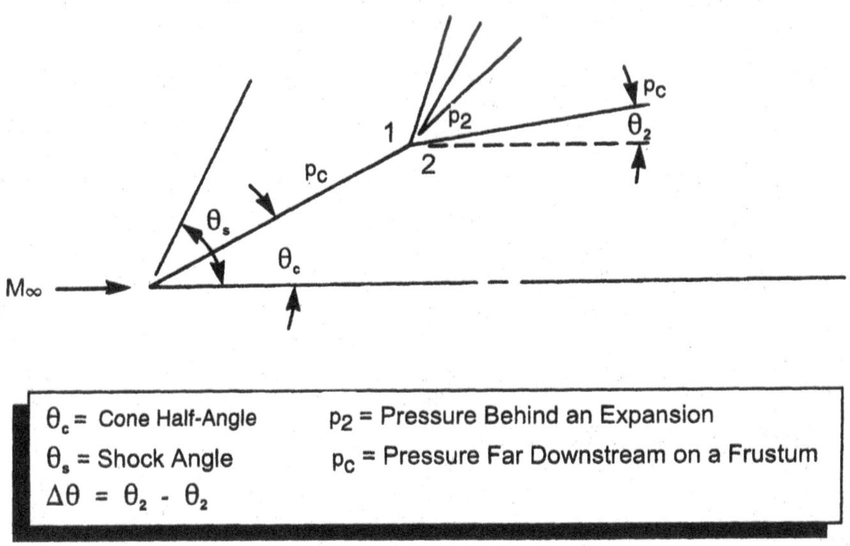

Fig. 4.8 Biconic configuration geometry for investigating pressure gradient around a corner (see Table 4.1).

results of this comparative study are presented in Table 4.1, which shows M_∞ varying from 3 to 15, cone half-angle from 5 to 20, and $\Delta\theta$ of 2 and 5 deg. Note first of all that the Ref. 3 approximation for

$$\left(\frac{r}{P_\infty}\frac{\partial p}{\partial s}\right)_2$$

agrees very closely with the exact integration of Ref. 11. As seen in Table 4.1, the approximate method of Ref. 10 gives reasonable results (within 10%) in most cases shown where $\Delta\theta$ is small.

The most important aspect of Table 4.1 results is the fact that η becomes negative quite often for Mach numbers of 5 and greater. Therefore, η must be defined and SOSET of Eq. (19) reverts back to generalized shock expansion theory (GSET) or tangent cone theory (TCT), that is, $p = p_2$ or $p = p_c$. As a result of this dilemma, Moore et al.[3] investigated a different approach that would allow one to take advantage of both p_c and p_2. This approach defined p in the SOSET as

$$p = p_c - (p_c - p_2)\eta_1 \tag{25}$$

Syvertson and Dennis[10] would revert back to first-order shock expansion theory when η was negative, which automatically gives p_2 from the value of Prandtl–Meyer expansion. Equation (25) allows p_2 to be the value of p if $\eta_1 = 1.0$ or p_c from the tangent cone solution to be the value of p if $\eta_1 = 0$. Or

Table 4.1 Estimates of $(r_2/p_\infty)(\partial p/\partial s)_2$ for various conditions

M_∞	θ_c, deg	$\Delta\theta$	Exact[11]	Syvertson and, Dennis[10]	Present[3]	$r_2(\partial p/\partial s)_2/p_\infty(p_c-p_2)$	p_2
3	15	2	0.022	0.021	0.022	0.657	1.86
		5	0.048	0.044	0.048	0.622	1.50
5	20	2	−0.015	−0.014	−0.015	−0.143	4.68
		5	−0.013	−0.011	−0.012	−0.077	3.56
5	10	2	0.015	0.015	0.015	0.389	1.88
		5	0.034	0.031	0.034	0.359	1.34
10	15	2	−0.088	−0.082	−0.087	−0.209	8.29
		5	−0.095	−0.079	−0.090	−0.188	5.20
10	10	2	−0.014	−0.013	−0.014	−0.097	3.96
		5	0.001	0.003	0.003	0.010	2.20
15	15	2	−0.260	−0.242	−0.258	−0.206	16.73
		5	−0.284	−0.231	−0.267	−0.193	9.77
15	10	2	−0.072	−0.065	−0.071	−0.141	7.17
		5	−0.042	−0.029	−0.035	−0.122	3.43
15	5	2	0.006	0.006	0.006	0.068	2.04
		5	0.017	0.014	0.017	0.051	0.72

a blend of p_c and P_2 could be obtained by setting η_1 somewhere 0 and 1.0. Moore et al.[3] found that tangent cone and Prandtl–Meyer expansion gave similar results for pointed bodies. However, for other than 100% blunt bodies, $\eta_1 = 0$ (TCT) gave better results for blunt bodies.

To illustrate the capacity of the SOSET, a couple of examples will be shown. The first is a pressure model test taken from Ref. 12. This test was conducted on cones and tangent ogive noses of 3- and 5-caliber nose length from 0- to 15-deg AOA. Test were conducted at $M_\infty = 3.0$, 4.25, and 5.05. Data were measured in 45-deg increments around the body and at five stations along the body, including $x = 0$, 1.0, 2.0, 3.0, and 4.5. Figure 4.9a illustrates the model tested and Fig. 4.9b shows the results of comparisons of the SOSET theory to experimental data. Comparisons are shown at $M_\infty = 3.0$ and 5.05 for the 3-caliber nose case at a 15-deg AOA. Only three planes of data ($x = 0$, 2.0, and 4.5) are shown to keep the figure easily discernible. As seen in the figure, SOSET gives excellent agreement to experiment with the exception of the leeward plane area at the $x = 4.5$ station. Here, flow separation is occurring, increasing the nonlinearity of the normal force, but having little effect on axial force. While no axial force results were given by Ref. 12, it is expected excellent agreement with axial force would be obtained due to the excellent agreement in pressure data. Savin[12] did give values of $(C_{N_\alpha})_{\alpha=0}$ for the $M_\infty = 5.05$ case. The value computed by the SOSET was 2.35 and the value from experiment was 2.29, or less than a 3% difference. As seen from Figs. 3.6 to 3.9 and Fig. 4.9, the combination of second-order Van Dyke (SOVD) and SOSET provide a powerful tool to get good estimates of low AOA aerodynamics in a fairly rapid manner. All of these computations in Figures 3.6 through 3.9 and 4.9 required less than 1 s on a personal computer with a 200-megahertz chip.

A similar configuration was tested[13] at $M_\infty = 2$, except the pressure database was more extensive. The configuration was exactly like that in Fig. 4.9a, except that the total length was 10 versus 3.67 calibers. However, the

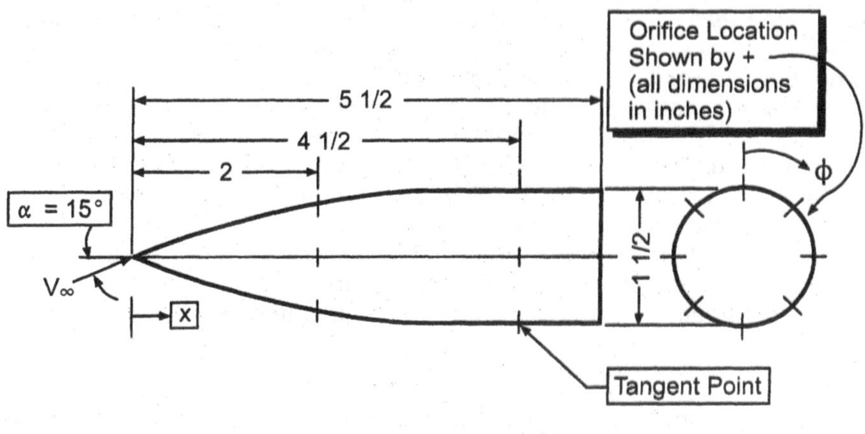

Fig. 4.9a Model used for pressure test[12] ($\ell_n = 3.0$ caliber).

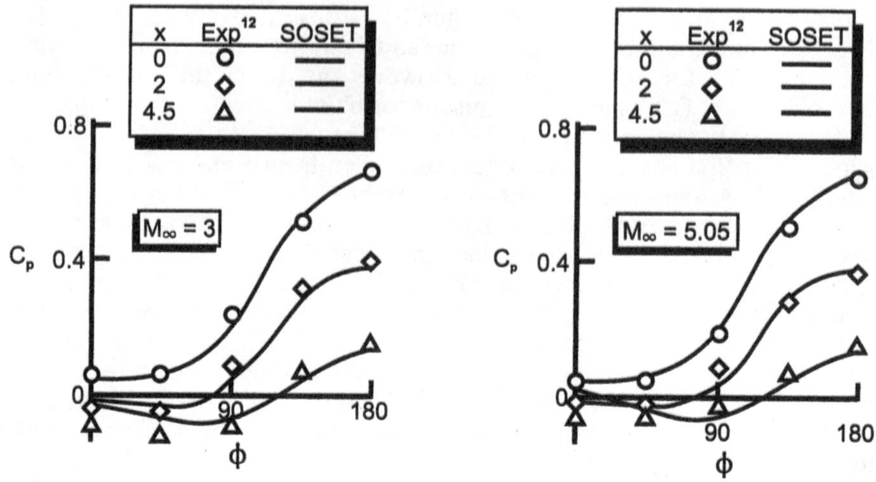

Fig. 4.9b Pressure distributions on 3-caliber tangent ogive nose at $\alpha = 15°$.

nose was a 3.0-caliber tangent ogive in both cases. The Ref. 13 pressure data were taken in increments of φ of 15 deg and at x stations of 0.56, 2.06, 4.06, 5.84, 7.61, and 9.39 calibers from the nose tip. At $M_\infty = 2.0$, both the SOSET and hybrid theory of Van Dyke (HTVD) methods are applicable, so both theories can be used for comparison to test data. Again, the 15-deg AOA case will be considered, in analogy to Fig. 4.9. Figure 4.10 presents the comparison of theory to experiment in terms of pressure coefficient for x locations of 0.56, 2.06, 4.06, 5.84, and 9.39 calibers. Note the scale on the C_P axis is enlarged for the $x = 5.84$ and 9.39 caliber cases.

In general, both the SOSET and HTVD give quite acceptable results, even for the AOA of 15 deg, which is about 5 deg higher than LT should be considered. In general, SOSET gives slightly better predictions in the windward plane than HTVD, whereas the reverse is true in the leeward plane. Experience has shown that, at $M_\infty = 2$, the average of forces and moments computed from the HTVD and the SOSET is closer to experiment than forces and moments from either theory on its own. It is also of interest to note the separated flow that occurs past the nose–cylinder juncture of 3.0 calibers. This separated flow creates nonlinear aerodynamics, which will be discussed in Chapter 5. Unfortunately, the *Handbook of Supersonic Aerodynamics*[13] did not give forces and moments, so we could not compare theory to experiment. It should also be pointed out that for AOAs to 10 deg or less (even though not shown), theory and experiment are in quite good agreement, even for the large x stations, as the regions of separated flow are quite small.

IV. Newtonian Impact Theory

Thus far, the approximate methods discussed are limited to sharp leading edges or pointed bodies. However, as shown in Chapter 1, many weapons

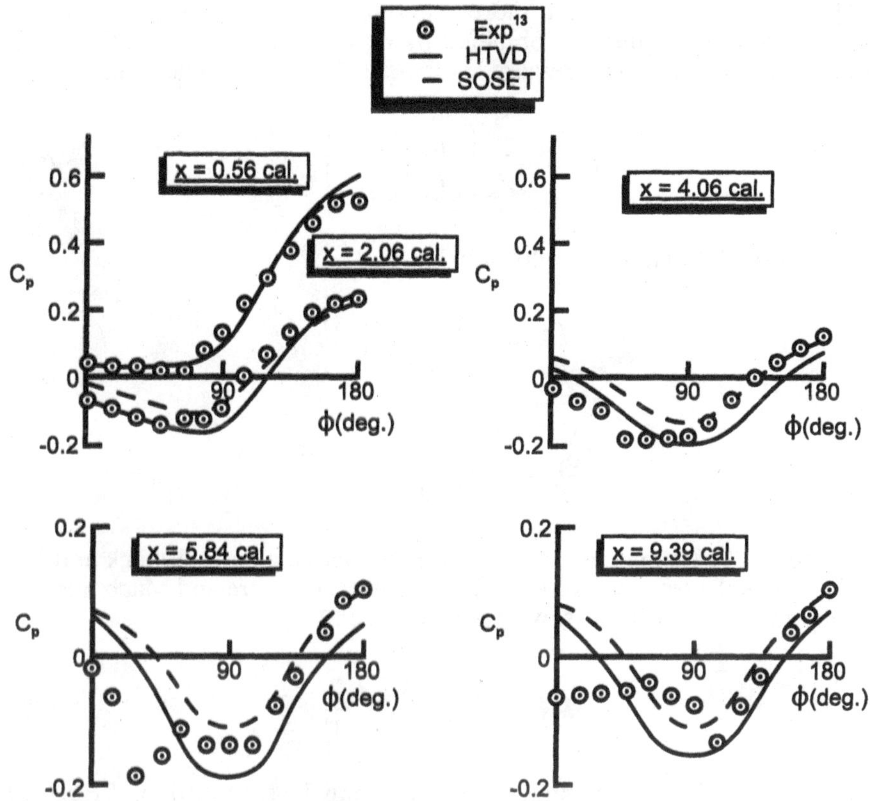

Fig. 4.10 Comparison of theory and experiment for 10-caliber tangent ogive cylinder (ℓ_n = 3 caliber) at various stations along body (M_∞ = 2.0 and α = 15°).

have blunt or truncated leading edges or tips for various design purposes. For these configurations, modified Newtonian theory (MNT) has proved to be a powerful, easy to use, and an accurate complement to perturbation methods discussed in Chapter 3 and local slope methods discussed in Chapter 4.

Newtonian impact theory assumes that, in the limit of high Mach number, the shock lies on the body, meaning that the disturbed flowfield lies in an infinitely thin layer between the shock and body. Applying the laws of conservation of mass and momentum across the shock yields the result that density behind the shock approaches infinite values and the ratio of specific heats approaches unity. The pressure coefficient on the surface becomes[14]

$$C_P = 2\sin^2\delta \tag{26}$$

where δ is the angle between the velocity vector and a tangent to the body at the point in question defined by Eq. (10).

Lees[14] noted that a much more accurate prediction of pressure on the blunt-nosed body could be obtained by replacing the constant "2" in Eq. (26) with the stagnation pressure coefficient C_{P_O}. C_{P_O} can be found from

$$C_{P_O} = \frac{2}{\gamma M_\infty^2}\left[\left(\frac{p_2}{P_\infty}\right)\left(\frac{p_{O_2}}{p_2}\right) - 1\right] \tag{27}$$

where p_2/p_∞ and p_{O_2}/p_2 are given by[1]

$$\frac{p_2}{p_\infty} = \frac{2\gamma M_\infty^2 - (\gamma - 1)}{\gamma + 1} \tag{28}$$

$$\frac{p_{O_2}}{p_2} = \left(1 + \frac{\gamma - 1}{2} M_2^2\right)^{\frac{\gamma}{\gamma-1}} \tag{29}$$

Equation (26) is simply the ratio of pressure across a normal shock and Eq. (29) relates the total pressure to the local static pressure and Mach number and is good for isentropic flow. Also, from Ref. 1,

$$M_2^2 = \frac{(\gamma - 1)M_\infty^2 + 2}{2\gamma M_\infty^2 - (\gamma - 1)} \tag{30}$$

so that Eq. (29) can be related to M_∞ and γ only by substitution of Eq. (30) into (29) to obtain

$$\frac{p_{O_2}}{p_2} = \left[\frac{4\gamma M_\infty^2 - 2(\gamma - 1)}{(\gamma + 1)^2 M_\infty^2}\right]^{\frac{-\gamma}{\gamma-1}} \tag{31}$$

Equations (28) and (30) hold only for a perfect gas. They are the relationships taken from Ref. 1 for flow across a normal shock wave. Combining Eqs. (31) and (28) into Eq. (27), we obtain

$$C_{P_O} = \frac{2}{\gamma M_\infty{}^2}\left\{\left[\frac{(\gamma + 1)M_\infty^2}{2}\right]^{\frac{\gamma}{\gamma-1}}\left[\frac{\gamma + 1}{2\gamma M_\infty^2 - (\gamma - 1)}\right]^{\frac{1}{\gamma-1}} - 1\right\} \tag{32}$$

and

$$C_P = C_{P_o} \sin^2 \delta \tag{33}$$

Equation (33) allows the calculation of the pressure coefficient all along the blunt surface of a missile nose or wing leading edge for a perfect gas where C_{P_O} is given by Eq. (32) and $\sin\delta$ by Eq. (10). Equation (33) and (32) are referred to as the MNT.

If the nose or leading edge of the configuration is truncated, then $\delta = 90$ deg and the author has found that a reasonable representation for the pressure coefficient on the truncated part of the configuration is

$$C_P = 0.9\ C_{P_O} \tag{34}$$

where C_{P_O} is computed by Eq. (32).

DeJarnette (of Ref. 3) derived a slightly more accurate expression for the pressure coefficient on a blunt nose by using a full Euler solution[15] in conjunction with MNT. This new methodology is given by

$$C_P = C_{P_{MN}} - k \cos^m \delta \left[\cos\delta - \cos(\delta)_m\right] \tag{35}$$

where $m = 2.78$, $(\delta)_m = 25.95$ deg, and

$$k = \frac{-\left(\dfrac{dC_p}{d\delta}\right)_{MN} + 1.5\left[\left(C_P + \dfrac{2}{\gamma M_\infty^2}\right)\left(\dfrac{dC_p}{d\delta}\right)\right]_{MN}^{1/2}}{\sin(\delta)_m \cos^m(\delta)_m}$$

Here

$$\left(\frac{dC_p}{d\delta}\right)_{MN} = C_{P_O}\ \sin(2\delta) \tag{36}$$

$C_{P_{MN}}$ of Eq. (35) is the value computed from MNT by Eq. (33).

This improved method has been applied to a sphere for $3.5 \leq M_\infty \leq 30$. The results are compared to computational fluid dynamics results from Ref. 15 for $x/R_N \leq 0.5625$. At $M_\infty = 3.5$ the maximum error in pressure is 3% at $x/R_N = 0.25$. For $M_\infty > 3.5$, the maximum error is less than 1% and, in most cases, less than 0.5%. In contrast, the modified Newtonian pressures had maximum errors from 7 to 10% compared to computational fluid dynamics calculations. Equation (35) can be used for x/R_N up to about 0.7, with errors less than 3%. A better prediction for $x/R_N > 0.5625$ can be obtained by using the second-order, shock-expansion method.

Figures 4.11 and 4.12 illustrate the accuracy of the improved modified Newtonian theory (IMNT) over MNT. The figures show the IMNT, MNT, and exact results from Ref. 15. Note that the IMNT gives virtually identical results to the exact computations. Although not shown, this excellent agreement in pressure prediction holds true all the way to $M_\infty = 30$.

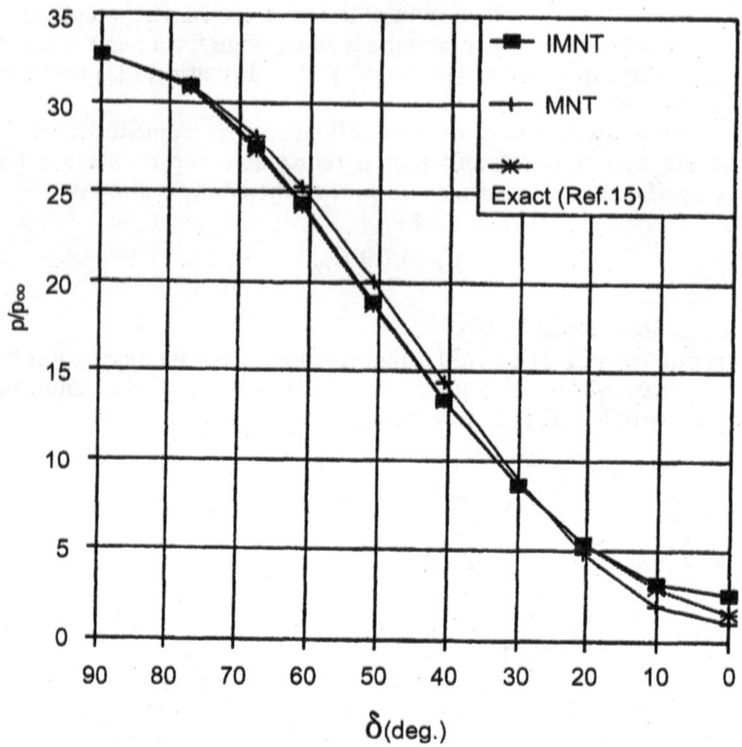

Fig. 4.11 Surface pressure distribution over a hemispherical forebody at M_∞ = 5.

V. Hybrid Theory of Van Dyke Plus Modified Newtonian Theory

Although the HTVD is a perturbation method and not a local slope technique such as the other methods discussed in this chapter, it was felt more appropriate to discuss the combined HTVD + MNT approach after the MNT method had been discussed. Section 4.6 will discuss the complementary SOSET plus MNT for higher Mach numbers. The SOVD method has an upper Mach number limit of about 3.0 before it is inappropriate to use it due to strong shocks. For short-nosed or blunt configurations, this upper limit is more like 2.0. The combination of SOVD + MNT for $1.2 \leq M_\infty < 2.0$ and SOSET + MNT for $M_\infty \geq 2.0$ provides for second-order accuracy for the wave component of the axial force calculation and first-order accuracy in normal force and center of pressure.

As noted in the discussion on the hybrid theory, it is limited to conditions where the body slope is less than the local Mach angle. The hybrid theory is, therefore, not applicable in the nose region of a blunt missile. On the other hand, MNT gives very acceptable estimates of pressure coefficients in the nose region, even for low supersonic Mach numbers where assumptions, inherent in the Newtonian impact theory, are violated. The author was the

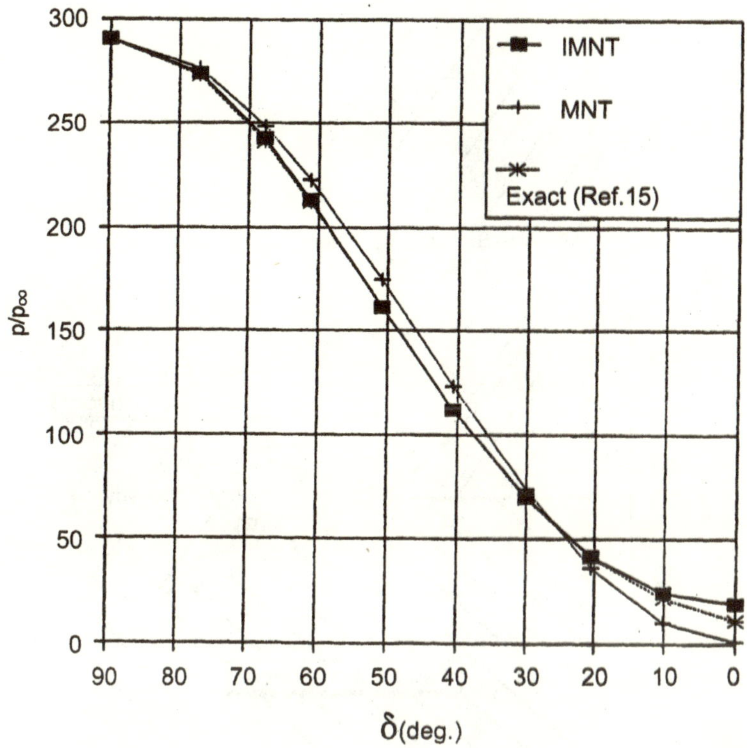

Fig. 4.12 Surface pressure distribution over a hemispherical forebody at $M_\infty = 15$.

first to recognize the possibility of combining these two theories.[16] The key to successful combination was in the starting solution. At low supersonic Mach numbers, the pressure overexpands on a blunt nose tip as it proceeds around the blunt portion from the stagnation point to the given portion of the nose. To capture this overexpansion, it was necessary to start the HTVD near its maximum acceptable slope and allow the pressure to expand around the surface.[16] Simultaneously, the MNT was started at the stagnation point and allowed to expand until the pressure coefficients of the MNT and the HTVD were equal, which was defined as the match point. Upstream of the match point, MNT was used in force and moment calculations, whereas, downstream, HTVD was used.

Figure 4.13 is an illustration of the boundaries of perturbation and Newtonian theories. As seen in Fig. 4.13, the upper limit on slope for starting the HTVD was found to be about 27.5 deg, regardless of how low the Mach number is. This upper limit was determined by the fact that even second-order accuracy was not acceptable for values of cone angles higher than 27.5 deg. (First-order theory provides acceptable accuracy only to about 5 to 10 deg cone half-angles.) As Mach number increases, this upper value of sur-

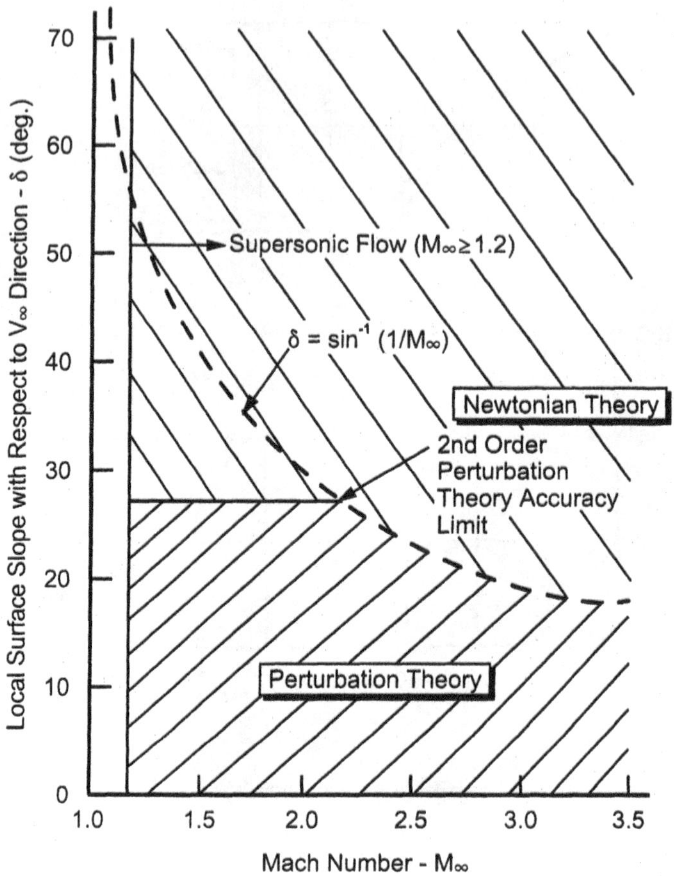

Fig. 4.13 Boundaries of perturbation and Newtonian theory.

face slope is reduced further below 27.5 deg due to the Mach angle limit. Fortunately, as shown in Section 4.4, MNT has its area of greatest accuracy for values of $\delta \geq 20$ deg, and is least accurate for $\delta < 10$ deg. Therefore, the HTVD and MNT and SOSET and MNT methods are very complementary to each other.

Another key to the successful combination of the HTVD with MNT was the treatment behind slope discontinuities. The author found that the HTVD gave oscillatory behavior behind a slope discontinuity. However, when small steps in the x direction were taken initially behind the discontinuity, and these steps were gradually increased in size as one proceeded downstream of the discontinuity, the solution was found to be stable. Also, the force and moments were found to be unaffected by oscillatory behavior of pressure coefficients immediately behind the discontinuity. This oscillatory behavior quickly died out after two to three steps behind the discontinuity.

As a result of the sensitivity of the HTVD behind discontinuities, two limiting constraints on its use are suggested. First, instead of allowing the method to be used all the way up to its upper limit where the body slope is equal to the Mach angle, a value of

$$\left(\frac{dr}{dx}\right)_{max} = 0.95 \sin^{-1} \mu \tag{37}$$

is suggested. Second, for pointed bodies, an upper Mach number of 2.4 is suggested and, for blunt bodies, an upper limit of 2.0 is suggested. These two upper limits have proven to minimize the sensitivity of the HTVD behind slope discontinuities.

While the sensitivity of the HTVD behind slope discontinuities is a nuisance that must be worked around through logic in a computer code or calculations, it remains the only approximate method the author is aware of to obtain accurate pressure predictions on blunt-nosed shapes at low supersonic Mach numbers. Figure 4.14 illustrates the accuracy of the pressure

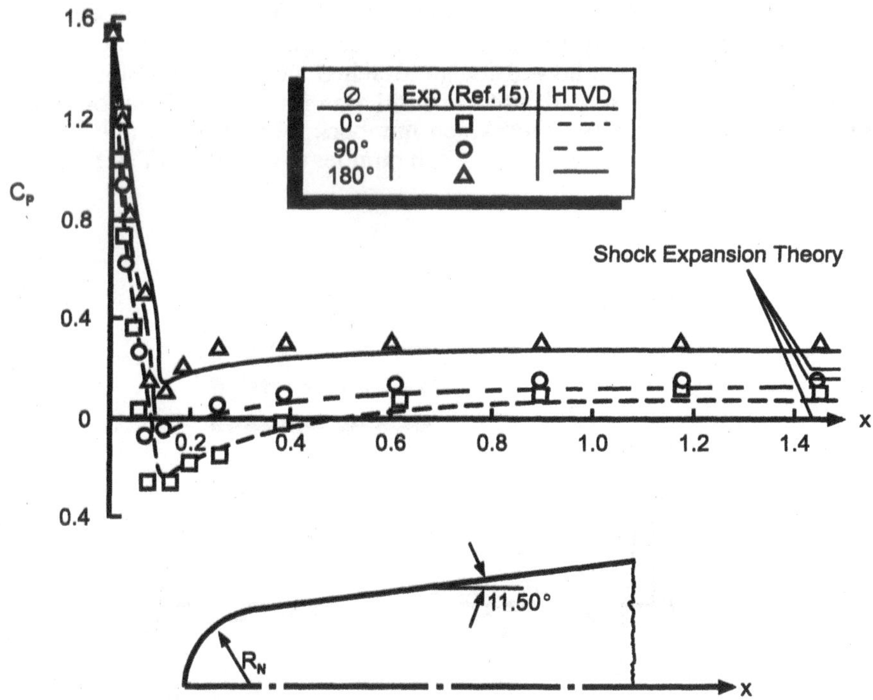

Fig. 4.14 Comparison of theory and experiment for blunted cone $\frac{R_N}{R_B} = 0.35, M_\infty = 1.5, \alpha = 8°$, and $\theta_c = 11.5°$.

coefficient prediction on a 35% blunt cone at $M_\infty = 1.5$ and $\delta = 8$ deg. Results are shown in the leeward, windward and $\varphi = 90$ deg planes and compared to experimental data of Ref. 15. Also shown on the figure are the shock expansion results at $x = 1.5$, showing this method to be inferior to the HTVD at low supersonic Mach numbers.

Figure 4.15 gives results of the HTVD compared to a numerical solution of the Euler equations[17] at $M_\infty = 1.2$ on a 40% blunt tangent ogive of length 3.0 calibers. Good agreement is obtained with the numerical solution except in a small region following the overexpansion. Integration of the axial force with the numerical solution gave a $C_{A_W} = 0.105$ versus a value of 0.095 for the HTVD. Both Figs. 4.14 and 4.15 give acceptable accuracy for wave drag calculations for engineering approximations.

VI. Second-Order Shock Expansion Plus Modified Newtonian Theory

This method combines the MNT of Section 4.4 with the SOSET method of Section 4.3 to allow computation of aerodynamics of blunt-nosed configurations for moderate to high supersonic Mach numbers. The major issue that must be dealt with is the match point between the MNT and the SOSET methods.

Several alternatives have been used in past literature to match the MNT to SOSET on the blunt nose part of the body. The first alternative was that of Jackson et al.[18], where the body slope was assumed to be that for the maximum wedge angle allowed for an attached shock wave. Their results showed good agreement with experimental data for $M_\infty \geq 2.3$ but only fair agreement for lower supersonic Mach numbers. DeJarnette et al.[11] developed an empirical equation for the Mach number to match MNT to SOSET.

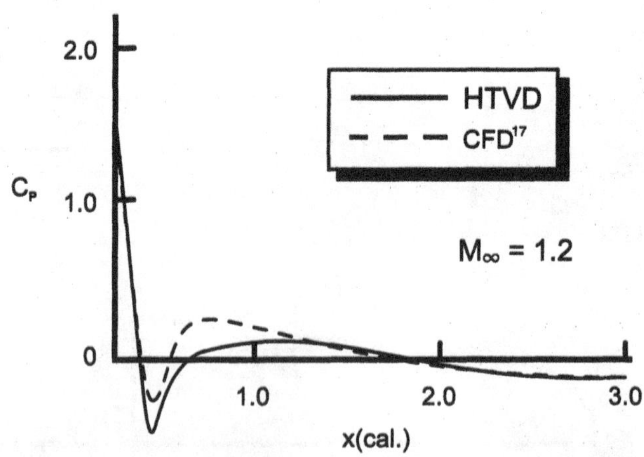

Fig. 4.15 Comparison of HTVD with exact numerical Euler solution (3-caliber tangent ogive with $r_n/r_b = 0.4$) (CFD = computational fluid dynamics).

This approach gave improved results over Ref. 18, particularly at low supersonic Mach numbers.

The final author of Ref. 3, which contains a derivation of the new methodology for pressures on spheres in hypersonic flows, found a constant value of a match point of $\theta_m = 25.95$ deg. The pressure at this point is defined by

$$p_1/p_\infty = 1 + \frac{\gamma_\infty M_\infty^2}{2} C_{p_1} \tag{38}$$

where C_{p_1} is given by Eq. (35). Differentiating Eq. (38), there is obtained

$$\frac{\mathrm{d}p_1}{\mathrm{d}s} = \frac{\gamma_\infty M_\infty^2 p_\infty}{2} \frac{\mathrm{d}C_{p_1}}{\mathrm{d}\delta} \frac{\mathrm{d}\delta}{\mathrm{d}s} \tag{39}$$

But, for a sphere,

$$\frac{\mathrm{d}\delta}{\mathrm{d}s} = -1/R_N \tag{40}$$

Also, Ref. 3 showed that

$$\frac{\mathrm{d}C_{p_1}}{\mathrm{d}\delta} = 1.5\left[\left(C_P + \frac{2}{\gamma M_\infty^2}\right)\frac{\mathrm{d}C_p}{\mathrm{d}\delta}\right]_{MN}^{\frac{1}{2}} \tag{41}$$

was accurate for $M_\infty \geq 3.5$ where

$$C_p = C_{p_O} \sin^2 \delta \tag{42}$$

Using Eqs. (39) through (42) and recognizing that $\delta = 25.95$ deg at the match point, Eq. (24) just past the match point becomes

$$\left(\frac{\partial p}{\partial s}\right)_2 = \left(\frac{\partial p}{\partial s}\right)_1 + \frac{\lambda_1}{R_N} = \frac{1}{R_N}\left[\frac{\rho_1 V_1^2}{\sqrt{M_1^2 - 1}} - 1.05 M_\infty^2 P_\infty \left(0.1676\, C_{P_O}^2 + \frac{1.251}{M_\infty{}^2} C_{P_O}\right)^{\frac{1}{2}}\right] \tag{43}$$

In Eq. (43), C_{P_O} is computed from Eq. (32). P_1 can be calculated from Eq. (38). Then, because total pressure is

$$P_O = \frac{1}{2} \rho_\infty V_\infty^2 C_{P_O} + p_\infty$$

local Mach number can be found from Eq. (29). That is

$$M_1 = \left\{\left(\frac{2}{\gamma - 1}\right)\left[\left(\frac{p_O}{p_1}\right)^{\frac{\gamma-1}{\gamma}} - 1\right]\right\}^{\frac{1}{2}} \tag{44}$$

Total and local temperatures can be found from Eq. (13) and then local density from the equation of state. Speed of sound is related to temperature through

$$a^2 = \gamma RT \tag{45}$$

V_1 is then

$$V_1 = M_1\, a_1 \tag{46}$$

In summary, at the match point, the pressure is computed from Eq. (35). Downstream of this point, the solution is continued by use of Eqs. (19), (20), and (24) or (43), depending on whether one is on the sphere part of the nose [Eq. (43) or some other part of the body]. P_C of Eq. (19) comes from a local tangent cone solution or an appropriate estimate of pressure such as Eq. (16). Also, p_2 comes from the Prandtl–Meyer solution discussed in Section 4.3.

For lifting properties, Jackson et al.[18] assumed that the body could be made up of several equivalent bodies of revolution as represented by the various meridians (see Fig. 4.16). This is basically equivalent to using Eqs. (19) and (20) in a local slope or tangent cone approach as discussed previously. Figure 4.17 is an example application of the combination of SOSE + MNT on the same blunt cone case of Fig. 4.14, except that the Mach number is 2.96 and the pressure coefficient is plotted along the body meridian versus the x axis. Note the excellent agreement of theoretical and experimental pressures along the surface.

A second example is chosen to illustrate the capability of both the SOSET and HTVD in combination with MNT to predict axial force. The configuration selected is shown in Fig. 4.18a and Table 4.2 and wind tunnel data is given in Ref. 20. This configuration is a body-alone wind tunnel model tested in the transonic and supersonic tunnels at Arnold Engineering and Development Center (AEDC) in Tullahoma, Tennessee. The configuration was tested without a transition trip, at Reynolds numbers varying from 1.8×10^6/ft to 5.3×10^6/ft and at Mach numbers 0.6 to 4.0 and AOAs from −6 to 14 deg. Interest here will focus on the zero AOA axial force comparison with various nose bluntness ratios. Only forebody axial force (wave plus skin friction) will be plotted because only forebody axial force was given in the related report by Butler et al.[20] Hence, Van Driest[21] is used to compute the viscous drag. This method will be discussed in Section 4.7. Figure 4.18b gives the theoretical and experimental forebody axial force comparisons for the

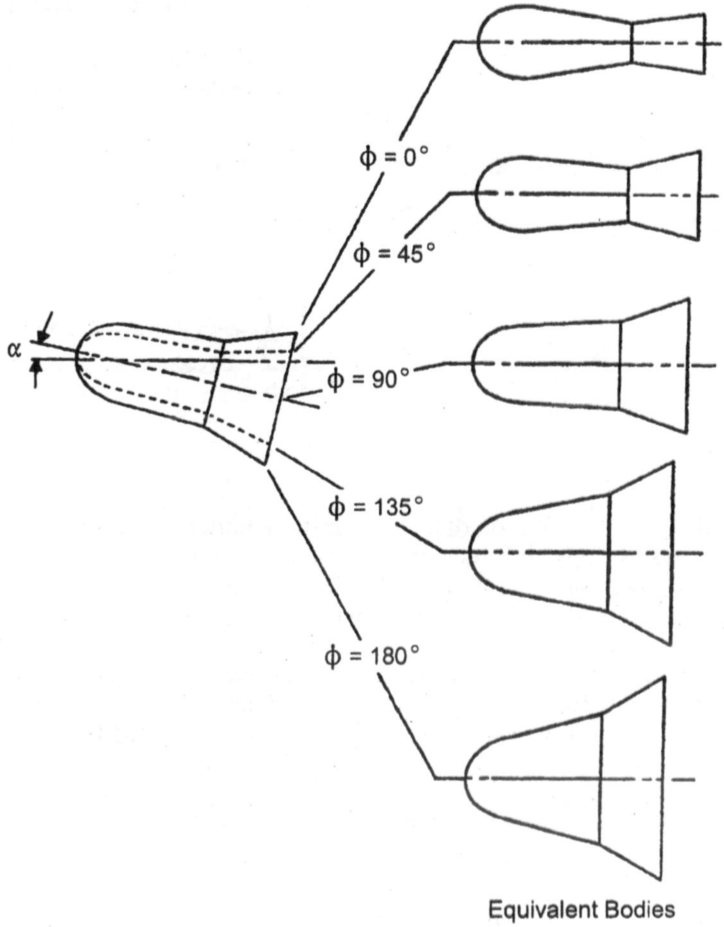

Fig. 4.16 Typical equivalent body shapes used for computing lifting properties with second-order shock expansion.

2-caliber ogive configuration, and Fig. 4.18c gives the 4-caliber ogive configuration comparisons. The only data available in the reference were for $M_\infty = 1.5$ and higher. The wave drag was computed by the HTVD + MNT at $M_\infty < 2$ and by SOSET + MNT for $M_\infty \geq 2.0$. Very good agreement with forebody axial force is obtained between the theory and experiment for all bluntness ratios and at all Mach numbers where data were available.

VII. Skin Friction Drag

For most practical configurations, the boundary layer begins in a laminar state and transitions at some point to turbulent flow. Most tactical weapons have a majority of the flow consisting of about 10 to 20% laminar flow

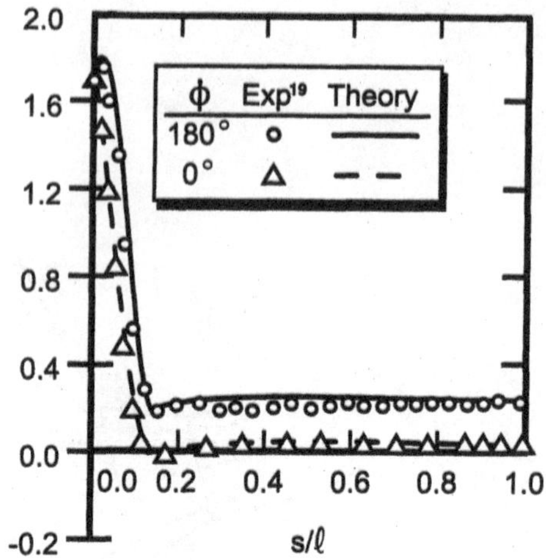

Fig. 4.17 Pressure distribution on a blunted cone $\frac{R_N}{R_B} = 0.35$, $M_\infty = 2.96$, $\alpha = 8°$, and $\theta_C = 11.5°$.

followed by 80 to 90% of the configuration under turbulent conditions. However, as altitude at which the weapon flies increases, the amount of laminar flow increases due to the lower flight Reynolds number.

For the turbulent boundary layer, Van Driest[21] is used to calculate the mean skin friction coefficient, $C_{f\infty}$, on a two-dimensional flat plate. The method of Van Driest assumes zero pressure gradient and Prandtl number equal to 1 to get a closed-form, two-dimensional solution of the turbulent boundary-layer equation. The solution for $C_{f\infty}$ is found by numerically solving Eq. (47).

$$\frac{0.242}{A\left(C_{f\infty}\right)^{1/2}}\left(T_W/T_\infty\right)^{1/2}\left(\sin^{-1} C_1 + \sin^{-1} C_2\right) = \log_{10}\left(R_{N\infty} C_{f\infty}\right) - \left(\frac{1+2n}{2}\right)\log_{10}\left(T_W/T_\infty\right) \tag{47}$$

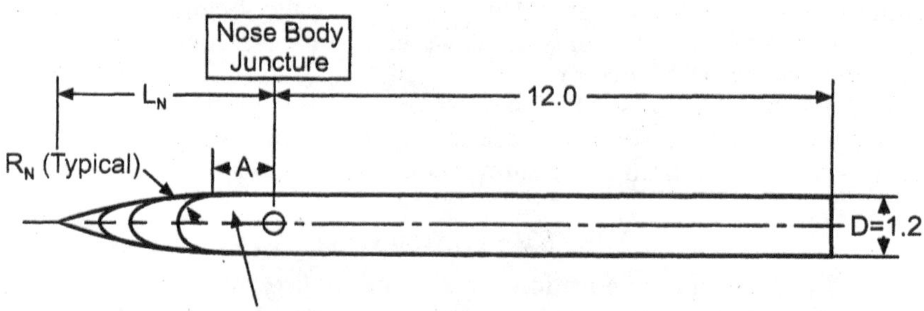

Fig. 4.18a Butler et al.[20] wind tunnel model tested.

Table 4.2 Butler et al.[20] wind tunnel model tested

Nose fineness ratio (L_N/D), caliber	Nose, R_N/R_B	R_N, in.	A, in.	D, in.	Total length, in.
2	0.00	0.00	2.40	1.2	14.4
2	0.25	0.15	2.06	1.2	14.06
2	0.50	0.30	1.67	1.2	13.67
2	0.75	0.45	1.17	1.2	13.17
4	0.00	0.00	4.80	1.2	16.80
4	0.25	0.15	4.15	1.2	16.15
4	0.50	0.30	3.38	1.2	15.38
4	0.75	0.45	2.38	1.2	14.38

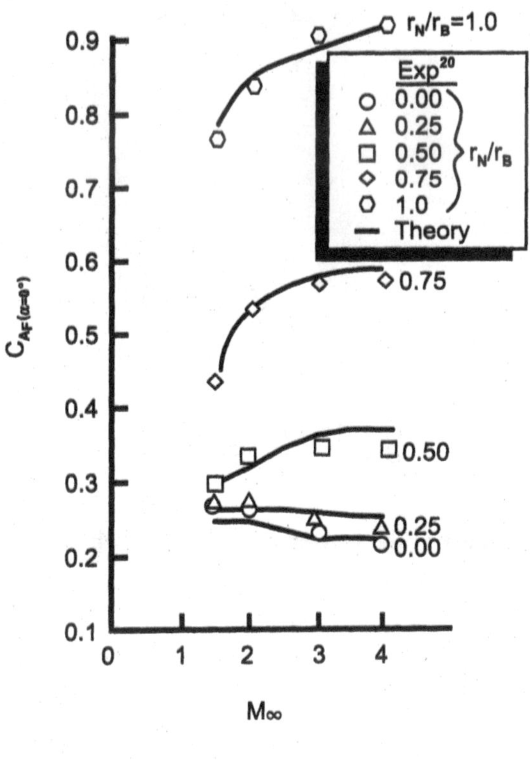

Fig. 4.18b Variation of forebody axial force coefficient with Mach number for various noses on 10-caliber afterbody (R_N/ft = 1.8 × 10^6, and ℓ_N = 2 caliber).

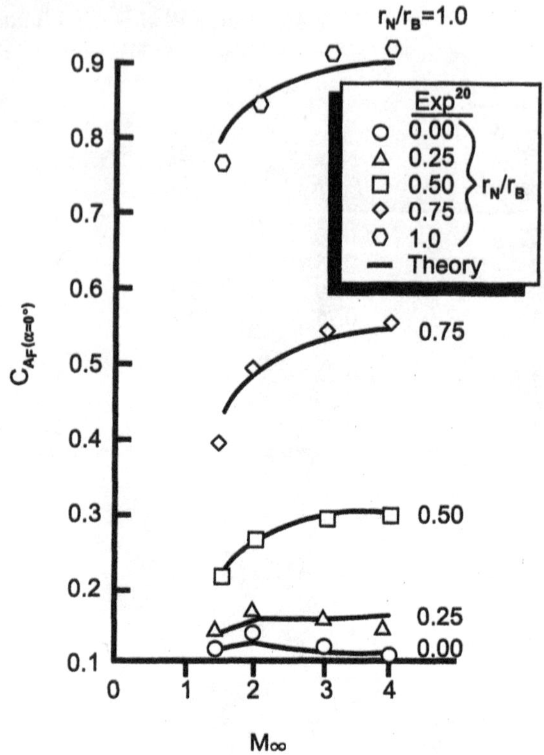

Fig. 4.18c Variation of forebody axial force coefficient with Mach number for various noses on 10-caliber afterbody (R_N/ft = 1.8 × 10^6, and ℓ_N = 4 caliber).

where

$$C_1 = \frac{2A^2 - B}{(B^2 + 4A^2)^{1/2}} \quad ; \quad C_2 = \frac{B}{(B^2 + 4A^2)^{1/2}}$$

and

$$A = \left[\frac{(\gamma - 1)M_\infty^2}{2T_W/T_\infty}\right]^{1/2} \quad ; \quad B = \frac{1 + (\gamma - 1)/2M_\infty^2}{T_W/T_\infty} - 1$$

The variable n of Eq. (47) is the power in the power viscosity law (which for air is 0.76):

$$\frac{\mu}{\mu_\infty} = \left(\frac{T_W}{T_\infty}\right)^n \tag{48}$$

To solve Eq. (47) for the mean skin friction coefficient $C_{f\infty}$, one must have values for T_W/T_∞, $R_{N\infty}$, and M_∞. The freestream Reynolds number is simply

$$R_{N_\infty} = \frac{\rho_\infty V_\infty \ell}{\mu_\infty} \tag{49}$$

To relate T_W/T_∞ to the freestream Mach number, assume the wall is adiabatic (no heat flux between the wall and the fluid). Defining a turbulent recovery factor R_T by

$$R_T = \left(\frac{T_W}{T_\infty} - 1\right)\frac{2}{(\gamma - 1)M_\infty^2}$$

then

$$\frac{T_W}{T_\infty} = 1 + R_T \frac{\gamma - 1}{2} M_\infty^2 \tag{50}$$

It has been shown that the recovery factor varies as the cube root of the Prandtl number[22] for turbulent flow so that

$$R_T = (\mathrm{Pr})^{1/3} \tag{51}$$

Recall that Van Driest's method assumes a Prandtl number of unity so, if this were used, then R_T would also be unity. However, the actual value of $Pr \cong 0.73$ so that the previous assumption of Prandtl number one can be compensated for somewhat by the above recovery factor, which for $Pr = 0.73$, would be 0.90. Thus Eq. (50) becomes

$$T_W/T_\infty = 1 + 0.9 \frac{\gamma - 1}{2} M_\infty^2 \tag{52}$$

Then, for a given set of freestream conditions (M_∞, ρ_∞, μ_∞ and V_∞), one can combine Eqs. (49) and (52) with (47) to solve for $C_{f\infty}$. The equation must be solved numerically however, because $C_{f\infty}$ cannot be explicitly determined. A procedure adaptable to equations of this type is the well-known Newton–Raphson method discussed in Ref. 23.

It has also been shown that, for conical bodies[22] a factor of 1.14 must be applied to the two-dimensional flat plate value from Eq. (20) to apply it to an axisymmetric body. This assumption will be made for any axisymmetric body.

For the laminar portion of the skin friction drag, incompressible flow over a flat plate is assumed, which gives

$$C_{f_\infty} = \frac{1.328}{\sqrt{R_{N_\infty}}} \tag{53}$$

where R_{N_∞} of Eq. (53) is the Reynolds number based on the x location where transition occurs. Equation (53) was later modified[24] for compressibility effects to yield

$$C_{f_\ell} = \frac{1}{\mathrm{Re}_\infty}\Big[1.328 - 0.0236M_\infty - 0.00335M_\infty^2 + 0.000349M_\infty^3 - 8.54 \times 10^{-6}M_\infty^4\Big] \tag{54}$$

The skin friction drag is simply the mean skin friction coefficient multiplied by the ratio of wetted area to reference area, that is

$$C_{A_F} = C_{f_\ell}\frac{(A_{\mathrm{WETTED}})_\ell}{A_{\mathrm{ref}}} + C_{f_T}\frac{(A_{\mathrm{WETTED}})_T}{A_{\mathrm{ref}}} \tag{55}$$

C_{f_ℓ} and C_{f_T} of Eq. (55) come from Eqs. (54) and (47), respectively. Again, if the skin friction drag is being computed for the body, Eq. (55) must be multiplied by 1.14; whereas, if it is for the wing surfaces, no factor is needed.

The final issue that must be addressed is transition Reynolds number. The author has found that a value of 1×10^6 for most practical bodies in flight and a value half that (0.5×10^6) for the wings give quite satisfactory results for most typical flight vehicles. However, for wind tunnel models, it has been found that most models are very smooth, and hence values of transition Reynolds number of 4×10^6 and 2×10^6 for the body and wing surfaces, respectively, are more realistic. Some wind tunnel models have boundary-layer trips. For these cases, flow should be assumed fully turbulent over the entire surface. Moore et al.[25] thus gave four alternatives for skin friction calculation. These alternatives and the associated transition Reynolds number are shown in Table 4.3. Again, referring back to a discussion of wind tunnel model roughness in Section 2.3, if the model is very rough, the transition Reynolds number will probably be somewhere between the value for flight vehicles and new wind tunnel models. Skin temperatures will be discussed in Chapter 7.

Experience has also shown that, for options 1 and 3 of Table 4.3, the region of turbulent flow moves toward the nose as AOA increases. An approximate way to account for this change for most vehicles is to assume fully turbulent flow at some AOA, say 30 deg. The transition Reynolds number can be assumed to vary from the values computed by options 1 and 3 to fully turbulent flow at $\alpha = 30$ deg in a linear fashion as a first approximation.

Table 4.3 Transition Reynolds number options for computing mean skin friction drag

Option	Transition Reynolds number	
	Body	Wings
1. Typical flight vehicle	1×10^6	0.5×10^6
2. Wind tunnel model with boundary-layer trip	1×10^2	1×10^2
3. Wind tunnel model with no boundary-layer trip	4×10^6	2×10^6
4. All laminar flow	1×10^{10}	1×10^{10}

VIII. Empirical Methods

It is fair to wonder why approximate aeroprediction codes are defined as semi-empirical with all the theoretical methods discussed so far. The truth is that, while these methods allow individual component forces and moments to be calculated fairly rigorously at a given Mach number or AOA, there are still many conditions where the analytical methods presented previously are either not applicable or not worth the effort because of the difficulty in applying them. In these cases, empirical methods are generally used. The combination of theoretical and empirical techniques in a code is thus why they are called semi-empirical codes. A few examples where empirical methods are typically used are transonic aerodynamics, body-alone subsonic aerodynamics, rotating band or protuberance aerodynamics, and base drag of the body and lifting surfaces. There are actually analytical methods available for transonic aerodynamic computations. However, most of the methods are inconsistent from a computational standpoint with the approximate codes. What is done in many cases is to use sophisticated analytical tools[16,26–29] to estimate transonic aerodynamics as a function of key geometric parameters, then to include these into an engineering code in a table lookup fashion. Obviously, for a vehicle that spends a large portion of its time in the transonic flow region, $0.8 < M_\infty < 1.2$, it would be justifiable to use a more sophisticated estimation process. Several examples of empirical prediction methods will be selected in this section to illustrate how empirical methods can be effectively combined with analytical methods to develop a complete set of aerodynamics.

A. Transonic Wave Drag Prediction

For transonic Mach numbers, the flowfield solution is nonlinear, even for slender bodies. The reason is that near $M_\infty = 1$, Eq. (45) from Chapter 3 is no longer valid and must be replaced by

$$\left[1 - M_\infty^2 - (\gamma + 1)M_\infty^2\Phi_x\right]\Phi_{xx} + \Phi_r/r + \Phi_{rr} = 0 \tag{56}$$

The term $(\gamma + 1) M_\infty^2 \Phi_x \Phi_{xx}$ makes Eq. (56) nonlinear and no linear small perturbation solutions exist, analogous to those discussed in Chapter 3. Hence, the alternatives to obtain transonic wave drag are (1) full numerical solution of Eq. (56), (2) numerical solution of the full potential equation, (3) numerical solution of the Euler equation, (4) numerical solution of the Navier–Stokes equations, or (5) experimental data. Moore[16] used the method of Wu and Aoyoma,[30] which solves Eq. (56) numerically, to develop tables of wave drag for tangent ogives of various lengths in transonic flow. Figure 4.19 gives these results for nose lengths 1.5 to 4.0. However, for fairly short nose lengths, the perturbation theory results are somewhat questionable. As a result, Chaussee[26] and Devan[29] extended these results to ogives with nose lengths down to 0.75 caliber through solution of the full potential equation, Eq. (10) from Chapter 3. Results of these computations are given in Table 4.4. Table 4.4 can be used to estimate wave drag for bodies of revolution for $M_\infty < 1.2$ by use of interpolation or extrapolation. Parameters included in the table include nose length, nose bluntness ratio, and Mach number.

The method of Wu and Aoyoma[30] is also used to compute the wave drag on boattails and flares, where the boattail or flare angle is fairly small in keeping with the perturbation flow assumption. For the convenience of the reader, results using the methods of Ref. 30 have been compiled in Table 4.5 for Mach numbers 0.9 to 1.1, boattail or flare lengths of 0.5 to 2.0 calibers,

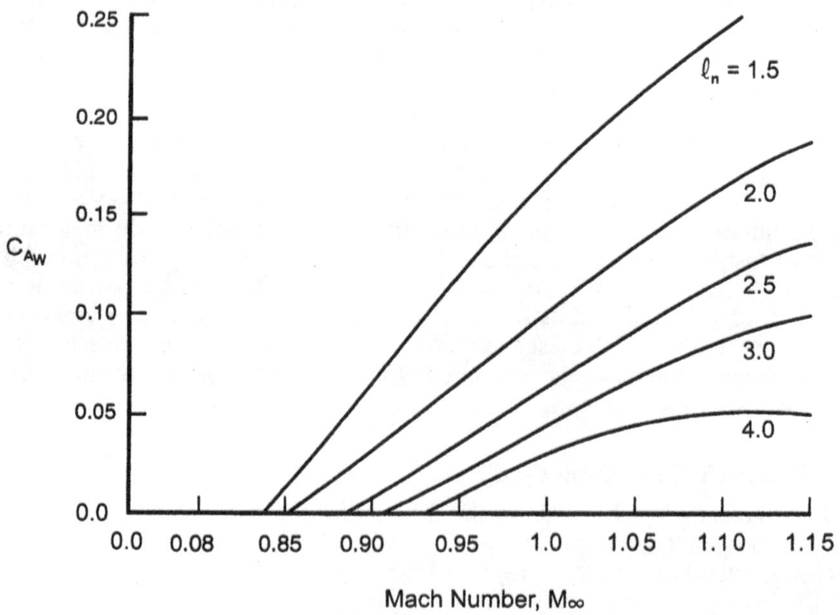

Fig. 4.19 Transonic wave drag of tangent ogives.

Table 4.4 C_{A_W} for blunted tangent ogives

M_∞	ℓ_n	r_n/r_b, 0	0.5	1.0
0.8	0.75	0.04	0.045	0.092
	1.0	0.025	0.030	0.092
	1.25	0.01	0.015	0.092
	1.5	0.01	0.015	0.092
	2.0	0.01	0.015	0.092
	3.0	0.01	0.015	0.092
	4.0	0.01	0.015	0.092
	5.0	0.01	0.015	0.092
0.95	0.75	0.16	0.189	0.279
	1.0	0.08	0.118	0.279
	1.25	0.04	0.070	0.279
	1.5	0.02	0.042	0.279
	2.0	0.02	0.035	0.279
	3.0	0.02	0.025	0.279
	4.0	0.02	0.020	0.279
	5.0	0.02	0.020	0.279
1.05	0.75	0.280	0.322	0.405
	1.0	0.200	0.252	0.405
	1.25	0.155	0.201	0.405
	1.5	0.135	0.152	0.405
	2.0	0.110	0.111	0.405
	3.0	0.078	0.062	0.405
	4.0	0.055	0.055	0.405
	5.0	0.036	0.050	0.405
1.2	0.75	0.419	0.460	0.55
	1.0	0.331	0.364	0.55
	1.25	0.283	0.286	0.55
	1.5	0.247	0.231	0.55
	2.0	0.194	0.155	0.55
	3.0	0.108	0.102	0.55
	4.0	0.065	0.095	0.55
	5.0	0.038	0.090	0.55

and boattail or flare angles varying from −7.5 to +7.5 deg in increments of 2.5 deg.

For wave drag of the wing alone, the assumption is made that the drag varies linearly between its value calculated at $M_\infty = 1.1$ from linear theory and a value of 0 at $M_\infty = 0.9$.

B. Viscous Separation and Rotating Band Drag

Figure 4.20a is a plot of forebody drag coefficients as a function of cone half-angle from data taken from Ref. 31. Because the skin friction drag coefficient is about 0.02 for this case, it can be subtracted from the curve of

Table 4.5 Wave drag of boattail and flares in transonic flow

M	C_{A_W}	Boattail (−) or flare (+) angle	Boattail or flare length, calibers	Base radius, calibers
1.1	0.0079	−2.5	0.5	0.478
1.0	0.0028	−2.5	0.5	0.478
0.9	0.0002	−2.5	0.5	0.478
1.1	0.032	−5.0	0.5	0.456
1.0	0.0109	−5.0	0.5	0.456
0.9	0.0004	−5.0	0.5	0.456
1.1	0.0692	−7.5	0.5	0.434
1.0	0.0234	−7.5	0.5	0.434
0.9	0.0005	−7.5	0.5	0.434
1.1	0.0088	+2.5	0.5	0.522
1.0	0.0033	+2.5	0.5	0.522
0.9	0.0000	+2.5	0.5	0.522
1.1	0.0418	+5.0	0.5	0.544
1.0	0.0139	+5.0	0.5	0.544
0.9	0.0000	+5.0	0.5	0.544
1.1	0.0941	+7.5	0.5	0.566
1.0	0.0314	+7.5	0.5	0.566
0.9	0.0000	+7.5	0.5	0.566
1.1	0.0089	−2.5	1.0	0.456
1.0	0.0034	−2.5	1.0	0.456
0.9	0.0004	−2.5	1.0	0.456
1.1	0.0436	−5.0	1.0	0.412
1.0	0.0149	−5.0	1.0	0.412
0.9	0.0006	−5.0	1.0	0.412
1.1	0.0986	−7.5	1.0	0.368
1.0	0.0333	−7.5	1.0	0.368
0.9	0.0007	−7.5	1.0	0.368
1.1	0.0122	+2.5	1.0	0.543
1.0	0.0041	+2.5	1.0	0.543
0.9	0.0000	+2.5	1.0	0.543
1.1	0.0606	+5.0	1.0	0.587
1.0	0.0202	+5.0	1.0	0.587
0.9	0.0000	+5.0	1.0	0.587
1.1	0.1476	+7.5	1.0	0.632
1.0	0.0492	+7.5	1.0	0.632
0.9	0.0000	+7.5	1.0	0.632
1.1	0.0096	−2.5	2.0	0.413
1.0	0.0036	−2.5	2.0	0.413
0.9	0.0006	−2.5	2.0	0.413
1.1	0.0460	−5.0	2.0	0.325
1.0	0.0159	−5.0	2.0	0.325
0.9	0.0008	−5.0	2.0	0.325
1.1	0.1093	−7.5	2.0	0.237
1.0	0.0370	−7.5	2.0	0.237
0.9	0.0008	−7.5	2.0	0.237
1.1	0.0122	+2.5	2.0	0.587
1.0	0.0041	+2.5	2.0	0.587

Table 4.5 Wave drag of boattail and flares in transonic flow *(continued)*

M	C_{A_W}	Boattail (−) or flare (+) angle	Boattail or flare length, calibers	Base radius, calibers
0.9	0.0000	+2.5	2.0	0.587
1.1	0.0662	+5.0	2.0	0.675
1.0	0.022	+5.0	2.0	0.675
0.9	0.0000	+5.0	2.0	0.675
1.1	0.178	+7.5	2.0	0.763
1.0	0.0593	+7.5	2.0	0.763
0.9	0.0000	+7.5	2.0	0.763

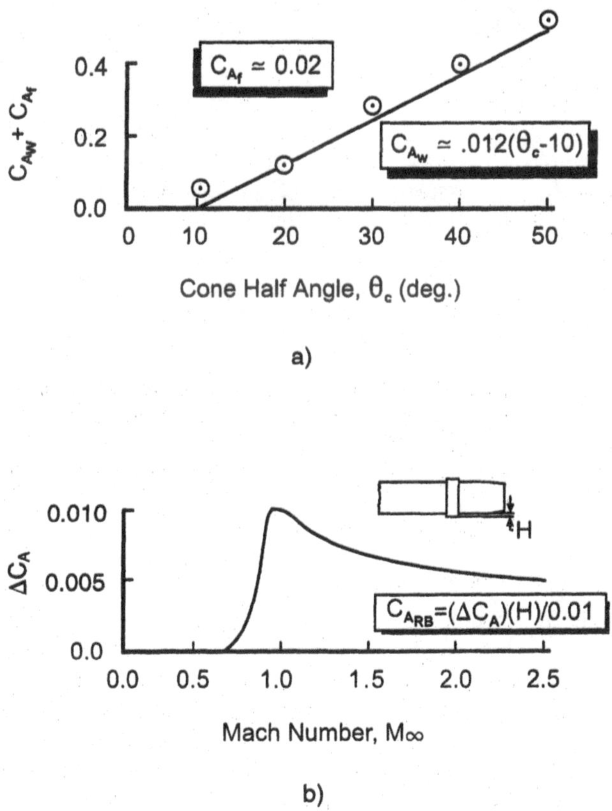

Fig. 4.20 Viscous separation and rotating band drag: a) viscous separation drag, $M_\infty = 0.5$, and b) rotating band drag, $C_{A_{RB}}$.

Fig. 4.20a to yield the pressure drag coefficient. Note that the freestream Mach number is 0.5, low enough so that no appreciable compressibility effects occur. The question, therefore, arises as to the origin of this type of drag because it is not compressibility or skin friction drag. It is, in fact, viscous separation drag. For very large cone half-angles, θ_c, the flow over the cone, instead of remaining attached, separates due to the very strong adverse pressure gradient and reattaches downstream. This separation prevents the pressure from decreasing as much as it would in inviscid flow and produces a drag. Oddly enough, this phenomenon does not occur on ogives or spherical surfaces, apparently due to body curvative effects on the boundary layer. As a result, one can derive an empirical expression for this viscous separation drag, where the important parameter is the angle δ^*, which the nose makes with the shoulder of the afterbody. Based on Fig. 4.20a, this relation is

$$C_{A_{vis}} = 0.012(\delta^* - 10^\circ) \quad ; \quad \delta^* \geq 10^\circ \tag{57}$$

$$C_{A_{vis}} = 0 \quad ; \quad \delta^* < 10^\circ$$

with δ^* in degrees and $\delta^* = \theta_c$ for a conical nose.

Moore[32] gives the measured effect of a rotating band on drag. The particular rotating band used in those wind tunnel tests had a mean height of about 0.024 calibers. An expression that functionalizes the above results for drag increment due to a rotating band is given by

$$C_{A_{RB}} = (\Delta C_A)\,(H)/0.01 \tag{58}$$

where H is the mean height of the band in calibers (on one side of the projectile only) and ΔC_A is the increment in axial force for an H of 0.01 caliber given in Fig. 4.20b. Although Eq. (58) was derived for a particular band, it checks well with the results of Charters[33] for a different band geometry.

C. Body-Alone Lift Properties for $M_\infty < 1.2$

At subsonic and transonic Mach numbers, lifting properties are more difficult to obtain. For subsonic velocities, the lift could be calculated by perturbation theory but, because weapons rarely fly at Mach numbers less than 0.7, a formulation on this basis was not justified. An alternative would be slender body theory, but the accuracy of this approach is inadequate in many cases. In light of the above reasoning, a semi-empirical method for normal force characteristics was derived[16] based on nose length, afterbody length, and boattail shape. This method was then extended through the transonic Mach number range because the state of the art in transonic flow does not allow one to handle the general body shapes or flow conditions.

The total inviscid normal force acting on the body may be written

$$C_{N_a} = \left(C_{N_a}\right)_n + \left(C_{N_a}\right)_a + \left(C_{N_a}\right)_B \tag{59}$$

where the subscripts n, a, and B stand for nose, afterbody, and boattail, respectively. The first term of Eq. (59) can be approximated by

$$\left(C_{N_a}\right)_n = C_1 \tan \delta^* + C_2 \tag{60}$$

where C_1 and C_2 are given in Fig. 4.21 as functions of Mach numbers. This relationship was determined empirically from the cone results of Owens.[31] It is approximately correct for $\ell_n \geq 1.5$, cone bluntness up to 0.5, and $M_\infty \leq$ 1.2. Note that the angle δ^* in Eq. (60) is the same as that discussed previously in Eq. (48).

The normal force coefficients of the afterbody and boattail can be obtained from Figs. 4.22 and 4.23, respectively. Figure 4.22 was derived analytically in the transonic Mach range from the method of Wu and Aoyoma[30] and in subsonic flow from the experimental data of Spring[36] and Gwin and Spring.[37] In the work of Gwin and Spring, the normal force of the nose plus

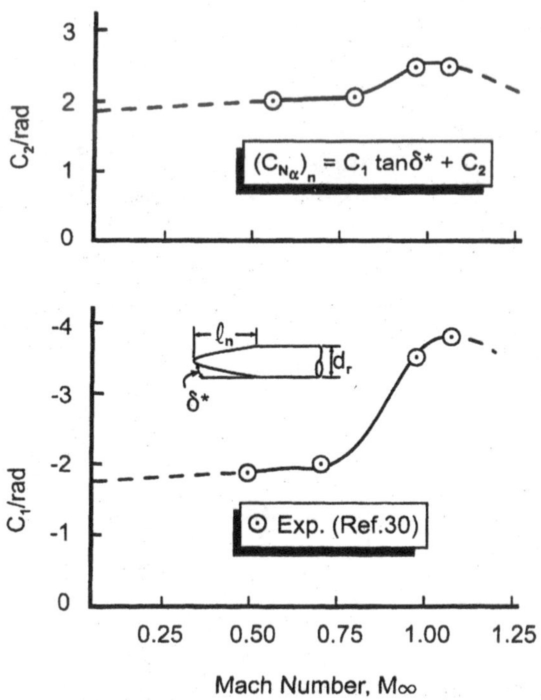

Fig. 4.21 Constants to determine $(C_{N_a})_n$ for $M_\infty < 1.2$.

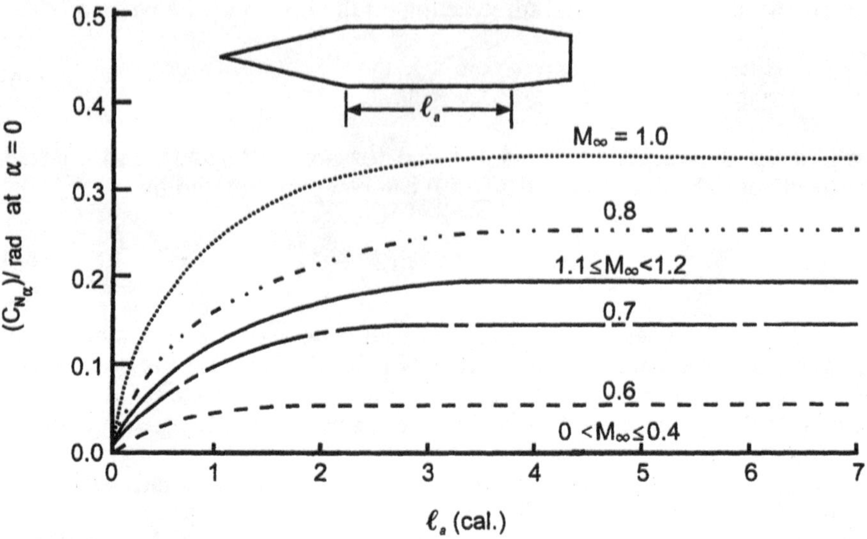

Fig. 4.22 Increase in C_{N_α} at subsonic and transonic Mach numbers due to afterbody.

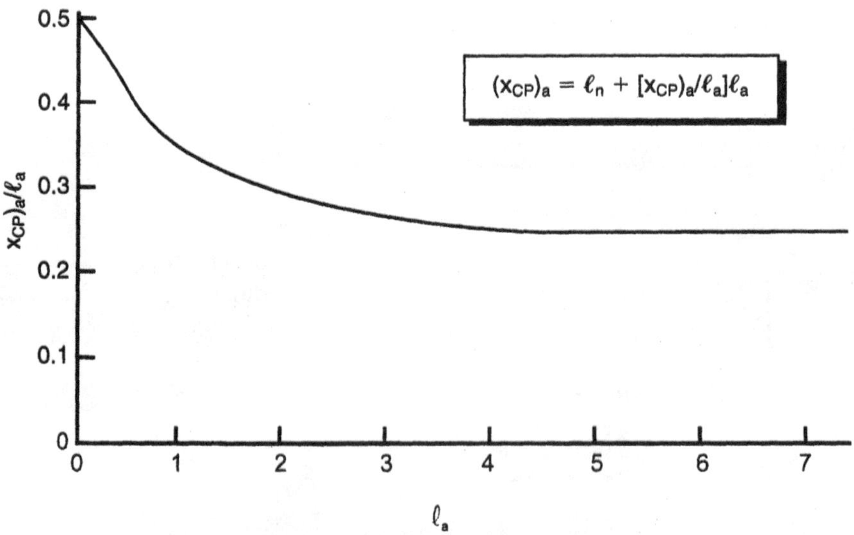

Fig. 4.23 Decrease in C_{N_α} due to boattail.

afterbody was given, but the nose component can be subtracted off by the use of Eq. (60). The boattail normal force coefficient (Fig. 4.23) was given by Washington and Pettis[34] but he stated that there were not enough data available in subsonic and transonic flow. Hence, the data of Washington and Pettis was supplemented by the 175-mm Army projectile and improved 5 in./54 Navy projectile data to derive the general curve of Fig. 4.23.

Although slender body theory may not be adequate for predicting the normal force coefficient, it appears to predict the center of pressure of the nose and boattail lift components quite adequately. According to slender body theory, the center of pressure of the nose is

$$\left(x_{CP}\right)_n = \ell_n - \frac{(\mathrm{Vol})_n}{\pi R_r^2} \tag{61}$$

and of the boattail

$$\left(x_{CP}\right)_B = \ell_n + \ell_a + \ell_B - \frac{(\mathrm{Vol})_B}{\pi R_r^2}$$

or

$$\left(x_{CP}\right)_B = \ell - \frac{(\mathrm{Vol})_B}{\pi R_r^2} \tag{62}$$

The center of pressure of the afterbody normal force was calculated analytically by the method of Wu and Aoyoma[30] in transonic flow and assumed to have the same value in subsonic flow. Figure 4.24 is a plot of $(x_{CP})_a/\ell_a$ versus afterbody length measured at the point where the afterbody begins. Now knowing the individual lift components and their center of pressure locations, one can compute the pitching moment about the nose as

$$C_{M_\alpha} = -\left[\left(C_{N_\alpha}\right)_n\left(x_{CP}\right)_n + \left(C_{N_\alpha}\right)_a\left(x_{CP}\right)_a + \left(C_{N_\alpha}\right)_B\left(x_{CP}\right)_B\right] \tag{63}$$

D. Wing-Alone Normal Force at Transonic Speeds

Airfoil thickness has only a small effect on wing lifting properties at subsonic and supersonic Mach numbers. However, at transonic speeds, we must account for the thickness effects. The preferred approach by the present author is the empirical approach. An approach developed by Douglas Aircraft[38] was adopted in Ref. 27 and carried throughout the aeroprediction code in all future versions. This procedure accounts, in a empirical manner, for sweep, Mach number, aspect ratio, and thickness ratio, but not airfoil section.

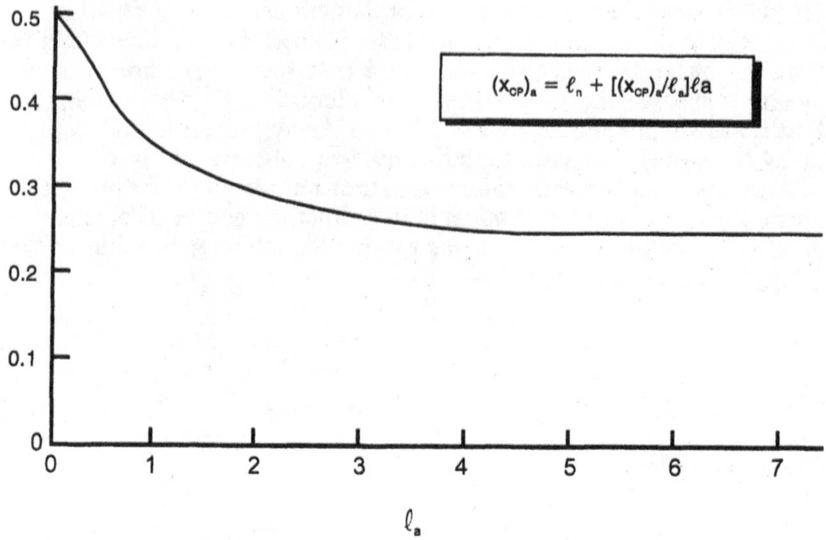

Fig. 4.24 Center of pressure of afterbody lift for $M_\infty < 1.2$.

To apply the empirical procedure of Ref. 38, the force break Mach number is found from Fig. 4.25a for a wing of zero sweep and corrected for sweep by Fig. 4.25b. The lift–curve slope at the force break Mach number is then computed by a simple expression derived from lifting line theory[39]:

$$\left(C_{N_a}\right)_{fb} = \frac{2\pi \mathrm{AR}}{2 + \left[\mathrm{AR}^2\left(\beta^2 + \tan^2 \Lambda_{1/2}\right) + 4\right]^{1/2}} \tag{64}$$

As stated earlier, the two-dimensional lift–curve slope of 2π can be replaced by a local wing section lift–curve slope if one is available. This value is corrected to agree with experiment according to Fig. 4.26a. The abrupt decrease in lift–curve slope associated with thick wings is approximated by the relation

$$\left(C_{N_a}\right)_a = (1 - a/c)\left(C_{N_a}\right)_{fb} \tag{65}$$

where a/c is given in Fig. 4.26b. The Mach number at point a is:

$$M_a = M_{fb} + 0.07 \tag{66}$$

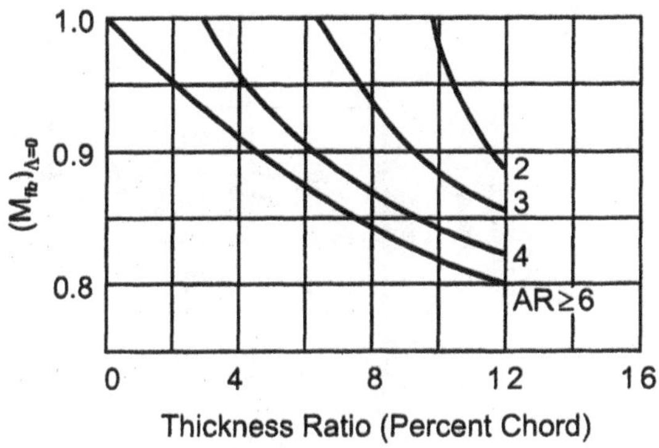

a)

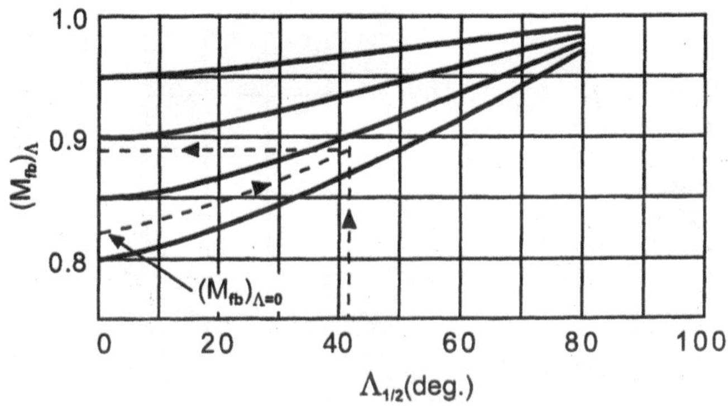

b)

Fig. 4.25 Transonic force break Mach number: a) zero sweep and b) with sweep correction.

The subsequent rise in C_{N_a} to a value at point b is

$$\left(C_{N_a}\right)_b = (1 - b/c)\left(C_{N_a}\right)_{fb} \tag{67}$$

where b/c is given in Fig. 4.26c. The Mach number at point b is

$$M_b = M_{fb} + 0.14 \tag{68}$$

The normal force curve slope at $M_\infty \geq 1.2$ is calculated by supersonic thin wing theory and for $M_\infty < M_{fb}$ by lifting surface theory. From this empirical correlation, one obtains $(C_{N_a})_{fb}$, $(C_{N_a})_a$, $(C_{N_a})_b$ and at Mach numbers M_{fb},

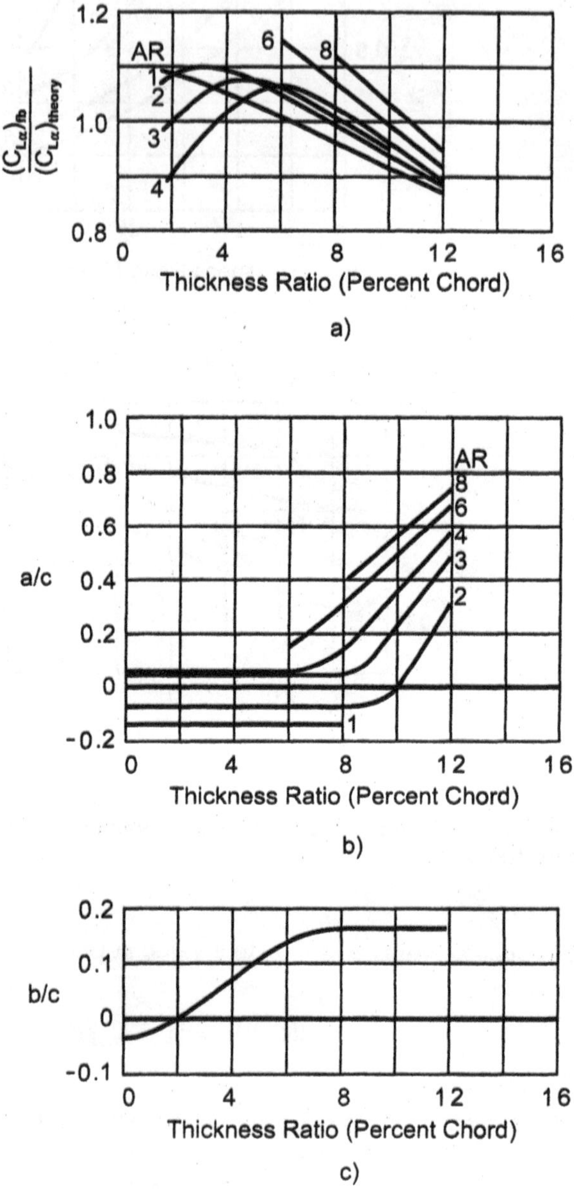

Fig. 4.26 Chart for determining transonic lift-curve slope at M_a and M_b. a) Correction to lift- curve slope at force-break Mach number, b) Chart for determining lift–curve slope at M_a, c) Chart for determining lift–curve slope at M_b

$M_{fb} + 0.07$, and $M_{fb} + 0.14$, respectively. For values of C_{N_a} in transonic flow in between these Mach numbers, interpolation is used.

The center of pressure of wing-alone lift is usually around the quarter-chord for subsonic flow and half-chord for supersonic flow. Transition from subsonic to supersonic flow is assumed to occur in a linear fashion between values calculated at M_∞ slightly less than M_{fb} and $M_\infty > 1.2$. The pitching moment coefficient derivative of the wing alone is then

$$C_{M_a} = -x_{CP} C_{N_a} \tag{69}$$

E. Base Drag

Base drag arises at the base of a weapon or at the rear of the blunt trailing edge of a fin due to flow separation. When flow separates, the flow in the base region typically drops below freestream values, giving a negative pressure coefficient. This, in turn, gives a positive contribution to axial force. There have been numerous experimental and theoretical efforts over the last 50 years to both analytically and experimentally estimate base drag. Some of the more important original works were the approximate models of Chapman[40] and Korst.[41] More recently, the Army Research Laboratory has made significant strides to predict base pressure with the full Navier–Stokes equation[42–45] as a function of several parameters of interest. Some of the more notable experimental data were taken by Love[46] for two-dimensional airfoils and Chapman[40] and Kurzweg[47] for bodies of revolution. More recent databases include those due to Lamb and Oberkampf[48] and Moore et al.[49]

In general, it is the author's feeling that, for an engineering approximation, a compilation of data is about as accurate as the full Navier–Stokes equation solutions and, in many cases, more accurate. As a result, the approach to predict base drag on both two-dimensional and axisymmetric surfaces relies on primarily data. Based on this data, approximate effects of AOA, fin location, and thickness effects; power on; and boattail effects are approximated. The major limitations of this approach are the following:

1) the data assume a fully developed turbulent boundary layer at the base so, if the flow is all laminar, the base drag prediction will be in error more than if the flow were all turbulent

2) Databases are for truncated bases and trailing edges. If round bases or trailing edges are used, errors may be higher.

3) Experimental data are more limited than desired to estimate a combination of effects, including fins, control deflection, AOA, and fin location. However, experience has shown that the empirical model of Ref. 50 works reasonably well based on many comparisons to date, so the empirical approach will be shown here. For those interested in other approaches, the references mentioned could be consulted.

Figure 4.27 is a summary of the empirical method of Ref. 50. Figure 4.27 was derived assuming the individual effects of AOA, tail fin thickness, deflection and location, and power on; and boattail or flare could be treated

A. Body Alone

$$\left(C_{P_B}\right)_{NF,\,\alpha} = \left(C_{P_B}\right)_{NF,\,\alpha=0}\left[1 + 0.01F_1\right]$$

F_1 = Body Alone α Effects

B. Body with Tail Fins

1. Deflection and Thickness Effects

$$\left(C_{P_B}\right)_{\alpha,\,\delta,\,t/c,\,x/c=0} = \left(1 + 0.01F_2\right)\left(C_{P_B}\right)_{NF,\,\alpha=0} + 0.01F_3\ (t/d)$$

$F_2 = |\alpha + \delta|$ Effects

F_3 = Additional Effects due to Thickness

2. Fin Location Effects when $x/c \neq 0$

$$\left(C_{P_B}\right)_{\alpha,\,\delta,\,t/c,\,x/c} = \left(C_{P_B}\right)_{NF,\,\alpha} + 0.01\left(\Delta C_{P_B}\right)_{\alpha,\,\delta,\,t/c,\,x/c=0}$$

C. Power On

ΔC_{P_B} due to Power-On by Modified Brazzel [51] Method Added to Above Values of C_{P_B}

D. Boattail

$$C_{A_B} = -\,C_{P_B}\left(d_B/d_{ref}\right)^3$$

E. Flare

$$C_{A_B} = -\,C_{P_B}\left(d_B/d_{ref}\right)^2$$

Fig. 4.27 Empirical base drag prediction model of Ref. 50.

individually and then combined together as shown in Fig. 4.27. That is, referring to Fig. 4.27, one first computes the body-alone base drag from Fig. 4.27a, then the tail fin effect from Fig. 4.27b, the power on, boattail or flare effects from Fig. 4.27c, d, and e. $(C_{P_B})_{NF,\alpha=0}$ of Fig. 4.27 is the average base pressure obtained for a body alone with no boattail at $\alpha = 0$. These data are given in Fig. 4.28, with data for two-dimensional base pressures that arise on the trailing edges of fins. Figure 4.28 was derived primarily from Refs. 40, 41, 46, 47, 48, and 49. It has been used extensively and appears to be fairly reliable. It is interesting to note the high values of base pressure coefficient at the subsonic, transonic, and low supersonic Mach numbers of Fig. 4.28. This is particularly true for the two-dimensional curve. It is not uncommon for base pressure to account for 40% or more of the total drag at low Mach number. It is also interesting to note that, at high Mach number, C_{P_B} approaches zero, the Newtonian flow assumption. The actual methodology of Ref. 50 is to use Fig. 4.28 up to M_∞ of about 4.3. Above $M_\infty = 4.3$, an empirical equation for $C_{P_{2\text{-}T}}$ and $C_{P_{3\text{-}D}}$ is

$$C_{P_B} = \frac{0.18}{M_\infty^2} + \frac{0.266}{M_\infty} - 0.018 \; ; \quad 4.3 < M_\infty < 16$$

and

$$C_{P_B} = 0 \qquad ; \quad M_\infty \geq 16 \tag{70}$$

The second parameter of Fig. 4.27 is F_1, the body-alone AOA effects. This parameter is given in Fig. 4.29a, where data for $M_\infty \geq 2$ were taken primarily from Ref. 49, and data for $M_\infty < 2$ were taken from Ref. 20 and are shown in Fig. 4.29b. As seen in Fig. 4.29, there is a fairly strong increase in the base drag from $\alpha = 0$ up to some small AOA, which is typically 15 deg or lower. Unfortunately, databases are limited to a 15-deg AOA, so the model of Ref. 49 extrapolates above $\alpha = 15$ deg.

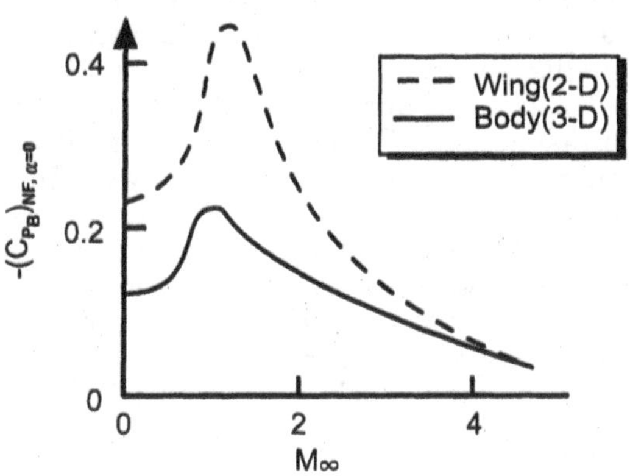

Fig. 4.28 Base pressure coefficient for blunt wing trailing edge and body base.

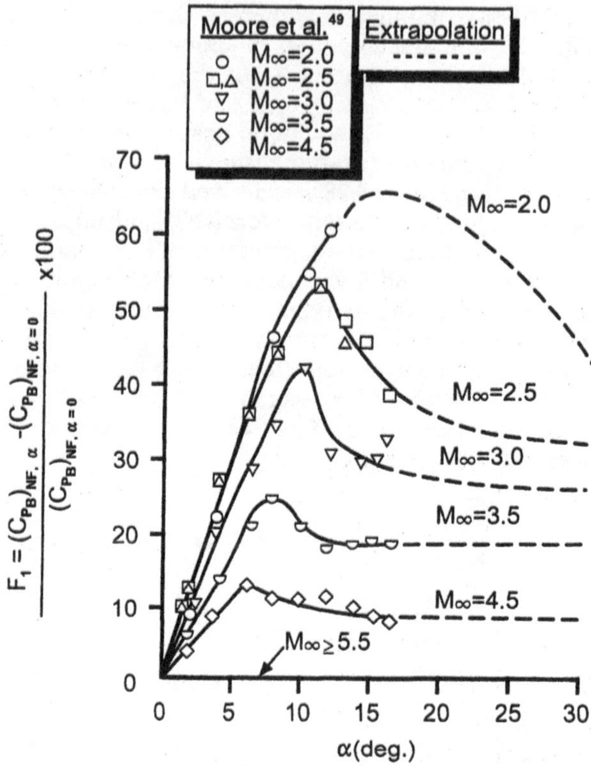

Fig. 4.29a Percent increase in body-alone base pressure coefficient due to AOA ($M_\infty \geq 2$).

The next two parameters of Fig. 4.27 are F_2 and F_3, which separate out the fin effects due to a combined AOA and control deflection ($|\alpha + \delta|$), and additional effects due to thickness. Values of these two parameters are defined in Figs. 4.30 and 4.31. Note that F_2, like F_1, measures the percent increase in base pressure of a given effect relative to the value of $(C_{P_B})_{NF,\alpha=0}$ of Fig. 4.28. Also, note that the values of F_3 are negative. However, when viewing the equation of Fig. 4.27 for combined effects of fin deflection and thickness, it is clear that both the F_2 and F_3 terms are additive because they both have negative signs. That is, they both tend to lower C_{P_B}, which in turn will increase base drag.

The next effect of Fig. 4.27 is fin location effects relative to the body base. Figure 4.32 shows the percent change in base pressure coefficient for various AOAs and t/d values as a function of x/c for $M_\infty = 2$. This figure was based on the wind tunnel data of Ref. 49. Note that at $\alpha = 0$, Fig. 4.32a, (taken here to be $|\alpha + \delta|$), ΔC_{P_B} goes close to zero about a caliber or so ahead of the base. Also note that the ΔC_{P_B} of Fig. 4.32 is the change in base pressure at a given AOA due to the presence of the fins.

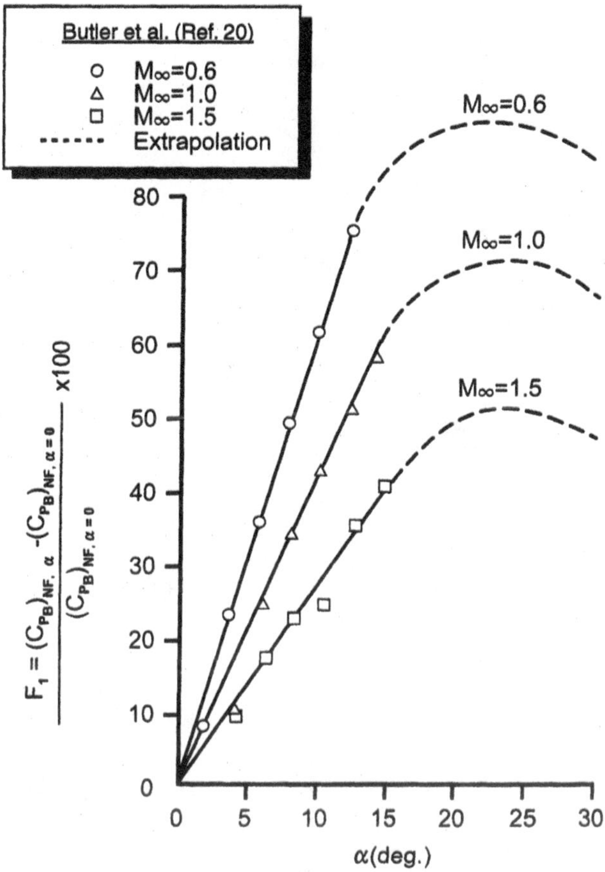

Fig. 4.29b Percent increase in body-alone base pressure coefficient due to AOA ($M_\infty < 2$).

On the other hand, examining Figs. 4.32b and 4.32c, it is seen that as a or $|a + \delta|$ becomes appreciable, the fins need to be located about 2.5 calibers ahead of the base before the fin effects are minimal and the body AOA effect is the dominant factor.

Also, it is noted that the percentage change in base pressure coefficient in Fig. 4.32 for $a = 5$ and 10 deg initially has a negative slope similar to the $a = 0$ case, but then reverses and approaches no change at $x/c = 2.5$. As a result, a numerical interpolation is used to compute the percent change in ΔC_{P_B} as a function of x/c, $|a + \delta|$, and t/d. This same percent change is assumed to occur at all Mach numbers because no data are available other than at $M_\infty = 2$.

The next step in the empirical method of Ref. 50 is to compute the change in C_{P_B} due to power on effects. When the engine is running, the base drag is

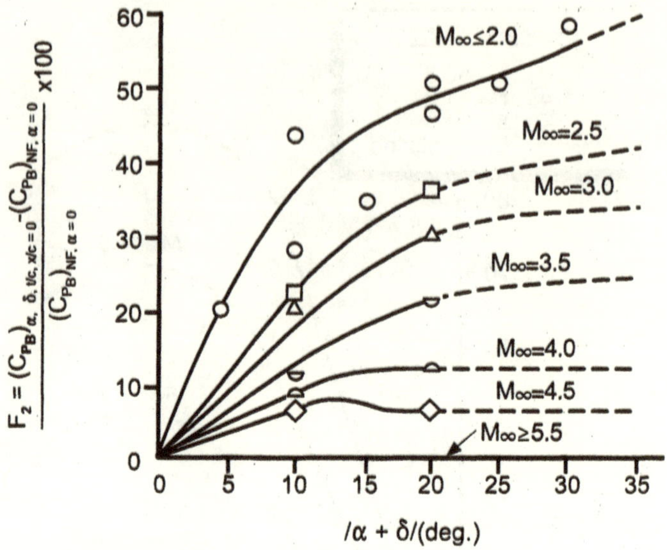

Fig. 4.30 Percent increase in base pressure coefficient due to combined effects of AOA and control deflection ($t/c \simeq 0$).

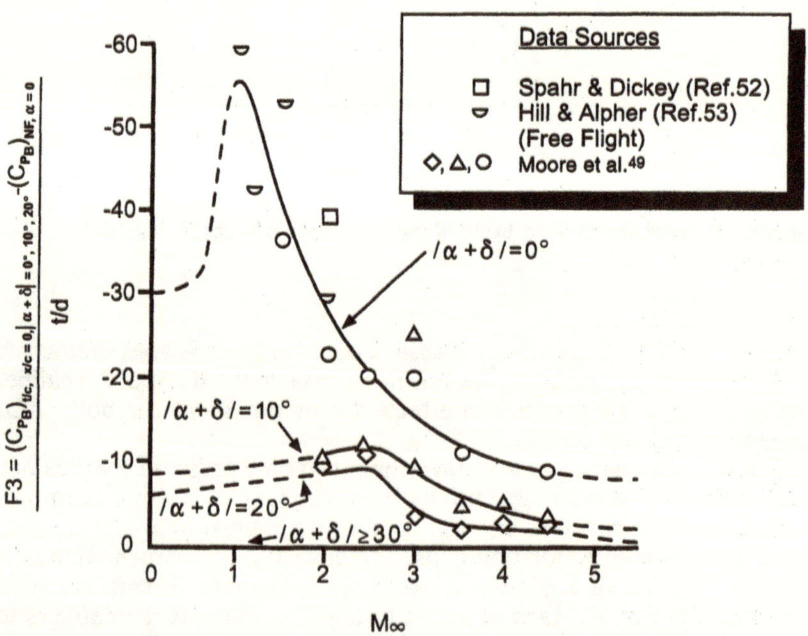

Fig. 4.31 Percent change in base pressure coefficient due to fin thickness at various values of $|\alpha + \delta|$.

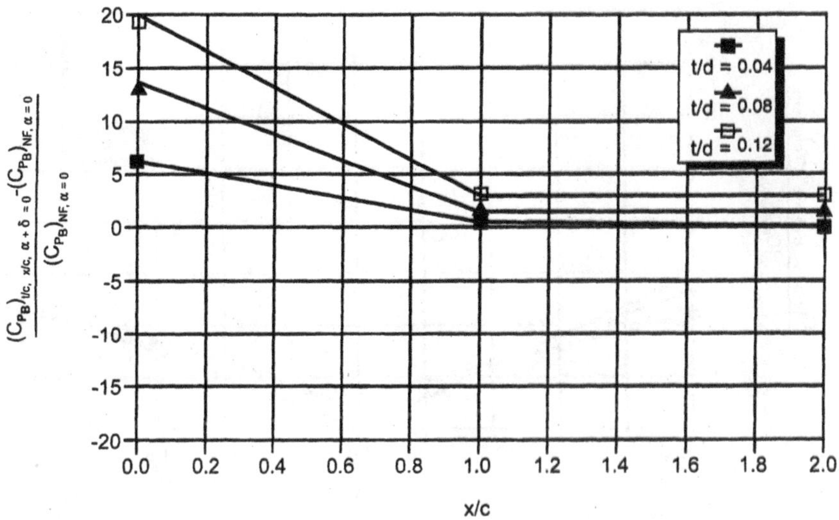

Fig. 4.32a Percent increase in base pressure coefficient due to fin location ($|\alpha + \delta|$ = 0° and M_∞ = 2.0).

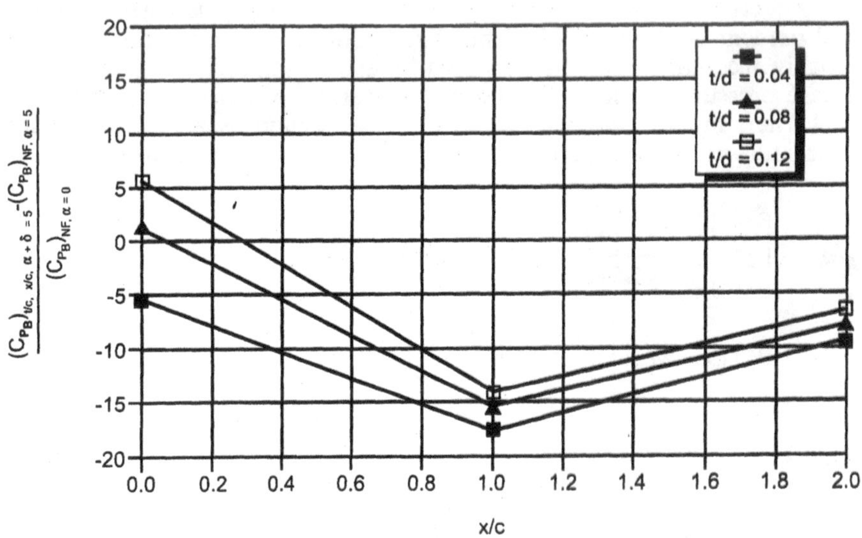

Fig. 4.32b Percent increase in base pressure coefficient due to fin location ($|\alpha + \delta|$ = 5.0° and M_∞ = 2.0).

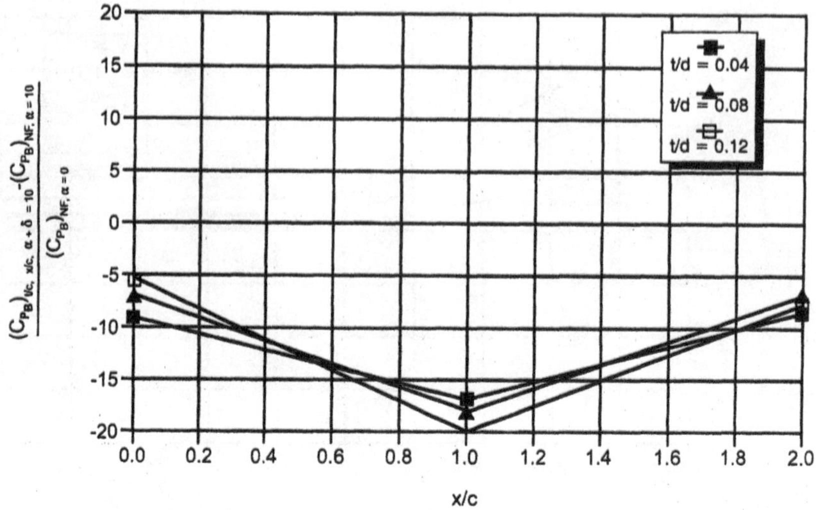

Fig. 4.32c Percent increase in base pressure coefficient due to fin location ($|\alpha + \delta| = 10.0°$ and $M_\infty = 2.0$).

usually lowered for two reasons. First, over the region of the base where the engine exit is located, the dead air region is replaced by an engine exhaust. Second, the dead air region outside the engine exhaust generally will have an increased base pressure coefficient due to increased mass flow. Both of these effects lower the base drag. However, if the engine is burning only a small percent of the time of flight of the vehicle, neglecting this effect is probably acceptable. The method used for estimating rocket engine effects on base pressure is a modification of the method of Brazzel and Henderson.[51] Their method defines the base pressure ratio as a function of jet temperature, jet nozzle location, and jet momentum flux ratio as

$$p_B/p_\infty = \left[\frac{T_j}{T_j^*}\right]\left[\frac{3.5}{1+2.5\left(A_B/A_{\text{ref}}\right)}\right] f(\text{RMF}) + 0.047\left[5 - M_\infty\right]\left[2\frac{x_j}{\text{d}_{\text{ref}}} + \left(\frac{x_j}{\text{d}_{\text{ref}}}\right)^2\right] \tag{71}$$

where

$$T_j/T_j^* = \frac{\left(\gamma_j + 1\right)/2}{1 + \dfrac{\gamma_j - 1}{2} M_j^2}$$

A_B = base area and x_j/d_{ref} = distance of nozzle exit plane in calibers from the missile base (positive aft) and

$$\text{RMF} = \frac{\gamma_j P_j A_j M_j^2}{\gamma_\infty P_\infty A_{ref} M_\infty{}^2}$$

The function f(RMF) of Eq. (71) is defined as follows:

$$\begin{aligned} f(\text{RMF}) &= 0.19 + 1.28\left(\frac{\text{RMF}}{1+\text{RMF}}\right) &&; \ \text{RMF} \leq 1.46 \\ f(\text{RMF}) &= 0.76 + 0.58\log(\text{RMF}) &&; \ 1.46 < \text{RMF} \leq 5.5 \\ f(\text{RMF}) &= 1.0 + (\text{RMF})^{0.33} &&; \ \text{RMF} > 5.5 \end{aligned} \tag{72}$$

The power on base pressure coefficient is then

$$\left(C_{P_B}\right)_{\text{on}} \frac{2}{\gamma M_\infty^2}(p_b/p_\infty - 1) \tag{73}$$

where p_b/p_∞ is defined by Eqs. (71) and (72) as a function of the various parameters of the engine.

The ΔC_{P_B} of Fig. 4.27c is then

$$\left(\Delta C_{P_B}\right)_{\text{eng}} = \left(C_{P_B}\right)_{\text{off}} - \left(C_{P_B}\right)_{\text{on}} \tag{74}$$

The total base pressure coefficient is then

$$C_{P_B} = \left(C_{P_B}\right)_{a,\delta,t/c,x/c} + \left(\Delta C_{P_B}\right)_{\text{eng}} \tag{74a}$$

Finally, if the configuration has a boattail, then the axial force is

$$C_{A_B} = -C_{P_B}\left(\text{d}_B/\text{d}_{\text{ref}}\right)^3 \tag{75}$$

If it has a flare, the axial force is

$$C_{A_B} = -C_{P_B}\left(\text{d}_B\text{d}_{\text{ref}}\right)^2 \tag{76}$$

Note that the cube of $\text{d}_B/\text{d}_{\text{ref}}$ of Eq. (75) comes from Stoney.[54]

IX. Configuration Aerodynamics at Low Angle of Attack

With the linearized theoretical methods discussed in Chapter 3, the local slope methods of Chapter 4, and the empirical methods of Section 4.8, we are now in a position to calculate total weapon configuration aerodynamics for low AOA. Several cases will be considered to investigate the accuracy of the methods considered up through Section 4.9 and as a lead-in to the nonlinear aerodynamic methods discussion of Chapter 5.

The first case considered is the spin-stabilized M33 projectile, which is shown in Fig. 4.33a and is taken from Ref. 55. Test data and parabolized Navier–Stokes (PNS) calculations were available from Ref. 55. Test data were for M33 and M8 rounds, which were basically the same aerodynamic shape. The M33 round has a fairly large nose bluntness of 36% and a boattail with a fairly large angle of 9 deg. Its total length is 4.46 calibers. Because data were all from sea level ballistics range tests, the boundary-layer option of "flight model" was used in computations for axial force. The zero yaw drag coefficient is compared to data and PNS in Fig. 4.33b. Excellent agreement with the test data is obtained. Due to the fact that PNS is a marching solution, the blunt nose drag and base drag were computed separately and added to the main computational fluid dynamics solution, which could have contributed to some of the error for the PNS calculation. The pitching moment coefficient derivative, center of pressure, and normal force coefficient derivative are shown in Figs. 4.33c, d, and e, respectively.

Reasonable agreement with experiment is obtained, particularly for pitching moment. The center of pressure is predicted to be slightly farther forward at low supersonic and transonic Mach numbers than data suggest, primarily due to the underprediction of normal force coefficient derivative, as shown in Fig. 4.33e. Figures 4.33f, g, and h present the dynamic derivative comparisons of theory to experiment. No experimental data were available for roll damping. Also note that the values of the dynamic derivatives taken from Ref. 55 were multiplied by 2 to convert from Army convention of d/V_∞ to Navy convention of using $d/2V_\infty$ in the nondimensionalization process. Notice the large scatter in data for both the magnus moment coefficient and

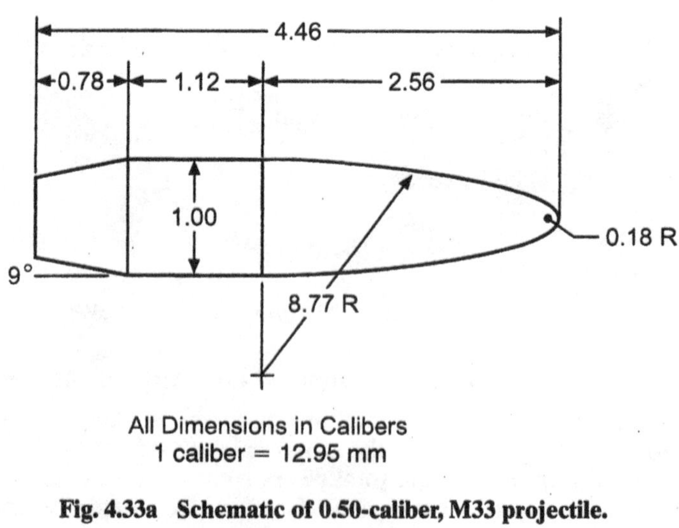

Fig. 4.33a Schematic of 0.50-caliber, M33 projectile.

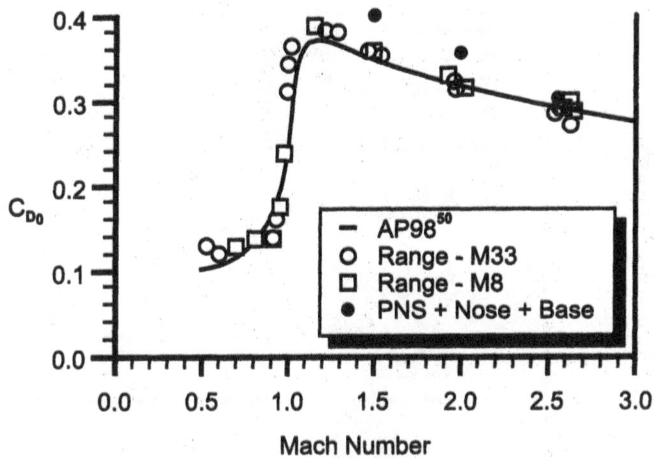

Fig. 4.33b Computed and measured zero-yaw drag coefficient versus Mach number.

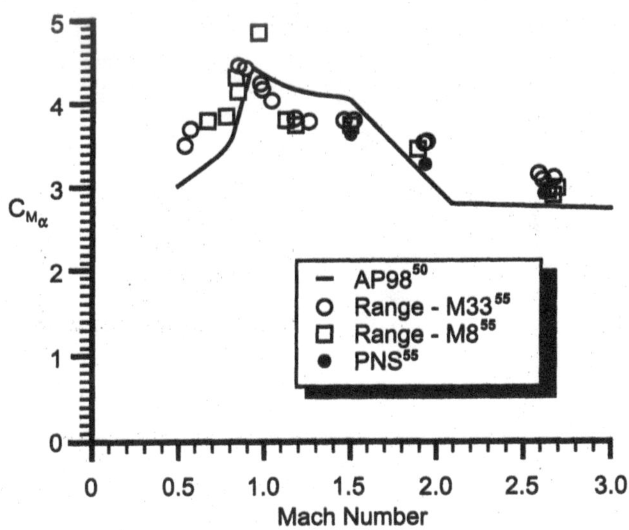

Fig. 4.33c Computed and measured pitching moment coefficient versus Mach number.

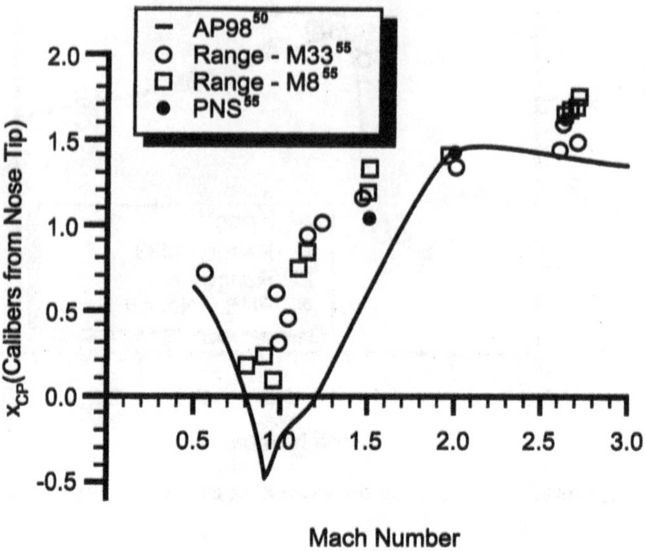

Fig. 4.33d Computed and measured normal force center of pressure versus Mach number.

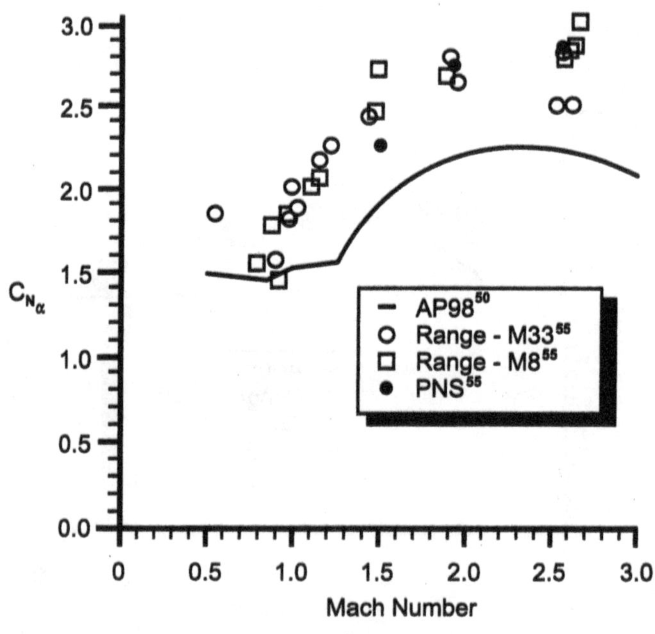

Fig. 4.33e Computed and measured normal force coefficient derivative versus Mach number.

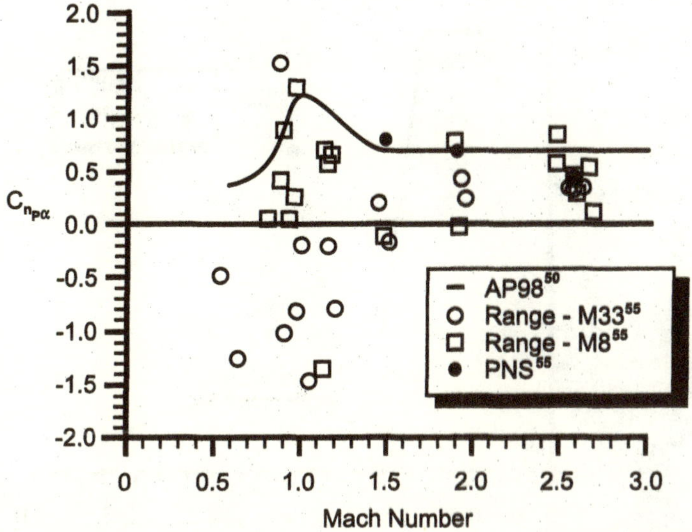

Fig. 4.33f Computed and measured magnus moment coefficient versus Mach number.

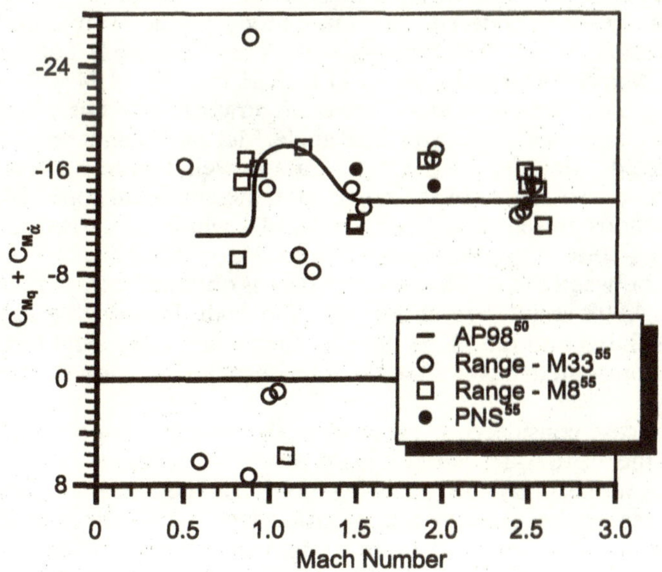

Fig. 4.33g Computed and measured pitch damping moment coefficient versus Mach number.

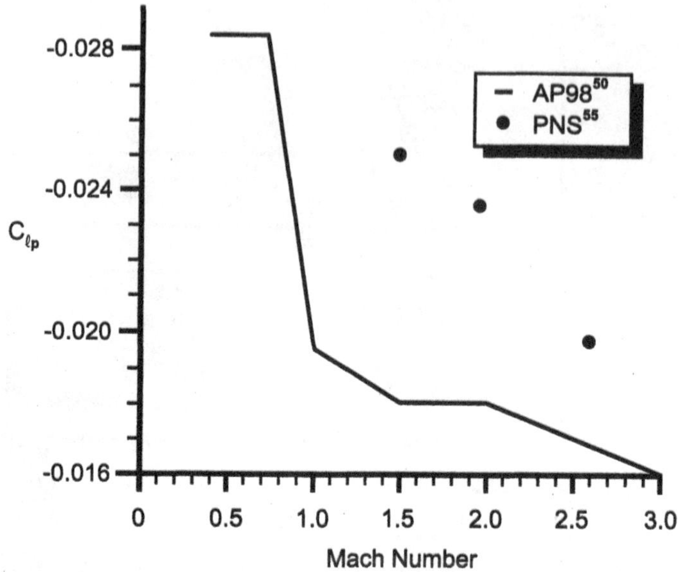

Fig. 4.33h Computed roll damping coefficient versus Mach number.

the pitch damping moment. Reasonable agreement between the empirical theory and experiment is obtained, considering the scatter in the data.

The second case considered is a 55-mm model of the Minuteman missile that was tested in the ballistic range at Army Research Laboratory in Aberdeen, Maryland, and described in Ref. 56 (see Fig. 4.34a). The model consists of a blunt cone–cylinder–flare configuration, with fairly high values of bluntness, cone half-angle, and flare angle. McCoy[56] compared an empirical code called "McDrag" to the range data as well. Figure 4.34b compares the axial force coefficients of the range data, the empirical code "McDrag", and the methods presented in Chapters 3 and 4, which are shown as AP98.[50] The McDrag code is applicable only up to $M_\infty = 4.0$. Note that excellent agreement between the AP98 and range data is obtained except at $M_\infty = 1.2$ where the AP98 results are about 10% too high. Reasonable agreement between McDrag and the range data is also shown. Only axial force information was presented in Ref. 56 so no other aerodynamics are shown for this case.

The third case considered is taken from the missile component database known as the NASA-Tri Service Data Base.[57] Configurations tested included body alone, body–tail, and wing alone. The body tested was a 12.33-caliber tangent–ogive cylinder configuration with a 3.0-caliber nose. Wings include aspect ratios from 0.25 up to 4.0 and taper ratios of 0 to 1.0, with $\lambda = 0.5$ having all aspect ratios in the test matrix. Mach numbers varied from 0.6 to 4.6, AOA from 0 to as high as 40 deg (depending on Mach number), roll angles 0 to 180 deg, and control deflections with some fins. Figure 4.35a

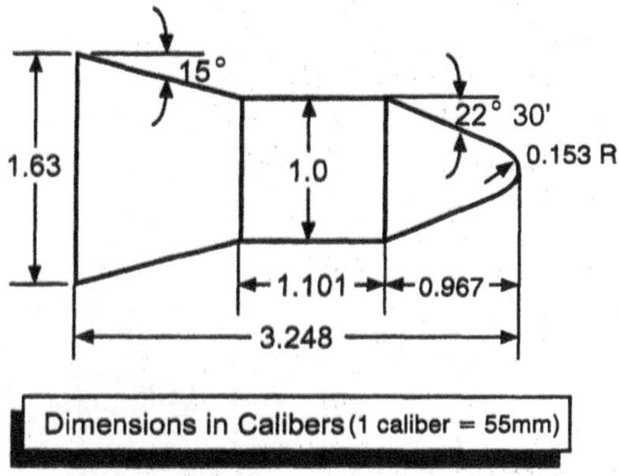

Fig. 4.34a 55-mm Minuteman model (Ref. 56).

shows the models tested for the wing–body test series[57] and Fig. 4.35b shows the wing-alone configurations tested.[58]

Two cases out of the many shown in Fig. 4.35a are considered. These are aspect ratio 0.5 and 1.0, both with a taper ratio of 0.5. The aspect ratio 0.5 case has a wing area to reference area of 2.54 and the leading edge is located 7.67 calibers from the nose tip. The aspect ratio 1.0 case has a wing area to reference area of 1.27 and the leading edge is located 9.86 calibers from the nose tip. Figures 4.36a and 4.36b present the normal force coefficient derivative and center of pressure as a function of Mach number for the AR = 0.5 and 1.0 cases, respectively. The combination of linear-theory, local-slope methods in conjunction with empirical methods gives excellent agreement compared to experimental data. Average errors are much less than 10% for C_{N_α} and less than 2% of the body length for center of pressure.

The last case considered is a canard–body–tail configuration shown in Fig.

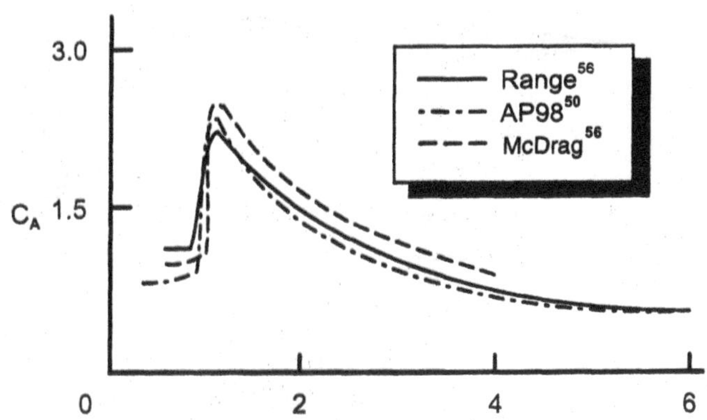

Fig. 4.34b Axial force coefficient versus Mach number for Minuteman model.

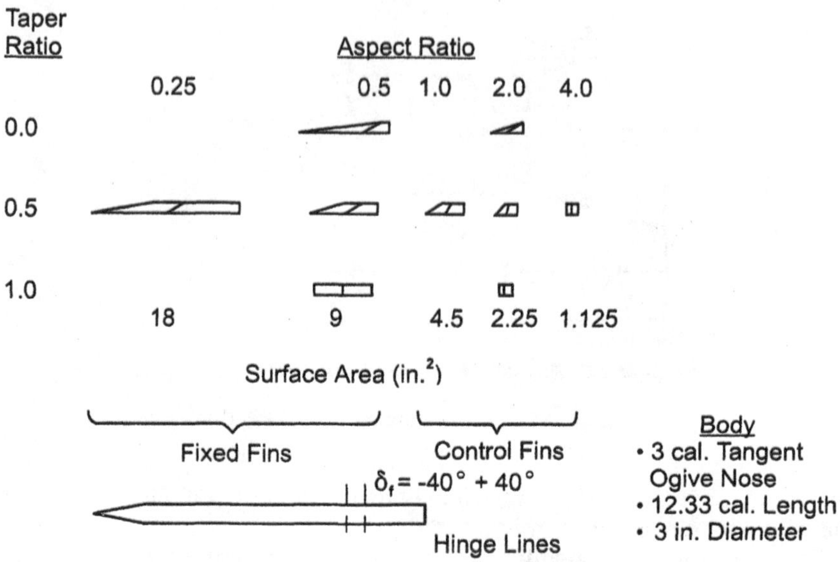

Fig. 4.35a Models used in NASA[57] wing–body tests.

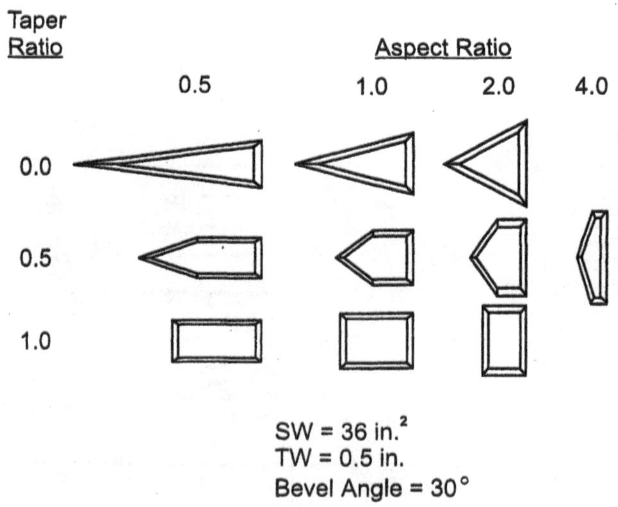

Fig. 4.35b Models used in Stallings and Lamb[58] wing-alone tests.

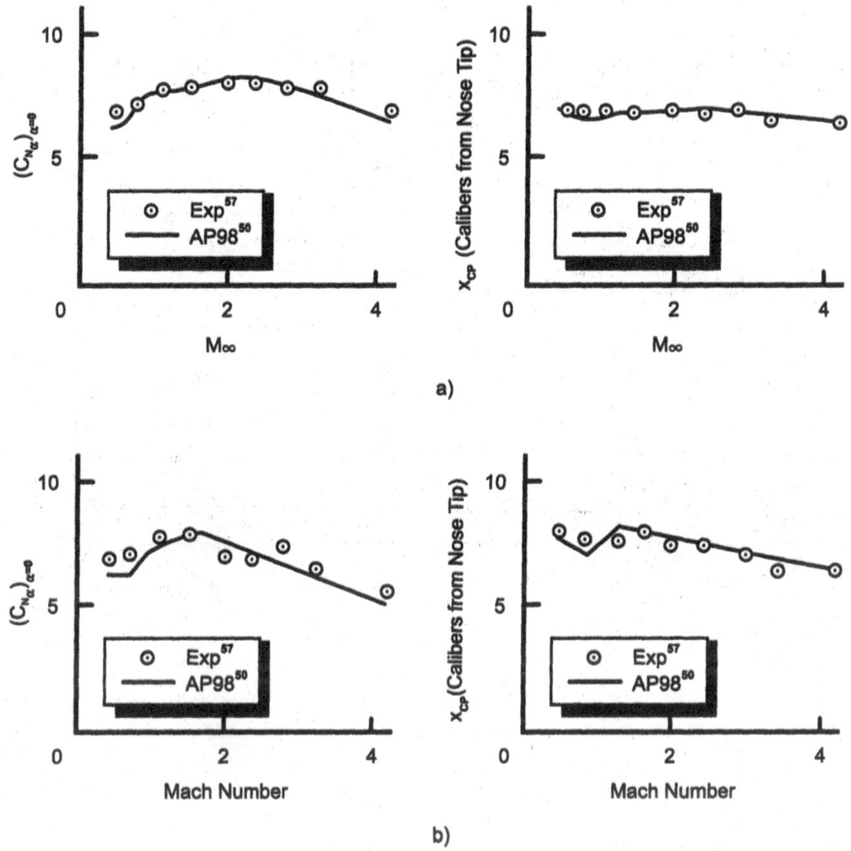

Fig. 4.36 Comparison of theory and experiment for normal force coefficient derivative and center of pressure near $\alpha = 0°$ for Fig. 4.35a configuration. a) AR = 0.5 λ = 0.5, X_{LE} = 7.67 caliber, b) AR = 1.0, λ = 0.5, X_{LE} = 9.86 caliber

4.37 that has a 100% blunt nose and is 22.2 calibers long. The canards have sharp leading and trailing edges and are 0.098 in. thick. The tail surfaces have sharp leading edges and truncated trailing edges. The tail thickness at the root chord and tip chord is 0.091 in. Experimental data[59] were taken over a range of Mach numbers and AOAs up to 20 deg. Shown at the top of Fig. 4.37 is the axial force coefficient and normal force coefficient derivative near zero AOA as a function of Mach number. As seen in the figure, excellent agreement with data on axial force is obtained, even for a configuration as complex as that in Fig. 4.37. However, at low AOA $(C_{N_\alpha})_{\alpha=0}$ is predicted to be too high at Mach numbers below about 3.0. It is possible this overprediction is due to a combination of fin thickness as well as trying to obtain data points from a large-scale chart from Ref. 59 for very small AOAs. It is estimated that this process could contribute as much as 5% to the error between data and

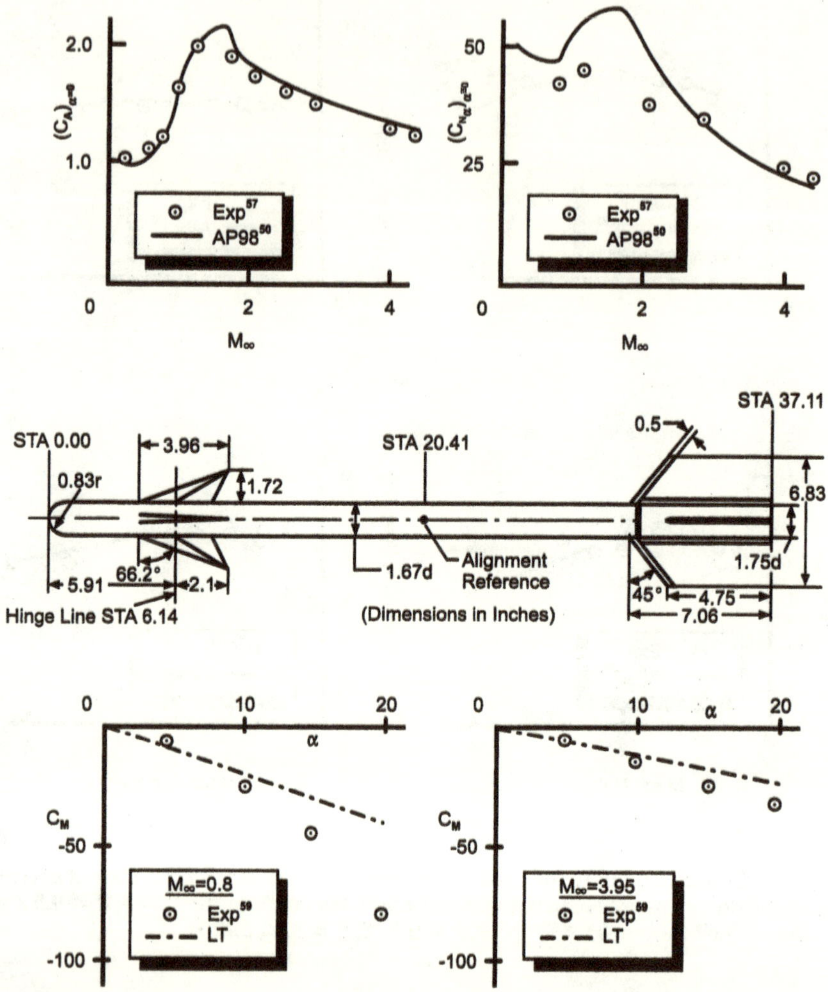

Fig. 4.37 Aerodynamics of a canard–body–tail configuration with a hemispherical nose.

theory. At the bottom of Fig. 4.37 is shown the pitching moment coefficient as a function of AOA for $M_\infty = 0.8$ and 3.95. The pitching moment is about the alignment reference point of 20.41 in. from the nose tip. Notice the comparisons of linear theory methods to data. For AOAs to about 5 to 10 deg, linear theory gives quite acceptable results at $M_\infty = 0.8$ and could actually be used all the way to 20 deg at $M_\infty = 3.95$. However, this is a fortuitous example and, in general, nonlinearities such as shown for the $M_\infty = 0.8$, C_M versus α case are more normal. As a result, nonlinear aerodynamics must be combined with linear theory or local slope methods to achieve accurate aerodynamics at AOAs above about 10 deg. There are configurations at certain flight conditions where the AOA can be extended higher than 10 deg for linear theory applications. However, these situations are rare. The next chapter will discuss the physics of the nonlinear aerodynamics followed by empirical and semi-empirical ways to model these physics.

References

[1]Ames Research Staff, "Equations, Tables, and Charts for Compressible Flow," NACA TR 1135, 1953.

[2]Taylor, G. I., and MacCall, J. W., "The Air Pressure on a Cone Moving at High Speed," *Proceedings of the Royal Society (London),* Ser. A., Vol. 139, 1933, pp. 278–311.

[3]Moore, F. G., Armistead, M. A., Rowles, S. H., and DeJarnette, F. R., "Second-Order Shock-Expansion Theory Extended to Include Real Gas Effects," Naval Surface Warfare Center Dahlgren Div. Rept. NAVSWC TR 90-683, Feb. 1992.

[4]Jones, D. J., "Numerical Solutions of the Flow Field for Conical Bodies in a Supersonic Stream," National Research Council Canada Rep. LR-507, July 1968.

[5]Eggers, A. J., Syvertson, C. A., and Kraus, S., "A Study of Inviscid Flow About Airfoils at High Supersonic Speeds," NACA Rep. 1123, 1953.

[6]Eggers, A. J., and Savin, R. C., "A Unified Two-Dimensional Approach to the Calculation of Three-Dimensional Hypersonic Flows, with Application to Bodies of Revolution," NACA Rep. 1249, 1995.

[7]Eggers, A. J., and Savin, R. C., "Approximate Methods for Calculating the Flow About Nonlifting Bodies of Revolution at High Supersonic Airspeeds," NACA TR 2579, 1951.

[8]Savin, R. C., "Application of the Generalized Shock Expansion Method to Inclined Bodies of Revolution Traveling at High Supersonic Airspeeds," NACA TN 3349, 1995.

[9]Liepmann, H. W., and Roshko, A., *Elements of Gas Dynamics,* Wiley New York, 1957.

[10]Syvertson, C. A., and Dennis, D. H., "A Second-Order Shock Expansion Method Applicable to Bodies of Revolution Near Zero Lift," NACA TR 1323, 1957.

[11]DeJarnette, F. R., Ford C. P., and Young, D. E., "Calculation of Pressure on Bodies at Low Angles of Attack in Supersonic Flow," *Journal of Spacecraft and Rockets* No. 79-1552, Vol. 17, No. 6, Nov–Dec. 1980.

[12]Savin, R. C., "Application of the Generalized Shock Expansion Method to Inclined Bodies of Revolution Traveling at High Supersonic Airspeeds," NACA TN 3349, April 1955.

[13]"Handbook of Supersonic Aerodynamics, Section 8—Bodies of Revolution," Defense Technical Information Center Naval Weapons Rep. 1488, Vol. 3, Fort Belvoir, VA, Oct. 1961, pp. 96–97.

[14]Lees, L., *Hypersonic Flow,* Institute Aeronautical Science, Preprint No. 554, 1955.

[15]Morrison, A. M., Solomon, J. M., Ciment, M., and Fergerson, R. E., "Handbook of Inviscid Sphere Cone Flowfields and Pressure Distributions," Vol. 1, Naval Surface Warfare Center, White Oak Lab. Rept. NSWC/WOL TR 75-45, Dec. 1975.

[16]Moore, F. G., "Body Alone Aerodynamics of Guided and Unguided Projectiles

at Subsonic, Transonic, and Supersonic Mach Numbers," Naval Weapons Laboratory TR-2976, Nov. 1972.

[17]Science Application, "Theoretical Analysis of the Flow Field Over a Family of Ogive Bodies," U.S. Army Missile Research Development TD-CR-77-5 Redstone Arsenal, AL, Aug. 1977.

[18]Jacksons, C. M., Sawyer, W. C., and Smith, R. S., "A Method for Determining Surface Pressures on Blunt Bodies of Revolution at Small Angles of Attack in Supersonic Flow," NASA TN P-4865, 1968.

[19]Owens, R. V., "Aerodynamic Characteristics of Spherically Blunted Cones at Mach Numbers 0.5 to 5.0," NASA TN D-3088, Dec. 1965.

[20]Butler, C. B., Sears, E. S., and Pallas, S. G., "Aerodynamic Characteristics of 2-, 3- and 4-Caliber Tangent Ogive Cylinders with Nose Bluntness Ratios of 0.00, 0.25, 0.50, and 0.75 at Mach Numbers from 0.6 to 4.0," Air Force Armament Test Lab. Rept. AFATL-TR-77-8, *Eglin Air Force Base, FL* Jan. 1977.

[21]Van Driest, E. R., "Turbulent Boundary Layers in Compressible Fluids," *Journal of Aeronautical Sciences,* Vol. 18, No. 3, 1951, pp. 145–160, 216.

[22]Truitt, R. W., *Fundamentals of Aerodynamic Heating,* Ronald, New York, 1960.

[23]Carahan, B., Luther, H. A., and Wilkes, J. O., *Applied Numerical Method,* Wiley New York, 1969.

[24]Moore, F. G., McInville, R. M., and Hymer, T. C., "The 1995 Version of the NSWC Aeroprediction Code: Part 1—Summary of New Theoretical Methodology," Naval Surface Warfare Center Dahlgren Div., Rept. NSWCDD/TR-94/379, Feb. 1995.

[25]Moore, F. G., Hymer, T. C., and McInville, R. M.,"Improved Aeroprediction Code: Part 1—Summary of New Methods and Comparison with Experiment," Rept. NSWCDD/TR-93/91, May 1993.

[26]Chaussee, D., "Improved Transonic Nose Drag Estimates for the NSWC Missile Computer Program," Naval Surface Warfare Center, Dahlgren Lab. Rept. NSWCDL TR 3030, Dahlgren, VA, April 1978.

[27] Moore, F. G., "Aerodynamics of Guided and Unguided Weapons: Part 1—Theory and Application," Naval Surface Warfare Center, Dahlgren Div., NWL TR-3018, Dahlgren, VA, Dec. 1973.

[28] Moore, F. G., and Swanson, R. C., "Aerodynamics of Tactical Weapons to Mach Number 3 and Angle of Attack 15 Degrees: Part 1—Theory and Application," NSWCDL TR-3584, Dahlgren, VA., Feb. 1977.

[29] Devan, L., "Aerodynamics of Tactical Weapons to Mach Number 8 and Angle of Attack 180°: Part 1, Theory and Application," Naval Surface Warfare Center, Dahlgren Div., Rept. NSWC TR 80-346, Dahlgren, VA, Oct. 1980.

[30] Wu, J. M., and Aoyoma, M., "Transonic Flow-Field Calculation Around Ogive Cylinders by Nonlinear-Linear Stretching Method," U.S. Army Missile Command TR No. RD-TR-70-12, April 1970.

[31] Owens, R. V., "Aerodynamic Characteristics of Spherically Blunted Cones at Mach Numbers from 0.5 to 5.0," NASA TN D-3088, Dec. 1965.

[32] Moore, F. G., "A Study to Optimize the Aeroballistic Design of Naval Projectiles," Naval Surface Warfare Center, Dahlgren Div., NWL TR-2337, Dahlgren, VA, Sept. 1969.

[33] Charters, A. C., "Some Aeroballistic Contributions to Aerodynamics," *14th Annual Meeting Inst. Aeronautical Sciences,* New York, Jan. 1946.

[34]Washington, W. D., and Pettis, W., Jr., "Boattail Effects on Static Stability at Small Angles of Attack," U.S. Army Missile Command Rep. No. RD-TM-68-5, 1968.

[35]Whyte, R. H., "Effects of Boattail Angle on Aerodynamic Characteristics of 175mm M437 Projectile at Supersonic Mach Numbers," U.S. Army Munitions Command TM 1646, Sept. 1965.

[36]Spring, D. J., "The Effect of Nose Shape and Afterbody Length on the Normal Force and Neutral Point Location of Axisymmetric Bodies at Mach Numbers from 0.80 to 4.50," U.S. Army Missile Command Rep. No. RF-TR-64-13, 1964.

[37]Gwin, H., and Spring, D. J., "Stability Characteristics of a Family of Tangent Ogive–Cylinder Bodies at Mach Numbers 0.2 to 1.5," U.S. Army Missile Command Rept. No. RG-TR-61-1, 1961.

[38]Douglas Aircraft Co., "USAF Stability and Control DATCOM," rev. by Wright Patterson Air Force Base, July 1963, 2 Vols.

[39]Lowry, J. G., and Polhamus, E. C., "A Method for Predicting Lift Increments Due to Flap Deflection at Low Angles of Attack in Incompressible Flow," NACA TN-3911, 1957.

[40]Chapman, D. R., "An Analysis of Base Pressure at Supersonic Velocities and Comparison with Experiment," NACA Rep. 1051, 1951.

[41]Korst, H. H., "A Theory for Base Pressure in Transonic and Supersonic Flow," *Journal of Applied Mechanics,* Dec. 1956, pp. 593–600.

[42]Sahu, J., Nietubicz, C. J., and Steger, J. L., "Navier–Stokes Computations of Projectile Base Flow With and Without Base Injection," U.S. Army Ballistics Research Lab. Rept. ARBRL-TR-02532, Aberdeen Proving Ground, MD, Nov. 1983 (AD A125738).

[43]Sahu, J., "Computations of Supersonic Flow Over a Missile Afterbody Containing an Exhaust Jet," *Journal of Spacecraft and Rockets,* Vol. 24, No. 5, Sept.–Oct. 1987, pp. 403–410.

[44]Sahu, J., "Three Dimensional Base Flow Calculation for a Projectile at Transonic Velocity," U.S. Army Ballistics Research Lab. Rept. BRL-MR-3610, Aberdeen Proving Ground, MD, Sept. 1987.

[45]Sahu, J., Nietubicz, C. J., and Heavey, K. R., "Computational Study for the M825 Projectile With Standard and Dome Bases," U.S. Army Ballistics Research Lab. Rept. BRL-MR-3662, Aberdeen Proving Ground, MD, March 1988.

[46]Love, R. S., "Base Pressure at Supersonic Speeds on Two-Dimensional Airfoils and on Bodies of Revolution With and Without Turbulent Boundary Layers," NACA TN 3819, 1957.

[47]Kurzweg, H. H., "New Experimental Investigations on Base Pressure in the NOL Supersonic Wind Tunnels at Mach Numbers 1.2 to 4.24," Naval Ordnance Lab. Memo 10113, Naval Surface Warfare Center, Dahlgren, Div., Dahlgren, VA 1950.

[48]Lamb, J. P., and Oberkampf, W. L., "A Review and Development of Correlations for Base Pressure and Base Heating in Supersonic Flow," Sandia National Lab., Rept. SAND 93-0280.UC-706, Albuquerque, NM, Nov. 1993.

[49]Moore, F. G., Wilcox, F. J., Jr., and Hymer, T., "Improved Empirical Model for Base Drag Prediction on Missile Configurations Based on New Wind Tunnel Data," NSWCDD/TR-92/509, Oct. 1992.

[50]Moore, F. G., McInville, R. M., and Hymer, T. C., "The 1998 Version of the NSWC

Aeroprediction Code: Part 1—Summary of New Theoretical Methodology," NSWCDD/TR-98/1, April 1998.

[51]Brazzel, C. E., and Henderson, J. H., "An Empirical Technique for Estimating Power-On Base Drag of Bodies-of-Revolution With a Single Jet Exhaust," *Proceedings of a Specialists' Meeting,* sponsored by AGARD Fluid Dynamics panel, Mulhouse, France, Sept. 1966.

[52]Spahr, J. R., and Dickey, R. R., "Effect of Tail Surfaces on the Base Drag of a Body of Revolution at Mach Numbers of 1.5 and 2.0," NACA TN-2360, 1951.

[53]Hill, F. K., and Alpher, R. A., "Base Pressure at Supersonic Velocities," Johns Hopkins Univ., Bumblebee Rept. 106, Nov. 1949.

[54]Stoney, W. E., Jr., "Collection of Zero-Lift Drag Data on Bodies of Revolution from Free-Flight Investigations," NASA TR R-100, 1961.

[55]Guidos, B. J., and Chung, S. K., "Computational Flight Design of 0.50 Caliber Limited Range Training Ammunition," U.S. Armament Research Development Command Rept. ARL-TR-662, Aberdeen Proving Ground, MD, Jan. 1995.

[56]McCoy, R. L., " 'McDrag'—A Computer Program for Estimating the Drag Coefficients of Projectiles," U.S. Armament Research Development Command Rept. ARBRL-TR-02293, Aberdeen Proving Ground, MD, Feb. 1981.

[57]NASA Langley Research Center Tri-service missile data base, transmitted from Jerry M. Allen to Naval Surface Warfare Center, Dahlgren Div., Nov. 1991 (Submitted for publication).

[58]Stallings, R. L., Jr., and Lamb, M., "Wing-Alone Aerodynamic Characteristics for High Angles of Attack at Supersonic Speeds," NASA TP 1889, July 1981.

[59]Graves, E. B., and Fournier, R. H., "Stability and Control Characteristics at Mach Numbers from 0.2 to 4.63 of a Cruciform Air-to-Air Missile With Triangular Canard Controls and a Trapezoidal Wing," NASA TM-X-3070, Nov. 1974.

Chapter 5

Nonlinear Aerodynamic Approximations

Nomenclature

AP98	= 1998 version of aeroprediction code, which contains the linear and nonlinear aerodynamics of Chapters 3 to 5
AR	= aspect ratio = b^2/A_W
LT	= linear theory
MNT	= modified Newtonian theory
SB, SBT	= slender body, slender-body theory
A_P	= two-dimensional planform area of the body in the crossflow plane (ft²) (for example, for a circular-cylinder of length ℓ, and diameter d, $A_P = \ell d$)
A_{ref}	= reference area (maximum cross-sectional area of body, if a body is present, or planform area of wing, if wing alone)(ft²)
A_W	= planform area of wing in crossflow plane (ft²)
b	= wing span (not including body)(ft)
C_A	= axial force coefficient
$C_{A_B}, C_{A_F}, C_{A_P}$	= base, skin-friction, and wave components, respectively, of axial force coefficient
C_{A_O}	= axial force coefficient at 0 deg AOA
C_{A_α}	= axial force coefficient term due to AOA
C_{A_δ}	= axial force coefficient term due to control deflection
C_{d_c}	= crossflow drag coefficient
C_M	= pitching moment coefficient (based on reference area and body diameter, if body present, or mean aerodynamic chord, if wing alone)
C_{M_L}	= linear component of pitching moment coefficient
$C_{M_{NL}}$	= nonlinear component of pitching moment coefficient
C_N	= normal-force coefficient
C_{N_B}	= normal-force coefficient of body alone
$C_{N_{B(V)}}$	= negative afterbody normal-force coefficient due to canard or wing-shed vortices
$C_{N_{B(W)}}, C_{N_{B(T)}}$	= normal-force coefficient of body, including interference effects from wing or tail

$\Delta C_{N_{B(W)}}, \Delta C_{N_{B(T)}}$ = additional normal-force coefficient on body as a result of the presence of wing or tail
C_{N_L} = linear component of normal-force coefficient
$C_{N_{NL}}$ = nonlinear component of normal-force coefficient
$C_{N_a}, (C_{N_a})_W, (C_{N_a})_T$ = normal-force coefficient slope of configuration, wing, or tail, respectively
$C_{N_{T(V)}}$ = negative normal-force coefficient component on tail due to wing or canard-shed vortex
C_{N_W} = normal-force coefficient of wing alone
$C_{N_{W(B)}}, C_{N_{T(B)}}$ = normal-force coefficient of wing or tail in presence of body
C_P = pressure coefficient $\left(\dfrac{p - p_\infty}{1/2\rho_\infty V_\infty^2}\right)$
c_r = root chord (ft)
c_t = tip chord (ft)
cal = caliber(s) (one body diameter)
d_B = body diameter (ft) at base
d_{ref} = reference body diameter (ft)
$\dfrac{\mathrm{d}K_{W(B)}}{\mathrm{d}a}, \dfrac{\mathrm{d}K_{B(W)}}{\mathrm{d}a}$ = rate at which $K_{W(B)}$ or $K_{B(W)}$ decreases
F, C_1, C_2, C_3 = dimensionless empirical factors used in nonlinear models of $k_{W(B)}$ and $C_{N_{T(V)}}$ to approximate effects due to high AOA or control deflection
f, h = lateral and vertical position of wing vortex
f_W, f_T = lateral location of wing or tail vortex (measured in feet from body center line)
i, i_1, i_4 = tail interference factor of a single fin, the windward plane, or leeward plane fin, respectively
$K_{B(W)}, K_{B(T)}$ = ratio of additional body normal-force coefficient in presence of wing, or tail to wing, or tail-alone normal-force coefficient at $\delta = 0$ deg
$k_{B(W)}, k_{B(T)}$ = ratio of additional body normal-force coefficient due to presence of wing or tail at a control deflection to that of wing or tail alone at $a = 0$ deg
$[K_{B(W)}]_{\min}$ = minimum value of $K_{B(W)}$ as percent of slender-body theory value
$K_{W(B)}, K_{T(B)}$ = ratio of normal-force coefficient of wing or tail in presence of body to that of wing or tail alone at $\delta = 0$ deg
$k_{W(B)}, k_{T(B)}$ = ratio of wing or tail normal-force coefficient in presence of body due to a control deflection to that of wing or tail alone at $a = 0$ deg
$\left.\begin{matrix}[\Delta K_{W(B)}]_{a=0} \\ [\Delta K_{B(W)}]_{a=0}\end{matrix}\right\}$ = amount that the experimental values of $K_{W(B)}$ and $K_{B(W)}$ exceed slender body theory at $a = 0$ deg

$\ell, \ell_n, \ell_a, \ell_b$	=	total length of body, length of nose, afterbody, and boattail, respectively
ℓ_{ref}	=	reference length (ft)
M_N	=	Mach number normal to body = $M_\infty \sin a$
M_{NC}	=	normal Mach number where flow transitions from subcritical to supercritical conditions
M_∞, M_L	=	freestream and local Mach number, respectively
p, p_L, p_{O_2}	=	pressure (lb/ft^2), local pressure, and total pressure behind shock, respectively
P_ℓ, P_W	=	loading factors in leeward and windward planes, respectively
r	=	local body radius (ft)
R_{NC}	=	crossflow Reynolds number where flow transitions from subcritical to supercritical conditions
R_{N_D}	=	Reynolds number based on body diameter
$R_{N_{\text{eff}}}$	=	an effective Reynolds number above which the flow transitions from subcritical to supercritical conditions
R_{N_N}	=	Reynolds number based on velocity normal to body
r_W, r_T	=	radius of body at wing or tail locations
s	=	wing or tail semispan plus the body radius in wing–body lift methodology
V_∞	=	freestream velocity
V_A, V_N	=	velocity along axis and normal to body, respectively
x_{CP}	=	center of pressure (in feet or calibers from some reference point that can be specified) in x direction
$(x_{\text{CP}})_L, (x_{\text{CP}})_{NL}$	=	center of pressure of linear and nonlinear terms of normal force
x, y, z	=	axis system fixed with x along centerline of body
y_{CP}	=	center of pressure of wing normal force in direction of wing span
a	=	AOA (deg)
a_C	=	AOA where wing–body interference factor starts decreasing (deg)
a_D	=	AOA where the wing–body interference factor reaches a minimum (deg)
a_M	=	AOA where $K_{W(B)}$ reaches a constant value
a_N	=	value of a where $C_{N_{T(V)}}$ goes to zero
a_{N_O}	=	value of a where $C_{N_{T(V)}}$ goes to zero for $A_W/A_{\text{ref}} = 5.5$
a_W, a_T	=	local AOA of wing or tail ($a_W + \delta$ or $a_T + \delta$, respectively, in degrees)
a_1, a_2	=	AOAs used in nonlinear model for $K_{B(W)}$
δ	=	control deflection (deg), positive leading edge up
δ_{eq}	=	angle between velocity vector and tangent to body surface

δ_W, δ_T	=	deflection of wing or tail surfaces (deg), positive leading edge up
η	=	parameter used in viscous crossflow theory for nonlinear body normal force (in this context, it is the normal force ratio of a circular cylinder of given length-to-diameter ratio to that of a cylinder of infinite length)
η_0	=	value of η at $M_N = 0$
Γ	=	vortex circulation, positive counterclockwise facing upstream (ft²/s)
φ	=	roll position of missile fins ($\varphi = 0$ deg corresponds to fins in the plus (+) orientation; $\varphi = 45$ deg corresponds to fins rolled to the cross (×) orientation), or a point on body surface ($\varphi = 0$ deg is leeward plane)
λ	=	taper ratio of a lifting surface = c_t/c_r
θ	=	local surface slope of body with respect to body axis
μ_L	=	local Mach angle = $\sin^{-1}\left(\dfrac{1}{M_L}\right)$

SECTION 1.2 briefly discussed nonlinear aerodynamic regions as a function of angles of attack (AOAs) and Mach number and the physical phenomena that cause those nonlinearities. This chapter will discuss these nonlinearities in detail and derive or define empirical or semi-empirical approaches to treat these nonlinearities for each of the force and moment components. Discussion of the nonlinearities will be part of the discussion of each configuration component; that is, body alone, wing or tail alone, and interference aerodynamics between the wing, tail, and body.

I. Nonlinear Aerodynamics Phenomena

The following are some of the physical phenomena that cause aerodynamic nonlinearities:

1) Moderate to high AOA,
2) Mach number,
3) Crossflow Reynolds number for $M_\infty < 2.0$,
4) Internal shock interactions for $M_\infty > 2.0$ and $\alpha > 25$ deg,
5) Asymmetric vortices for $M_\infty < 2.0$ and $\alpha > 25$ deg,
6) Real gas effects for $M_\infty > 6.0$, and
7) Configuration geometry.

By far the two largest sources of nonlinear aerodynamics phenomena are due to AOA and Mach number. As seen in Fig. 1.6 and as evidenced by comparison to data in Chapters 3 and 4, for AOAs less than about 10 deg, the methods discussed in Chapters 3 and 4 will yield reasonable aerodynamics. As AOA increases above about 10 deg, aerodynamics become increasingly nonlinear. This is true for all the configuration components of body, wing, tail, and mutual aerodynamic interference effects between the configuration components. Increased AOA is also the major source that generates other nonlinear phenomena listed previously including internal shock interactions and asymmetric vortices.

Mach number aerodynamic nonlinearities occur at all AOAs. By including analytical methods to compute aerodynamics at low AOA, many of these nonlinearities can be resolved. However, axial force at AOA is particularly strongly dependent on Mach number. For example, at subsonic Mach numbers, flow separation that occurs in the leeward plane region of the body can move to the nose region, causing a negative axial force due to pressure. This term can more than offset the positive skin friction and base drag components, causing a total negative axial force coefficient. As Mach number increases, the dynamic pressure in the windward plane more than offsets the leeward plane separation in the nose region so a total positive axial force coefficient is obtained. Mach number is one of the main contributors to the internal shock interaction phenomena as well.

Crossflow Reynolds number plays a role primarily in the nonlinear normal force of the body alone for Mach numbers less than about 2.0. Physically what happens is that, when the crossflow Reynolds number is supercritical, the boundary layer stays attached to the body in the crossflow plane of a two-dimensional, circular-cylinder (see Fig. 5.1) past the normal separation point of the shoulder.[1] This causes the crossflow drag coefficient to drop and the nonlinear normal force term to be a strong function of crossflow Reynolds number. In Fig. 5.1b, a supercritical value of R_{N_C} causes the flow to remain attached to the circular cylinder, causing a drop in the crossflow drag coefficient. This in turn causes a drop in the nonlinear normal force term. This problem occurs all along the body alone. However, when wings or tails are placed on the body, low crossflow drag phenomena cannot occur in the vicinity of the wing or tail surfaces. The crossflow Reynolds number nonlinearity is difficult to account for in semi-empirical methods, but will be discussed in Section 5.2.

Figure 5.2 illustrates the nonlinear phenomena described as internal shock interactions (here "internal" refers to shocks generated internal to the bow shock but outside the body). These interactions are from the nose bow shock intersecting a wing or tail surface, a wing or tail surface shock intersecting an adjacent wing or tail, or a shock from a forward lifting surface impacting a rear tail surface. The internal shock interactions appear to affect the normal force only slightly but can have more significant effects on the pitching moment. This is particularly true for forward shocks impacting tail surfaces, where even a small loss in normal force can have a long moment arm and hence fairly significant effects on pitching moment and static stability.

Asymmetric vortices occur at AOAs of 20 to 25 deg and higher and generally for freestream Mach numbers of 2.0 and lower. The mechanism of the asymmetric shedding is not clear, but it is suspected to be caused by small asymmetries in the body nose shape, wind tunnel turbulence levels, or

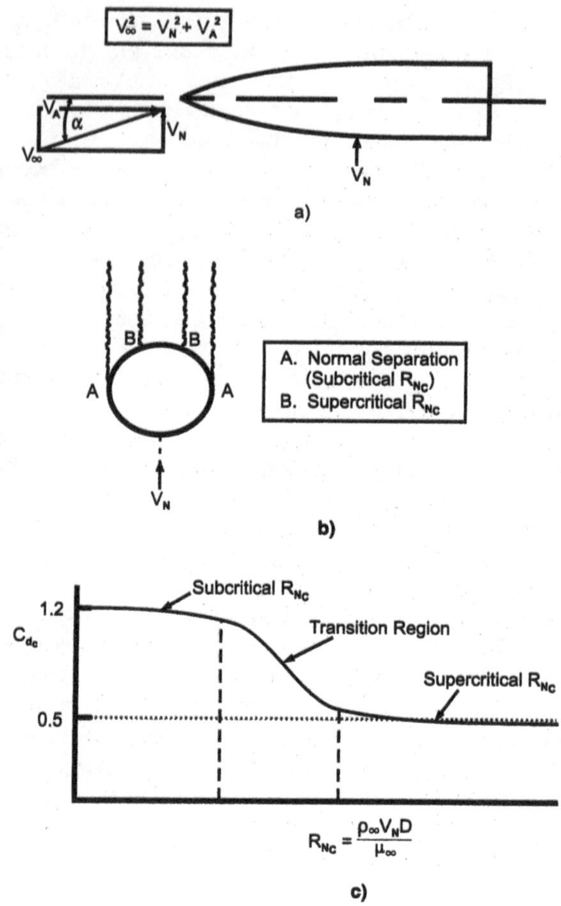

Fig. 5.1 Crossflow separation on a two-dimensional circular cylinder at AOA. a) axial and normal velocity components, b) normal and supercritical R_{N_C} separation, c) variation of crossflow drag coefficient with R_{N_C}.

body roughness effects. Blunt nose shapes and canards in the nose region tend to minimize the asymmetric vortex shedding problem. Asymmetric vortices have only about a ±10% effect on normal force but can cause side forces to be as large as normal forces.

Real gas effects occur in hypersonic aerodynamics as a result of high temperatures. These high temperatures excite internal energy modes of vibration, electronic excitation, dissociation of molecules, or ionization of atoms. These internal energy modes have only secondary effects on forces and moments but major effects on inviscid surface temperature and hence convective heat transfer. Therefore, for semi-empirical codes interested only in forces and moments and not heat transfer, this nonlinear aerodynamic effect can be neglected.

Configuration geometry is partly accounted for in the theoretical low AOA aerodynamic methods. However, configuration geometry affects non-

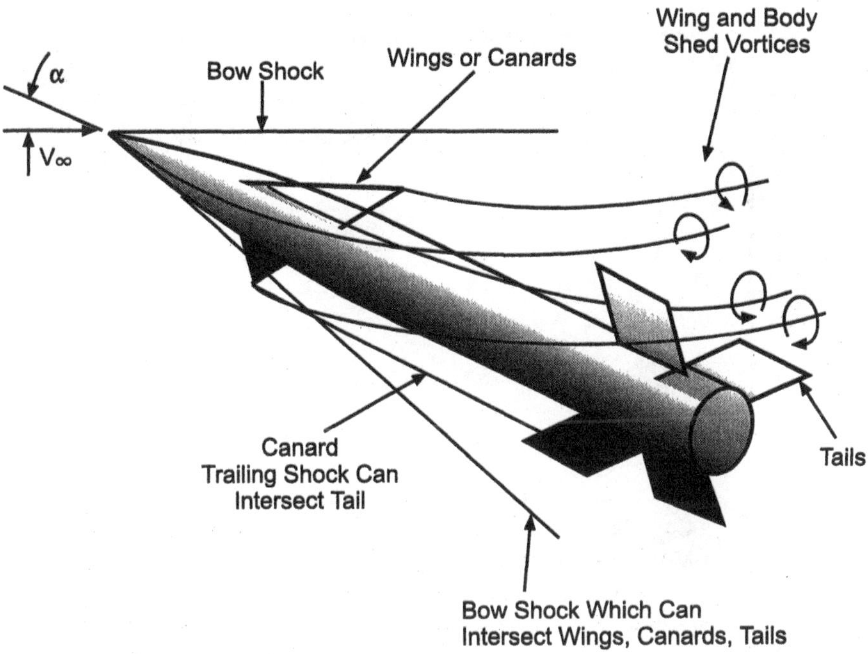

Fig. 5.2 Illustration of internal shock interactions and wing-shed vortices.

linear aerodynamic terms as well. For the body alone, the nose shape has already been mentioned as a partial cause of asymmetric vortex shedding. The planform area in the crossflow plane also directly affects the nonlinear normal force of the body alone. As far as lifting surfaces are concerned, the aspect ratio has a major effect on the nonlinear aerodynamics. High aspect ratio wings tend to have negative nonlinear terms in comparison to the linear aerodynamic term. Low aspect ratio wings, on the other hand, have just the opposite effect. Moderate aspect ratio wings are in between these two extremes. Figure 5.3 illustrates qualitatively these results for wings. The reason for these phenomena is the fact that high aspect ratio wings have a large value of linear lift and, as a result, stall occurs quicker as a function of AOA. On the other hand, low aspect wings have a very low value of linear normal force and, as AOA increases, the nonlinear normal force increases significantly as AOA increases.

Before leaving this discussion of the physical phenomena that cause aerodynamic nonlinearities, it should also be pointed out that the understanding of these phenomena is much more critical for semi-empirical codes than for the full Navier–Stokes code or even Euler codes because many of the nonlinearities come out as part of the solution in numerical codes. In semi-empirical codes, each nonlinear phenomenon must be understood and a physical model described to approximate the nonlinearity. This chapter will

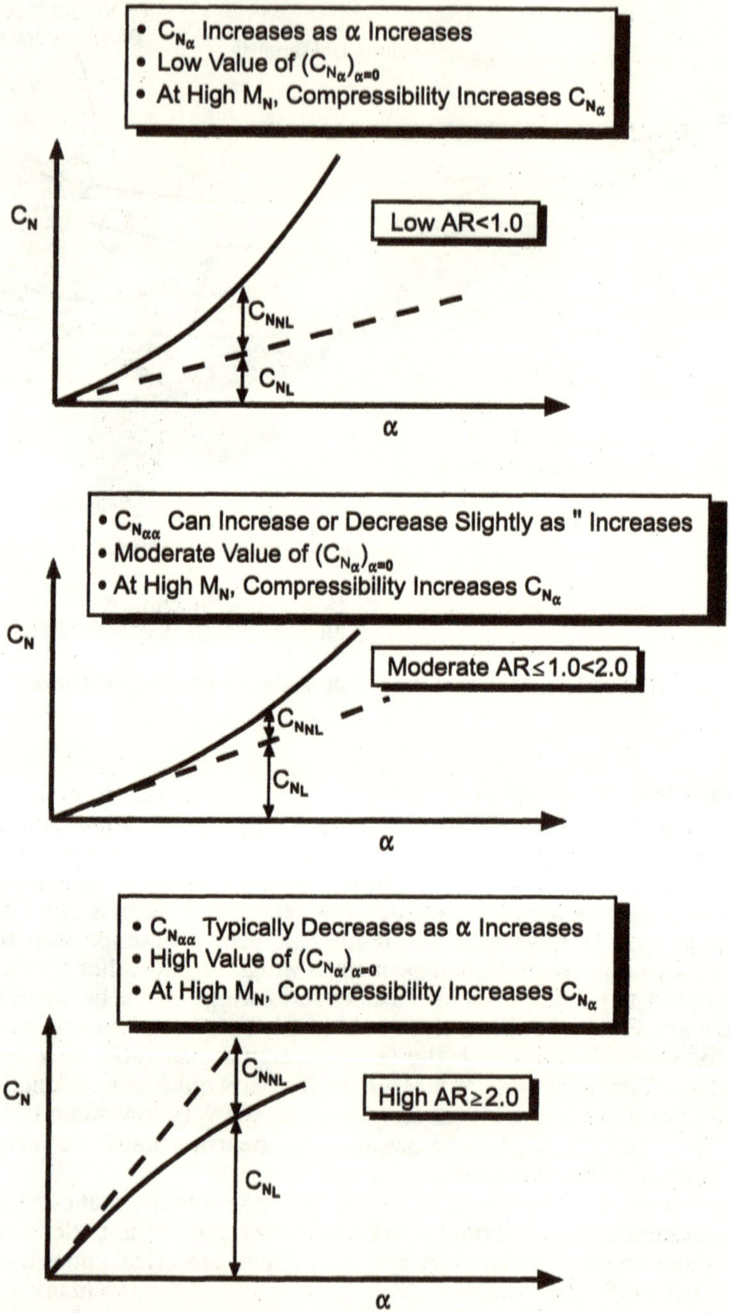

Fig. 5.3 Qualitative characteristics of normal force coefficient of wings alone.

describe some of the approximate physical models the author has developed or found to be quite useful to describe nonlinear aerodynamic terms.

Equations (1) to (4) in Chapter 3 represent the static aerodynamics of a weapon configuration that consists of a wing–body–tail configuration. Each of the terms in Eqs. (2) and (3) in Chapter 3 will be treated separately in this chapter in terms of defining a nonlinear counterpart to the low AOA linear aerodynamics described in Chapters 3 and 4.

II. Body-Alone Normal Force and Center of Pressure

Section 5.1 discussed some of the major nonlinearities that occur in weapon aerodynamics. Of those listed in Fig. 5.1, the ones that have the most influence on the body alone are AOA, Mach number, crossflow Reynolds number, and asymmetric vortices. All of these phenomena can be modeled in an approximate sense except for the asymmetric shedding of vortices. The asymmetric vortex phenomenon is the strongest in the subsonic Mach number range and starts to dissipate at crossflow Mach numbers of about 0.5. While the phenomenon is reasonably well understood, it is predictable in the general sense only in terms of upper bounds on side forces created by asymmetric vortex shedding. Because the phenomenon is not truly predictable, weapon designs that require flight at AOA > 25 deg for $M_\infty < 2.0$ (and definitely for $M_\infty < 1.5$) should design around this problem. Design alternatives that help to alleviate the problem and make vortices more symmetric include a blunt nose, strakes, or canards in the nose region, or even air blown symmetrically in the nose region. As a result of the inability to quantitatively predict side forces due to asymmetric vortex shedding, it will not be considered. The other nonlinear phenomena discussed previously will be treated.

The body term of Eq. (2) in Chapter 3 can be expanded as

$$C_{N_B} = \left(C_{N_B}\right)_L + \left(C_{N_B}\right)_{NL} \tag{1}$$

where $(C_{N_B})_L$ is calculated by the methods described in Chapters 3 and 4. The term $(C_{N_B})_{NL}$ can be calculated by a revised method of Allen and Perkins.[2]

In developing this theory, Allen reasoned that the total force of an inclined body of revolution is equal to the potential term discussed in Chapters 3 and 4, plus a crossflow term. This crossflow term is based on the drag force experienced by an element of a circular cylinder of the same diameter in a stream moving at the cross component of the stream velocity, $V_\infty \sin a$. This crossflow term is primarily created by the viscous effects of the fluid as it flows around the body, often separating and creating a nonlinear force coefficient. In equation form, the so-called viscous crossflow theory is

$$C_{N_{NL}} = \eta C_{d_c}\left(\frac{A_P}{A_{\text{ref}}}\right)\sin^2 a \tag{2}$$

Here, η is the drag proportionality factor or the ratio of crossflow drag of a cylinder of finite length to one of infinite length. C_{d_c} is the crossflow drag coefficient.

As explained in detail in a study by Ericsson and Reding[1] and one by Jorgensen,[3] C_{d_c} varies in a complex fashion for $M_C > 0.5$. Below a certain critical Reynolds number $R_{N_{crit}}$, the body boundary layer is laminar and the flow separates at about the 90- to 100-deg circumferential location (where $\varphi = 0$ deg is the leeward plane), producing a large wake region and a relatively constant value of C_{d_c} (see Fig. 5.1). At $R_{N_{crit}}$, the separated boundary-layer transitions to turbulent and reattaches, remaining attached until approximately the 40-deg circumferential location (140 deg from the windward plane). The resulting smaller wake produces a sudden decrease in the value of C_{d_c}. As the Reynolds number increases further, transition moves forward and the laminar separation bubble disappears, leading to an eventual stabilization of C_{d_c} as the turbulent separation becomes fixed at about the 80-deg location. At crossflow Mach numbers greater than 0.5, localized supersonic regions develop as the flow expands around the body. The shocks that are produced by these supersonic pockets are strong enough to separate the boundary layer regardless of its laminar or turbulent state, and the "drag bucket" disappears, leaving C_{d_c} virtually independent of the state of the boundary layer.

As pointed out in the Ericsson and Reding study,[1] the observed value of $R_{N_{crit}}$ can vary drastically from test to test and from one wind tunnel to another because of a number of factors, including tunnel turbulence, wall and model roughness, and model vibration, In fact, over an order-of-magnitude difference has been found using the same model in different wind tunnels.

The C_{d_c} model that will be discussed was developed primarily using data from a NASA Langley Research Center study[4] with added input from the Jorgensen study.[3] The Langley body-alone data were obtained using a model with a 3.0-caliber tangent ogive nose and an overall length of 12.33 calibers. Body diameter was 3.0 in. The Jorgensen data represent measurements on two configurations. One has a 2.5-caliber ogive nose and a total length of 9.5 calibers. The other has a 3.5-caliber ogive nose and a total length of 10.5 calibers. The body diameter in both cases is 1.5 in. Freestream Mach numbers ranged from 0.6 to 1.2, and AOA varied from 5 to 60 deg. Because only total C_N values were given, C_{d_c} was extracted by first subtracting the linear component as determined by slender body theory, for simplicity. Equation (2) was then used to compute C_{d_c}, with the value of η coming from Jorgensen.

The resulting C_{d_c} values were plotted in two different fashions. First, a plot was generated showing the relationship of the C_{d_c} values of Reynolds number. Instead of the crossflow Reynolds number, an effective value intended to produce superior correlation of data was used.[1] This Reynolds number was defined by

$$R_{N_{eff}} = \frac{1}{2} R_{N_D} \left\{ \frac{\cos a \cot a + 2 \sin a}{2} \right\} \left\{ \frac{1 + \left[1 + (1/2 \cot a)^2\right]^{1/2}}{\left[1 + (1/2 \cot a)^2\right]^{1/2}} \right\} \tag{3}$$

where R_{N_D} is the Reynolds number based on the body diameter. This plot is shown in Fig. 5.4. With the exception of the Jorgensen[3] data for the 2.5-caliber nose case, there is fairly good correlation, as indicated by the superimposed dashed curve, to a critical Reynolds number of approximately 400,000. The degree of scatter present is not unexpected given the conditions discussed previously. The lack of general agreement of the one data set could be an extreme case of the expected scatter or it could be the result of errors in the C_{d_c} extraction process. Available information[4] was sketchy and some assumptions were made that may not have been correct. In any event, the $R_{N_{crit}}$ value of 400,000 is higher than that indicated in Ericsson and Reding[1] by a factor of 4 and about twice that shown in Jorgensen. It is, however, well within the possible order-of-magnitude variation referenced previously. In addition, it is fairly close to the theoretical two-dimensional value of 330,000. Thus, it would seem that the extracted C_{d_c} values represent a reasonable database from which to work, at least for the NASA wind tunnel.

The second plot of extracted C_{d_c} values shows their variation with crossflow Mach number and is shown in Fig. 5.5. Note the presence of a well-defined "drag bucket" at the lower crossflow Mach numbers. There are few if any, data from the present cases to define the upper branch of the C_{d_c} curve because it applies only for subcritical Reynolds number, and the conditions of the Langley[4] and Jorgensen[3] studies produced, for the most part, supercritical situations. The upper branch of the C_{d_c} curve can be drawn on the plot with reasonable certainty because it is known to be basically a constant value of 1.2 for crossflow Mach numbers below 0.4. This resulted in a location for the lower branch of the curve that is shifted slightly toward lower M_{N_C} values, more so than that shown in Jorgensen, extending from $0.03 \leq M_{N_C} \leq 0.4$ rather than from $0.1 \leq M_{N_C} \leq 0.5$.

Because the critical crossflow Reynolds number varies from wind tunnel to wind tunnel, from model to model, it is believed average values for R_{N_C} of 180,000 and M_C of 0.1 should be used as approximate values of these

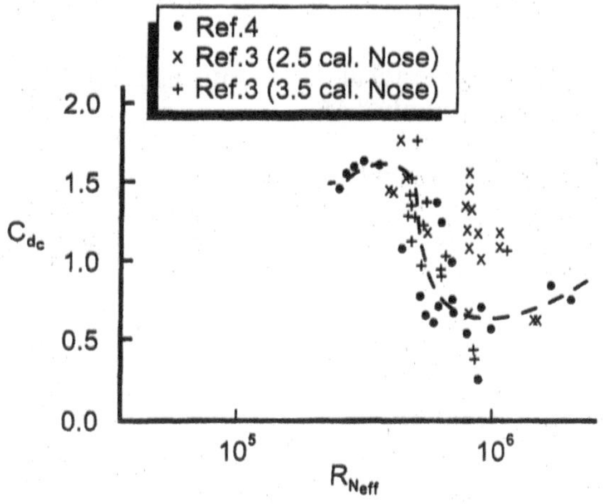

Fig. 5.4 Crossflow drag coefficient as a function of effective Reynolds number.

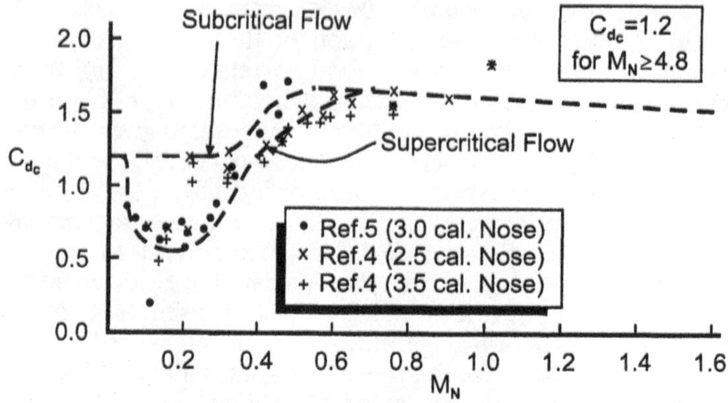

Fig. 5.5 Variation of crossflow drag coefficient with crossflow Mach number.

parameters unless the values of these parameters are known for a given tunnel. If they are known for a given tunnel, those values should be used as starting points. Thus the process for determining C_{d_c} at a given Mach number consists of the following steps:

1) Compute value of $(R_N)_{\text{eff}}$ from Eq. (3) for a given α, M_∞.
2) Define values of R_{N_C} and M_{N_C} based on experience of a given wind tunnel or wind tunnel data if available. If you do not know these values, use average values of $R_{N_C} = 180{,}000$ and M_{N_C} of 0.1.
3) Determine if the flow is supercritical by comparing $(R_N)_{\text{eff}}$ and M_N with R_{N_C} and M_{N_C}. For supercritical flow,

$$\begin{aligned}(R_N)_{\text{eff}} &> R_{N_C} \\ M_N &> M_{N_C}\end{aligned} \tag{4}$$

4) Depending on whether the flow is supercritical or subcritical, define C_{d_c} from Fig. 5.5 for a given crossflow Mach number M_N.

It is also recommended that one allow a gradual transition from the subcritical to supercritical curve. That is, in transitioning from the subcritical to the supercritical drag curves of Fig. 5.5, if C_{d_c} is allowed to vary from its subcritical value to the supercritical value as

$$R_{N_{\text{eff}}} = R_{N_C} + 25{,}000 \tag{5}$$

a smooth versus abrupt change in C_{d_c} will occur.

The final physical consideration in calculating C_{d_c} for a given configuration is to recognize the fact that all of the data of Figs. 5.4 and 5.5 are for body-alone configurations. When wings are attached, the physical mechanism that causes the flow to be supercritical certainly cannot occur in the vicinity of the wings. An approximate way to account for this is to subtract out the planform area of the body in the vicinity of the wings. That is, one breaks down the planform area of the body [A_P of Eq. (2)] into that not influenced by the wings and that influenced by the wings. If the planform area influenced by the wings is assumed to be two root chord lengths times the body diameter, and if $(C_{d_c})_1$ and $(C_{d_c})_2$ are the sub- and supercritical values of C_{d_c}, respectively, then a modified nonlinear normal force of the body alone for supercritical flow (where wings are present) is

$$\frac{\left(C_{N_{NL}}\right)_{\text{mod}}}{C_{N_{NL}}} = 1 + \frac{2c_r}{\ell}\left[\frac{\left(C_{d_c}\right)_1}{\left(C_{d_c}\right)_2} - 1\right] \tag{6}$$

Before leaving the discussion on computation of the crossflow drag coefficient of Eq. (2), it should be pointed out that, from a practical standpoint, the issue of subcritical flow versus supercritical flow is most important for subsonic flow. At $M_\infty \geq 2.0$, one goes through the drag bucket at low AOA and it is generally not discernible at high AOA.

The other parameter that must be defined in Eq. (2) is η, the drag proportionality factor. Jorgensen[3] gives values of η_0 for the incompressible case where $M_N = 0$ and also shows data for the value of η as crossflow Mach number increases. These results for both η_0 and η are shown in Figs. 5.6 and 5.7 respectively, both taken from Ref. 3. The dashed line in Fig. 5.7 is a linear approximation to the compressibility effects of the Ref. 3 results.

Thus, to compute η for a given configuration, first compute η_0 from Fig. 5.6 for a configuration with a given ℓ/d ratio. Then, assuming the dashed line of Fig. 5.7 is a reasonable approximation of compressibility effects,

$$\begin{aligned} \eta &= \left(\frac{1-\eta_0}{1.8}\right)M_N + \eta_0 \quad &&\text{for } M_N \leq 1.8 \\ \eta &= 1 &&\text{for } M_N > 1.8 \end{aligned} \tag{7}$$

It has also been recommended by some references that, as freestream Mach number approaches moderate values, η approaches 1.0 regardless of M_N. A value of M_∞ of 2.5 to 3.0 is recommended if this option is used.

Knowing the value of C_{d_c} and η for a given Mach number and AOA, the nonlinear normal force coefficient component of the body can be computed from Eq. (2). The total body-alone normal force is then computed from Eq. (1), where C_{N_L} comes from the process described in Chapters 3 and 4. The center of pressure of the nonlinear body-alone normal force term is assumed to lie at the centroid of the planform area, A_P. The overall body-alone center

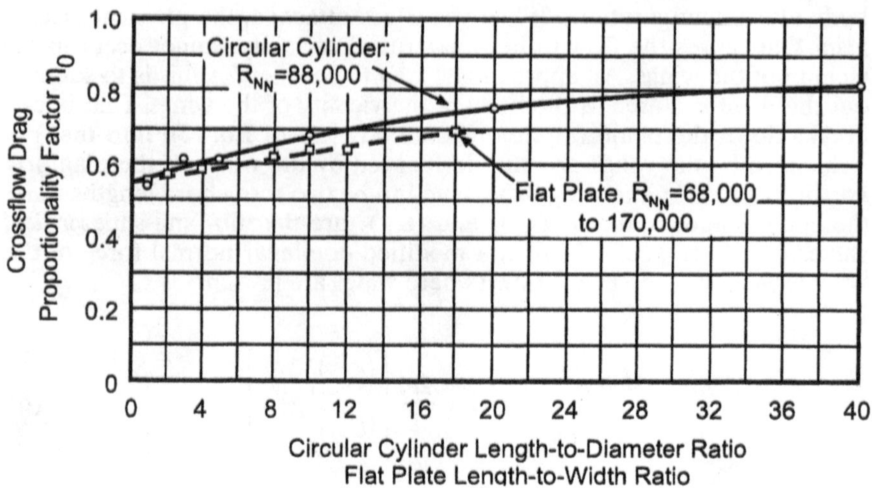

Fig. 5.6 Ratio of crossflow drag coefficient for a finite length cylinder (or flat plate) to that for an infinite-length cylinder (or flat plate) at $M_N = 0$ (figure taken from Ref. 3).

of pressure is then computed by an averaging process of the linear and nonlinear components as follows:

$$x_{CP} = \frac{C_{N_L}(x_{CP})_L + C_{N_{NL}}(x_{CP})_{NL}}{C_{N_L} + C_{N_{NL}}} \tag{8}$$

Again, $(x_{CP})_L$ comes from the analytical computations of Chapters 3 and 4.

Two other considerations need to be made concerning Eq. (8). The first is the variation of C_{N_L} as a function of a. In the limit when a = 90 deg, it

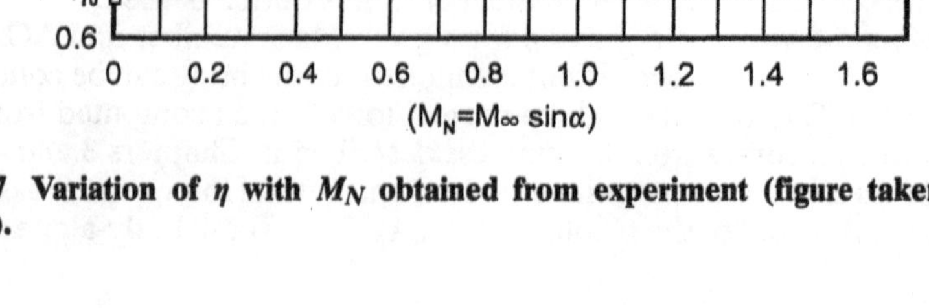

Fig. 5.7 Variation of η with M_N obtained from experiment (figure taken from Ref. 3).

appears reasonable to assume that the entire normal force of the body comes from the nonlinear Eq. (2) and the linear term is zero. The author currently assumes, in his calculations, that C_{N_L} varies with a as follows:

$$C_{N_L} = \left(C_{N_a}\right)_L a \; ; \; a \leq 30^\circ$$

$$C_{N_L} = \left(C_{N_L}\right)_{a=45^\circ}\left[1 - \frac{a - 45^\circ}{45^\circ}\right] ; \; 45^\circ < a \leq 90^\circ \tag{9}$$

The second comment regarding use of Eq. (8) for center of pressure has to do with the fact that two nonlinear phenomena are present that Eq. (9) does not fully resolve. One phenomenon is the shift in center of pressure that occurs as a result of asymmetric body vortices that typically occur above AOA = 25 deg and below $M_\infty = 2.0$. This phenomenon appears to affect the normal force only slightly but shifts the center of pressure forward. The second phenomenon occurs in the transonic flow range where a shock forms on the body as the flow approaches sonic conditions. The center of pressure shift tends to be forward from this phenomenon as well.

To estimate the center of pressure shifts, two different databases[4,5] were used and nondimensionalized with respect to body length. In implementing these results into complete missile configurations, shifts in the transonic region were found to be slightly too large and were reduced somewhat. Also, an upper bound on maximum allowable shift was used based on a 15-caliber body length. In other words, if a 20-caliber body were being considered, the maximum allowable shift in the center of pressure is based on a 15-caliber body. The final tabulation of shifts in the body-alone center of pressure as a function of AOA and Mach number is given in Table 5.1. Again, Table 5.1 is in percent of body length with an upper body length shift based on a 15-caliber body. Positive values in the table indicate a shift toward the body base, and negative values, a shift toward the nose.

In examining Table 5.1 one other point is worth noting. The point is that at low AOA, most of the shifts are in the positive direction. It is speculated that this is because the center of pressure of the linear term of normal force is assumed constant, whereas it probably moves rearward like the nonlinear component of normal force.

Knowing the total body-alone normal force and center of pressure, the pitching moment about some point x_o is then

$$C_M = - \, C_N \left(x_{CP} - x_o\right) \tag{10}$$

Several examples will be chosen to illustrate the capability of the body-alone prediction methodology. The first example is a 20-caliber-long tangent ogive cylinder with a 2.7-caliber ogive nose. Test data for normal force coefficient[6] were available at a low Mach number of 0.15. This case was chosen to illustrate the criticality of the value of R_{N_C} and M_C of Eqs. (3) through (5). An R_{N_C} value of 180,000 was used as a default, and an M_C value

Table 5.1 Shift in body-alone center of pressure as a function of Mach number and AOA (as a fraction of body length)

$M \backslash \alpha$	0	10	20	30	40	50	60	70	80	90
0.00	0.00	0.01	0.01	0.000	−0.025	−0.040	−0.040	−0.030	−0.010	0.00
0.20	0.00	0.02	0.02	0.010	−0.025	−0.040	−0.045	−0.030	−0.010	0.00
0.40	0.00	0.03	0.03	0.010	−0.025	−0.040	−0.050	−0.030	−0.015	0.00
0.60	0.00	0.03	0.03	0.010	−0.035	−0.055	−0.070	−0.050	−0.030	0.00
0.80	0.00	0.02	0.02	−0.015	−0.050	−0.070	−0.070	−0.050	−0.015	0.00
0.90	0.00	0.00	0.00	−0.015	−0.050	−0.070	−0.070	−0.040	−0.015	0.00
1.00	0.00	0.00	0.00	−0.015	−0.040	−0.040	−0.040	−0.030	−0.005	0.00
1.15	0.00	0.00	0.00	−0.015	−0.020	−0.025	−0.030	−0.025	−0.005	0.00
1.30	0.00	0.00	0.00	−0.005	−0.010	−0.010	−0.010	−0.005	0.000	0.00
1.50	0.00	0.00	0.00	0.000	0.000	0.000	−0.005	−0.005	0.000	0.00
2.00	0.00	0.02	0.02	0.020	0.015	0.010	0.005	0.000	0.000	0.00
2.50	0.00	0.03	0.03	0.030	0.015	0.010	0.005	0.000	0.000	0.00
6.00	0.00	0.03	0.03	0.030	0.015	0.010	0.005	0.000	0.000	0.00
20.00	0.00	0.00	0.00	0.00	0.00	0.00	0.00	0.000	0.000	0.00

of 0.03. Figure 5.8 shows the results of these calculations in the form of the dotted line. The dotted line assumed that criteria for supercritical flow were satisfied. However, data suggested that flow was subcritical. By shifting the value of M_C from 0.03 to 0.13 through numerical experiments, the results indicated by the solid line were obtained and agree with experimental data.

The second case considered is the 12.33-caliber tangent ogive of Fig. 4.35a, which has a 3.0-caliber nose. Figure 5.9a gives a comparison of theoretical and experimental normal force coefficients for M_∞ = 0.6, 2.0, and 4.5. The value of R_{NC} is 330,000 and M_C is zero for this case. These values were determined through trial and error for this database. Note that excellent agreement is obtained between experiment and theory. It is also interesting to note that the case most affected by the selection of R_{NC} and M_C is the M_∞ = 0.6 case. Transition from supercritical flow to subcritical begins occurring at a = 25 deg and is complete at a = 35 deg for this case. Figure 5.9b presents the comparison of center of pressure predictions and experiment for this same case. Quite acceptable agreement with experiment is obtained. The worst comparisons to experiment are at low AOAs, where it is most difficult to get accurate wind tunnel data for this parameter. In general, the average center of pressure error for these three Mach numbers and at the six AOAs shown (18 cases) is less than 2% of the body length, or about 0.25 calibers.

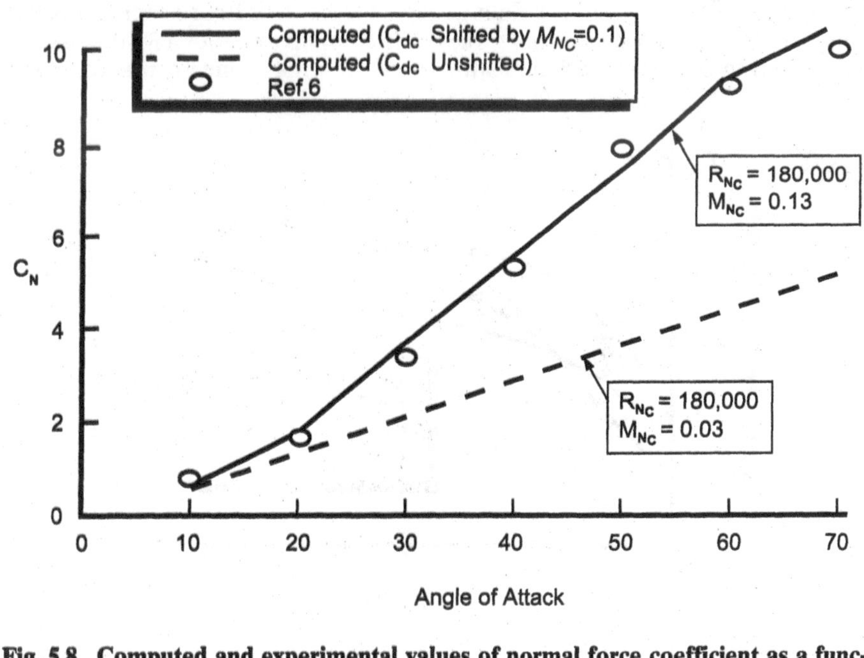

Fig. 5.8 Computed and experimental values of normal force coefficient as a function of AOA for a body-alone configuration at M_∞ = 0.15.

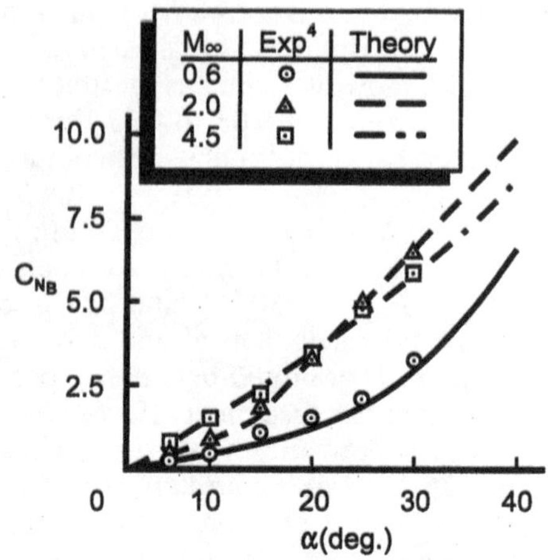

Fig. 5.9a Comparison of theoretical and experimental normal force coefficients for 12.33-caliber tangent ogive cylinder (ℓ_n = 3.0 caliber, R_{N_C} = 330,000 and M_{N_C} = 0).

The worse condition is at $a = 5$ deg, $M_\infty = 2.0$, were the error is 0.75 calibers or 6% of the body length.

III. Wing-Alone Normal Force and Center of Pressure

Figure 5.3 showed a qualitative trend of the normal force coefficient as a function of AOA for various values of aspect ratio. This figure represents the trends of the wing-alone normal force coefficient up to about a 30-deg AOA. However, Fig. 5.10 is more representative of the wing-alone normal force for

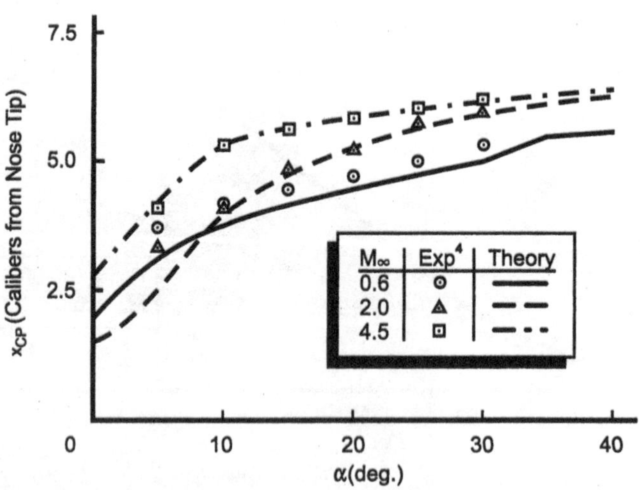

Fig. 5.9b Comparison of theoretical and experimental center of pressure for 12.33-caliber tangent ogive cylinder (ℓ_n = 3.0 caliber, R_{N_C} = 180,000, and M_{N_C} = 0).

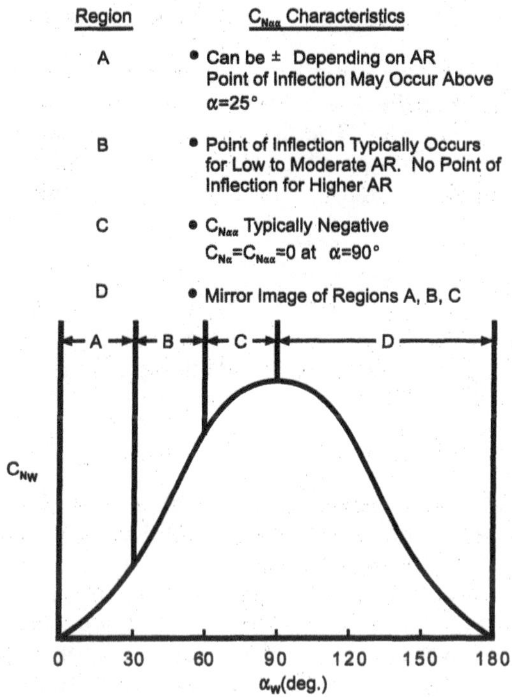

Fig. 5.10 Qualitative characteristics of normal force coefficient of wings alone at all AOAs.

AOAs of 0 to 180 deg. As seen in Fig. 5.10, the fact that the curve of C_N versus a for a given wing may have both a concave up and concave down trend at various a values requires a higher than second-order equation to represent the wing normal force. Moore and McInville[7] tried a second-, third-, and fourth-order equation in AOA to represent the wing-alone normal force coefficient as a function of AOA. The fourth-order equation for the wing-alone normal force was found to be the most accurate for all AOAs.

The wing-alone nonlinear normal force coefficient is thus defined by

$$C_{NW} = a_0 + a_1 a_W + a_2 a_W^2 + a_3 a_W^3 + a_4 a_W^4 \tag{11}$$

The a_W term in Eq. (11) is defined as the AOA of the unperturbed freestream on the wing itself. That is,

$$a_W = | \, a + \delta \, | \tag{12}$$

where δ is the control deflection and a the AOA of the wing-body configuration. Here, only positive AOAs (a_W) are considered because it is assumed that missile wing planforms have no camber and, as a result, the normal force at a negative AOA is simply the negative of that at the same positive value of AOA.

To predict the wing-alone normal force using Eq. (11) now requires that five constants be evaluated. Because there are five constants, five independent equations or conditions are needed. The first condition has already been mentioned. That is, most weapon lifting surface planforms are symmetric and have zero camber. As a result, $(C_N)_{\alpha=0} = 0$ and, therefore, from Eq. (11),

$$a_0 = 0 \tag{13}$$

Second, we will assume that at $\alpha = 0$, C_{N_α} can be estimated accurately enough by the linearized theories of Chapter 3. Once again, Eq. (11) yields

$$a_1 = (C_{N_\alpha})_{\alpha=0} \tag{14}$$

These values of $(C_{N_\alpha})_{\alpha=0}$ are already available from Chapter 3 and, as shown in Fig. 3.20, give reasonably accurate results for planforms where the thickness is not too great and the AOA is fairly small.

To determine the remaining three conditions, several alternatives are available. The first alternative is to take advantage of the fact that at $\alpha = \pi/2$, $C_{N_\alpha} = 0$ because the C_{N_W} is a maximum at $\alpha = \pi/2$. Using this condition and two values of C_{N_W} at two different AOAs, the remaining three constants can be determined. The question here is which two AOAs to use. One set that warrants consideration is $\alpha_1 = \pi/6$ and $\alpha_2 = \pi/3$, because the AOA range is divided into equal $\pi/6$ increments where the conditions are defined for evaluating the constants a_i.

Examining Fig. 5.10, it is seen that a fairly strong nonlinearity occurs for $\alpha < 30$ deg for low aspect ratio wings. Hence, an argument could be made that a lower value of α would be more appropriate than $\alpha_1 = \pi/6$. Hence, values of $\alpha_1 = 20$ deg and $\alpha_2 = 60$ deg and $\alpha_1 = 20$ deg and $\alpha_2 = 45$ deg can be used to compare against the $\alpha_1 = 30$ deg and $\alpha_2 = 60$ deg predictions.

A fourth option for defining the three remaining constants using the fourth-order Eq. (11) is to use three values of α versus using two values of α and the condition that $C_{N_\alpha} = 0$ at $\alpha = \pi/2$. Using the rationale of the strong nonlineriarity below $\pi = 30$ deg for low aspect ratio configurations and the fact that $\alpha \leq 30$ deg is the most important part of the AOA range, values of $\alpha_1 = 15$ deg, $\alpha_2 = 30$ deg, and $\alpha_3 = 60$ deg, seem appropriate. To provide more equal spacing between $\alpha_1 = 15$ deg and $\alpha_3 = 60$ deg, yet focus more on the $\alpha = 0$ to 30 deg range, another alternative for α_2 is 35 deg. Above $\alpha = 60$ deg, approximate equations can be used to extend the C_{N_W} from $\alpha = 60$ deg to $\alpha = 90$ deg.

Each of the five alternatives outlined in the previous discussion for computing the three coefficient a_2, a_3, and a_4 of Eq. (11) require three different sets of equations. Moore and McInville[7] evaluated all five of these alternatives and found the most accurate of the alternatives to be the option where we use the three AOAs in the range 0 to 60 deg, $\alpha = 15$, 35, and 60 deg, to approximate the normal force of the wing alone. Above α_W of 60 deg, approximate equations were used to extend the range of applicability to a 90-deg AOA.

The equations that govern the three coefficients a_2, a_3, and a_4 of Eq. (11) are

$$a_2 = 34.044(C_N)_{\alpha=15^\circ} - 4.824(C_N)_{\alpha=35^\circ} + 0.426(C_N)_{\alpha=60^\circ} - 6.412\ a_1 \quad (15)$$

$$a_3 = 88.240(C_N)_{\alpha=15^\circ} + 23.032(C_N)_{\alpha=35^\circ} - 2.322(C_N)_{\alpha=60^\circ} + 11.464\ a_1 \quad (16)$$

$$a_4 = 53.219(C_N)_{\alpha=15^\circ} - 17.595(C_N)_{\alpha=35^\circ} + 2.661(C_N)_{\alpha=60^\circ} - 5.971\ a_1 \quad (17)$$

The terms $(C_N)_{\alpha=15^\circ}$, $(C_N)_{\alpha=35^\circ}$, and $(C_N)_{\alpha=60^\circ}$ are the normal force coefficients of a wing alone at α_W = 15 deg, 35 deg, and 60 deg, respectively as a function of Mach number, taper ratio, and aspect ratio. Knowing values of these parameters for a given wing platform, a_0, a_1, a_2, a_3, and a_4 can be computed from Eqs. (13) to (17), and C_{N_W}, from Eq. (11).

To evaluate Eqs. (15) to (17), requires tables of data for α = 15, 35, and 60 deg. These data are given in Tables 5.2 through 5.4 based on databases of Refs. 5, 8, and 9. To form Tables 5.2 through 5.4 requires plots of C_{N_W} versus Mach number for each value of α noted above and for each wing aspect and taper ratio. In general, a great deal of consistency exists between the databases at the lower values of AOA. However, as AOA increased above α = 30 deg, there were some inconsistencies in the databases as well as some gaps. To fill the gaps, plots were made of C_{N_W} versus aspect ratio for a given

Table 5.2 Values of $(C_N)_{\alpha=15^\circ}$

Aspect ratio	Taper ratio	Mach number 0	0.6	0.8	1.0	1.2	1.6	2.0	3.0	4.5	≥6.0
≤0.5	0.0	0.28	0.29	0.30	0.32	0.33	0.33	0.32	0.24	0.175	0.125
	0.5	0.39	0.41	0.415	0.42	0.43	0.42	0.39	0.28	0.22	0.18
	1.0	0.34	0.34	0.36	0.40	0.42	0.42	0.40	0.30	0.23	0.19
1.0	0.0	0.43	0.44	0.46	0.49	0.53	0.47	0.43	0.33	0.26	0.21
	0.5	0.47	0.50	0.60	0.62	0.625	0.55	0.50	0.39	0.29	0.22
	1.0	0.46	0.48	0.52	0.58	0.60	0.55	0.51	0.39	0.29	0.22
2.0	0.0	0.55	0.59	0.65	0.67	0.68	0.58	0.48	0.33	0.26	0.22
	0.5	0.56	0.59	0.66	0.76	0.80	0.66	0.54	0.40	0.31	0.26
	1.0	0.56	0.59	0.66	0.76	0.70	0.62	0.55	0.38	0.29	0.23
≥4.0	0.0	0.65	0.66	0.71	0.83	0.86	0.73	0.59	0.43	0.31	0.23
	0.5	0.69	0.71	0.75	0.88	0.92	0.75	0.62	0.45	0.34	0.28
	1.0	0.69	0.71	0.75	0.88	0.92	0.75	0.62	0.45	0.34	0.28

Table 5.3 Values of $(C_N)_{\alpha=35°}$

Aspect ratio	Taper ratio	Mach number 0	0.6	0.8	1.0	1.2	1.6	2.0	3.0	4.5	≥6.0
≤0.5	0.0	0.89	0.91	0.93	0.95	0.98	0.95	0.88	0.72	0.65	0.61
	0.5	1.10	1.13	1.16	1.25	1.20	1.09	1.0	0.84	0.76	0.72
	1.0	1.06	1.08	1.13	1.16	1.19	1.12	1.03	0.86	0.76	0.72
1.0	0.0	1.18	1.20	1.22	1.24	1.18	1.09	1.0	0.80	0.70	0.66
	0.5	1.2	1.22	1.24	1.33	1.40	1.20	1.15	0.95	0.82	0.76
	1.0	1.10	1.11	1.16	1.26	1.36	1.20	1.16	0.95	0.82	0.76
2.0	0.0	0.95	1.01	1.13	1.20	1.28	1.18	1.08	0.93	0.86	0.81
	0.5	1.0	1.07	1.18	1.3	1.4	1.32	1.17	1.0	0.90	0.85
	1.0	0.98	1.05	1.17	1.27	1.39	1.32	1.21	1.0	0.90	0.85
≥4.0	0.0	0.97	1.05	1.17	1.21	1.34	1.22	1.10	0.95	0.87	0.83
	0.5	1.03	1.09	1.22	1.32	1.44	1.35	1.25	1.05	0.96	0.92
	1.0	1.03	1.09	1.21	1.32	1.44	1.35	1.25	1.05	0.96	0.92

taper ratio and Mach number and engineering judgment was exercised in making consistent data sets.

It is fair to say that, even though the databases of Refs. 5, 8, and 9 are extensive, additional wing-alone data are needed at low Mach numbers ($M_\infty \leq 1.2$) and AOAs greater than 30 deg and at all Mach numbers for aspect ratio 4 and $\lambda = 0$ and 1.0.

In forming Tables 5.2 to 5.4, in some cases, one database was relied upon more than another. The reason for this decision goes back to the way the data were collected for the three data sets. Stallings and Lamb[8] data were

Table 5.4 Values of $(C_N)_{\alpha=60°}$

Aspect ratio	Taper ratio	Mach number 0	0.6	0.8	1.0	1.2	1.6	2.2	3.0	4.5	≥6.0
≤0.5	0.0	1.10	1.11	1.15	1.26	1.33	1.37	1.31	1.25	1.21	1.18
	0.5	1.26	1.27	1.30	1.40	1.54	1.64	1.54	1.44	1.39	1.36
	1.0	1.26	1.27	1.30	1.40	1.51	1.58	1.54	1.46	1.40	1.36
1.0	0.0	1.44	1.46	1.49	1.53	1.56	1.61	1.50	1.42	1.38	1.36
	0.5	1.40	1.42	1.45	1.53	1.58	1.70	1.64	1.54	1.48	1.45
	1.0	1.33	1.34	1.35	1.44	1.62	1.72	1.67	1.57	1.50	1.46
2.0	0.0	1.26	1.27	1.34	1.48	1.59	1.74	1.68	1.54	1.48	1.45
	0.5	1.30	1.31	1.37	1.48	1.63	1.84	1.80	1.63	1.57	1.54
	1.0	1.30	1.31	1.37	1.48	1.63	1.76	1.73	1.64	1.57	1.54
≥4.0	0.0	1.27	1.28	1.37	1.50	1.64	1.80	1.70	1.56	1.50	1.47
	0.5	1.31	1.32	1.40	1.52	1.70	1.89	1.82	1.66	1.60	1.56
	1.0	1.31	1.32	1.40	1.52	1.70	1.78	1.75	1.66	1.60	1.57

taken on a full wing planform at $1.6 \le M_\infty \le 4.63$. Measurements of pressure on the wing planforms were actually taken at many points on the wing and these values were then integrated to obtain normal force. This meant the thickness of the wings was larger than most practical wing planforms due to all the pressure taps and lines inside the wing. This was particularly true for the larger values of λ where the root thickness was higher. It is suspected that this increased thickness contributed to a slightly lower value of normal force for some cases of Ref. 8 compared to Ref. 5. Moore and McInville[5] data were taken on fairly thin wings mounted on a splitter plate. This meant that only one-half of the wing was used in the wind tunnel which, apparently, resulted in some fictitious stall results at higher AOA. At these conditions, Ref. 8 data were relied on primarily. Finally, the Ref. 9 database was made primarily from a combination of Refs. 8 and 9 data, with other available data. Moore and McInville[10] data are important in comparing the Refs. 8 and 9 data and in making a decision between the Refs. 8 and 9 databases on occasion.

Two other points need to be considered before leaving the wing-alone nonlinear normal force discussion. The first has to do with the prediction of normal force coefficient above $a_W = 60$ deg and the second has to do with center of pressure. For the AOA range $60 < a_W \le 90$, approximate formulas were developed consistent with the trends of the data in Refs. 5, 8, and 9. These approximate equations are

For $M_\infty \le 1.2$; $60^\circ < a_W \le 90^\circ$

$$C_{N_W} = \left(C_{N_W}\right)_{a_W=60^\circ}\left(\frac{\sin a_W}{\sin 60^\circ}\right)^{1/3}$$

For $M_\infty \ge 2.0$; $60^\circ < a_W \le 90^\circ$

$$C_{N_W} = \left(C_{N_W}\right)_{a_W=60^\circ}\left(\frac{\sin a_W}{\sin 60^\circ}\right)$$

For $1.2 < M_\infty < 2.0$; $60^\circ < a_W \le 90^\circ$

Extrapolate between values at $M = 1.2$ and 2.0 (18)

$$C_{N_W} = \left(C_{N_W}\right)_{M=1.2} + \left[\left(C_{N_W}\right)_{M=2.0} - \left(C_{N_W}\right)_{M=1.2}\right]\frac{(M-1.2)}{0.8}$$

Finally, for a_W greater than 90 deg, we define an AOA a^* by

$$a = \pi/2 + a^*$$

where a^* is the value of a greater than $\pi/2$. Then, from symmetry considerations,

$$\left(C_{N_W}\right)_{a=\pi/2+a^*} = \left(C_{N_W}\right)_{a=\pi/2-a^*} \tag{19}$$

The wing-alone and wing-body normal force center of pressure at roll position $\varphi = 0$ is based on the linearized theory at small AOA and the center of pressure moves to the centroid of the wing planform area at large AOA ($\alpha_W = 60$ deg).

This transition is accomplished in a quadratic fashion. If A and B are the centers of pressure of the linear and nonlinear normal force terms (in percent of mean geometric chord), and $\alpha_W = \alpha + \delta$, then the center of pressure of the wing-body or wing-alone lift is

$$\left(x_{CP}\right)_{WB} = \left(x_{CP}\right)_{w} = A + \frac{1}{36}\left|\alpha_{w}\right| (B - A) + \frac{1}{5400}\alpha_{w}^{2}(A - B) \tag{20}$$

Equation (20) is the methodology used for roll position of 0 deg.

When the fins are rolled to a non-zero roll orientation, the center of pressure equation (20) will change because of the geometry of the wings and an asymmetric effect on the wing loading. To visualize this effect, imagine a missile rolled to $\varphi = 45$ deg and increasing in AOA. As AOA increases, two things occur. First, the windward plane fins carry more and more of the load compared to the leeward plane fins. Second, the local Mach number in the windward plane is different and typically lower than the leeward plane. This has the effect of shifting the wing-alone center of pressure forward in the windward plane. Because the normal force and wing centers of pressure are different on the windward and leeward plane fins, this results in a net forward shift in the center of pressure for $\varphi = 45$ deg roll compared to the $\varphi = 0$ deg computation of Eq. (20). This shift appears to occur for all Mach numbers and is largest at moderate AOA and goes to zero at AOA of 0 and 90 deg. At 90 deg, the windward plane fins carry almost all the load compared to the leeward plane fins but, geometrically, the fins are aligned perpendicular to the AOA plane.

Mathematically, this geometrical shift can be approximated by

$$\left(\Delta x_{CP}\right)_{WB} \cong y_{CP} \cos^{2} \varphi \sin (2\alpha) \frac{\Delta C_{N_{W(B)}}}{C_{N_{W(B)}}} \tag{21}$$

y_{CP} of Eq. (21) is the centroid of the planform area in the y direction of the wing planform. If the wing is assumed to be represented by a trapezoidal shape, then

$$y_{CP} = r + \left(\frac{b}{c_r + c_t}\right)\left(\frac{c_r}{2} - \frac{c_t}{3}\right) \tag{22}$$

The $\cos^2 \varphi$ factor of Eq. (21) rotates the ΔC_N term from being normal to the wing planform, to the plane of the normal force vector. It also rotates the y_{CP} term from being out the wing to a plane normal to the velocity vector. The $\sin 2\alpha$ term allows the center of pressure shift to vary between 0 at AOA of 0 and 90 deg to a maximum at AOA of 45 deg. The term $\Delta C_{N_{W(B)}}/C_{N_{W(B)}}$ represents the difference between the load of the windward and leeward plane fins as a fraction of the total wing-body load. Using the NASA database,[4] an estimate was made for several wings. An approximate linear trend with AOA was determined. While data are not available above a of 40 deg, it will be assumed this trend is linear between AOA of 0 to 65 deg. At $a \geq$ 65 deg, it is assumed the windward plane fins carry 80% of the load of the wings and the leeward plane fins carry only 20%. With this assumption,

$$\frac{\Delta C_{N_{W(B)}}}{C_{N_{W(B)}}} = 0.8\left(\frac{a}{65}\right); \; a < 65^\circ \tag{23}$$

In Eq. (23), a is in degrees. Combining Eqs. (21), (22), and (23), there is obtained

$$\begin{aligned}(\Delta x_{CP})_W &= (\Delta x_{CP})_{WB} \\ &= -\left[r + \left(\frac{b}{c_r + c_t}\right)\left(\frac{c_r}{2} - \frac{c_t}{3}\right)\right]\cos^2(\varphi)\sin(2a)\left(\frac{0.8a}{65}\right); a \leq 65 \\ &= -0.8\left[r + \left(\frac{b}{c_r + c_t}\right)\left(\frac{c_r}{2} - \frac{c_t}{3}\right)\right]\cos^2(\varphi)\sin(2a); a > 65^\circ\end{aligned} \tag{24}$$

Equation (24) is added to Eq. (20) for the roll orientation of 45 deg.

A couple of examples will be given to show the robustness and overall accuracy of the wing-alone methodology. The first case is an aspect ratio 0.5 wing with a taper ratio of 0.5. Figure 5.11 gives the comparison of theory and experiment for $M_\infty = 0.8$ and 4.5 cases for AOAs of 0 to 90 deg. Experimental data are available only to $a = 60$ deg for C_{N_W} and only for the $M_\infty = 4.5$ cases for x_{CP}. As seen in the figures, the nonlinear model represented by Eqs.

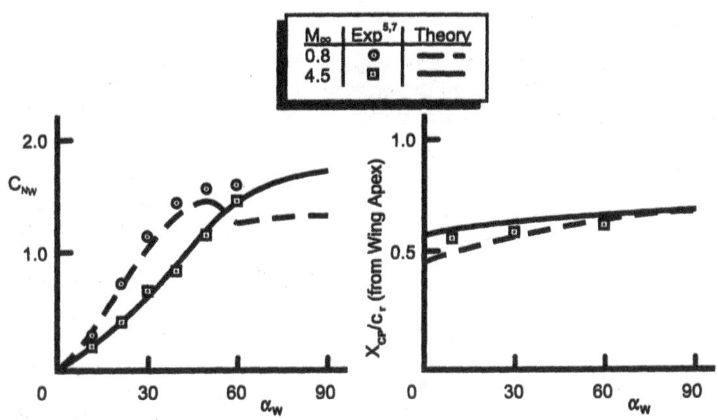

Fig. 5.11 Comparison of theory and experiment for wing normal force coefficient and center of pressure (AR = 0.5 and λ = 0.5).

(11) and (20) for the normal force coefficient and center of pressure at $\varphi = 0$ deg roll does a good job of predicting aerodynamics of the wing alone. The theory for the wing alone at $M_\infty = 0.8$ is intentionally low compared to the data of Ref. 10; however, it is high compared to the data of Ref. 5, which is not shown.

The second example (Fig. 5.12) compares the theory and experiment as a function of aspect ratio for $M_\infty = 4.6$ and $\lambda = 0.5$ at $\alpha = 30, 45$ and 60 deg. The theory is intentionally slightly higher than the data[8] due to the fairly thick wings tested. However, results of both normal force and center of pressure are in reasonably good agreement with the data. Figures 5.11 and 5.12 are just two examples considered for illustrative purposes. Many more examples are given in Ref. 7. In general, it has been found that the nonlinear wing-alone model presented in this section does a very good job of predicting wing-alone characteristics. An exception is in the stall region where the current method smooths out this region, as opposed to yielding an abrupt decrease in wing-alone normal force at some AOA.

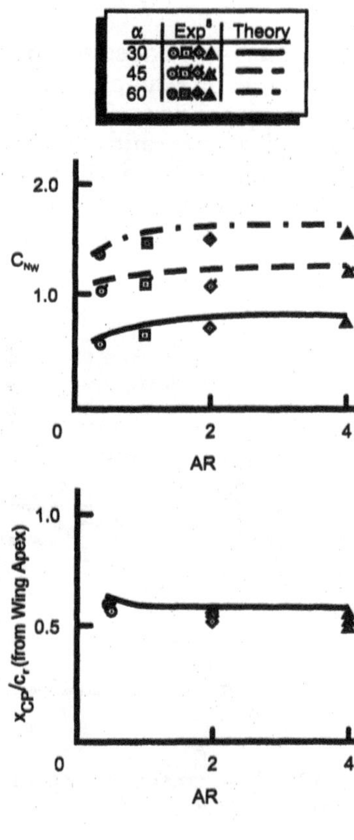

Fig. 5.12 Comparison of theory and experiment for wing normal force coefficient and center of pressure $M_\infty = 4.6$ and $\lambda = 0.5$).

IV. Wing–Body and Body–Wing Interference Due to Angle of Attack

Equations (2) and (3) in Chapter 3 define normal force and pitching moment coefficient components of force and moment associated with the body, wing, and mutual interference effects. Equation (136) in Chapter 3 further defined Chapter 3's Eq. (2) in terms of individual normal force components of the body and wing and interference effects. Equation (136) in Chapter 3 is the equation most used in the 1970s and 1980s before semi-empirical nonlinear aerodynamic prediction codes were available. A code that was developed for low AOA could be made to fit within the linear context of Eq. (136) in Chapter 3 by defining a linear and nonlinear term for each of the components in that equation. Sections 5.2 and 5.3 defined the nonlinear methods of C_{N_B} and C_{N_W} to complement the low AOA methods described in Chapters 3 and 4. This section will define the nonlinear methods for the interference factors $K_{W(B)}$ and $K_{B(W)}$. These are the interference factors associated with the normal force of the wing in the presence of the body and the additional normal force on the body as a result of the wing being present due to AOA.

Mathematically, they are defined as

$$K_{W(B)} = \frac{C_{N_{W(B)}}}{C_{N_W}} \; ; \; K_{B(W)} = \frac{\Delta C_{N_{B(W)}}}{C_{N_W}} \tag{25}$$

where $\Delta C_{N_{B(W)}} = C_{N_{B(W)}} - C_{N_B}$. The term $C_{N_{W(B)}}$ of Eq. (2) in Chapter 3 can be rewritten in a nonlinear form to correspond to a part of Chapter 3's Eq. (136) as follows:

$$C_{N_{W(B)}} = \left[\left(C_{N_\alpha}\right)_L + \left(C_{N_\alpha}\right)_{NL}\right]_W \left\{\left[\left(K_{W(B)}\right)_{\text{SBT}} + \left(\Delta K_{W(B)}\right)_{NL}\right] \alpha + \left(C_1\left[k_{W(B)}\right]_{\text{SBT}} + C_2\right)\delta_W\right\}\left(\frac{A_W}{A_{\text{REF}}}\right) \tag{26}$$

The term $[(C_{N_\alpha})_L]_W$ of Eq. (26) is the value of the linear normal force curve–slope determined for the wing-alone methods from Chapter 3. The nonlinear term, $[(C_{N_\alpha})_{NL}]_W$, can be determined from the value of total wing-alone normal force computed in Section 5.3 minus the linear theory value. In other words, C_{N_W} of Section 5.3 is the total value of $[(C_{N_\alpha})_L + (C_{N_\alpha})_{NL}]_W$ of Eq. (26). There are two types of interference terms in Eq. (26), one due to AOA and the other due to control deflection. The interference term due to AOA will be treated here and the nonlinear term due to control deflection will be treated in the next section.

Each of the other terms of Eq. (2) in Chapter 3 could be treated in a similar fashion to Eq. (26) to modify Eq. (136) from Chapter 3 to include nonlinear aerodynamics. If one is developing an all empirical computer code from scratch to include linear and nonlinear aerodynamics, Eq. (2) from Chapter 3 is the approach one should use. This is the approach a computational fluid dynamics (CFD) tool would use that solves nonlinear equations of motion (either Euler or Navier–Stokes). We seek here the nonlinear definition of each of the terms in Eq. (2) in Chapter 3 at the roll-stable positions of $\varphi = 0$ deg and $\varphi = 45$ deg. These are the roll positions of most weapons' flight during the early and midcourse portions and, therefore, are of the most interest from a semi-empirical method standpoint. During terminal maneuvers, the weapon does not fly in a roll-stable position of $\varphi = 0$ or 45 deg in many cases, due to maneuver requirements. However, the planar aerodynamics can still be used as a benchmark for aerodynamics at roll positions other than $\varphi = 0$ or 45 deg.

While roll dependence is not an issue for the body, it is an issue for the interference terms. Because we would like to maintain the use of slender body theory and linear theory at low AOA, Moore et al.[10] investigated the implications of roll dependence on aerodynamics using slender body theory. A summary of key findings from that study are as follows:

1) For cruciform wings alone or a wing–body combination, the total normal force is independent of roll.
2) For a planar wing–body combination at roll, the loading on the windward plane panel is greater by an equal amount to that on the leeward plane panel. Therefore, if one were trying to design a code for lateral aerodynamics, roll dependence of each fin planform must be considered. On the other hand, if longitudinal aerodynamics are of primary interest, the total normal force on the entire wing planform can be considered.
3) For a cruciform wing–body–tail configuration at roll, eight vortices are shed in the wing–body region, which adversely affects the tail lift. This is as opposed to four vortices at $\varphi = 0$ deg.
4) The planar theory developed for wing–tail interference can be used to approximate the loss of lift on the tails at $\varphi = 45$ deg.
5) The aerodynamics of a cruciform wing–body–tail combination with zero control deflections are independent of roll position.

These findings for roll dependence from linearized or slender body theory are quite useful in helping plan how to develop a nonlinear aeroprediction code for both $\varphi = 0$ and 45 deg roll. While the conclusions of linear theory roll dependence may not translate to the nonlinear case, we will still use the findings to help guide the nonlinear methods development. In particular, the item (1) conclusion implies use of the $\varphi = 0$ deg, wing-alone data for $\varphi = 45$ deg. Notably, available wing-alone databases are all at $\varphi = 0$ deg. This means that any nonlinear wing-alone roll dependence will be included in the interference factors rather than the wing-alone solution, which is independent of φ.

The second major result of the key slender body/linear theory roll dependence findings is that, for cruciform missiles, we can use the same interference approaches for $\varphi = 0$ and 45 deg, except that the constants need to be changed because of a different roll angle. The combination of these two conclusions is quite important because they basically allow the direct use of linear theories with different constants for the nonlinear interference terms at $\varphi = 45$ deg versus $\varphi = 0$ deg.

The third significant conclusion is that, for small AOA, wing–body–tail aerodynamics are independent of roll position. This allows the use of wing–tail interference methodology designed for planar computations for different roll orientations, as long as the proper number of vortices are considered. Again, different nonlinear corrections are expected for the $\varphi =$ 45 deg versus the $\varphi = 0$ deg roll position.

To define $K_{W(B)}$ and $K_{B(W)}$ in the nonlinear sense for the $\varphi = 0$ and 45 deg roll position will require an extensive component wind tunnel database or the equivalent using a CFD code. Fortunately, there are several missile component databases that have tested the body alone and then the body close to the wing, with separate measurements for the wing and body. The largest and most complete of these databases are defined by the Langley[4], Stallings and Lamb,[8] and Baker[5] studies. The body tested in the Langley study was a 12.33-caliber-long tangent ogive cylinder with a 3.0-caliber sharp nose, whereas that in the Baker study was a 10-caliber body with a 2.5-caliber nose. The wings tested include aspect ratios from 0.5 to 4.0 and taper ratios of 0, 0.5, and 1.0 (see Fig. 4.35). Mach numbers included 0.6 to 4.6, AOAs included 0 to 40 deg, and roll angles included 0 to 90 deg. Not all conditions were tested at the maximum AOA ($M_\infty \leq 1.2$, for example).

The third wing–body database is from a more recent study by Meyer[11] and complements the Langley database.[4,8] Meyer tested a 10-caliber-long tangent ogive cylinder with a 2-caliber nose and three different sets of fins. Measurements were made for body alone, body in proximity to the fins, and fins in proximity to the body. Data were taken for AOA to 80 deg for various roll positions and at $M_\infty = 0.1$. Because the lower limit of the Langley database was only $M_\infty = 0.6$ and $\alpha = 40$ deg, the Meyer database should provide insight to the interference factor behavior above $\alpha = 40$ deg for subsonic speeds as well as insight to the Mach number range of $M_\infty < 0.6$.

The databases measured the following quantities:

$$C_N, C_M, x_{CP}, C_{N_{W(B)}}, C_{N_W}, C_A, C_{N_B}, C_{M_B}, (x_{CP})_B$$

as a function of α, M_∞, φ, AR, and λ. As a result, $K_{W(B)}$ can be calculated directly from two measured quantities, $C_{N_{W(B)}}$ and C_{N_W}. The main sources of errors are wind tunnel measurement errors, any error associated with a different thickness wing alone and wing in conjunction with the body, and different tunnels and tunnel entries for the wing alone and wing in conjunction with the body measurement. On the other hand, $C_{N_{B(W)}}$ must be calculated by subtracting numbers of the same order of magnitude from other numbers that were measured in different tests; that is

$$\Delta C_{N_{B(W)}} = C_N - C_{N_B} - C_{N_{W(B)}} \tag{27}$$

As long as there is a significant magnitude to $C_{N_{B(W)}}$ its value can be computed with reasonable accuracy. This reasonable accuracy is possible when Mach number is low and the percent of wing planform area to body planform area is fairly high (AR = 0.5 and $M_\infty < 2$). However, as this percent gets low or Mach number gets high, $C_{N_{B(W)}}$ gets small and the accuracy of $K_{B(W)}$, which is defined by Eq. (25) in conjunction with Eq. (27), gets worse. An accuracy analysis of Ref. 10 pointed out that, in defining nonlinear factors $K_{W(B)}$ and $K_{B(W)}$, engineering judgment and wind tunnel test data were necessary. Also, once nonlinear factors were developed, it was found necessary to fine-tune these factors after being applied to configurations outside wind tunnel databases. This fine-tuning process helped to reduce errors due to wind tunnel measurements and judgment.

An example of nonlinear interference factors taken from Refs. 10 and 12 is shown in Fig. 5.13 for the aspect ratio 0.5 case at $M_\infty = 1.5$ and for $\varphi = 0$ and $\varphi = 45$ deg. Also shown on the curves are the slender body theory values of $K_{W(B)}$ and $K_{B(W)}$ for $r/s = 0.5$, which are about 1.45 and 0.8, respectively. Wind tunnel data are shown for taper ratios of 0, 0.5, and 1.0. Several points are worthy of note. First of all, slender body theory gives a reasonable representation of the data near $a = 0$ for both $K_{W(B)}$ and $K_{B(W)}$. However, as AOA increases, slender body theory becomes inadequate to predict the behavior of either $K_{W(B)}$ or $K_{B(W)}$. Second, there is a dependence on λ and φ, although the dependence on φ is more obvious for $K_{B(W)}$ than $K_{W(B)}$ at this Mach number. While not shown here, Moore and McInville[10] and Moore et al.[12] showed a clear dependence on AR and Mach number as well.

In general, it was found that $K_{W(B)}$ and $K_{B(W)}$ followed the general shape of Fig. 5.14 for both $\varphi = 0$ and 45 deg, at all Mach numbers, aspect ratios, and taper ratios. Inspecting Fig. 5.14a, $K_{W(B)}$ can be described in terms of the value given by slender body theory and some deviation from that at $a = 0$ and four other parameters. Likewise, Fig. 5.14b shows that $K_{B(W)}$ can be generally described in terms of the value given by slender body theory and five other parameters. Referring to Fig. 5.14, definitions of all the parameters in the figure are

$[\Delta K_{W(B)}]_{a=0}$ = difference between slender body theory and data at $a = 0$

a_c = AOA where $K_{W(B)}$ starts decreasing

$\dfrac{dK_{W(B)}}{da}$ = rate of decrease of $K_{W(B)}$ between $a = a_c$ and $a = a_D$

a_D = AOA where $K_{W(B)}$ reaches an initial minimum

a_M = AOA where $K_{W(B)}$ reaches a constant value

$[K_{W(B)}]_{a=a_M}$ = value of $K_{W(B)}$ where a constant value with AOA has been attained

$[\Delta K_{B(W)}]_{a=0}$ = difference between slender body theory/linear theory and data at $a = 0$

$\dfrac{dK_{B(W)}}{da}$ = rate of change of $K_{B(W)}$ between $a = 0$ and $a = a_1$

a_1 = AOA where $dK_{B(W)}/da$ changes sign

a_2 = AOA where $K_{B(W)}$ reaches a constant

$[K_{B(W)}]_{\min}$ = constant value of $K_{B(W)}$ above $a = a_2$ as a percent of linear theory or slender body theory

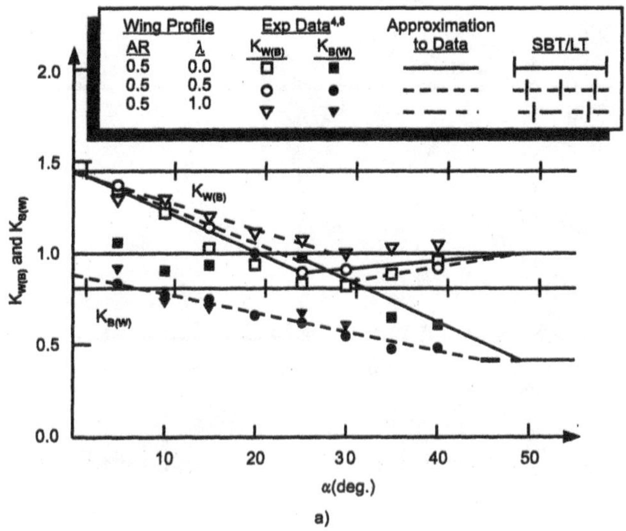

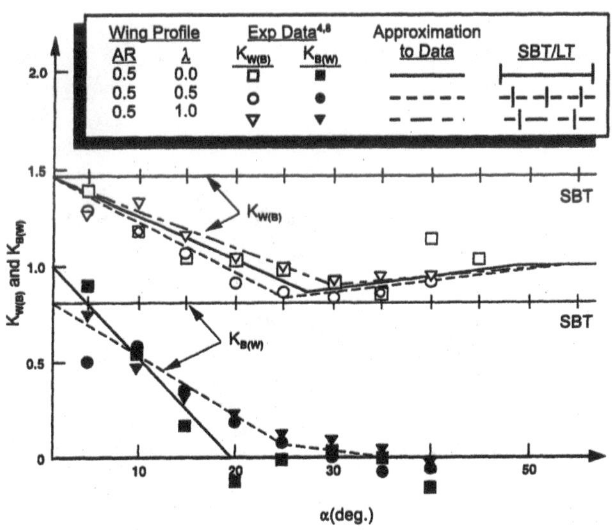

Fig. 5.13 Wing–body and body–wing interference factors as functions of AOA (AR = 0.5, M_∞ = 1.5, and r/s = 0.5): a) φ = 0° and b) φ = 45°.

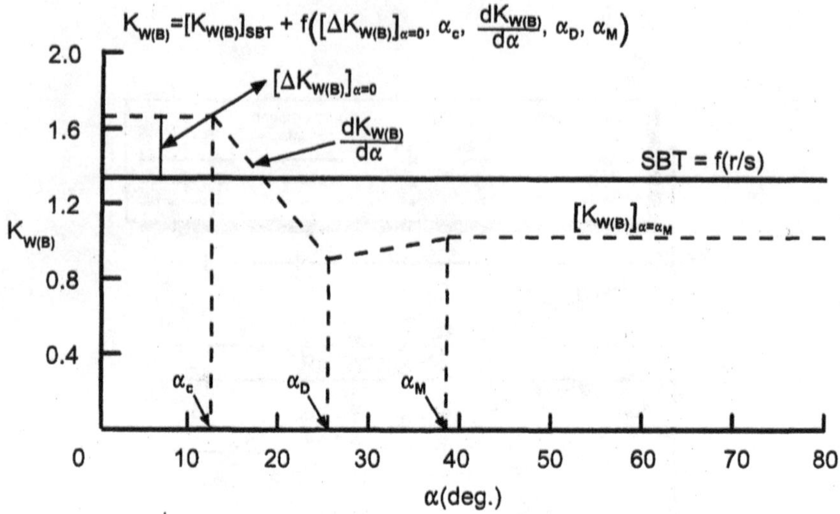

Fig. 5.14a Generic representation of $K_{W(B)}$ with AOA.

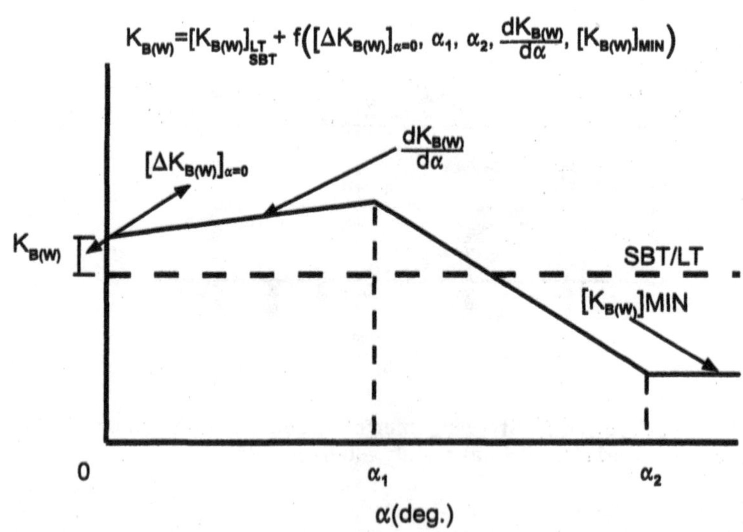

Fig. 5.14b Generic representation of $K_{B(W)}$ with AOA.

The mathematical models for $K_{W(B)}$ and $K_{B(W)}$ are once again defined based on slender body theory and linear theory. The specific equations for $K_{W(B)}$ are

$$
\begin{aligned}
K_{W(B)} &= \left[K_{W(B)}\right]_{\mathrm{SBT}} + \left[\Delta K_{W(B)}\right]_{a=0} \; ; a \le a_C \\
K_{W(B)} &= \left[K_{W(B)}\right]_{\mathrm{SBT}} + \left[\Delta K_{W(B)}\right]_{a=0} + \left| \left(|a| - a_C \right) \right| \frac{\mathrm{d}K_{W(B)}}{\mathrm{d}a} \; ; a_C < a \le a_D \\
K_{W(B)} &= 1 - \left(\frac{a_M - |a|}{a_M - a_D} \right) \left(1 - \left[K_{W(B)}\right]_{a=a_D} \right) ; \; a_D < a \le a_M \\
K_{W(B)} &= \left[K_{W(B)}\right]_{a=a_M} ; a > a_M
\end{aligned}
\tag{28}
$$

Here,

$$
\frac{\mathrm{d}K_{W(B)}}{\mathrm{d}a} = \left(\left[K_{W(B)}\right]_{aD} - \left[K_{W(B)}\right]_{\max} \right) / \left(a_D - a_C \right)
$$

where

$$
\left[K_{W(B)}\right]_{\max} = \left[K_{W(B)}\right]_{\mathrm{SBT}} + \left[\Delta K_{W(B)}\right]_{a=0}
$$

The equations for the body–wing interference factor are

$$
\begin{aligned}
K_{B(W)} &= \left[K_{B(W)}\right]_{\substack{LT\\ \mathrm{SBT}}} + \left[\Delta K_{B(W)}\right]_{a=0} + |a| \frac{\mathrm{d}K_{B(W)}}{\mathrm{d}a} \; ; a \le a_1 \\
K_{B(W)} &= \left[K_{B(W)}\right]_{a=a_1} + \left(\frac{a_1 - |a|}{a_2 - a_1} \right) \left\{ \left[K_{B(W)}\right]_{a=a_1} - \left[K_{B(W)}\right]_{\min} \right\} ; \; a_1 < a \le a_2 \\
K_{B(W)} &= \left[K_{B(W)}\right]_{\min} ; a \ge a_2
\end{aligned}
\tag{29}
$$

Data for the parameters defined in Eqs. (28) and (29) for the roll position of $\varphi = 0$ are given in Tables 5.5 to 5.13 and, for the roll position of $\varphi = 45$ deg, in Tables 5.14 to 5.23. The term $[K_{B(W)}]_{\min}$ is defined in Fig. 5.15 for both $\varphi = 0$ and 45 deg.

Linear interpolation is used in Tables 5.5 to 5.23 to define the parameters in Eqs. (28) and (29) other than $[K_{W(B)}]_{\mathrm{SBT}}$ and $[K_{B(W)}]^{LT}_{\mathrm{SBT}}$, as a function of M_∞, a, AR, and λ. These two terms are calculated analytically using slender body theory or linear theory as described in Chapter 3.

In examining the empirical model for $K_{W(B)}$ and $K_{B(W)}$ defined by Eqs. (28) and (29) in conjunction with Tables 5.5 to 5.23 and Fig. 5.15, some

Table 5.5 Data for $[\Delta K_{W(B)}]_{\alpha=0}$ for $\varphi = 0$ deg

Aspect ratio	Taper ratio	Mach number ≤0.1	0.6	0.8	1.2	≥1.5
≤0.25	0, 0.5, 1.0	−0.30	0	0	0	0
0.5	0.5	0.30	0.27	0.23	0.05	0
1.0	0.5	0.54	0.25	0.10	0	0
≥2.0	0.5	0	0.20	0.20	0.10	0
0.5	0	0.30	0.35	0.42	0.18	0
1.0	0	0.54	0.29	0.16	0.06	0
≥2.0	0	0	0.20	0.20	0.10	0
0.5	1.0	0.30	0.27	0.29	0.05	0
1.0	1.0	0.54	0.31	0.19	0.06	0
≥2.0	1.0	0	0.20	0.20	0.10	0

discussion of the physics that this model tries to represent is in order. The wing–body interference factor is somewhat easier to understand than the body–wing interference. Wing–body experimental data show that, at low Mach number, slender body theory slightly underpredicts experimental data. As AOA is increased, $K_{W(B)}$ starts decreasing and, in some cases, decreases below its wing-alone value. As AOA increases, $K_{W(B)}$ approaches its wing-alone value. As Mach number increases, the positive interference lift on the wing, caused by the presence of the body, is lost faster and faster as AOA increases. That is, the wing-alone solution is recovered much faster at high Mach number, as AOA increases, than at low Mach number. This is believed to be the result of the Newtonian impact mechanism where, at high Mach number, the momentum of the air particle is lost almost entirely on

Table 5.6 Data for α_c (deg) for $\varphi = 0$ deg

Aspect ratio	Taper ratio	Mach number ≤0.1	0.6	0.8	1.2	≥1.5
≤0.25	0, 0.5, 1.0	0	22.0	22.0	0	0
0.5	0.5	30.0	17.3	11.5	10.0	0
1.0	0.5	30.0	15.0	11.0	10.0	0
≥2.0	0.5	10.0	20.0	20.0	15.0	0
0.5	0	30.0	12.0	10.0	10.0	0
1.0	0	30.0	13.0	10.0	10.0	0
≥2.0	0	10.0	20.0	20.0	15.0	0
0.5	1.0	30.0	17.3	15.0	10.0	0
1.0	1.0	30.0	15.0	12.5	10.0	0
≥2.0	1.0	10.0	20.0	20.0	15.0	0

Table 5.7 Data for $[\Delta K_{W(B)}]_{\alpha=\alpha_D}$ for $\varphi = 0$ deg

Aspect ratio	Taper ratio	Mach number ≤0.1	0.6	0.8	1.2	1.5	2.0	2.5	3.0	3.5	4.5	≥5.0
≤0.25	0, 0.5, 1.0	1.0	1.0	1.0	1.0	1.0	1.0	1.0	1.0	1.0	1.0	1.0
0.5	0.5	1.0	1.0	1.0	0.90	0.90	1.0	0.95	1.0	0.97	1.0	1.0
1.0	0.5	1.0	1.0	1.0	0.95	1.0	1.0	1.0	1.0	1.0	1.0	1.0
≥2.0	0.5	1.0	1.0	0.95	0.95	1.0	1.0	1.0	1.0	1.0	1.0	1.0
0.5	0	1.0	1.0	1.0	1.05	0.90	0.90	0.90	0.90	0.90	0.90	1.0
1.0	0	1.0	1.0	1.0	0.95	1.0	1.0	1.0	1.0	1.0	1.0	1.0
≥2.0	0	1.0	1.0	0.95	1.0	1.0	1.0	1.0	1.0	1.0	1.0	1.0
0.5	1.0	1.0	1.0	1.0	1.0	1.0	1.0	1.05	1.15	1.13	1.15	1.0
1.0	1.0	1.0	1.0	1.0	0.95	0.95	0.95	1.0	1.0	1.0	1.0	1.0
≥2.0	1.0	1.0	1.0	1.0	1.0	1.0	1.0	1.0	0.93	0.90	0.95	1.0

Table 5.8 Data for α_D (deg) for $\varphi = 0$ deg

Aspect ratio	Taper ratio	Mach number										
		≤0.1	0.6	0.8	1.2	1.5	2.0	2.5	3.0	3.5	4.5	≥5.0
≤0.25	0, 0.5, 1.0	44.0	40.0	38.0	35.0	30.0	25.0	16.3	15.1	13.9	13.1	12.3
0.5	0.5	50.0	33.0	31.4	27.5	30.0	16.8	17.8	17.0	15.0	15.0	14.0
1.0	0.5	50.0	32.5	39.0	22.0	20.0	22.5	17.5	18.0	10.0	17.0	15.0
≥2.0	0.5	42.0	35.0	35.0	30.0	25.0	16.5	17.0	16.0	10.0	17.0	15.0
0.5	0	50.0	30.0	30.0	21.2	25.0	15.0	14.0	15.0	15.0	12.0	11.5
1.0	0	50.0	31.0	39.0	20.0	18.0	21.5	16.0	17.0	11.0	13.0	13.0
≥2.0	0	42.0	35.0	35.0	30.0	25.0	20.0	17.7	17.0	12.0	12.6	11.5
0.5	1.0	50.0	33.0	34.2	26.0	30.0	14.2	17.0	13.4	11.8	12.2	11.5
1.0	1.0	50.0	33.0	40.0	21.0	20.0	22.0	17.0	16.0	9.0	14.0	12.0
≥2.0	1.0	42.0	35.0	35.0	30.0	25.0	18.0	15.0	15.5	12.0	12.6	11.5

Table 5.9 Data for α_M (deg) for $\varphi = 0$

Aspect ratio	Taper ratio	Mach number ≤0.1	0.6	0.8	1.2	1.5	2.0	2.5	3.0	3.5	4.5	≥5.0
≤0.25	0, 0.5, 1.0	50.0	45.0	45.0	40.0	44.0	38.0	50.0	46.0	50.0	50.0	46.0
0.5	0.5	50.0	33.0	31.4	40.0	50.0	17.0	40.0	17.0	40.0	15.0	14.0
1.0	0.5	50.0	33.0	39.0	45.0	50.0	50.0	50.0	36.0	33.0	17.0	17.0
≥2.0	0.5	50.0	43.0	45.0	30.0	50.0	50.0	50.0	36.0	33.0	17.0	17.0
0.5	0	50.0	30.0	30.0	40.0	50.0	48.0	50.0	50.0	50.0	50.0	50.0
1.0	0	50.0	31.0	40.0	50.0	42.0	50.0	50.0	50.0	44.0	40.0	40.0
≥2.0	0	50.0	43.0	45.0	45.0	50.0	50.0	50.0	50.0	50.0	50.0	35.0
0.5	1.0	50.0	33.0	34.2	50.0	31.0	50.0	50.0	50.0	50.0	50.0	50.0
1.0	1.0	50.0	33.0	40.0	50.0	42.0	50.0	50.0	50.0	44.0	40.0	40.0
≥2.0	1.0	50.0	43.0	45.0	45.0	25.0	18.0	15.0	36.0	33.0	37.0	30.0

Table 5.10 Data for $[\Delta K_{B(W)}]_{\alpha=0}$ for $\varphi = 0$ deg

Aspect ratio	Taper ratio	Mach number ≤0.1	0.6	0.8	1.2	1.5	2.0	2.5	3.0	3.5	4.5	≥5.0
≤0.25	0, 0.5, 1.0	0.0	0.0	0.0	0.0	0.0	0.0	0.0	0.0	0.0	0.0	0.0
0.5	0.5	0.0	−0.28	−0.15	0.16	0.10	−0.02	0.0	0.0	0.0	0.0	0.0
1.0	0.5	0.0	−0.20	−0.20	0.15	0.20	0.05	0.0	0.0	0.0	0.0	0.0
≥2.0	0.5	0.0	−0.20	−0.07	0.17	0.18	0.10	0.0	0.0	0.0	0.0	0.0
0.5	0	0.0	−0.33	−0.30	0.28	0.20	0.10	0.08	0.0	0.0	0.0	0.0
1.0	0	0.0	−0.24	−0.25	0.13	0.28	0.05	0.0	0.0	0.0	0.0	0.0
≥2.0	0	0.0	−0.20	−0.07	0.17	0.0	0.05	0.0	0.0	0.0	0.0	0.0
0.5	1.0	0.0	−0.28	−0.15	0.25	0.0	0.10	0.0	0.0	0.0	0.0	0.0
1.0	1.0	0.0	−0.20	−0.20	0.22	0.10	0.05	0.0	0.0	0.0	0.0	0.0
≥2.0	1.0	0.0	−0.20	−0.07	0.17	0.20	0.10	0.15	0.0	0.0	0.0	0.0

Table 5.11 Data for $dK_{B(W)}/d\alpha$ (per deg) for $\varphi = 0$ deg

Aspect ratio	Taper ratio	Mach number										
		≤ 0.1	0.6	0.8	1.2	1.5	2.0	2.5	.30	3.5	4.5	≥ 5.0
≤ 0.25	0, 0.5, 1.0	0.0	0.0	0.0	0.0	0.0	0.0	−0.007	−0.014	−0.015	−0.02	−0.024
0.5	0.5	0.006	0.023	0.023	−0.009	−0.012	−0.010	−0.015	−0.014	−0.015	−0.016	−0.020
1.0	0.5	0.006	0.012	0.011	−0.003	−0.003	−0.005	−0.006	−0.008	−0.010	−0.012	−0.015
≥ 2.0	0.5	0.0	0.012	0.011	0.0	0.0	−0.001	−0.012	−0.014	−0.015	−0.016	−0.020
0.5	0	0.006	0.043	0.058	0.0	0.0	0.0	−0.004	−0.014	−0.015	−0.016	−0.020
1.0	0	0.006	0.020	0.0225	−0.003	−0.003	−0.005	−0.006	−0.008	−0.010	−0.012	−0.015
≥ 2.0	0	0.0	0.012	0.011	0.0	0.0	−0.002	−0.012	−0.014	−0.015	−0.016	−0.020
0.5	1.0	0.006	0.038	0.033	−0.013	−0.012	−0.010	−0.015	−0.014	−0.015	−0.016	−0.020
1.0	1.0	0.006	0.007	0.005	−0.003	−0.010	−0.010	−0.015	−0.016	−0.016	−0.016	−0.018
≥ 2.0	1.0	0.0	0.012	0.011	0.0	−0.002	−0.007	−0.012	−0.014	−0.015	−0.016	−0.020

Table 5.12 Data for α_1 (deg) for $\varphi = 0$ deg

Aspect ratio	Taper ratio	Mach number										
		≤0.1	0.6	0.8	1.2	1.5	2.0	2.5	3.0	3.5	4.5	≥5.0
≤0.25	0, 0.5, 1.0	30.0	21.1	16.5	45.0	37.0	33.3	23.3	20.5	18.0	15.0	14.0
0.5	0.5	30.0	22.2	16.7	62.0	43.0	40.0	25.0	25.0	25.0	20.0	20.0
1.0	0.5	30.0	25.0	20.0	70.0	30.0	25.0	28.6	23.0	20.4	26.0	26.0
≥2.0	0.5	30.0	25.0	20.0	40.0	66.0	58.0	30.0	24.0	20.4	26.0	26.0
0.5	0	30.0	24.2	17.2	25.0	25.0	20.0	20.0	10.0	27.0	20.0	20.0
1.0	0	30.0	25.0	20.0	70.0	61.0	18.0	27.0	18.0	24.0	24.0	24.0
≥2.0	0	30.0	25.0	20.0	40.0	48.5	49.0	30.0	32.0	30.0	26.0	26.0
0.5	1.0	30.0	17.0	15.5	48.5	43.0	40.0	25.0	26.5	21.6	20.0	20.0
1.0	1.0	30.0	25.0	20.0	70.0	54.0	22.0	29.5	23.5	18.0	22.0	22.0
≥2.0	1.0	30.0	25.0	20.0	40.0	48.0	47.0	32.0	26.0	20.0	26.0	26.0

Table 5.13 Data for α_2 (deg) for $\varphi = 0$ deg

Aspect ratio	Taper ratio	Mach number										
		≤0.1	0.6	0.8	1.2	1.5	2.0	2.5	3.0	3.5	4.5	≥5.0
≤0.25	0, 0.5 1.0	90.0	80.0	65.0	63.4	45.0	43.3	42.5	31.5	37.3	40.0	40.0
0.5	0.5	90.0	80.0	65.0	62.0	43.0	41.0	42.5	25.0	42.0	40.0	40.0
1.0	0.5	90.0	80.0	80.0	80.0	65.0	46.0	40.0	36.0	40.0	40.0	40.0
≥2.0	0.5	90.0	80.0	80.0	80.0	90.0	90.0	42.0	40.0	40.0	40.0	40.0
0.5	0	90.0	80.0	80.0	80.0	49.0	47.8	42.5	43.0	26.5	40.0	40.0
1.0	0	90.0	80.0	80.0	80.0	59.0	46.0	40.0	40.0	34.0	40.0	40.0
≥2.0	0	90.0	80.0	80.0	80.0	90.0	90.0	41.0	35.0	40.0	43.0	43.0
0.5	1.0	90.0	80.0	53.2	48.7	43.0	41.0	42.5	26.5	43.5	40.0	40.0
1.0	1.0	90.0	80.0	74.0	72.0	55.0	46.0	40.0	32.0	40.0	40.0	40.0
≥2.0	1.0	90.0	80.0	80.0	80.0	90.0	90.0	45.0	30.0	40.0	43.0	43.0

Table 5.14 Data for $[\Delta K_{W(B)}]_{\alpha=0}$ at $\varphi = 45$ deg

Aspect ratio	Taper ratio	Mach number ≤0.1	0.6	0.8	1.2	1.5	2.0	2.5	3.0	3.5	4.5	≥5.0
≤0.25	0, 0.5, 1.0	−0.20	0.00	0.00	0.00	0.00	0	0	0	0	0	0
0.5	0.5	0.00	0.05	0.00	−0.13	0.00	0	0	0	0	0	0
1.0	0.5	0.13	0.05	0.00	0.00	−0.10	0	0	0	0	0	0
≥2.0	0.5	0.00	0.00	0.00	0.00	0.00	0	0	0	0	0	0
0.5	0	0.00	0.10	0.14	0.00	−0.22	0	0	0	0	0	0
≥2.0	0.00	0.00	0.00	0.00	0.00	−0.18	0	0	0	0	0	
0.5	1.0	0.00	0.10	0.05	−0.23	0.00	0	0	0	0	0	0
≥2.0	1.0	0.00	0.00	0.00	0.00	0.00	0	0	0	0	0	0

Table 5.15 Data for α_c at $\varphi = 45$ deg

Aspect ratio	Taper ratio	Mach number ≤0.1	0.6	0.8	1.2	1.5	2.0	2.5	3.0	3.5	4.5	≥5.0
≤0.25	0, 0.5, 1.0	10.0	22.0	22.0	0.0	0.0	0	0	0	0	0	0
0.5	0.5	45.0	11.5	11.0	10.0	0.0	0	0	0	0	0	0
1.0	0.5	45.0	13.3	0.0	6.5	0.0	0	0	0	0	0	0
≥2.0	0.5	20.0	10.0	0.0	6.5	2.2	0	0	0	0	0	0
0.5	0	39.0	15.0	11.5	10.0	0.0	0	0	0	0	0	0
≥ 2.0	0	20.0	10.0	0.0	6.5	0.0	0	0	0	0	0	0
0.5	1.0	45.0	15.0	15.0	15.0	0.0	0	0	0	0	0	0
≥2.0	1.0	20.0	10.0	0.0	6.5	1.5	0	0	0	0	0	0

Table 5.16 Data for $[\Delta K_{W(B)}]_{\alpha=\alpha_D}$ at $\varphi = 45$ deg

Aspect ratio	Taper ratio	Mach number ≤0.1	0.6	0.8	1.2	1.5	2.0	2.5	3.0	3.5	4.5	≥6.0
≤0.25	0, 0.5, 1.0	1.0	1.0	1.00	1.00	1.00	1.00	1.00	1.00	1.00	1.00	1.0
0.5	0.5	1.0	1.0	1.00	0.90	0.90	1.00	0.95	1.00	0.97	1.00	1.0
1.0	0.5	1.0	1.0	1.00	0.95	1.00	1.00	1.00	1.00	1.00	1.00	1.0
≥2.0	0.5	1.0	1.0	0.95	0.95	1.00	1.00	1.00	1.00	1.00	1.00	1.0
0.5	0	1.0	1.0	1.00	1.00	0.90	0.90	0.90	0.90	0.90	0.90	1.0
≥ 2.0	0	1.0	1.0	0.95	1.00	1.00	1.00	1.00	1.00	1.00	1.00	1.0
0.5	1.0	1.0	1.0	1.00	1.00	1.00	1.00	1.00	1.00	1.00	1.00	1.0
≥2.0	1.0	1.0	1.0	1.00	1.00	1.00	1.00	1.00	0.93	0.90	0.95	1.0

Table 5.17 Data for α_D at $\varphi = 45$ deg

Aspect ratio	Taper ratio	Mach number ≤ 0.1	0.6	0.8	1.2	1.5	2.0	2.5	3.0	3.5	4.5	≥ 6.0
≤ 0.25	0, 0.5, 1.0	20.0	40.0	38.0	35.0	30.0	25.0	16.3	15.1	13.9	13.1	10.0
0.5	0.5	59.0	33.0	30.0	25.6	25.0	15.0	15.0	10.0	15.0	12.0	10.0
1.0	0.5	59.0	38.0	32.0	26.0	24.0	17.0	15.0	14.4	10.0	10.0	10.0
≥ 2.0	0.5	39.0	31.5	30.0	28.0	25.0	16.5	15.0	14.4	10.0	13.0	10.0
0.5	0	39.0	35.5	33.0	39.5	29.5	15.0	25.0	15.0	15.0	10.0	10.0
≥ 2.0	0	39.0	31.5	30.0	28.0	24.7	17.0	13.5	11.4	10.0	10.0	10.0
0.5	1.0	59.0	35.5	33.0	25.6	29.5	15.0	15.0	15.0	12.0	13.0	10.0
≥ 2.0	1.0	39.0	31.5	30.0	28.0	23.3	14.0	16.0	15.0	11.8	12.0	10.0

Table 5.18 Data for a_M at $\varphi = 45$ deg

Aspect ratio	Taper ratio	Mach number										
		≤0.1	0.6	0.8	1.2	1.5	2.0	2.5	3.0	3.5	4.5	≥6.0
≤0.25	0, 0.5, 1.0	40.0	45.0	45.0	40.0	44.0	43.0	38.0	28.0	25.0	29.0	20.0
0.5	0.5	65.0	33.0	30.0	49.0	52.0	40.0	40.0	30.0	25.0	25.0	20.0
1.0	0.5	65.0	38.0	47.0	49.5	66.0	48.5	45.0	41.0	40.0	10.0	20.0
≥2.0	0.5	40.0	31.5	40.0	56.0	57.0	45.0	45.0	41.0	40.0	28.0	20.0
0.5	0	40.0	35.5	33.0	65.0	48.0	50.0	46.0	30.0	30.0	50.0	20.0
≥2.0	0	40.0	31.5	40.0	56.0	55.0	58.5	49.8	44.2	41.5	28.5	20.0
0.5	1.0	65.0	35.5	33.0	49.0	52.0	40.0	28.0	24.0	21.0	13.0	20.0
≥2.0	1.0	40.0	31.5	40.0	56.0	49.5	44.0	40.0	33.0	32.0	28.0	20.0

Table 5.19 Data for $[\Delta K_{W(B)}]_{\alpha=\alpha_M}$ at φ 45 deg

Aspect ratio	Taper ratio	Mach number ≤0.1	0.6	0.8	1.2	1.5	2.0	2.5	3.0	3.5	4.5	≥6.0
≤0.25	0, 0.5, 1.0	0.80	0.95	1.0	1.0	1.0	1.0	1.0	1.0	1.0	1.0	1.0
0.5	0.5	0.80	0.95	1.0	1.0	1.0	1.0	1.0	1.0	1.0	1.0	1.0
1.0	0.5	0.80	0.90	1.0	1.0	1.0	1.0	1.0	1.0	1.0	1.0	1.0
≥2.0	0.5	0.80	0.90	1.0	1.0	1.0	1.0	1.0	1.0	1.0	1.0	1.0
0.5	0	0.85	0.95	1.0	1.0	1.0	1.0	1.0	1.0	1.0	1.0	1.0
≥2.0	0	0.85	0.95	1.0	1.0	1.0	1.0	1.0	1.0	1.0	1.0	1.0
0.5	1.0	0.80	0.95	1.0	1.0	1.0	1.0	1.0	1.0	1.0	1.0	1.0
≥2.0	1.0	0.80	0.95	1.0	1.0	1.0	1.0	1.0	1.0	1.0	1.0	1.0

Table 5.20 Data for $[\Delta K_{B(W)}]_{\alpha=0}$ for $\varphi = 45$ deg

Aspect ratio	Taper ratio	Mach number										
		≤0.1	0.6	0.8	1.2	1.5	2.0	2.5	3.0	3.5	4.5	≥6.0
≤0.25	0, 0.5, 1.0	−0.10	−0.18	0.00	0.00	0.0	0	0	0	0	0	0
0.5	0.5	0.0	−0.22	0.00	0.00	0.2	0.1	0.08	0	0	0	0
1.0	0.5	0.0	−0.07	−0.18	0.20	0.2	0	0	0	0	0	0
≥2.0	0.5	0.0	−0.23	−0.18	0.20	0.0	0	0	0	0	0	0
0.5	0	0.0	−0.12	0.00	0.25	0.2	0	0	0	0	0	0
≥2.0	0	0.0	−0.23	−0.18	0.20	0.0	0	0	0	0	0	0
0.5	1.0	0.0	−0.22	0.00	0.25	0.2	0.1	0.08	0	0	0	0
≥2.0	1.0	0.0	−0.23	−0.18	0.20	0.0	0	0	0	0	0	0

Table 5.21 Data for $dK_{B(W)}/d\alpha$ (per deg) for $\varphi = 45$ deg

Aspect ratio	Taper ratio	Mach number ≤0.1	0.6	0.8	1.2	1.5	2.0	2.5	3.0	3.5	4.5	≥6.0
≤0.25	0, 0.5, 1.0	−0.0050	−0.00557	−0.00925	−0.0215	−0.0238	−0.0450	−0.0268	−0.0650	−0.0650	−0.060	−0.06
0.5	0.5	0.0015	−0.00210	−0.00720	−0.0100	−0.0140	−0.0440	−0.0275	−0.0330	−0.0620	−0.060	−0.06
1.0	0.5	−0.0030	0.00000	0.00750	−0.0150	−0.0150	−0.020	−0.025	−0.0220	−0.0143	−0.013	−0.01
≥2.0	0.5	0.0030	0.00670	0.01330	−0.0150	−0.0170	−0.0300	−0.045	−0.054	−0.060	−0.062	−0.065
0.5	0	−0.0020	−0.00830	−0.01000	−0.0170	−0.0260	−0.005	−0.0125	−0.0100	−0.005	−0.005	−0.06
≥2.0	0	0.0030	0.00670	0.01330	−0.0150	−0.0170	−0.0300	−0.045	−0.054	−0.060	−0.062	−0.065
0.5	1.0	0.0015	−0.00120	−0.00100	−0.0140	−0.0140	−0.0425	−0.0275	−0.0400	−0.0720	−0.060	−0.06
≥2.0	1.0	0.0030	0.00670	0.01330	−0.0150	−0.0170	−0.0300	−0.045	−0.054	−0.060	−0.062	−0.065

Table 5.22 Data for α_1 (deg) for $\varphi = 45$ deg

Aspect ratio	Taper ratio	Mach number									
		≤ 0.1	0.6	0.8	1.2	1.5	2.0	2.5	3.0	3.5	4.5	≥6.0
≤.025	0, 0.5, 1.0	10.0	45.5	35.0	30.0	23.0	22.0	20.8	20.0	20.0	15.0	8.0
0.5	0.5	10.0	57.0	45.0	30.0	25.0	16.0	20.0	15.0	10.0	10.0	8.0
1.0	0.5	10.0	34.5	35.0	35.0	36.6	10.0	15.0	17.5	42.0	40.0	30.0
≥2.0	0.5	10.0	15.0	30.0	35.0	20.0	20.0	18.0	17.5	30.0	35.0	35.0
0.5	0	10.0	35.0	45.0	30.0	19.0	20.0	22.5	15.0	15.0	15.0	8.0
≥2.0	0	20.0	15.0	30.0	35.0	20.0	20.0	18.0	17.5	30.0	35.0	35.0
0.5	1.0	10.0	57.0	45.0	30.0	25.0	16.0	20.0	15.0	10.0	10.0	8.0
≥2.0	1.0	10.0	15.0	30.0	35.0	20.0	10.0	15.0	17.5	30.0	35.0	35.0

Table 5.23 Data for α_2 (deg) for $\varphi = 45$ deg

Aspect ratio	Taper ratio	Mach number ≤0.1	0.6	0.8	1.2	1.5	2.0	2.5	3.0	3.5	4.5	≥6.0
≤0.25	0, 0.5, 1.0	55.0	55.0	50.0	50.0	45.0	37.0	44.0	29.5	29.5	29.5	25.0
0.5	0.5	75.0	65.0	55.0	43.0	40.0	38.0	44.0	44.0	36.0	30.0	20.0
1.0	0.5	75.0	65.0	60.0	60.0	60.0	60.0	62.0	80.0	42.0	40.0	30.0
≥2.0	0.5	75.0	65.0	60.0	60.0	50.0	60.0	62.0	80.0	42.0	45.0	45.0
0.5	0	75.0	60.0	60.0	52.0	40.0	35.0	44.0	50.0	36.0	30.0	20.0
≥2.0	0	75.0	65.0	60.0	60.0	50.0	60.0	62.0	80.0	42.0	45.0	45.0
0.5	1.0	75.0	65.0	55.0	42.0	40.0	38.0	44.0	40.0	36.0	30.0	20.0
≥2.0	1.0	75.0	65.0	60.0	60.0	60.0	60.0	62.0	80.0	42.0	45.0	45.0

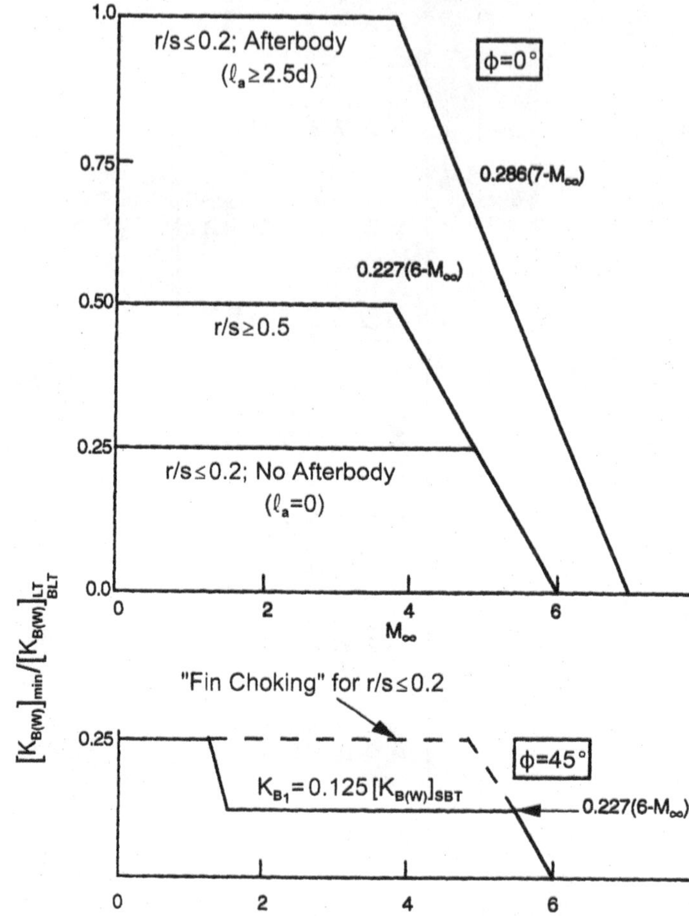

Fig. 5.15 Minimum value of body–wing interference factor at high AOAs.

direct impact of a surface, as opposed to wrapping around the surface and carrying some of the momentum with it, as at low Mach numbers.

The $K_{B(W)}$ model contains body vortex effects, nose- and wing-to-wing shock effects, as well as the usual added dynamic pressure of the body caused by the presence of the wing. While some of the trends in Tables 5.10 to 5.13 and 5.20 to 5.23 and Fig. 5.15 can be rationalized, others cannot, except in light of these combination effects. The stronger variations in the tables for $K_{B(W)}$ than in the tables for $K_{W(B)}$ are caused, in part, by trying to approximately model all three of these phenomena simultaneously. The alternative is to try to estimate the effects of the body vortices and nose- and wing-shock interactions and subtract them. This process not only complicates the methodology but adds additional inherent errors because these effects cannot be easily and accurately estimated. The present approach neglects some of the scale effects caused by the position of the wing on the body. However, this error is probably smaller than approximating analytically the other effects, subtracting them, and then adding them again for a different geometrical configuration.

In general, $K_{B(W)}$ actually increases with AOA at low Mach numbers to a certain point, where it starts decreasing analogously to $K_{W(B)}$. However, a certain amount of lift or force enhancement is gained all the way to $a = 90$ deg for low Mach numbers. This phenomenon is assumed to occur all the way to $M_\infty = 6.0$, based on extrapolated data from the point where experimental data end, which is AOA of 25 to 40 deg depending on Mach number and to $a = 90$ deg, shown in Fig. 5.15 as $[K_{B(W)}]_{\min}$.

Additional higher AOA data above $a = 40$ deg are needed for both $K_{W(B)}$ and $K_{B(W)}$ to modify assumed extrapolations of the models for $K_{W(B)}$ and $K_{B(W)}$ at high AOA. However, until additional data are available, approximate nonlinear models for $K_{W(B)}$ and $K_{B(W)}$ can be used to estimate aerodynamics for engineering use. This statement will be validated for a limited set of flight conditions in a later section.

Also notice that $[K_{B(W)}]_{\min}$ of Fig. 5.15 is significantly different for $\varphi = 0$ than $\varphi = 45$ deg. This is one of the main reasons that the normal force on a wing–body configuration is lower at $\varphi = 45$ deg than at $\varphi = 0$ deg when a goes above 20 to 25 deg. Apparently the windward plane fins shield the wrap around effect of the flow around the body and the fins in the leeward plane area do not produce enough body carryover lift in the $\varphi = 45$ deg roll position to make up for the carryover lift at $\varphi = 0$ deg. Again, referring to Fig. 5.15, note that for $\varphi = 45$ deg roll and small values of r/s, the minimum value of $K_{B(W)}$ is higher when "fin choking" is present. "Fin choking" is a phenomenon similar to that which occurs when an inlet becomes unstarted or a wind tunnel achieves its maximum rate of flow (an increase in power produces no more mass flow through the inlet). As the body increases in AOA with the fins oriented in the "x" or cross orientation, the flow between the fins will eventually "choke" at some AOA and at moderate to large supersonic Mach numbers. When this happens, a strong shock is formed just in front of the fin,[13,14] producing a high-pressure region on the fins and body. This high-pressure region is shifted forward from where it would be if supersonic flow occurred through the fins. While the absolute value of pressure on the body is higher than for the unchoked flow, it occurs over a much smaller region and hence gives only slightly higher body–wing interference normal force.

It should also be stated that, while many aerodynamic nonlinearities are included in the missile component configuration databases of Refs. 4, 5, and 11, several are not. These are associated with a combination of high Mach number and AOA. The "fin choking" phenomenon is partially accounted for in Fig. 5.15. However, when the bow shock intersects the wing, resulting in a trailing shock that intersects the tail, these effects are not accounted for in the present method. This is because wind tunnel databases had only one set of lifting surfaces present, and these were generally in the tail region of the body.

In summary, it is fair to say that, while the present approximate method takes into account some of the nonlinearities in missile aerodynamics due to internal shocks, it is far from complete. Additional new data, combined with numerical computations, are needed to develop an approximate engineering model to represent these nonlinearities accurately and hence completely cover the desired Mach number and AOA boundary for all configurations of interest to weapons designers.

The center of pressure of the wing–body normal force is assumed to occur at the same location as the wing-alone values defined by Eqs. (20) for $\varphi = 0$ deg in conjunction with Eq. (24) for $\varphi = 45$ deg roll. The body–wing center of pressure at low AOA is estimated by the linear theory methods of Section 3.9.1. As AOA increases, the local Mach number in the windward plane decreases for supersonic flow and the Mach angle that defines the Mach box of Fig. 3.28 decreases according to

$$\mu_L = \sin^{-1}\left(\frac{1}{M_L}\right) \tag{30}$$

The local Mach number M_L can be approximated by using modified Newtonian theory where

$$M_L = \left\{\left(\frac{2}{\gamma - 1}\right)\left[\left(\frac{p_{O_2}}{p_L}\right)^{\frac{\gamma-1}{\gamma}} - 1\right]\right\}^{1/2} \tag{31}$$

and

$$\frac{p_{O_2}}{p_L} = \frac{\frac{\gamma M_\infty^2}{2} C_{P_O} + 1}{\frac{\gamma M_\infty^2}{2} C_P + 1} \tag{32}$$

$$C_{P_O} = \frac{2}{\gamma M_\infty^2}\left\{\left[\frac{(\gamma + 1)M_\infty^2}{2}\right]^{\frac{\gamma}{\gamma-1}}\left[\frac{\gamma + 1}{2\gamma M_\infty^2 - (\gamma - 1)}\right]^{\frac{1}{\gamma-1}} - 1\right\} \tag{33}$$

$$\delta_{eq} = \sin^{1}(\sin\theta\cos a - \sin a\cos\varphi\cos\theta) \tag{34}$$

$$C_P = C_{P_O}\sin^2(\delta_{eq}) \tag{35}$$

Use of Eqs. (30) to (35) to determine the local Mach number for use in determining the area affected by the carryover pressure on the body has the effect of moving the center of pressure of the body carryover term forward toward the wing apex as AOA increases.

As an illustration of the nonlinear methodology for the wing–body and body–wing interference due to AOA, consider the configuration of Fig. 5.16a, which is the aspect ratio 0.5, taper ratio 0.5 case of the NASA database illustrated in Fig. 4.35. The Fig. 5.16a configuration was tested at several Mach numbers, to AOA 40 deg in most cases and at various roll orientations. Data were taken in component buildup fashion so $C_{N_{W(B)}}$ was available directly and $\Delta C_{N_{B(W)}}$ could be calculated from the data by use of Eq. (27). Examples are shown in Fig. 5.16b and 5.16c for $C_{N_{W(B)}}$ and $\Delta C_{N_{B(W)}}$ at the $\varphi = 0$ and 45 deg roll orientation, respectively, for the Mach numbers of 0.6, 2.0, and 4.5. Critical crossflow Reynolds number and Mach number of 330,000 and 0.0 were used in the nonlinear computations. Both linear theory and nonlinear theory are compared to experiment. As seen in the figures, for this case, the linear theory gives reasonable agreement with data up to at least a 5-deg AOA at all Mach numbers and to higher AOA for supersonic Mach numbers. The nonlinear theory does an excellent job in predicting the total wing–body load and additional load on the body due to the presence of the fins. Figures 5.16d and 5.16e present the total normal force and center of pressure for the Fig. 5.16a configuration, again at $\varphi = 0$ and 45 deg roll, respectively. Linear theory is independent of roll so it gives the same values of C_N and X_{CP}. However, note the nonlinear theory does a very good job in predicting the static aerodynamics at all AOAs for which data are available. While not shown, this statement applies to other configuration in the NASA database. However, in fairness, nonlinear methods were based on the NASA database primarily so they should do well for these configurations. The true test of the accuracy of nonlinear methods will come when comparisons are made to configurations outside this database, later in Chapter 5.

V. Wing–Body and Body–Wing Interference Due to Control Deflection

The initial attempt at developing a nonlinear model for $k_{W(B)}$ and $k_{B(W)}$ was to use slender body theory in conjunction with the database of Ref. 4. However, due to the small size of the fins deflected in Ref. 4, this approach

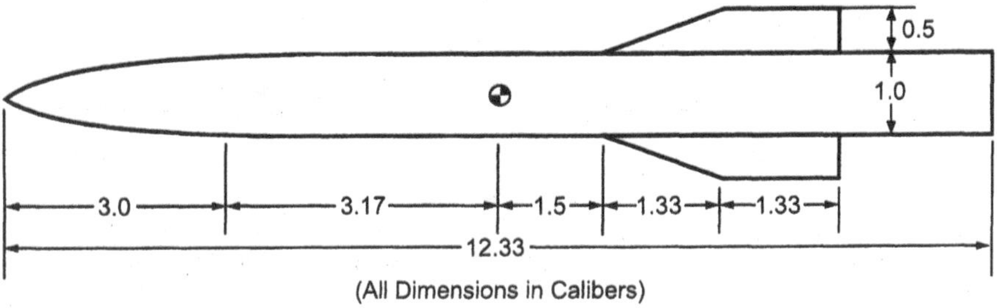

Fig. 5.16a Body–tail configuration of NASA database[4] (AR = 0.5 and λ = 0.5).

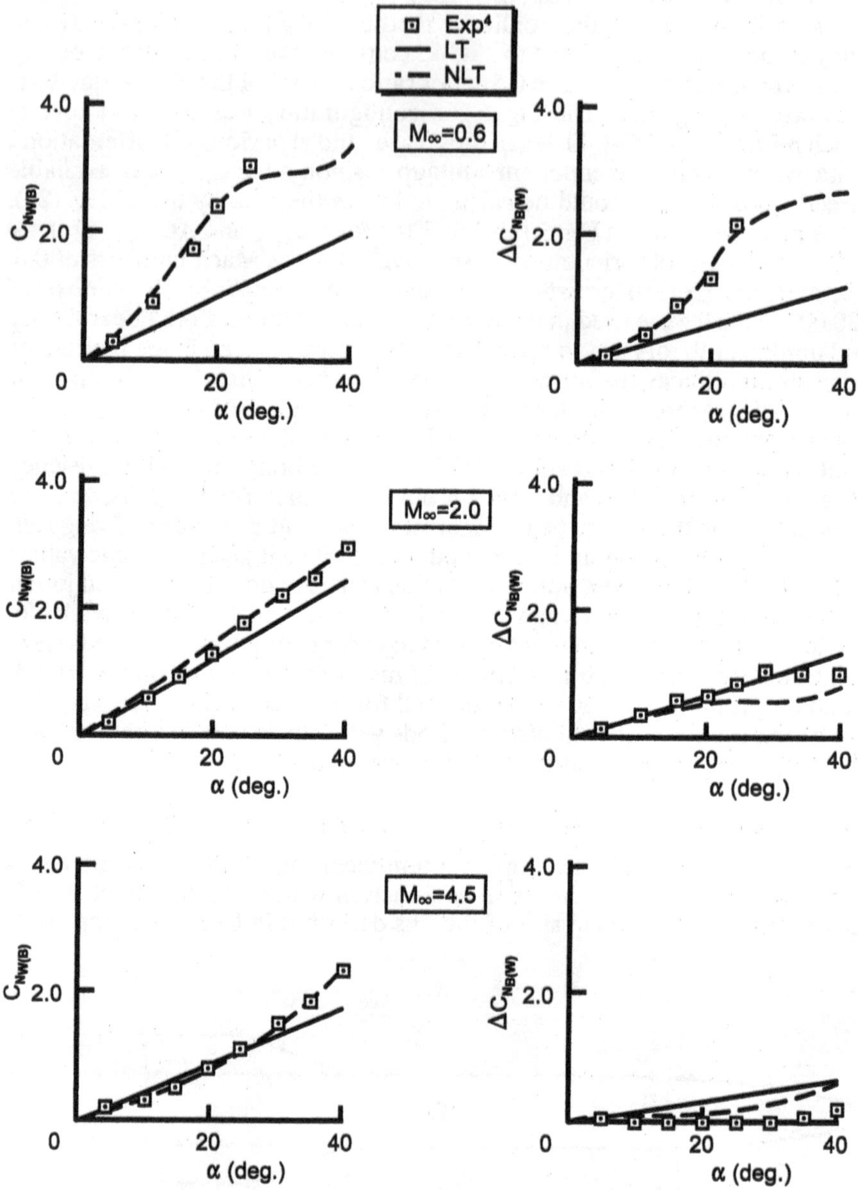

Fig. 5.16b Comparison of experiment, linear theory, and nonlinear theory for wing–body and body–wing normal force components due to AOA at roll position of $\varphi = 0°$ (AR = 0.5 and $\lambda = 0.5$).

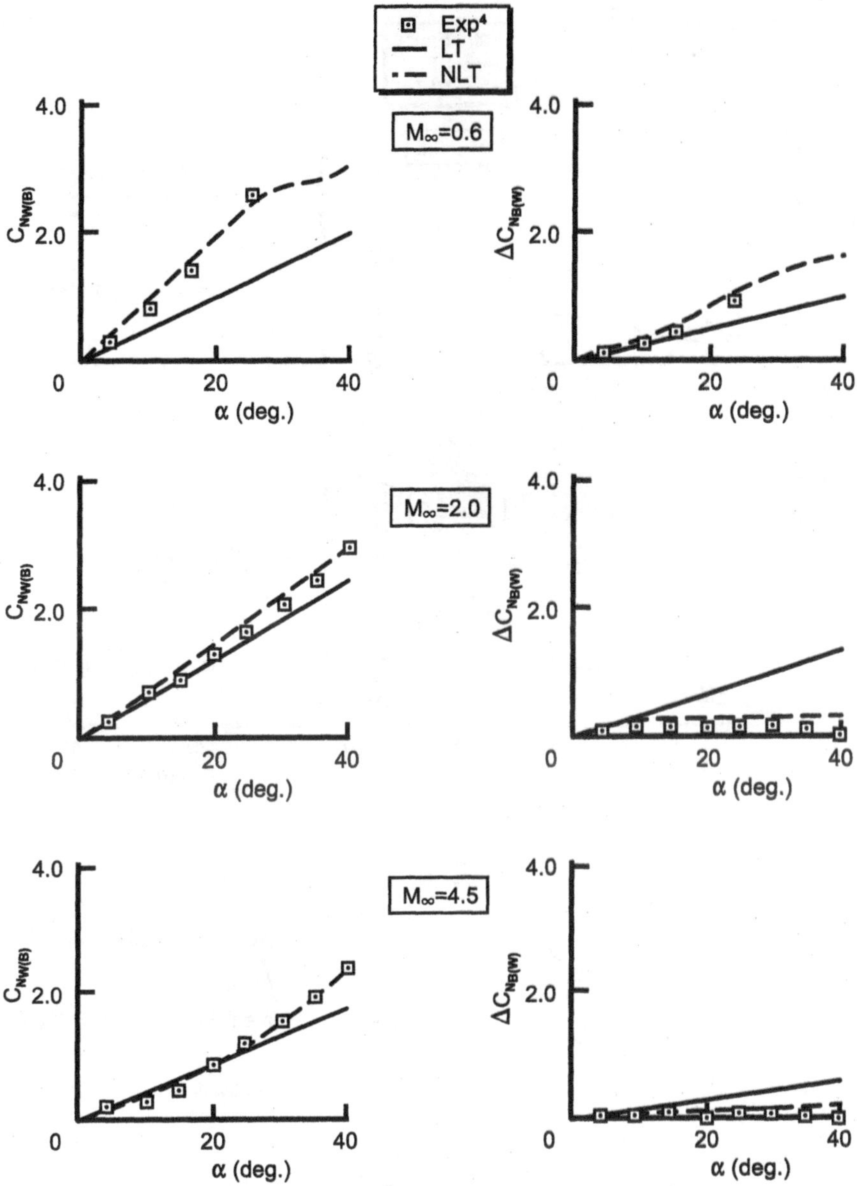

Fig. 5.16c Comparison of experiment, linear theory, and nonlinear theory for wing–body and body–wing normal force components due to AOA at roll position of $\varphi = 45°$ (AR = 0.5 and $\lambda = 0.5$).

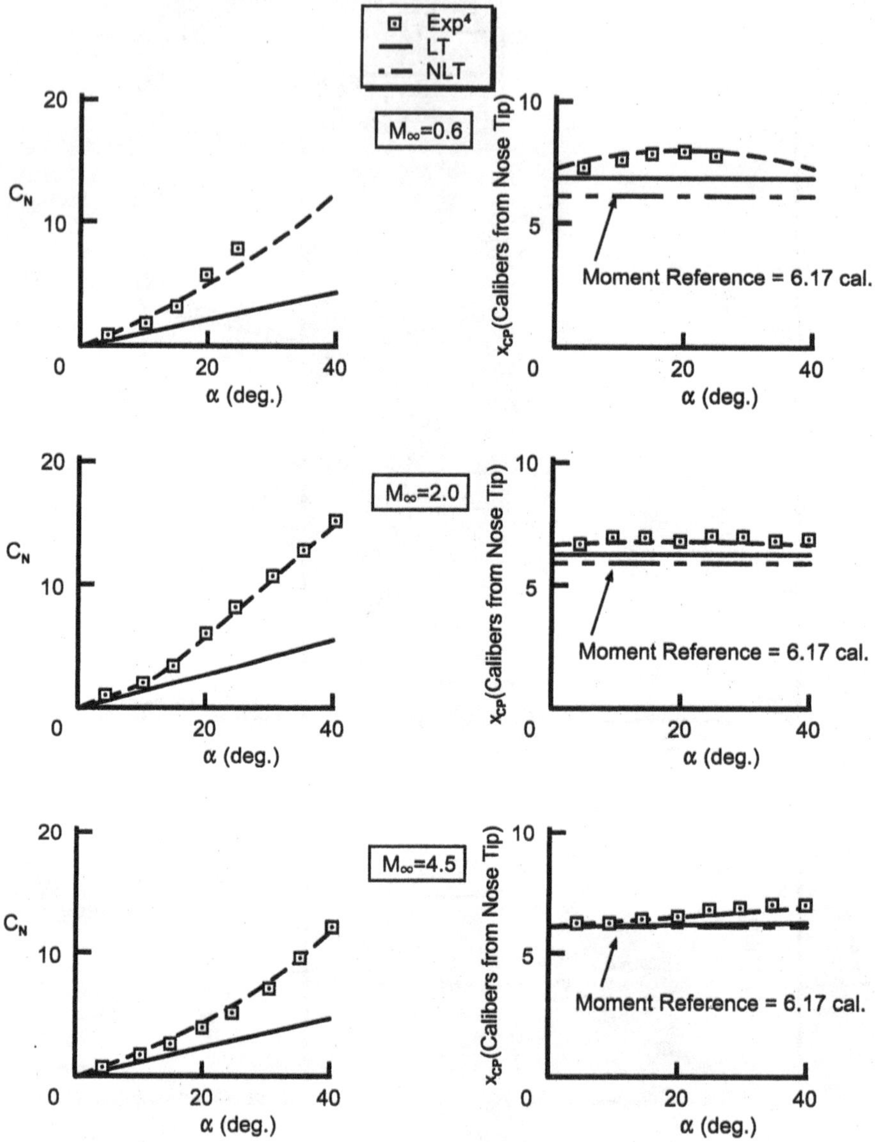

Fig. 5.16d Comparison of experiment, linear theory, and nonlinear theory for normal force coefficient and center of pressure for Fig. 5.16a configuration (AR = 0.5, $\lambda = 0.5$, and $\varphi = 0°$).

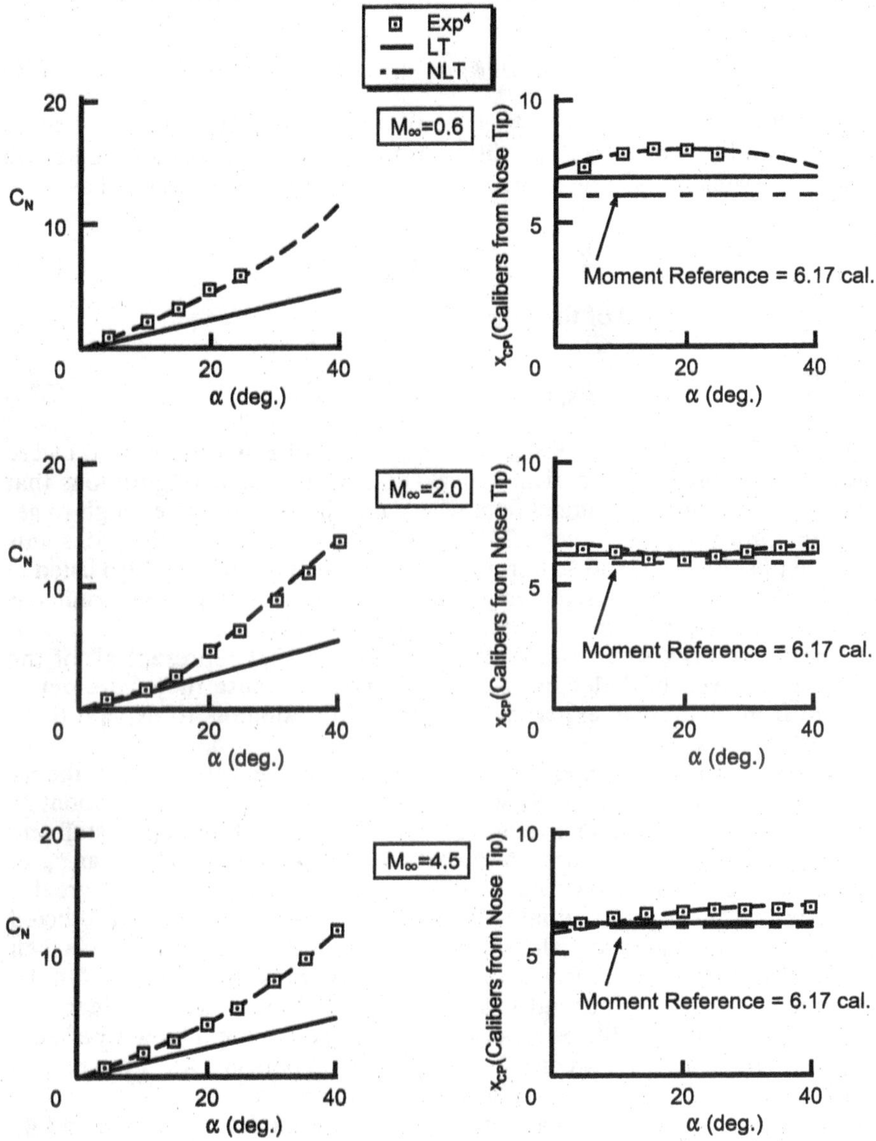

Fig. 5.16e Comparison of experiment, linear theory, and nonlinear theory for normal force coefficient and center of pressure for Fig. 5.16a configuration (AR = 0.5, λ = 0.5, and φ = 45°).

proved difficult. The second alternative the author has used with success is to model the wing–body interference due to control deflection as

$$k_{W(B)} = C_1(M_\infty)\,[k_{W(B)}]_{\mathrm{SBT}} + C_2(|\alpha_W|, M_\infty) \tag{36}$$

for both the $\varphi = 0$ and $\varphi = 45$ deg roll orientations. However, the values $C_1(M_\infty)$ and $C_2(|\alpha_W|, M_\infty)$ are different for the two roll positions. Likewise the body–wing interference due to control deflection was modeled as

$$k_{B(W)} = [k_{B(W)}]_{\mathrm{SBT}} + C(\alpha_W) \tag{37a}$$

for $\varphi = 0$ deg roll and of the form

$$k_{B(W)} = [k_{B(W)}]_{\mathrm{SBT}} + C(\delta_W) \tag{37b}$$

for $\varphi = 45$ deg roll. The constants C_1, C_2, and C of Eqs. (36) and (37) were determined through numerical experiments on missile configurations that had deflected fin wind tunnel data where the fins were large enough to get accurate information. Tables 5.24 and 5.25 give the value of these constants for roll positions of 0 and 45 deg, respectively. The parameter F also listed in Tables 5.24 and 5.25 is an empirical parameter used in the nonlinear wing–tail interference model discussed in the next section.

The constants C_1, C_2, and C of Eqs. (36) and (37) represent all of the physics involved in deflecting a control surface because they were determined from numerical experiments. Figure 5.17 attempts to explain these physics in a qualitative sense.

At low Mach number, Fig. 5.17a indicates that the slender body theory gives a low value of $k_{W(B)}$ for small values of α_W. At a value of α_W of about 25 deg, the control loses effectiveness as a result of a combination of stall and blow-by effects due to the separation between the wing and body. At an α_W of about 55 deg, controls have lost all effectiveness. At Mach numbers greater than about 4, the controls initially generate less effectiveness than is predicted by slender body theory for values of α_W up to about 20 or 25 deg. Controls then become more effective because of nonlinear compressibility effects. On the other hand, at an α_W of around 45 to 50 deg, controls once again begin to lose effectiveness, presumably because of shock interactions and blow-by effects. For Mach numbers in between subsonic and high supersonic, $k_{W(B)}$ has behavior in between the two extremes illustrated in Fig. 5.17a.

In comparing the nonlinear control deflection models for $\varphi = 0$ and 45 deg roll in Tables 5.24 and 5.25, a lot of similarity is seen. The constants for the $\varphi = 45$ deg are slightly different from those for $\varphi = 0$ deg, and the values of α_W where nonlinearities begin are somewhat different. However, by and large, Eq. (36) holds for both the $\varphi = 0$ and 45 deg roll cases. It should be pointed out that, in Ref. 12, mostly linear variations of $k_{W(B)}$ with α_W were used. However, these have been improved upon for the $\varphi = 45$ deg case with cubic fits of control deflection data as seen in Table 5.25. As such, all

Table 5.24 Semiempirical nonlinear control deflection model (φ = 0 deg)

Mach	Nonlinear model
$M_\infty \le 0.8$	If $\lvert a_W \rvert \le 24.0 \rightarrow k_{W(B)} = 1.1[k_{W(B)}]_{SB}$
	If $\lvert a_W \rvert > 24.0 \rightarrow k_{W(B)} = 1.1[0.000794\lvert a_W \rvert^2 - 0.0933\lvert a_W \rvert + 2.71]$
	If $\lvert a_W \rvert > 52.0 \rightarrow k_{W(B)} = 0.04$
	$F = 1.1$
$M_\infty = 1.1$	If $\lvert a_W \rvert \le 15.0 \rightarrow k_{W(B)} = 1.0[k_{W(B)}]_{SB}$
	If $\lvert a_W \rvert > 15.0 \rightarrow k_{W(B)} = 1.0[1.33 - \lvert a_W \rvert/45]$
	If $\lvert a_W \rvert > 58.0 \rightarrow k_{W(B)} = 0.04$
	$F = 1.1$
$M_\infty = 1.5$	If $\lvert a_W \rvert \le 10.0 \rightarrow k_{W(B)} = 1.0\ [k_{W(B)}]_{SB}$
	If $\lvert a_W \rvert > 10.0 \rightarrow k_{W(B)} = 1.0\ [k_{W(B)}]_{SB} - 0.005[\lvert a_W \rvert - 10.0]$; $[k_{W(B)}]_{\min} = 0.6$
	If $\lvert a_W \rvert \le 20.0 \rightarrow F = 0.8$
	If $\lvert a_W \rvert > 20.0 \rightarrow F = 0.8 + 0.10[\lvert a_W \rvert - 20.0]$
$M_\infty = 2.0$	If $\lvert a_W \rvert \le 10.0 \rightarrow k_{W(B)} = 0.9\ [k_{W(B)}]_{SB}$
	If $\lvert a_W \rvert > 10.0 \rightarrow k_{W(B)} = 0.9\ [k_{W(B)}]_{SB} - 0.003[\lvert a_W \rvert - 10.0]$; $[k_{W(B)}]_{\min} = 0.63$
	If $\lvert a_W \rvert \le 20.0 \rightarrow F = 0.8$
	If $\lvert a_W \rvert > 20.0 \rightarrow F = 0.8 + 0.17[\lvert a_W \rvert - 20.0]$
$M_\infty = 2.3$	If $\lvert a_W \rvert \le 40.0 \rightarrow k_{W(B)} = 0.9[k_{W(B)}]_{SB}$
	If $\lvert a_W \rvert > 40.0 \rightarrow k_{W(B)} = 0.9[k_{(B)}]_{SB} + 0.005[\lvert a_W \rvert - 40.0]$; $[k_{W(B)}]_{\max} = 0.76$
	If $\lvert a_W \rvert \le 30.0 \rightarrow F = 0.9$
	If $\lvert a_W \rvert > 30.0 \rightarrow F = 0.9 + 0.15[\lvert a_W \rvert - 30.0]$
$M_\infty = 2.87$	If $\lvert a_W \rvert \le 40.0 \rightarrow k_{W(B)} = 0.9[k_{W(B)}]_{SB}$
	If $\lvert a_W \rvert > 40.0 \rightarrow k_{W(B)} = 0.9[k_{W(B)}]_{SB} + 0.005[\lvert a_W \rvert - 40.0]$; $[k_{W(B)}]_{\max} = 0.96$
	If $\lvert a_W \rvert \le 30.0 \rightarrow F = 0.9$
	If $\lvert a_W \rvert > 30.0 \rightarrow F = 0.9 + 0.17[\lvert a_W \rvert - 30.0]$
$M_\infty = 3.95$	If $\lvert a_W \rvert \le 20.0 \rightarrow k_{W(B)} = 0.8[k_{W(B)}]_{SB}$
	If $\lvert a_W \rvert > 20.0 \rightarrow k_{W(B)} = 0.8[k_{W(B)}]_{SB} + 0.007[\lvert a_W \rvert - 20.0]$; $[k_{W(B)}]_{\max} = 1.07$
	If $\lvert a_W \rvert \le 30.0 \rightarrow F = 0.9$
	If $\lvert a_W \rvert > 30.0 \rightarrow F = 0.9 + 0.2[\lvert a_W \rvert - 30.0]$
$M_\infty \ge 4.6$	If $\lvert a_W \rvert \le 20.0 \rightarrow k_{W(B)} = 0.75[k_{W(B)}]_{SB}$
	If $\lvert a_W \rvert > 20.0 \rightarrow k_{W(B)} = 0.75[k_{W(B)}]_{SB} + 0.013[\lvert a_W \rvert - 20.0]$; $[k_{W(B)}]_{\max} = 1.33$
	If $\lvert a_W \rvert \le 35.0 \rightarrow F = 0.9$
	If $\lvert a_W \rvert > 35.0 \rightarrow F = 0.9 + 0.2[\lvert a_W \rvert - 35.0]$

Where $a_W = a + \delta$ and for large a_W

$$k_{B(W)} = [k_{B(W)}]_{SB} \text{ for } \lvert a_W \rvert \le 70 \text{ deg}$$

$$k_{B(W)} = [k_{B(W)}]_{SB}\left[1 - \left(\frac{\lvert a_W \rvert - 70}{20}\right)\right] \text{ for } 70 < \lvert a_W \rvert \le 90$$

$$k_{B(W)} = 0 \text{ for } \lvert a_W \rvert > 90 \text{ deg}$$

Table 5.25 Semiempirical nonlinear control deflection model (φ = 45 deg)

Mach	Nonlinear model
$M_\infty \leq 0.8$	If $\lvert \alpha_W \rvert \leq 40.0$ deg$\rightarrow k_{W(B)} = 1.15[k_{W(B)}]_{SB}$ If $\lvert \alpha_W \rvert > 40.0$ deg$\rightarrow k_{W(B)} = 1.15[-0.0455 \lvert \alpha_W \rvert + 2.74]$; $[k_{W(B)}]_{\min} = 0.01$ $F = 1.1$
$M_\infty = 1.1$	If $\lvert \alpha_W \rvert \leq 15.0 \rightarrow k_{W(B)} = 0.95[k_{W(B)}]_{SB}$ If $\lvert \alpha_W \rvert > 15.0 \rightarrow k_{W(B)} = 0.95[1.33 - \lvert \alpha_W \rvert /45$; $[k_{W(B)}]_{\min} = 0.04$ $F = 1.1$
$M_\infty = 1.5$	If $\lvert \alpha_W \rvert \leq 35.0 \rightarrow k_{W(B)} = 0.95[k_{W(B)}]_{SB}$ If $35.0 \leq \lvert \alpha_W \rvert \leq 55.0 \rightarrow k_{W(B)} = [-8.067 \times 10^{-5} (\lvert \alpha_W \rvert - 35.0)^3 + 0.00201 (\lvert \alpha_W \rvert - 35.0)^2 - 0.0295 (\lvert \alpha_W \rvert - 35.0) + 0.94] [k_{W(B)}]_{SB}$ If $\lvert \alpha_W \rvert > 55.0 \rightarrow k_{W(B)} = [k_{W(B)}]_{\alpha_W=55} - ((\lvert \alpha_W \rvert - 55.0)/35.0) ([k_{W(B)}]_{\alpha_W=55} - 0.2[k_{W(B)}]_{SB})$; $[k_{W(B)}]_{\min} = 0.2 [k_{W(B)}]_{SB}$ If $\lvert \alpha_W \rvert \leq 20.0 \rightarrow F = 0.8$ If $\lvert \alpha_W \rvert > 20.0 \rightarrow F = 0.8 + 0.17[\lvert \alpha_W \rvert - 20.0]$
$M_\infty = 2.0$	If $\lvert \alpha_W \rvert \leq 32.5 \rightarrow k_{W(B)} = 0.95[k_{W(B)}]_{SB}$ If $32.5 < \lvert \alpha_W \rvert \leq 55.0 \rightarrow k_{W(B)} = [-0.8 \times 10^{-5} (\lvert \alpha_W \rvert - 32.5)^3 - 0.00091 (\lvert \alpha_W \rvert - 32.5)^2 + 0.011 (\lvert \alpha_W \rvert - 32.5) + 0.95] [k_{W(B)}]_{SB}$ If $\lvert \alpha_W \rvert > 55.0 \rightarrow k_{W(B)} = [k_{W(B)}]_{\alpha_W=55} - ((\lvert \alpha_W \rvert - 55)/35) ([k_{W(B)}]_{\alpha_W=55} - 0.2[k_{W(B)}]_{SB})$ If $\lvert \alpha_W \rvert \geq 75.0$, $k_{W(B)} = 0.2[k_{W(B)}]_{SB}$ If $\lvert \alpha_W \rvert \leq 20.0 \rightarrow F = 0.8$ If $\lvert \alpha_W \rvert > 20.0 \rightarrow F = 0.8 + 0.17[\lvert \alpha_W \rvert - 20.0]$
$M_\infty = 2.35$	If $\lvert \alpha_W \rvert \leq 30.0 \rightarrow k_{W(B)} = 0.95[k_{W(B)}]_{SB}$ If $30.0 < \lvert \alpha_W \rvert \leq 40.0 \rightarrow k_{W(B)} = [4.257 \times 10^{-5} (\lvert \alpha_W \rvert - 30.0)^3 - 0.00291 (\lvert \alpha_W \rvert - 30.0)^2 + 0.0388 (\lvert \alpha_W \rvert - 30.0) + 0.976] [k_{W(B)}]_{SB}$ If $40.0 < \lvert \alpha_W \rvert \leq 78.0 \rightarrow k_{W(B)} = [k_{W(B)}]_{\alpha_W=40} - \left(\frac{\lvert \alpha_W \rvert - 40.0}{38.0}\right)([k_{W(B)}]_{\alpha_W=40} - 0.20 [k_{W(B)}]_{SB})$; $[k_{W(B)}]_{\min} = 0.2 [k_{W(B)}]_{SB}$ If $\lvert \alpha_W \rvert > 78.0$, $k_{W(B)} = 0.2[k_{W(B)}]_{SB}$ If $\lvert \alpha_W \rvert \leq 30.0 \rightarrow F = 0.9$ If $\lvert \alpha_W \rvert > 30.0 \rightarrow F = 0.9 + 0.15[\lvert \alpha_W \rvert - 30.0]$
$M_\infty = 2.87$	If $\lvert \alpha_W \rvert \leq 30.0 \rightarrow k_{W(B)} = 0.95[k_{W(B)}]_{SB}$ If $30.0 < \lvert \alpha_W \rvert \leq 50.0 \rightarrow k_{W(B)} = [6.526 \times 10^{-5} (\lvert \alpha_W \rvert - 30.0)^3 - 0.00405 (\lvert \alpha_W \rvert - 30.0)^2 + 0.0575 (\lvert \alpha_W \rvert - 30.0) + 0.947] [k_{W(B)}]_{SB}$ If $50.0 < \lvert \alpha_W \rvert \leq 78.0 \rightarrow k_{W(B)} = [k_{W(B)}]_{\alpha_W=40} - \left(\frac{\lvert \alpha_W \rvert - 50.0}{28.0}\right)([k_{W(B)}]_{\alpha_W=50} - 0.20 [k_{W(B)}]_{SB})$; $[k_{W(B)}]_{\min} = 0.2 [k_{W(B)}]_{SB}$ If $\lvert \alpha_W \rvert \leq 30.0 \rightarrow F = 0.9$ If $\lvert \alpha_W \rvert > 30.0 \rightarrow F = 0.9 + 0.17[\lvert \alpha_W \rvert - 30.0]$

Table 5.25 Semi-empirical nonlinear control deflection model (φ = 45 deg) *(continued)*

Mach	Nonlinear model
$M_\infty = 3.95$	If $\lvert\alpha_W\rvert \leq 35.0 \rightarrow k_{W(B)} = 0.88[k_{W(B)}]_{SB}$
	If $35.0 \leq \lvert\alpha_W\rvert \leq 48.0 \rightarrow k_{W(B)} = [-8.84 \times 10^{-5}(\lvert\alpha_W\rvert - 35.0)^3 + 0.000173(\lvert\alpha_W\rvert - 35.0)^2 + 0.0397(\lvert\alpha_W\rvert - 35.0) + 0.884][k_{W(B)}]_{SB}$
	If $48.0 \leq \lvert\alpha_W\rvert \leq 80.0 \rightarrow k_{W(B)} = [k_{W(B)}]_{\alpha_W=48} - \left(\frac{\lvert\alpha_W\rvert - 48.0}{32.0}\right)([k_{W(B)}]_{\alpha_W=48} - 0.20[k_{W(B)}]_{SB})$; $[k_{W(B)}]_{\min} = 0.2[k_{W(B)}]_{SB}$
	If $\lvert\alpha_W\rvert \leq 30.0 \rightarrow F = 0.9$
	If $\lvert\alpha_W\rvert > 30.0 \rightarrow F = 0.9 + 0.2[\lvert\alpha_W\rvert - 30.0]$
$M_\infty \geq 4.6$	If $\lvert\alpha_W\rvert \leq 35.0 \rightarrow k_{W(B)} = 0.83[k_{W(B)}]_{SB}$
	If $35.0 \leq \lvert\alpha_W\rvert \leq 55.0 \rightarrow k_{W(B)} = [4.697 \times 10^{-5}(\lvert\alpha_W\rvert - 35.0)^3 - 0.00463(\lvert\alpha_W\rvert - 35.0)^2 + 0.0739(\lvert\alpha_W\rvert - 35.0) + 0.83][k_{W(B)}]_{SB}$
	If $50.0 \leq \lvert\alpha_W\rvert \leq 77.0 \rightarrow k_{W(B)} = [k_{W(B)}]_{\alpha_W=50} - \left(\frac{\lvert\alpha_W\rvert - 50.0}{27.0}\right)([k_{W(B)}]_{\alpha_W=50} - 0.20[k_{W(B)}]_{SB})$; $[k_{W(B)}]_{\min} = 0.2[k_{W(B)}]_{SB}$
	If $\lvert\alpha_W\rvert \leq 35.0 \rightarrow F = 0.9$
	If $\lvert\alpha_W\rvert > 35.0 \rightarrow F = 0.9 + 0.2[\lvert\alpha_W\rvert - 35.0]$
$0 \leq M_\infty \leq \infty$	$k_{B(W)} = [k_{B(W)}]_{SB} - \frac{\lvert\delta_W\rvert}{30.0}(0.75[k_{B(W)}]_{SB})$ for $\lvert\delta_W\rvert \leq 30.0$
	$k_{B(W)} = 0.25[k_{B(W)}]_{SB}$ for $\lvert\delta_W\rvert > 30.0$
	If $\lvert\alpha_W\rvert \geq 55$, $k_{B(W)} = 0$, linearly goes to zero from α_W = 50 to 55 deg

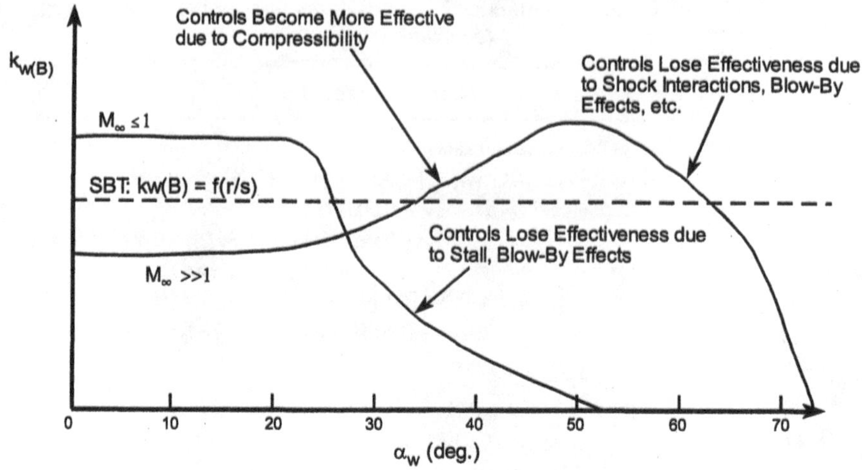

Fig. 5.17a Qualitative trend of wing–body interference due to control deflection as a function of M_∞, α_W.

nonlinear effects are included in variations of $k_{W(B)}$ as a function of Mach number and $|\alpha + \varphi|$.

Figure 5.17b assumes that $k_{B(W)}$ can be represented by slender body theory up to some value of $|\alpha_W|$, at which point it decays to a percent of slender body theory. For $\varphi = 0$ deg, these values are $|\alpha_W| = 70$ deg and 50% of slender body theory as a minimum. For $\varphi = 45$ deg roll, the model $k_{B(W)}$ is based on $|\delta|$ only and begins decaying at $|\delta| = 0$. It reaches a minimum of $k_{B(W)}$ of 25% of slender body theory analogous to the $[K_{B(W)}]_{\min}$ of Fig. 5.15.

It can be also noted that $k_{W(B)}$ and $k_{B(W)}$ of Table 5.25 are multiplied by 1.414 to indicate that all four fins are assumed to be deflected by an equal amount in the $\varphi = 45$ deg roll position. This multiplication is done internal to the 1998 version of the Aeroprediction Code[15] (which contains linear and nonlinear aerodynamics of chapters 3 to 5). Finally, for Mach numbers in between the values on Tables 5.24 and 5.25, linear interpolation is used. Several examples will be shown later in the configuration aerodynamics section to illustrate the overall accuracy and robustness of the nonlinear control deflection model.

It should also be pointed out that the centers of pressure of $k_{W(B)}$ and $k_{B(W)}$ are assumed to be at the same locations as $K_{W(B)}$ and $K_{B(W)}$.

VI. Nonlinear Wing–Tail Interference Model

The final nonlinear term of Eq. (2) in Chapter 3 to be modeled is the wing–tail interference $C_{N_{T(V)}}$. Section 3.9.2 gave a fairly thorough description of the linearized theory approach to computing the term $C_{N_{T(V)}}$. The discussion here will build upon the linear theory to incorporate nonlineari-

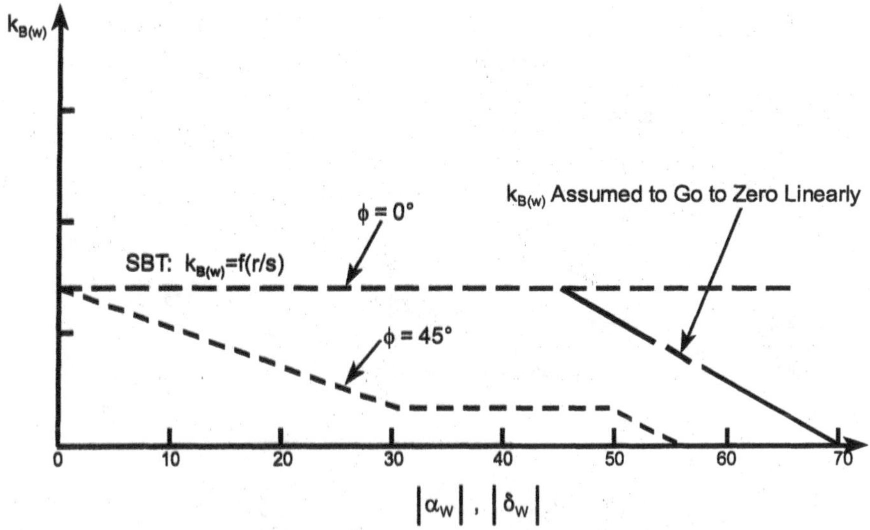

Fig. 5.17b Qualitative trend of body–wing interference due to control deflection as a function of α_W, δ_W.

ties in AOA, Mach number, and control deflection. Separate semi-empirical models will be developed for the $\varphi = 0$ and 45 deg roll positions.

To improve on the present slender body theory/linear theory wing–tail methodology, the literature was searched for wing–tail interference data. Washington and Spring[16] and Aiello and Bateman[17] were found in this process. These references documented wind tunnel tests on two different missile configurations at which actual wing–tail normal force measurements were made. Washington and Spring[16] measured these results on individual fins as a function of roll position at Mach numbers of 2, 3, and 4. Aiello and Bateman[17] gave results for only the $\varphi = 0$ deg roll position at $M_\infty = 1.1$. Figure 5.18 shows the configuration tested in Ref. 16 with results at $M_\infty = 1.96$ and 4.02 at roll position of 0 deg. Also reported in Ref. 16 were the theoretical computations from analytical codes referred to as AERODSN,[18] MISSILE 2A,[19] and LRCDM.[20] According to Washington and Spring[16], the AERODSN code has the same wing–tail interference model as the older version of the Aeroprediction Code and is based on slender body theory from Ref. 21. The MISSILE 2A program is based on theory and experiment and, therefore, should contain some of the nonlinearity associated with the wing–tail interference. The LRCDM code is based on paneling methods and databases and, therefore, should also contain some of the nonlinearities of the wing–tail interference. For comparison purposes, $C_{N_{T(V)}}$ computed by the AP95[12] for a single fin is also shown in Figs. 5.18b and 5.18c. The AP95, although it contains some nonlinearities, still resembles the slender body theory, as seen from the close proximity to AERODSN results. It is fair to say, in viewing the comparison of the

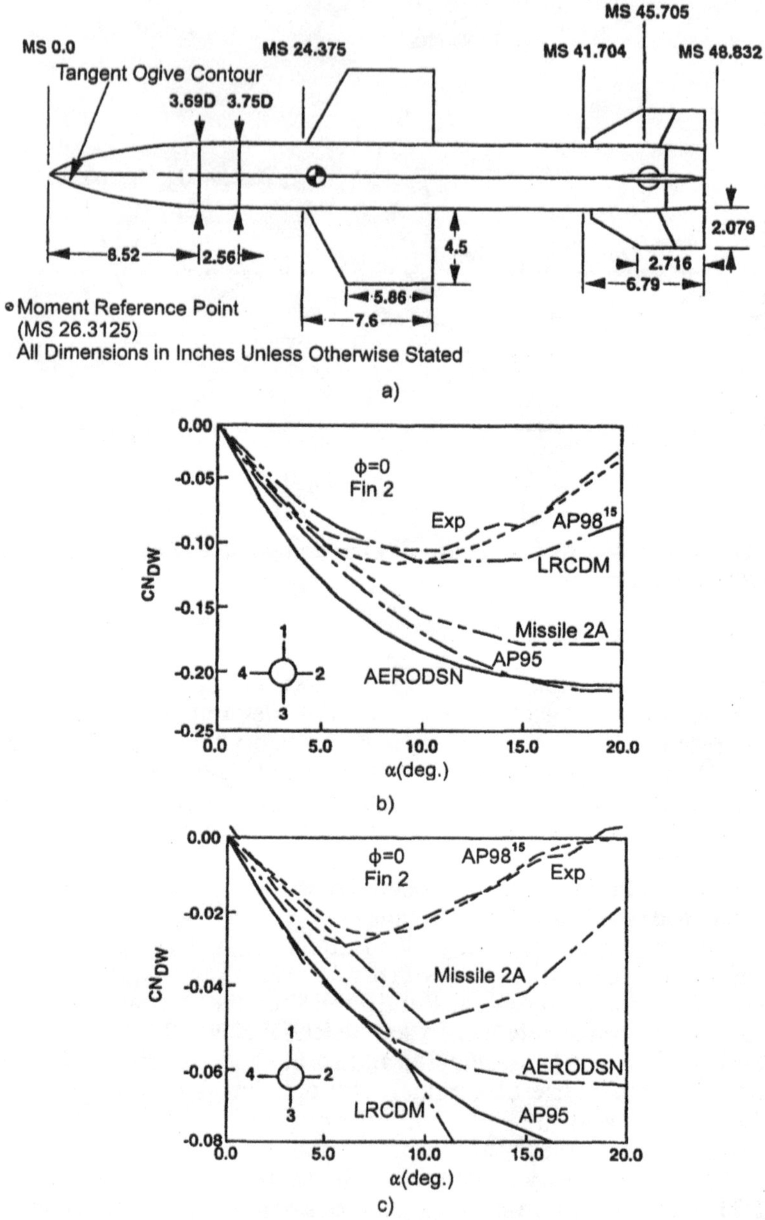

Fig. 5.18 Configuration and single fin data from Ref. 16 for wing-tail interference normal force. a) Configuration tested in Ref. 16, b) Single fin wing–tail interference (M_∞ = 1.96), c) Single fin wing–tail interference (M_∞ = 4.02)

various theoretical approaches for predicting wing–tail interference in Fig. 5.18, that improvements in the theory are needed.

One might argue that the value of $C_{N_{T(V)}}$ is so small that it can be neglected. However, in most cases the moment arm is fairly large so even that a small loss of lift on the tail is magnified in the pitching moment. As a result of this and the fact that $C_{N_{T(V)}}$ is not predicted well by any of the theoretical approaches, it was decided to try to develop a new semi-empirical approach to predict $C_{N_{T(V)}}$. The new approach will still be anchored by the traditional slender body theory[21] but will be modified according to data from Refs. 16 and 17.

The wing–tail interference equation for $\varphi = 0$ deg roll is

$$C_{N_{T(V)}} = \frac{\left(C_{N_\alpha}\right)_W \left(C_{N_\alpha}\right)_T \left[K_{W(B)}\alpha + k_{W(B)}\delta_W\right] i(s_T - r_T)A_W}{2\pi (AR)_T (f_W - r_W) A_{\text{ref}}} \tag{38}$$

Equation (38) can be rewritten in the form

$$C_{N_{T(V)}} = \left[C_{N_{T(V)}}\right]_\alpha + F\left[C_{N_{T(V)}}\right]_\delta \tag{39}$$

where F is an empirical constant or parameter derived from numerical experiments of Eq. (39) compared to data for configurations with control deflection. This parameter was defined in Tables 5.24 and 5.25 as a part of the nonlinear interference model due to control deflection. Also $\left(C_{N_\alpha}\right)_W$, $\left(C_{N_\alpha}\right)_T$, $K_{W(B)}$, and $k_{W(B)}$ of Eqs. (38) and (39) are nonlinear terms determined from earlier sections of Chapter 5.

While Eq. (38) was helpful in defining some of the nonlinearities of missile aerodynamics not predicted by slender body theory or linear theory, it still had two major weaknesses. First of all, the tail interference factor methodology was independent of Mach number and, second, there was no parameter for the first term of Eq. (39), similar to the parameter F of the second term, to help control the quasilinear character of the equation. These two problems were partially accounted for in the AP95[12] by placing upper limits on $C_{N_{T(V)}}$ as a percent of total tail lift and with the upper limits a function of Mach number. However, these upper limits mainly affected large wing configurations where the wing–tail interference could exceed the tail lift.

To address these two issues, a model for no control deflection, based on a third-order equation in AOA is postulated to fit data from both Refs. 16 to 17 at $\varphi = 0$ deg roll. This model is defined by

$$\left[C_{N_{T(V)}}\right]_\alpha = A + B\alpha + C\alpha^2 + D\alpha^2 \tag{40}$$

The four constants of Eq. (40) require four conditions to define them. These conditions are as follows:

1) The wings are assumed to be thin and disturb the flow slightly so that $C_{N_{T(V)}} = 0$ at $\alpha = 0$.

2) The slope of slender body theory value for $C_{N_{T(V)}}$ near $\alpha = 0$ is required (using this value and a value of the real slope from data, a modified value of this slope can be obtained).
3) The AOA at which $C_{N_{T(V)}}$ goes to zero based on data is needed.
4) Finally, the maximum value of $C_{N_{T(V)}}$ as a percent of slender body theory at AOA where $C_{N_{T(V)}}$ is a maximum is required.

Note that in the four conditions chosen, slender body theory is used twice. Based on these conditions, the four constants of Eq. (40) become

$$A = 0 \tag{41a}$$

$$B = \left[\left(\frac{dC_{N_{T(V)}}}{d\alpha}\right)_{\alpha=0}\right]_{\text{SBT}} E_1 \tag{41b}$$

$$C = \frac{-B - D\alpha_N^2}{\alpha_N} \tag{41c}$$

$$D = \frac{E_2\alpha_N - B\alpha_N\alpha_F + B\alpha_F^2}{\alpha_N\alpha_F^3 - \alpha_F^2\alpha_N^2} \tag{41d}$$

The parameters in the constants *B*, *C*, and *D* are defined as follows:

α_{N_o} = value from Fig. 5.19a

α_F = (value from Fig. 5.19b) × $\dfrac{\alpha_N}{100}$

E_1 = value from Fig. 5.19c

E_2 = (value from Fig. 5.19d) × $\left(\left[C_{N_{T(V)}}\right]_{\text{SBT}}\right)_{\alpha=\alpha_F}$

The values of the parameters α_{N_O}, α_F, E_1, and E_2 of Fig. 5.19 were all determined through the use of Refs. 16 and 17 databases, with linear theory values of $C_{N_{T(V)}}$ computed from Eq. (38). Figure 5.19, in essence, represents the amount slender body theory deviates from actual experimental data.

The question arises regarding how to account for fins that are different sizes and locations than those tested in Refs. 16 and 17. Fortunately, both fins tested in these references were significantly larger than the tail surfaces and were located in a wing versus canard location. Hence, the present approach will be to use these results directly for wings or canards of less area to body reference area than those tested. For wings of greater area, it is assumed that the AOA, α_N, at which the $C_{N_{T(V)}}$ becomes negligible is increased according to

$$\alpha_N = \alpha_{N_O} \text{ for } A_W/A_{\text{ref}} \leq 5.5$$

$$\alpha_N = \alpha_{N_O}\left(\frac{A_W/A_{\text{ref}}}{5.5}\right) \text{ for } A_W/A_{\text{ref}} > 5.5 \tag{42}$$

with an upper limit on α_N of 2.5 α_{N_O}.

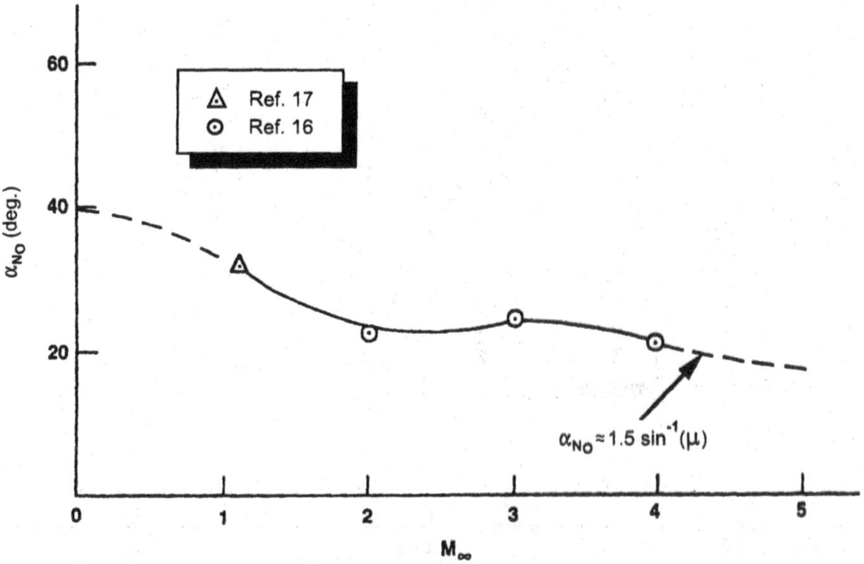

Fig. 5.19a Angle of attack at which wing–tail interference is negligible (parameter used in nonlinear wing–tail interference model at $\varphi = 0°$).

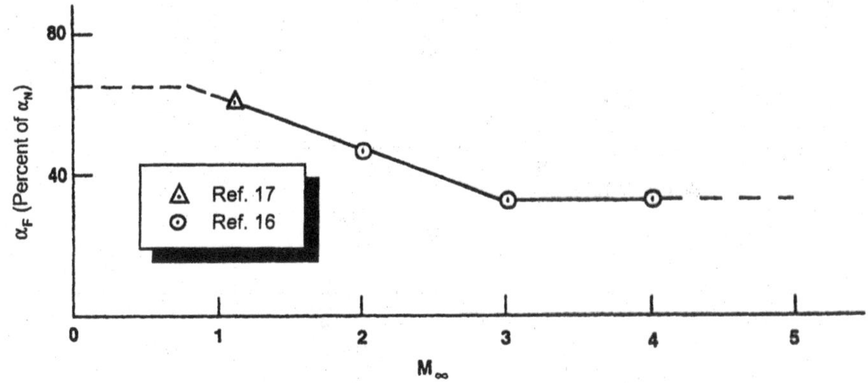

Fig. 5.19b Angle of attack at which wing–tail interference is a maximum (percent of α_N) (parameter used in nonlinear wing–tail interference model at $\varphi = 0°$).

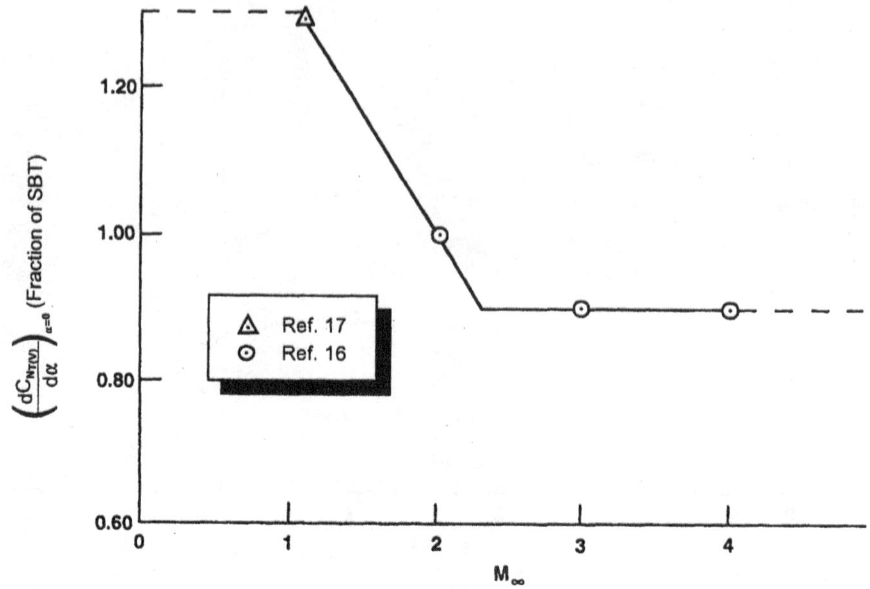

Fig. 5.19c Initial slope at $\alpha = 0$ of wing–tail interference as a function of M_∞ (parameter used in nonlinear wing–tail interference model at $\varphi = 0°$).

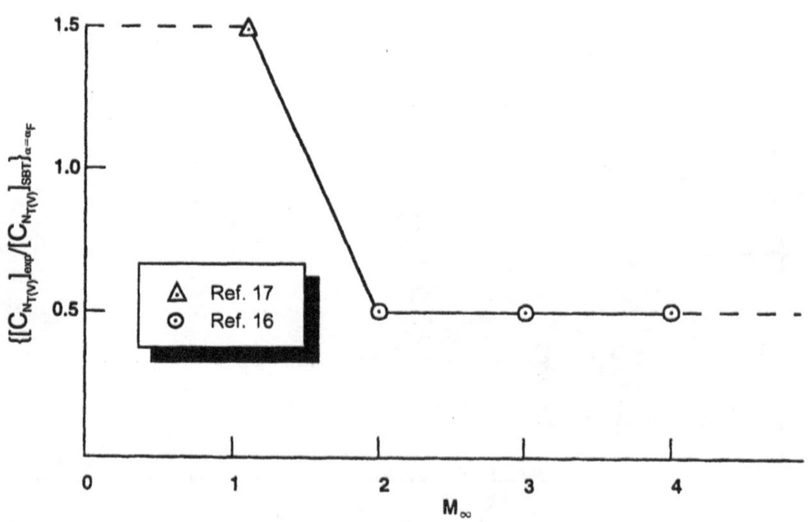

Fig. 5.19d Slender body theory prediction of wing–tail interference at AOAs at which $C_{N_{T(V)}}$ reaches a maximum as a fraction of experimental data (parameter used in nonlinear wing–tail interference model at $\varphi = 0°$).

The value of A_W/A_{ref} of 5.5 corresponds to the wing area ratio of Ref. 16. Also, for wings larger than those of Ref. 16, an upper limit on the amount of lift loss on the tail will remain in effect. This upper limit is defined by the following methodology:

For $M_\infty \leq 1.5$

$$\frac{|C_{N_{T(V)}}|}{C_{N_T}} = 1.0 \; ; a \leq 5$$

$$\frac{|C_{N_{T(V)}}|}{C_{N_T}} = 1.0 - 0.04125\,(a - 5)\,; a > 5 \tag{43a}$$

For $1.5 < M_\infty \leq 2.5$

$$\frac{|C_{N_{T(V)}}|}{C_{N_T}} = 0.9 - 0.025a \; ; a \leq 10$$

$$\frac{|C_{N_{T(V)}}|}{C_{N_T}} = 0.65 - 0.0235\,(a - 10)\,; a > 10 \tag{43b}$$

For $M_\infty > 2.5$

$$\frac{|C_{N_{T(V)}}|}{C_{N_T}} = 0.8 - 0.025a \tag{43c}$$

where a is AOA in degrees. Equation (43a) says that, at $a = 0$ deg, the maximum lift loss on the tail is limited to 100% of the tail lift, regardless of the size of the wings. The percent lift loss then decreases linearly with AOA as defined by Eqs. (43a) to (43c).

Admittedly, this is conservative [overpredicts $C_{N_{T(V)}}$ for values of A_W/A_{ref} < 5.5 and is simply a judgment based on numerical experiments for values of A_W/A_{ref} > 5.5. It does accomplish the objective of making the wing–tail interference with no control deflection more closely approximate data than available approaches, including the AP95. This is illustrated by the results shown in Figs. 5.18b and 5.18c for AP98.[15]

Before moving to the wing–tail interference methodology for the $\varphi = 45$ deg roll position, a comment could be valuable on the Fig. 5.19 results, which basically compare slender body theory to data. First of all, it is clear that, at low AOA, slender body theory underpredicts $C_{N_{T(V)}}$ for low Mach number and overpredicts it at high Mach number, for the Ref. 16 configuration. The point of optimum prediction appears to be around Mach 2 (see Fig. 5.19c). Second, the $C_{N_{T(V)}}$ term decays much faster at high AOA than does slender body theory, a fact that is increasingly true as Mach number increases. This again highlights the Newtonian impact assumptions at high Mach number where any vortices in the leeward plane are completely dominated by the dynamic pressure in the windward plane. Figures 5.19a and 5.19b illustrate this fact, showing that the AOA at which $C_{N_{T(V)}}$ is negligible gets smaller as M_∞ increases; also, the maximum magnitude as a percent of slender body theory gets smaller with increasing M_∞.

The wing–tail interference normal force for the roll position of $\varphi = 45$ deg is defined by

$$C_{N_{T(V)}} = \frac{A_W (C_{N_\alpha})_W (C_{N_\alpha})_T \left[K_{W(B)}\alpha + F k_{W(B)}\delta_W\right]}{2\pi (AR)_T (f_W - r_W) A_{\text{ref}}} (s_T - r_T)\left[i_1 \cos\varphi + i_4 \sin\varphi\right] \tag{44}$$

Again, $(C_{N_\alpha})_W, (C_{N_\alpha})_T, K_{W(B)}$ are all nonlinear quantities described in Sections 5.3 to 5.5. The wing–tail interference factor for $\varphi = 45$ deg is different for the leeward plane fin (i_4) than the windward plane fin (i_1). However, examining Fig. 3.33, from symmetry, i_1 and i_2 are equal in magnitude as are i_3 and i_4. Hence, only one plane of the flowfield (0 to 180 deg) needs to be considered assuming the vortices from the 180- to 360-deg plane are symmetric to those from 0 to 180 deg.

As pointed out in Section 3.9.2, the assumption is made that the strength of the vortex shed from the windward plane fins is equal in magnitude and opposite in direction to the leeward plane fins. At small AOA, this assumption is quite reasonable and is partially what leads to the fact that cruciform wing–body–tail missile aerodynamics are independent of roll position. As AOA increases, this assumption becomes less and less valid. In fact, the lift of the windward plane fin is much larger than the leeward plane fin as AOA increases. This difference was modeled approximately in the center of pressure shift discussed earlier by approximating a linear variation in the shift of normal force to the windward plane fin from the leeward plane fin up to AOA of 65 deg. At that point, the ratio of the windward to leeward plane load remained constant.

If we define the factors,

$$\begin{gathered} P_W = \left(1.0 + 0.6\frac{\alpha}{65}\right), P_\ell = \left(1.0 - 0.6\frac{\alpha}{65}\right); \alpha \le 65 \\ P_W = 1.6, P_\ell = 0.4; \alpha > 65 \end{gathered} \tag{45}$$

then the interference factors can be weighted by Eq. (45) depending on where the vortex is shed. If it is shed in the windward plane, then the P_W

factor is appropriate; whereas if it is shed from the leeward plane, then the P_ℓ factor in Eq. (45) is appropriate. This is an approximate way to represent the nonlinear nature of the load in the windward to leeward plane fin and the strength of the vorticity shed from each fin.

Using Eq. (45) with Eqs. (A-42) and (A-43) of Ref. 10, an approximate nonlinear representation of the wing–tail interference factors for the $\varphi = 45$ deg roll position for use in Eq. (44) is

$$
\begin{aligned}
i_1 = {} & \left(\frac{2\cos\varphi}{1+\lambda_T}\right)\left\{P_W\left[L\left(\lambda_T, \frac{r_T}{s_T}, \frac{y_1}{s_T}, \frac{z_1}{s_T}\right) - L\left(\lambda_T \frac{r_T}{s_T}, \frac{y_2}{s_T}, \frac{z_2}{s_T}\right)\right.\right. \\
& -L\left(\lambda_T, \frac{r_T}{s_T}, \frac{y_5}{s_T}, \frac{z_5}{s_T}\right) + L\left.\left(\lambda_T, \frac{r_T}{s_T}, \frac{y_6}{s_T}, \frac{z_6}{s_T}\right)\right] \\
& +P_\ell\left[-L\left(\lambda_T, \frac{r_T}{s_T}, \frac{y_3}{s_T}, \frac{z_3}{s_T}\right) + L\left(\lambda_T, \frac{r_T}{s_T}, \frac{y_4}{s_T}, \frac{z_4}{s_T}\right)\right. \\
& \left.\left.+L\left(\lambda_T, \frac{r_T}{s_T}, \frac{y_7}{s_T}, \frac{z_7}{s_T}\right) - L\left(\lambda_T, \frac{r_T}{s_T}, \frac{y_8}{s_T}, \frac{z_8}{s_T}\right)\right]\right\}
\end{aligned}
\tag{46}
$$

$$
\begin{aligned}
i_4 = {} & \left(\frac{2\sin\varphi}{1+\lambda_T}\right)\left\{P_W\left[L\left(\lambda_T, \frac{r_T}{s_T}, \frac{y_1}{S_T}, \frac{z_1 - f_W\cos\varphi}{s_T}\right)\right.\right. \\
& -L\left(\lambda_T, \frac{r_T}{s_T}, \frac{y_2}{s_T}, \frac{z_1 - f_W\cos\varphi}{s_T}\right) \\
& -L\left(\lambda_T, \frac{r_T}{s_T}, \frac{y_5}{s_T}, \frac{z_5}{s_T}\right) + L\left.\left(\lambda_T, \frac{r_T}{s_T}, \frac{y_5}{s_T}, \frac{z_6}{s_T}\right)\right] \\
& +P_\ell\left[-L\left(\lambda_T, \frac{r_T}{s_T}, \frac{y_3}{s_T}, \frac{z_3}{s_T}\right) + L\left(\lambda_T, \frac{r_T}{s_T}, \frac{y_4}{s_T}, \frac{z_4}{s_T}\right)\right. \\
& \left.\left.+L\left(\lambda_T, \frac{r_T}{s_T}, \frac{y_7}{s_T}, \frac{z_7}{s_T}\right) - L\left(\lambda_T, r_T, \frac{y_8}{s_T}, \frac{z_8}{s_T}\right)\right]\right\}
\end{aligned}
\tag{47}
$$

For small AOA and no control deflection, Eq. (44) reverts back to slender body theory. Refer to Figs. 3.32 and 3.33 for definitions of y and z.

The final modification for inclusion of nonlinearities to the wing–tail interference model for the $\varphi = 45$ deg roll position is to adjust Eq. (44) based on experimental data analogous to the $\varphi = 0$ deg methodology of Fig. 5.19. For the 45-deg roll position, the only data found on individual fins were from Ref. 16. Figure 5.20 gives the $C_{N_{T(V)}}$ values for M_∞ values of 1.96, 3.01, and 4.02. In developing an analogous model to Fig. 5.19 for $\varphi = 45$ deg, qualitative use will be made of the $\varphi = 0$ deg results at lower Mach number because at least one set of data[17] existed for $M_\infty = 1.1$. These results will be used to compare trends of data as functions of Mach number and AOA, not their absolute values.

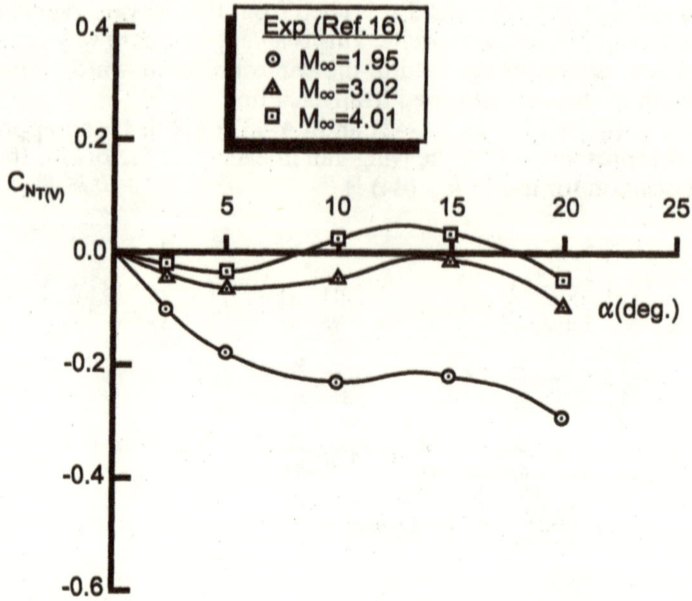

Fig. 5.20 Wing–tail interference (φ = 45° and δ_W = 0°).

Note that the curves of Fig. 5.20 show a point of inflection in experimental data between a 10- and 15-deg AOA. This is because, at a very low AOA, windward and leeward plane vortices shed from the wings adversely affect the tail normal force. However, at a slightly higher AOA, the windward plane wing-shed vortex has a positive effect on the leeward plane tail because of the counterclockwise vortex hitting the windward side of the leeward tail surface. As the AOA is increased higher, the wing-shed windward plane vortex rises above the leeward plane tail fin, at which point both wing-shed vortices again have an adverse affect on the tail normal force.

While a model such as that derived for the φ = 0 roll position could be derived for the φ = 45 deg plane, it is more difficult because of a lack of data below M_∞ = 2.0 and the shape of the curves in Fig. 5.20. As a result, modified slender body theory was used to calculate $C_{N_{T(V)}}$ at various AOAs and at the three Mach numbers where data were available. The results were compared to the experimental data of Fig. 5.20, and the semi-empirical model of Fig. 5.21 was defined for no control deflection.

Referring to Eq. (39), the first part of the equation for φ = 45 deg, δ_w = 0 is

$$\left[C_{N_{T(V)}}\right]_a = G_1\left[C_{N_{T(V)}}\right]_{\text{SBT}} \tag{48}$$

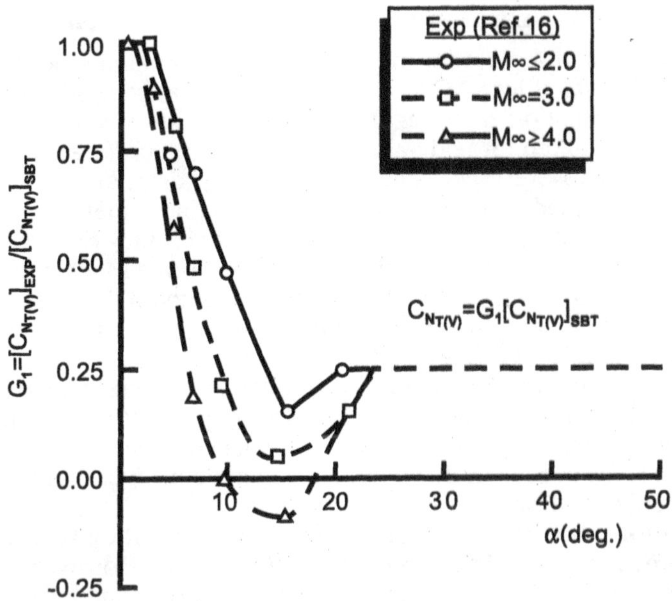

Fig. 5.21 Parameter used in wing–tail interference model for no control deflection at $\varphi = 45°$.

with the same maximum constraints placed on Eq. (48) as for the $\varphi = 0$ deg roll methodology. This constraint was given by Eq. (43). Equation (43a) basically allows the negative tail normal force to not exceed a certain percent of the normal force of the tail regardless of slender body theory predictions. Equations (48) and (43), in conjunction with G_1 from Fig. 5.21, and modified slender body theory Eqs. (46), (47), and (44) define the wing–tail interference model used for $\varphi = 45$ deg.

To summarize the semi-empirical model for wing–tail interference, it uses the slender body/linear theory formulation but uses nonlinear values for all the variables in Eqs. (38) and (44). These nonlinear variables are defined based on limited component wing–tail interference data and numerical experiments of the wing–tail interference model to configurations outside the databases. In these numerical experiments, generally good agreement has been obtained with the $\varphi = 45$ deg roll wing–tail model and mixed success with the $\varphi = 0$ deg roll model. By mixed success is meant that, for some cases, the complete nonlinear model discussed here is best, whereas for some cases, the older quasi-non-linear model of the AP95, which resembles slender body theory, works best. It is believed additional wind tunnel or CFD data need to be gathered to fine-tune the wing–tail interference models. However, in general, semi-empirical models presented here are superior to low AOA slender body and linear theories.

VII. Axial Force Coefficient at Angle of Attack

Chapters 3 and 4 placed a major emphasis on obtaining accurate axial force coefficients at low AOA. It was found that, in general, second-order accurate perturbation methods in conjunction with modified Newtonian theory were required for body-alone wave drag; first-order accurate perturbation methods were acceptable for wing wave drag; and all other drag components needed to be treated individually. These methods have been shown to give quite good axial force coefficient accuracy for over 20 years. However, axial force changes with AOA and for some Mach numbers substantially. The low AOA methods of Chapters 3 and 4 do not account for these changes. We will, therefore, define an empirical mathematical model to estimate axial force AOA effects.

Before developing a mathematical model to address changes in axial force coefficient as a function of AOA, it is appropriate to discuss the physics of the flow that causes these changes. To visualize the changes, the axial force will be once again broken down into its components due to pressure, skin-friction, and base drag.

For subsonic Mach numbers and spherically blunt- or ogive-shaped bodies, the pressure drag at low to moderate AOA is zero. However, as AOA is increased or for bodies with truncated noses or large cone half-angles, the flow forms a separation bubble in the vicinity of the nose region. This separation bubble has a negative pressure coefficient, which means that the axial force decreases with AOA. On the other hand, at high Mach number, the pressure coefficient on the windward side of the body is a function of the sine squared of the angle between a tangent to the body surface and the velocity vector. This means the increase in axial force at high Mach number with AOA is positive due to compressibility effects of the air. On the leeward surface of the body, the pressure coefficient approaches zero at high Mach number so it has little effect on the axial force coefficient. None of the theoretical methods being used for computing the pressure component of the axial force coefficient accurately account for these changes above an AOA of about 10 deg at moderate supersonic Mach numbers and below.

The approach that will be discussed here[22] to define C_A is

$$C_A = C_{A_O} + f(M_\infty, a) \tag{49}$$

where

$$f(M_\infty,a) = Aa + Ba^2 + Ca^3 + Da^4 \tag{50}$$

To evaluate the coefficients A, B, C, and D requires four independent conditions. These conditions are.

i) $\left.\dfrac{\partial f}{\partial a}\right|_{a=0} = f'(M_\infty,0)$

ii) to iv) The value of $f(M_\infty,a)$ at $a = 30$ deg, 60 deg, and 90 deg, respectively.

Using these four conditions, and putting α in radians versus degrees, four equations are arrived at to solve simultaneously. Solving these equations simultaneously for A, B, C, and D yields

$$\begin{aligned}
A &= f'(M_\infty,0) \\
B &= -3.509\,f'(M_\infty,0) + 11.005\,f(M_\infty,30) - 2.757\,f(M_\infty,60) + 0.41\,f(M_\infty,90) \\
C &= 3.675\,f'(M_\infty,0) - 17.591\,f(M_\infty,30) + 7.041\,f(M_\infty,60) - 1.179\,f(M_\infty,90) \\
D &= -1.181\,f'(M_\infty,0) + 6.771\,f(M_\infty,30) - 3.381\,f(M_\infty,60) + .752\,f(M_\infty,90)
\end{aligned} \tag{51}$$

Knowing $f'(M_\infty,0)$, $f(M_\infty,30)$, $f(M_\infty,60)$, and $f(M_\infty,90)$ in conjunction with Eqs. (49), (50), and (51), improved estimates of C_A at AOAs up to 90 deg can be obtained. In reality, if the body is symmetric, then this AOA should be $\pm$ 90 as long as Eq. (50) is viewed in terms of absolute values for AOA.

The question that must be addressed is how to determine values of the parameters $f'(M_\infty,0)$, $f(M_\infty,30)$, $f(M_\infty,60)$, and $f(M_\infty,90)$. To do this, the two large wind tunnel databases of Refs. 4 and 5 were utilized. The Ref. 4 database was used primarily for the lower AOA information, $f'(M_\infty,0)$ and $f(M_\infty,30)$, and Ref. 5 was used for the higher AOA variables, $f(M_\infty,60)$ and $f(M_\infty,90)$. The reason for this was that Ref. 4 data were available in tabular form, making the estimates of the parameters more accurate than from the Ref. 5 information, which was available only in graphs. For some data points in the transonic flow region, the parameter $f(M_\infty,30)$ was obtained by averaging the data from both Refs. 4 and 5.

Figure 5.22 gives the values of the parameters $f'(M_\infty,0)$, $f(M_\infty,30)$, $f(M_\infty,60)$, and $f(M_\infty,90)$ for the body alone. Ingram et al.[23] indicated that analysis showed that the first parameter [$f'(M_\infty,0)$] was independent of nose length, nose shape, and total body length and dependent only on Mach number. This assumption will also be made for the other three parameters. Comparison of the new method on configurations different from those within the database will determine the validity of this assumption.

Figure 5.23 gives the values of the same parameters for a configuration with one set of lifting surfaces. These parameters were derived for mostly body–tail configurations. In comparing values of the parameters in Fig. 5.23 with those of Fig. 5.22, it is seen that similar values exist for each parameter, but they are slightly different. The Aiello and Bateman[24] method shows no difference between configurations that are bodies alone versus those with lifting surfaces, and Ingram et al.[23] shows a difference only as AOA approaches 90 deg. Thus, the present method will also give slightly different results for body-alone and wing–body or body–tail missile configurations than those of Refs. 23 and 24 due to differences in the values of the parameters used in defining nonlinearity with AOA. This is in addition to differences arising from the use of a missile equation for all cases in the present approach versus three different equations used in Refs. 23 and 24.

Unfortunately, neither of the databases given in Refs. 4 and 5 tested configurations with two sets of lifting surfaces. While the coefficients defined

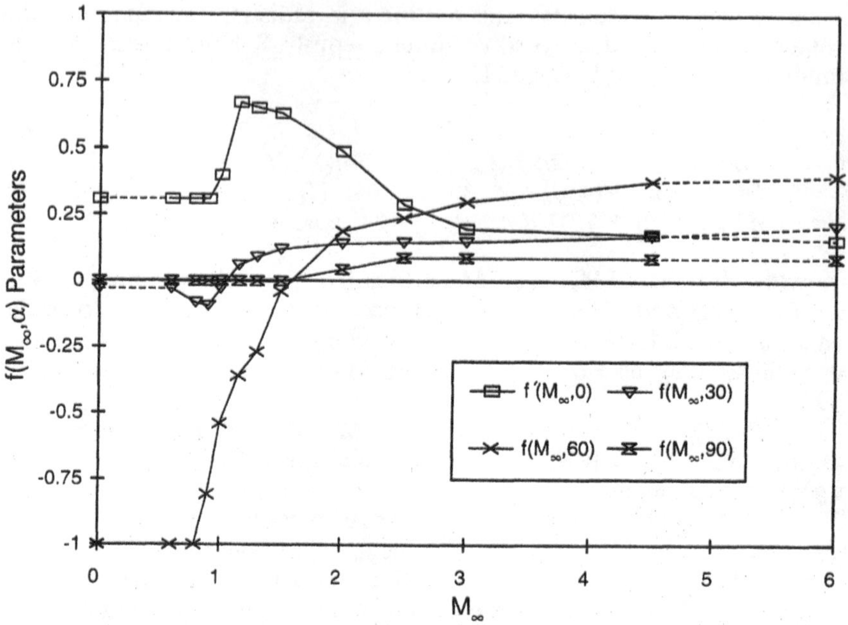

Fig. 5.22 Axial force AOA variation parameters for body alone.

by Fig. 5.23 could be used as an approximation for configurations with two sets of lifting surfaces, this approach can be improved upon. To this end, Refs. 25 through 28 were utilized. References 25 and 26 were very helpful in defining $f'(M_\infty,0)$ and $f(M_\infty,30)$ for Mach numbers 0.8 to 4.6. Above Mach number 4.6, values of these parameters were extrapolated as was done in Figs. 5.22 and 5.23. References 27 and 28 were utilized for low Mach number values of $f'(M_\infty,0)$, $f(M_\infty,30)$, and $f(M_\infty,60)$. The value $f(M_\infty,60)$ for high Mach numbers could be extrapolated reasonably well based on the Ref. 26 data in conjunction with Fig. 5.23. The $f(M_\infty,90)$ data are simply a best guess based on $f(M_\infty,60)$ values of wing–body–tail cases and $f(M_\infty,60)$ and $f(M_\infty,90)$ values of Fig. 5.23.

Values of the four parameters of Eq. (5) are given in Fig. 5.24 for configurations having two sets of lifting surfaces. In comparing Figs. 5.23 and 5.24, some similarities are seen, but also some differences. At low Mach numbers, trends of all four parameters are similar, which means that Fig. 5.23 could be used successfully for configurations with more than one set of lifting surfaces and still get reasonably accurate estimates of axial force change with AOA, which is, in fact, what Refs. 23 and 24 do. However, as Mach number increases, values of the parameters $f'(M_\infty,0)$ and $f(M_\infty,30)$ tend to be higher for two sets of lifting surfaces compared to one. That is, C_A increases faster and reaches a higher peak value with AOA for a wing–body–tail case than for a wing–body or body–tail case.

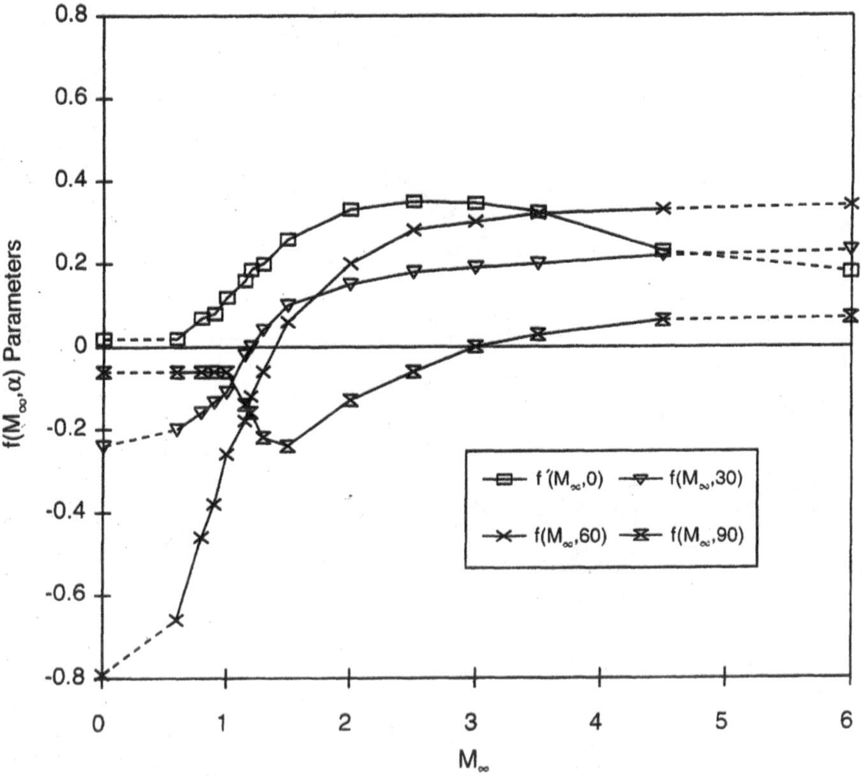

Fig. 5.23 Axial force AOA variation parameters for a body–tail configuration.

The last physical phenomenon to be modeled is the change in axial force coefficient with control deflection as AOA increases. The normal way to define the axial force term due to control deflection is

$$C_{A\delta W} = C_{N_{W(B)}} \sin \delta_W \tag{52a}$$

for the forward lifting surface and as

$$C_{A\delta T} = \left[C_{N_{T(B)}} + C_{N_{T(V)}}\right] \sin \delta_T \tag{52b}$$

for a rearward lifting surface if two sets of lifting surfaces are present. If only one set of lifting surfaces are present, Eq. (52a) applies regardless of the fin location. Equation (52b) contains the tail interference term as a result of the forward set of fins, whereas Eq. (52a) does not have this term. Both $C_{N_{W(B)}}$ and $C_{N_{T(B)}}$ of Eq. (52) are the normal force coefficient on the wing or tail due to both AOA and control deflection.

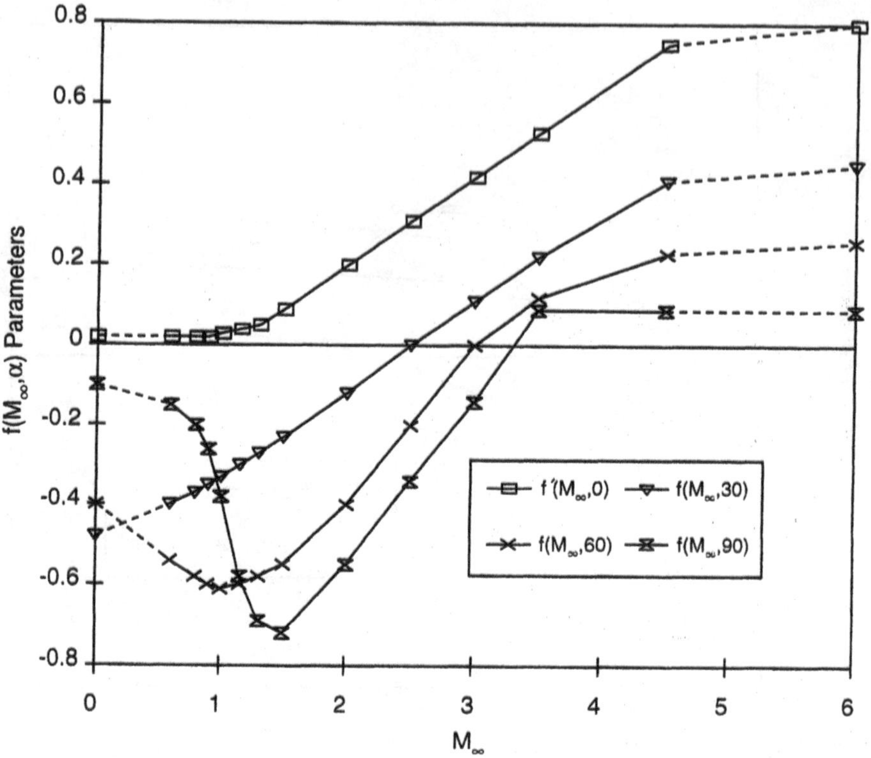

Fig. 5.24 Axial force AOA variation parameters for a wing–body–tail configuration.

In analyzing Eqs. (52a) and (52b), they both basically assume that the axial force due to control deflection is simply the normal force of the wing in conjunction with the body times the sine of the control deflection angle δ. In examining comparisons of this approach to data, it was found that, when a and δ were of the same sign, Eq. (52) gave agreement with experimental data that was quite acceptable in most cases. However, when a and δ were of opposite signs, it was found the agreement was not as good as desired for higher Mach numbers. It is suspected that part of the reason for poor accuracy when a and δ are of opposite signs is that the nonlinear models of $k_{W(B)}$ and $k_{B(W)}$ are accurate for a and δ of the same sign but, when a and δ are of opposite signs, the total values of $k_{W(B)}$ and $k_{B(W)}$ are correct, but each term may be in error and the errors tend to compensate. It is speculated that the fundamental source of this discrepancy is wing gap effects when the wings are deflected. This cancellation of errors could give accurate normal force, but possibly inaccurate axial force combinations due to AOA. The other possible source of the inaccuracy in the C_{A_a} term is the fact that $k_{W(B)}$ does not include nonlinearities from control deflection directly (only a_W and

M_∞ nonlinearities are included). These nonlinear wing gap effects could be accounted for adequately when a and δ are of the same sign from the a alone contribution but inadequately when they are of opposite sign.

Because no direct data measurements are available and the control deflection matrix of Ref. 12 gives quite acceptable values of normal force and pitching moment, the approach taken here to model control deflection effects is to define a term $f(M_\infty, a_W)$ to multiply Eq. (52) by when a and δ are of opposite signs. Then

$$C_{N\delta_W} = \left[C_{N_{W(B)}} \sin\delta_W\right] f(M_\infty, a_W) \tag{53a}$$

and

$$C_{A\delta_T} = \left[C_{N_{T(B)}} + C_{N_{T(V)}}\right] \sin\delta_T f(M_\infty, a_T) \tag{53b}$$

a_W and a_T of Eq. (53) are defined by

$$a_W = a + \delta_W$$

$$a_T = a + \delta_T \tag{54}$$

The a of Eq. (54) is the AOA of the body, and δ_W and δ_T are the control deflections with respect to the body axis of symmetry. Because generic wind tunnel databases available have fins that are too small to accurately determine $f(M_\infty, a)$, use other databases,[26,29] in conjunction with the AP95[12] code, will be used instead. Figure 5.25a gives values of $f(M_\infty, a_W)$ for Mach numbers 1.5 and 4.60 at $\varphi = 0$, and Fig. 5.25b gives the complementary values at $\varphi =$ 45 deg. For Mach numbers above 4.6, the values at 4.6 are used. For Mach numbers below 1.5, $f(M_\infty, a_W)$ is assumed to go to its value of 1.0 at $M_\infty = 0.8$ and remain at the value below that Mach number. This assumption again is based on comparisons to data at low speeds. For a and δ of the same sign, $f(M_\infty, a_W)$ is always 1.0.

It should also be pointed out that the value of $f(M_\infty, a_W)$ at $a_W = 90$ deg has been assumed to be 0.5 for both the $\varphi = 0$ and $\varphi = 45$ deg planes. If the wing were not in the presence of the body, then a value of 1.0 would be natural at least at $\varphi = 0$. However, due to body interference, the full effect of control deflection on $f(M_\infty, a_W)$ at a_W does not appear to be obtained at $a_W = 90$ deg based on extrapolated experimental data at lower values of a_W, hence the assumed value of 0.5. It is hoped that higher AOA experimental data will become available that can be used to improve upon this assumption.

The final issue that must be resolved is how to break down the changes in C_A with AOA into the individual components. The present approach is to break these components down into those due to axial pressure drag, skin-friction, and base drag. The present AP98[15] methodology uses Section VIII of Chapter 4 to account for changes in C_{A_B} to an AOA of 30 deg. From an

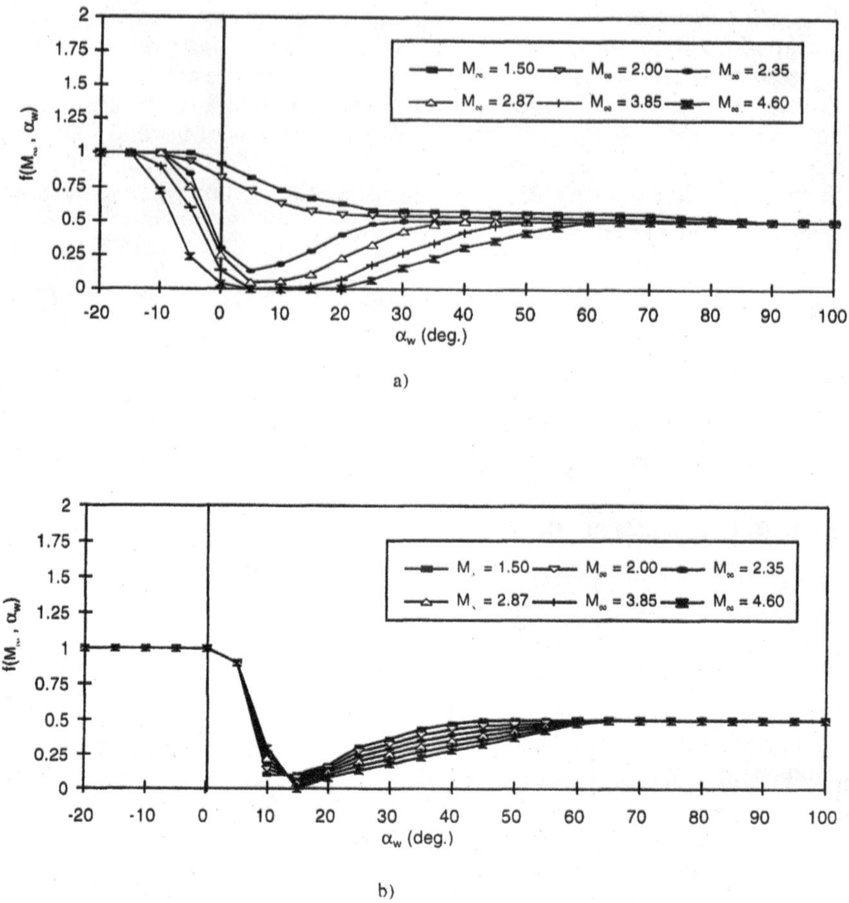

Fig. 5.25 Axial force variation parameters for control deflection when α and δ are opposite signs: a) $\varphi = 0°$ and b) $\varphi = 45°$.

AOA of 30 to 90 deg, C_{A_B} is assumed to go linearly to zero from its value at an AOA of 30 deg. The approach to compute C_{A_B} will not change from Ref. 15 because Ref. 15 was based on wind tunnel data. It will be assumed that C_{A_B} is independent of AOA. This means that the remaining change in C_A with AOA is accounted for in the pressure term. Thus

$$\left(\Delta C_{A_P}\right)_a = f(M_\infty, a) - \left(\Delta C_{A_B}\right)_a \tag{55}$$

Then, the individual components of axial force of the body are

$$\left(C_{A_P}\right)_a = \left(C_{A_P}\right)_{a=0} + \left(\Delta C_{A_P}\right)_a \tag{56a}$$

$$\left(C_{A_f}\right)_a = \left(C_{A_f}\right)_{a=0} \tag{56b}$$

$$\left(C_{A_B}\right)_a = \left(C_{A_B}\right)_{a=0} + \left(\Delta C_{A_B}\right)_a \tag{56c}$$

Of course,

$$(C_A)_a = \left(C_{A_P}\right)_a + \left(C_{A_F}\right)_a + \left(C_{A_B}\right)_a = C_{A_O} + f(M_\infty, a) \tag{57}$$

It will also be assumed that there is no change in the zero lift axial force coefficient of a lifting surface with AOA. This assumption will also apply to the components of axial force on the lifting surface. Thus all the axial force change with AOA for bodies alone or those with lifting surfaces at $\delta = 0$ will be accounted for by the body change in C_A with the angle of attack. These changes are defined by Eq. 55 through 57. The lifting surface change in C_A due to δ is accounted for by Eq. (53) with $f(M_\infty, a_W)$ being 1.0 when a and δ are the same sign and varying according to Fig. 5.25 when of opposite signs.

The only example case that will be shown here is for a body-alone configuration. C_A as a function of AOA, for wing–body and wing–body–tail cases, with and without control deflection, will be shown as a part of overall forces and moments in the next section. The case chosen is the body-alone axial force configuration of Ref. 5. It consists of a 10-caliber pointed tangent ogive with a 2.5-caliber nose length. Data were gathered to a 90-deg AOA.

Figure 5.26 compares the new theoretical approach with the body-alone axial force data of Ref. 5. Also shown on Fig. 5.26 for the lower Mach number cases is the fourth-order axial force method of Ref. 23 and the third-order method of Ref. 24 for the higher Mach number cases. Note that the new method achieves one of its objectives of being as accurate as the Refs. 23 and 24 methods, while using a single versus dual method. One of the reasons Refs. 23 and 24 methods compare as well as they do to data is that the C_{A_O} prediction uses the Baker[5] experimental database. The new methods use AP98[15] theoretical methodology to predict C_{A_O} and still give as good or better comparison of the Baker[5] data than either Ref. 23 or 24.

As seen in Fig. 5.26, axial force changes substantially with AOA. While it is true that C_A is still small in general compared to normal force, it still needs to be accounted for to have accurate static forces and moments at AOA.

VIII. Configuration Aerodynamics

Chapters 3 and 4 developed low AOA methods based primarily on analytical techniques with some empiricism where necessary. Earlier sections of this chapter developed empirical and semi-empirical approximations to various aerodynamic nonlinearities associated with AOA, Mach number, control deflection, and missile configuration. These nonlinearities were developed in component form. The total aerodynamics of a configuration are, therefore, the sum of linear and nonlinear terms. This section of Chapter 5 will show the robustness and accuracy of these nonlinear aerodynamic

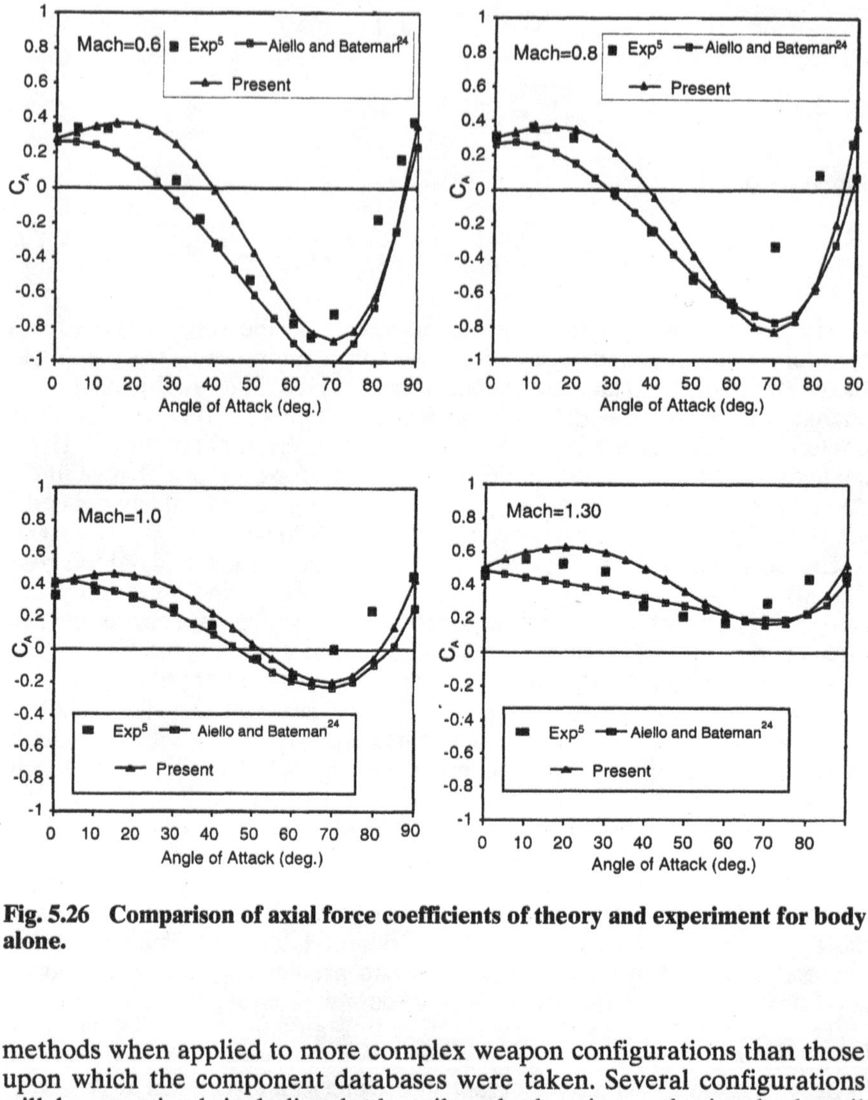

Fig. 5.26 Comparison of axial force coefficients of theory and experiment for body alone.

methods when applied to more complex weapon configurations than those upon which the component databases were taken. Several configurations will be examined, including body–tail or body–wing and wing–body–tail with and without control deflection. Throughout the example cases, the methods discussed in Chapters 3 to 5 will be shown as the AP98.[15]

The first case considered is the canard–body–tail case of Fig. 4.37, which has a 100% blunt nose. Figure 4.37 illustrated how the linear theory methods degrade with AOA. Application of the nonlinear aerodynamics methods of Chapter 5, in conjunction with the low AOA methods of Chapters 3 and 4 is shown in Fig. 5.27. Figure 5.27a is for $\varphi = 0$ deg roll and Fig. 5.27b is for $\varphi = 45$ deg roll. Results are shown for $a = 0$ and 20 deg and $\delta_C = 0$ and 20 deg in terms of axial and normal force coefficients and pitching moment coefficient. Of course, for $a = \delta = 0$, $C_N = C_M = 0$ so no curves exist for this condition. In

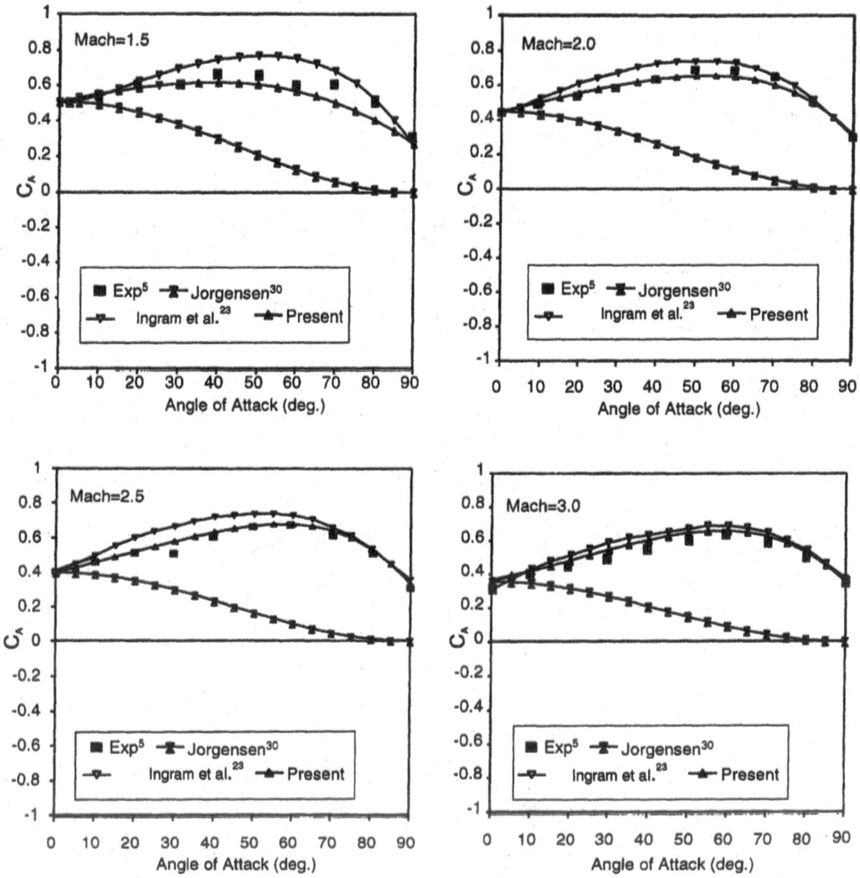

Fig. 5.26 *(Continued)*

examining Fig. 5.27 for both $\varphi = 0$ and 45 deg, it is seen that acceptable prediction accuracy of aerodynamics is obtained at most all Mach numbers for all of the aerodynamic coefficients. An exception is for subsonic Mach numbers where the experimental data predict a lower normal force and less stable airframe than the theory. The other exception is for axial force coefficient at combined α and δ for $\varphi = 45$ deg where the $\pm$ 10% accuracy criterion is exceeded. Even with these errors, when all of the errors on axial force and normal force are averaged, the $\pm$ 10% criterion is met. This also holds true for the $\pm$ 4% of body length criterion for average center of pressure accuracy.

The next example considered for comparison purposes is taken from Ref. 31. Body-alone, body–tail, and wing–body–tail configurations were all part of this test series. Figure 5.28a shows one of the configurations tested and considered here for validation of the AP98 results. The model is 13.5 calibers

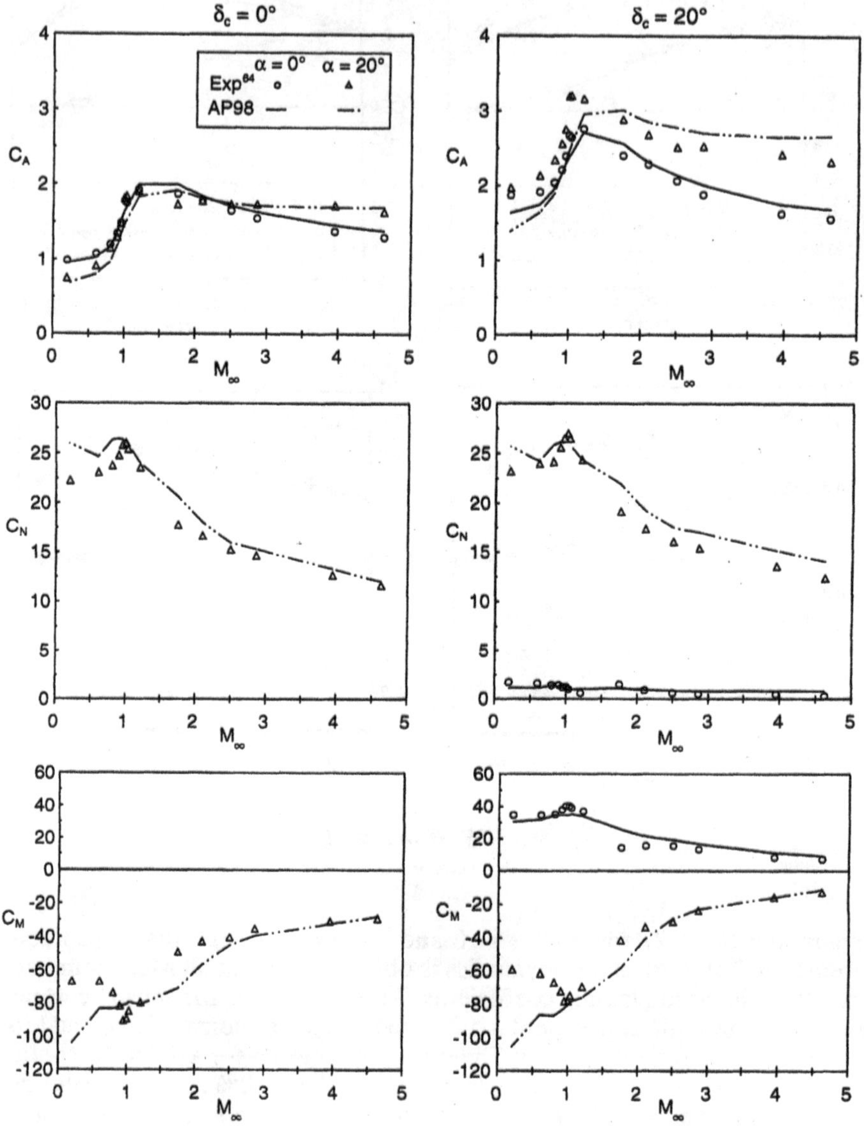

Fig. 5.27a C_A, C_N, and C_M versus Mach number for configuration of Fig. 4.37 ($\varphi = 0°$).

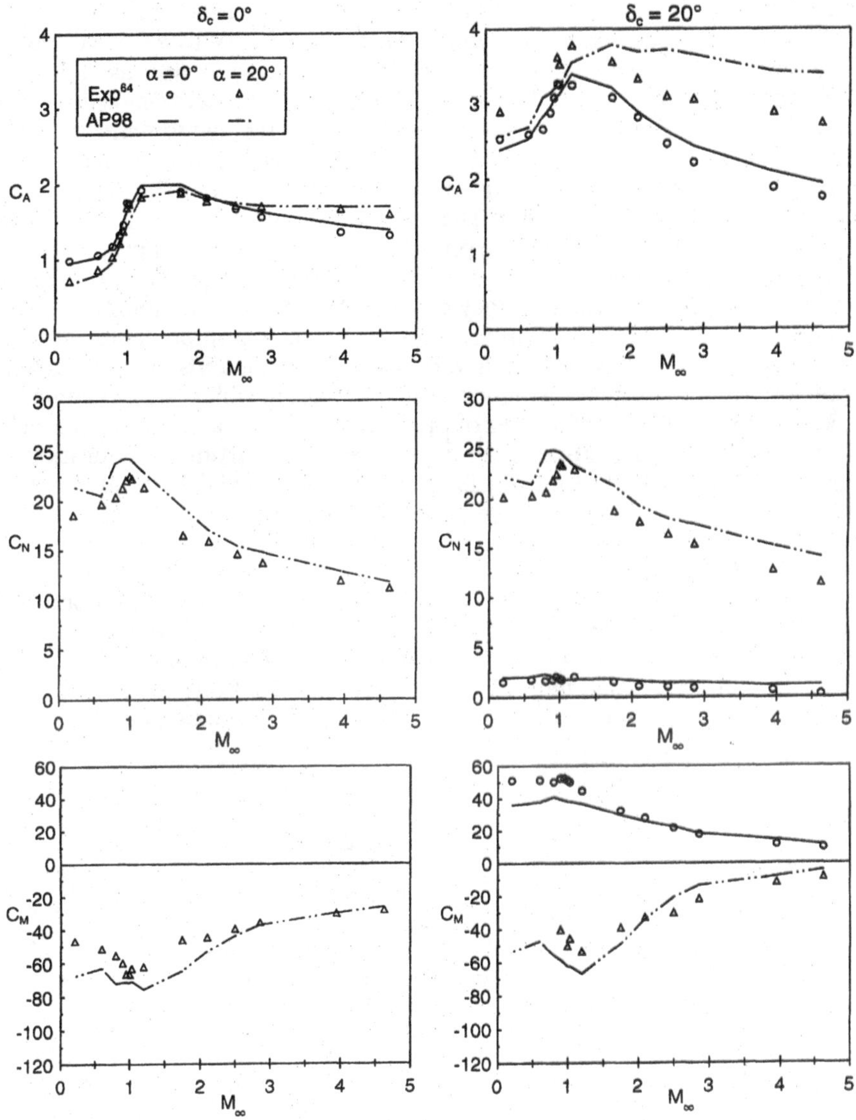

Fig. 5.27b C_A, C_N, and C_M **versus Mach number for configuration of Fig. 4.37** (φ = 45°).

in length with a 1.5-caliber tangent ogive nose. The wing surfaces are fairly large with a thickness of $t/c_r = 0.0178$ and wedge angles on the leading and trailing edges of 15 deg. The tail surfaces have thicknesses of $t/c_r = 0.05$ and wedge angles of 20 deg. The tests were conducted at Mach numbers 0.7 to 3.08 with Reynolds number varying from about 2×10^6 to 4.6×10^6 per ft. The smooth model without boundary-layer trip option was used for the AP98 calculations. Angles of attack to 25 deg were considered in wind tunnel tests. For comparison purposes, normal force and pitching moments are compared to data at $M_\infty = 1.42$ and 3.08 for the $\varphi = 0$ deg roll orientation. Figure 5.28b presents these results. As seen in the figure, the AP98 gives quite acceptable comparisons to data. The average normal force error here is well under the $\pm 10\%$ criterion. While pitching moment coefficient predictions at first glance do not appear to be as good as desired, when looked at in terms of center of pressure, the accuracy criteria of $\pm 4\%$ of body length is easily met. One reason the author prefers an accuracy criterion on center of pressure as a percent of body length versus pitching moments, is that pitching moment accuracy is dependent upon the point about which the moments are taken. On the other hand, center of pressure prediction error as a percent of body length can be compared on a consistent basis from configuration to configuration. Gudmundson and Tangren[31] also gave axial force information where the base pressure had been subtracted. Unfortunately, only a single side camber tap was used, so the AOA information was not believed to be accurate. Hence, no axial force comparisons with AOA are shown.

The third case considered in the validation process of linear and nonlinear aerodynamics methodology of Chapters 3 to 5 is shown in Fig. 5.29a. This configuration was tested[32] with a body alone, body with flare, and body with fins. The options chosen here for comparison are the body alone and body with a 10-deg flare and body with 15-deg fins. The basic body of Fig. 5.29a

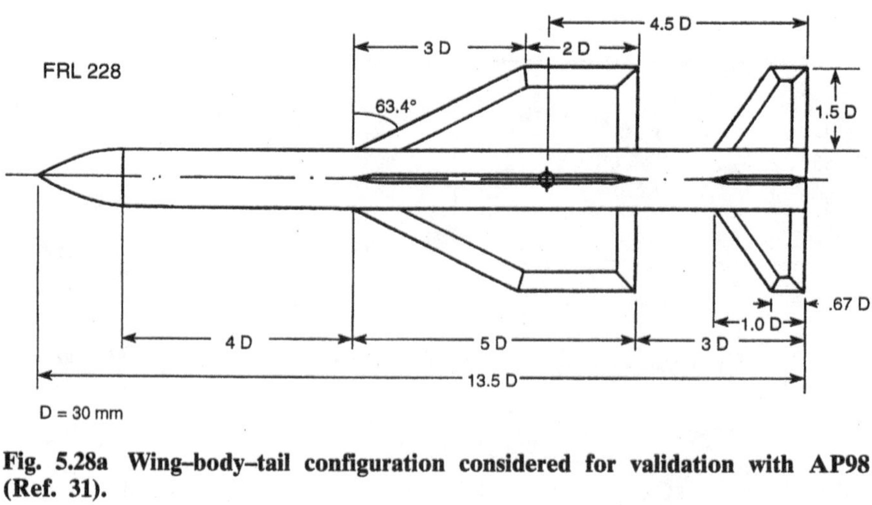

Fig. 5.28a Wing–body–tail configuration considered for validation with AP98 (Ref. 31).

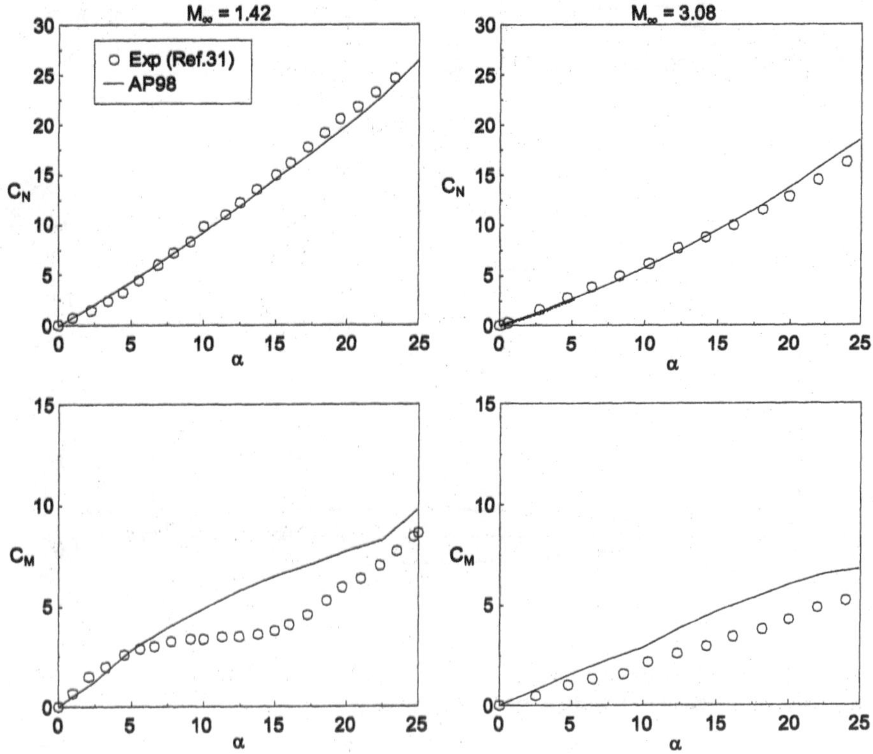

Fig. 5.28b Normal force and pitching moment comparisons of theory and experiment for Fig. 5.28a configuration ($\varphi = 0°$).

consists of a 5-caliber ogive forebody with a 20% blunt nose. The fins were flat plates having rounded leading edges and blunt trailing edges. Tests were conducted at $M_\infty = 2.01$ and R_N/ft of 2×10^6. Base axial force measurements were subtracted out of the experimental data so the axial force will not include the base drag term. No boundary-layer trip was mentioned in Ref. 32 so the "smooth model with no boundary layer trip" option was used in the computations. Data were given to AOAs of 24 to 28 deg.

Figure 5.29b presents the AP98 computations of $C_A - C_{A_B}$, C_N, and C_M compared to the data of Ref. 32. Comparison of the theory to data is acceptable, except for C_A. It is suspected that the process of subtracting out the base pressure from experiment and theory is the primary source of the disagreement. To get accurate base pressure measurements from experimental data generally requires at least a 3:1 ratio of base to sting diameter plus several base pressure taps. Also, the values of wave plus skin friction drag given by the experiment for the body-alone case of 0.065 to 0.085 appear low. If the base pressure component were too high by 0.02 to 0.04, this could account for the error between experiment and theory for the axial

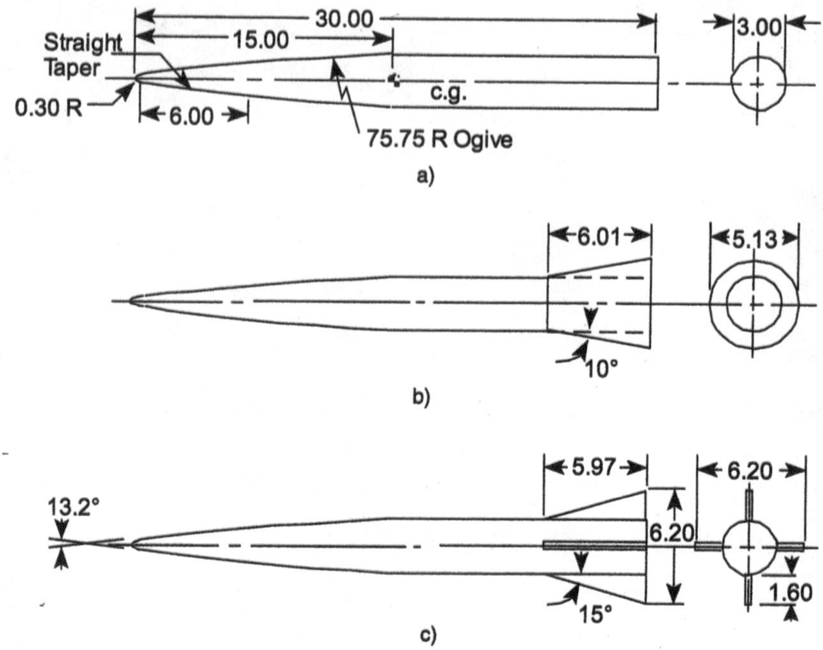

Fig. 5.29a Some configurations tested at $M_\infty = 2.01$ (Ref. 32): a) basic body, b) body with 10-deg flare, and c) body with 15-deg fins.

force coefficient on Fig. 5.29b. Figure 5.29b gives the same static aerodynamic results for the 10-deg flare case of Fig. 5.29a, as well as the results for the 15-deg fin case of Fig. 5.29a. Comparison of the AP98 and experiment is quite good, with the exception of the base pressure issue discussed previously. While at first sight, agreement between theory and experiment for pitching moment does not appear to be as good as desired, this is misleading. The reason, as mentioned before, is that the accuracy criteria we have strived to meet is on center of pressure. This accuracy criteria sets a goal of ±4% of body length as the average error criterion for center of pressure. The average center of pressure error is

$$\left(\frac{x_{CP}}{\ell}\right)_{\substack{\text{av}\\\text{error}}} = \frac{1}{N\ell}\sum_{i=1}^{N}\frac{C_{M_T} - C_{M_{\text{exp}}}}{C_{N_{\text{exp}}}} \times 100 \tag{58}$$

Equation (58) is in percent of body length ℓ, where ℓ is in calibers. Applying Eq. (58) to any of the three configurations of Fig. 5.29 shows the average center of pressure error to be less than 3% of the body length, or 0.3 caliber. This is well within the ± 4% goal on center of pressure error and is the reason for the comment of good agreement on static aerodynamics, even

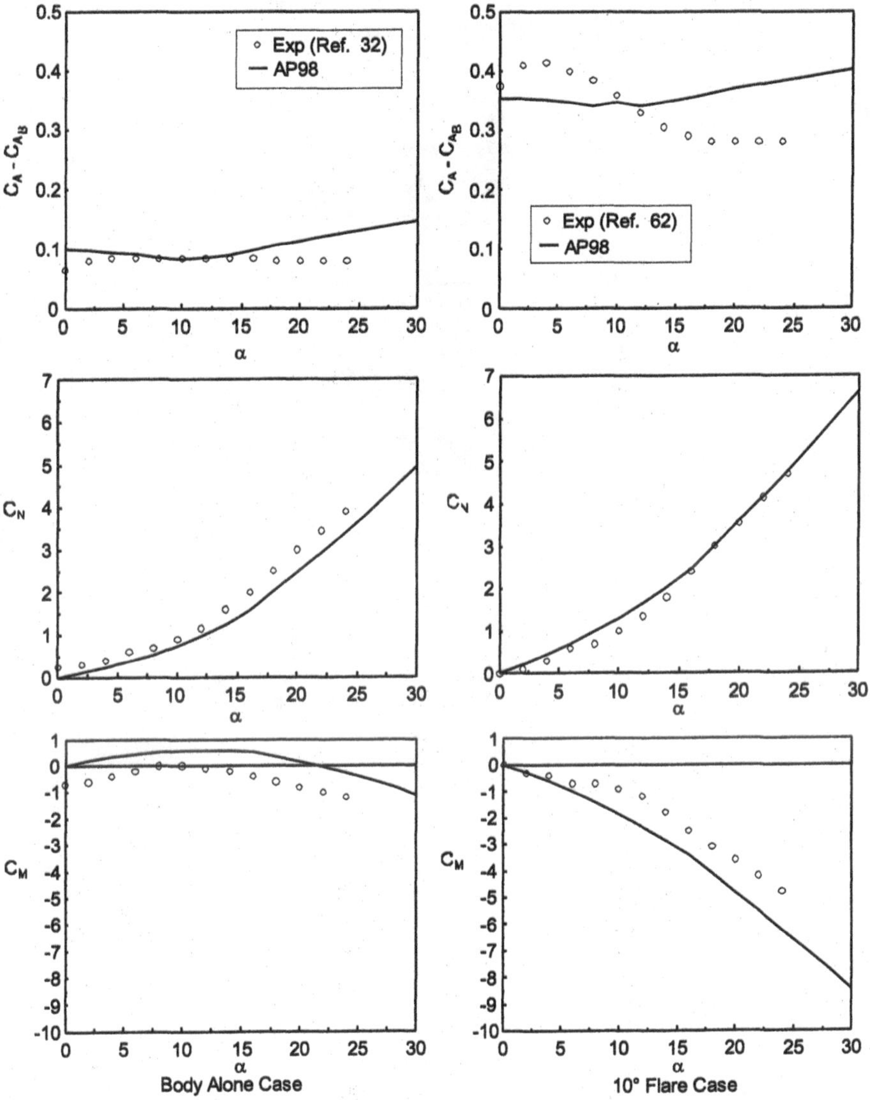

Fig. 5.29b Static aerodynamics of configuration of Fig. 5.29a.

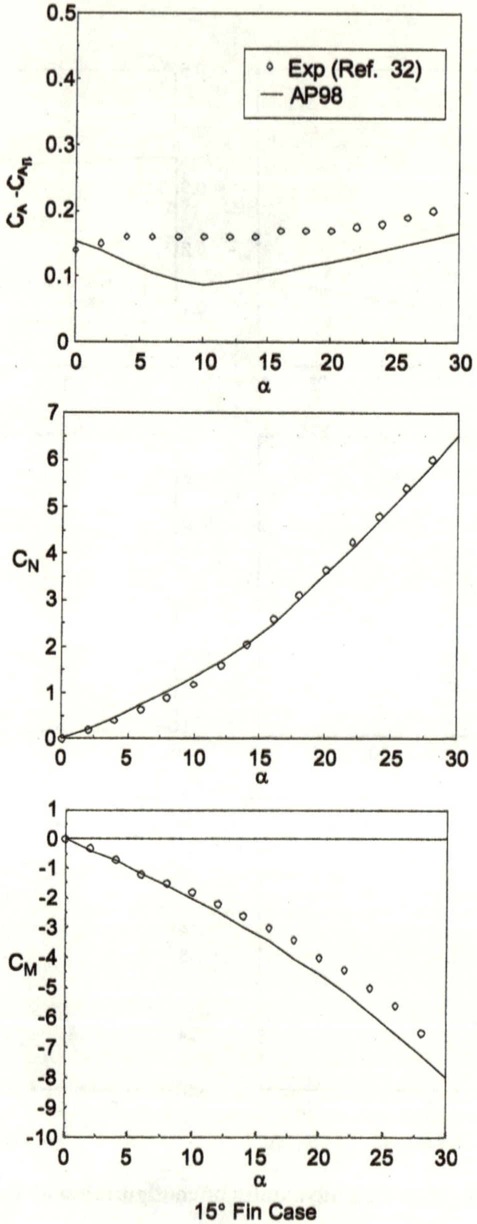

Fig. 5.29b ***(Continued)***

though pitching moments at first glance do not appear to be predicted as well as might be desired.

The fourth case considered is a canard–body–tail configuration shown in Fig. 5.30a. This case was tested[33] at $M_\infty = 0.2$. The configuration tested has aspect ratio canards of 1.59 and tails of 0.9. The body was 22.6 calibers in length. Test data were available at a 50-deg AOA for both $\varphi = 0$ and 45 deg roll orientation with canards deflected 0 and ±20 deg. Smith et al.[33] pointed out that the sting balance was designed for the large normal forces at high AOA, and hence the axial force measurements are not expected to be accurate. The wind tunnel model with boundary-layer trip option was chosen for aeroprediction computations. Sea level conditions were also assumed. Figure 5.30b gives the axial and normal force and pitching moment coefficients for the roll position of $\varphi = 0$ and control deflections of 0 and ±20 deg. Results are given for the experimental data, AP98, and Missile DATCOM[34] (which was taken from Ref. 33). As seen in the figures, the AP98 gives results for normal force and pitching moments that are in reasonable agreement with data. The AP98 appears to give better results compared to data than the DATCOM results presented in Ref. 33.

The next case is taken from Ref. 35. This data set consisted of various length-to-diameter canard–body–tail configurations tested at $M_\infty = 2.01$ to an AOA of 25 to 30 deg. Data were taken at a R_N/ft of 3.47×10^6 and with no boundary-layer trip present. Figure 5.31 shows configurations considered for comparison to data. These cases consist of 19.1-, 16.7-, and 14.8-caliber

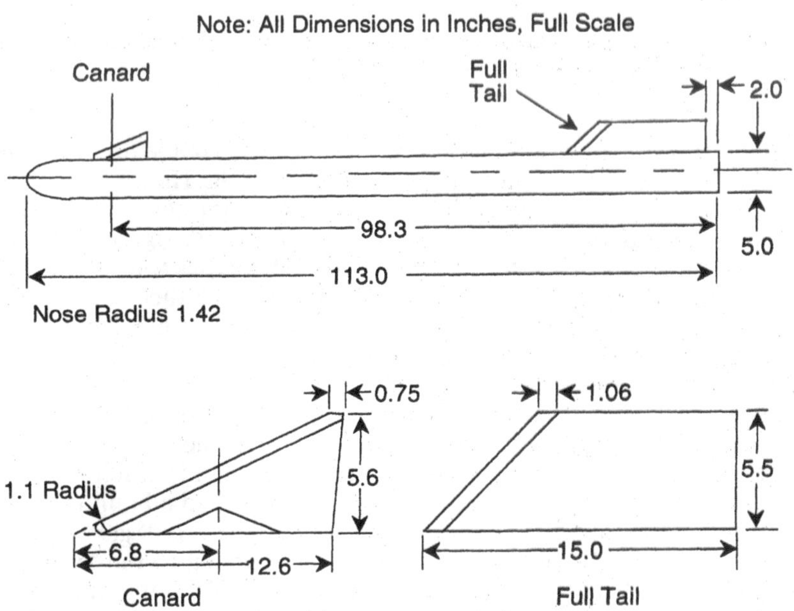

Fig. 5.30a Canard-controlled missile configuration.[33]

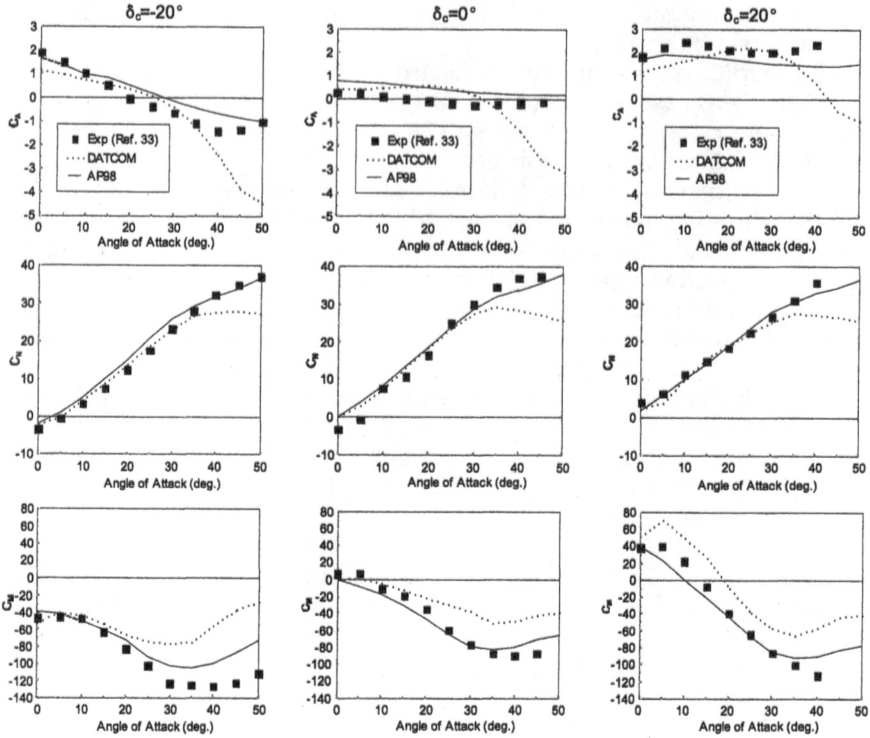

Fig. 5.30b Comparison of static aerodynamics between experiment and theory for Fig. 5.30a configuration ($\varphi = 0°$ and $M_\infty = 0.2$).

canard–body–tail configurations. The tail has about 2% of the tip eliminated, but this was not considered in the computational results. Hence, it should be expected that the theory should be slightly high on normal force and have slightly more stable pitching moments compared to data. Figure 5.31 shows the normal force and pitching moment coefficient comparisons of the AP98 to data. Even with the slightly larger tail fins used in the theory, the normal force and center of pressure average errors are well within the ±10 and ±4% of body length goals, respectively.

The next case considered is taken from Ref. 36 and is shown in Fig. 5.32. The model was about 22 calibers in length with a sharp nose of 2.25 calibers. The canards had an aspect and taper ratio of 2.0 and 0.3, respectively. Various tail fin spans were considered. This model was tested at Mach numbers 1.6 to 3.5 at AOAs to about 18 to 20 deg. It had a boundary-layer trip present and was tested at a R_N/ft of 2.0×10^6. Blair et al.[36] gave separate values of base axial force coefficient, which were added to the axial force values given in the reference to compare to the AP98 computations. To compare the experimental data to theory, Mach numbers of 2.5 and 3.5 are selected at roll angle 45 deg. Also, values of the tail-to-canard semispan of

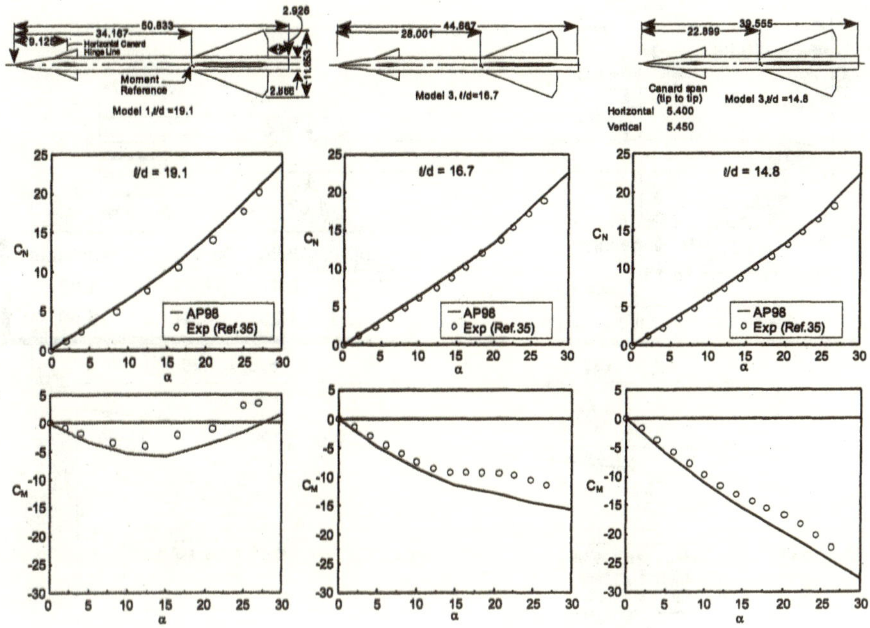

Fig. 5.31 Canard–body–tail configuration aerodynamics of various length-to-diameter ratios; M_∞ = 2.01 (Ref. 35).

0.47 and 1.25 are considered. Figure 5.32 presents the comparison of theory to experiment for b_t/b_c = 0.47 and b_t/b_c = 1.25. Results are shown for C_A, C_N, and C_M. As seen in Fig. 5.32, comparison of theory to experiment is quite acceptable and meets the average accuracy goal of ±10% on axial and normal force and ±4% of body length for center of pressure. The worst case error on pitching moment at M_∞ = 2.5 and b_t/b_c = 1.25 represents an error of less than 3% of the body length in terms of center of pressure. Also shown on Fig. 5.32 are M3HAX computations taken from Ref. 37 for the b_t/b_c = 1.25 case at M_∞ = 2.5. The C_A results for φ = 0 deg from Ref. 37 were assumed to apply to the φ = 45 deg case because no control deflections were assumed for this example. As seen in the figure, M3HAX gives reasonable comparison to data at φ = 45 deg. Lesieutre et al.[37] shows M3HAX giving improved comparisons to data at φ = 0 deg roll, compared to φ = 45 deg roll, particularly for normal force and pitching moment coefficients.

The next case considered is a wing–body–tail case with a low aspect ratio wing and tested at M_∞ = 0.1[38]. The configuration is shown at the top of Fig. 5.33. Note that the model tested in the wind tunnel at the top of the figure is slightly different from that where calculations were performed with the AP98. This is because the aeroprediction code cannot handle the detailed dorsal and wing geometry shown at the top of the figure. The configuration

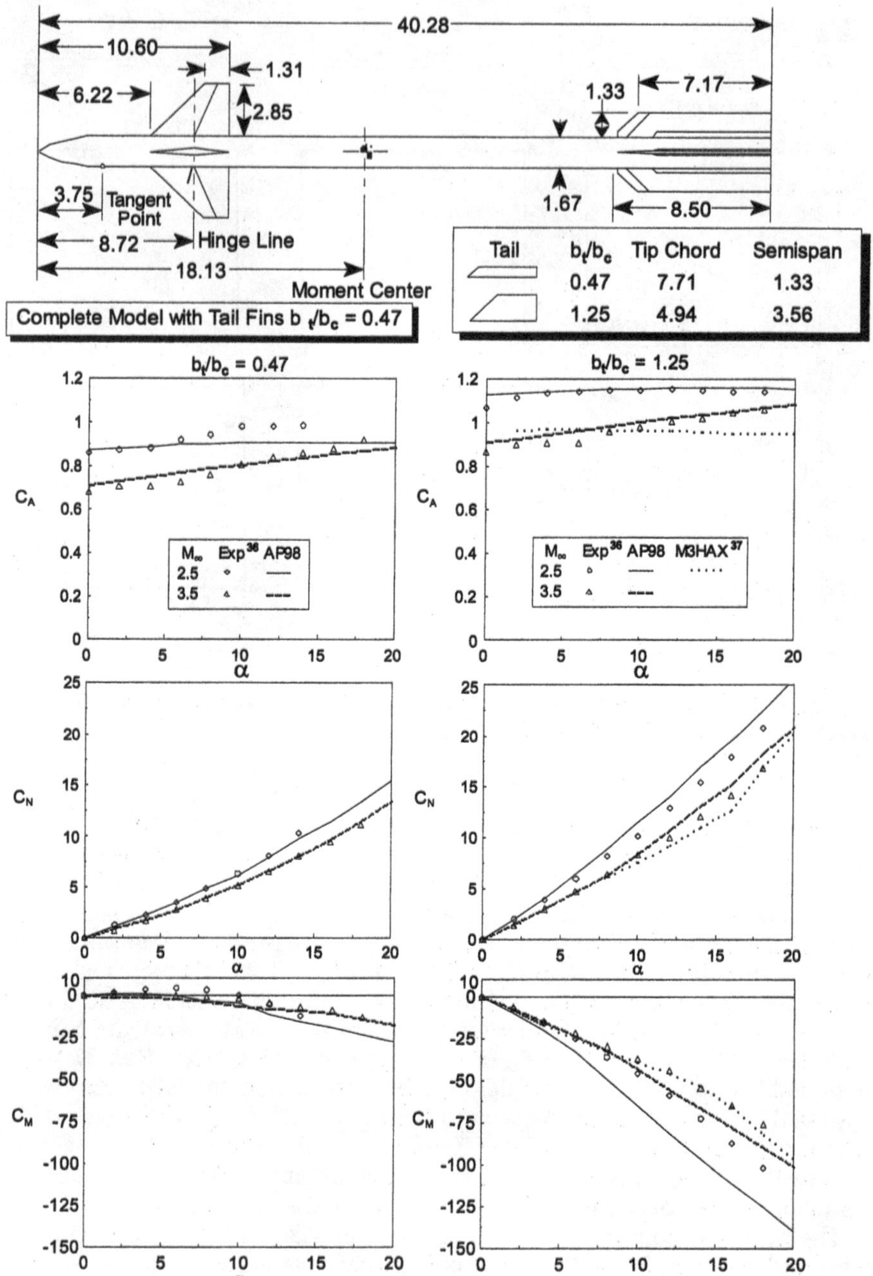

Fig. 5.32 Static aerodynamics of a canard–body–tail configuration with varying tail span (from Ref. 36).

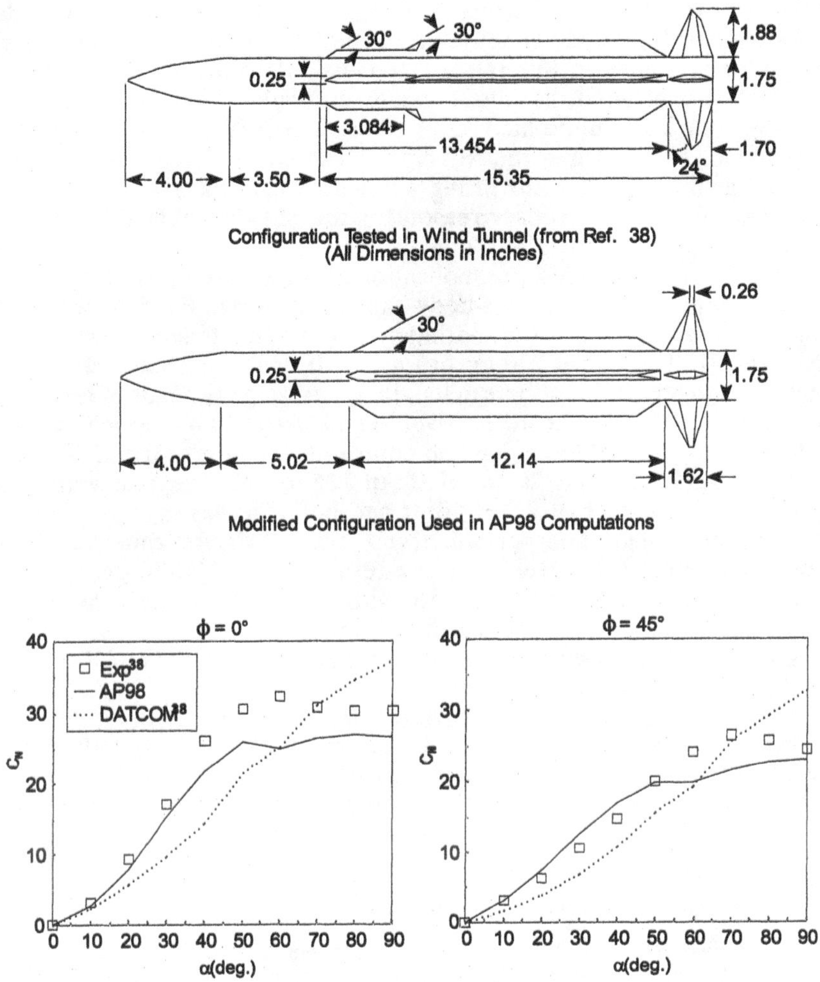

Fig. 5.33 Normal force coefficient comparisons of theory and experiment for body–dorsal–tail ($M_\infty = 0.1$).

shown in the middle of Fig. 5.33 has the same aspect ratio, span, taper ratio, leading edge sweep angle, and area of both the dorsal and tail as the actual model at the top of Fig. 5.33. Also, although not shown, the distance to the centroid of the planform areas is also held constant. Of course, the body is also the same between the two cases.

The only results given in Ref. 38 were for normal force. Experimental data and Missile DATCOM[34] results were both given in Ref. 38 for roll positions of $\varphi = 0$ and 45 deg. These results are compared to the AP98 at the bottom of Fig. 5.33 for $\varphi = 0$ deg and $\varphi = 45$ deg. Note that comparisons to data are reasonable and quite good except at an AOA of 40 to 70 deg. It is possible that wind tunnel sting to model interference increased the experimental results in this range somewhat. Previous wind tunnel studies[39,40] have concluded that model sting can increase model normal force loads by 10 to 20% in the high AOA range. The AP98 gives slightly better comparisons to data than the Missile DATCOM[34] for this case. It is suspected the low aspect ratio lifting surfaces, which have a great deal of nonlinear lift, are the primary reason for the Missile DATCOM accuracy problems, particularly at $\varphi = 0$ deg.

The last case considered for application of the semi-empirical methodology of Chapters 3 to 5 is a wing–body–tail case shown in Fig. 5.34, where the wings are used for control. Experimental results for this case were taken from Ref. 41. This configuration has a length of about 18 calibers with a tangent ogive nose 2.25 calibers in length. It has wings and tails of fairly high aspect ratios of 2.8 and 2.6, respectively. Data were taken at Mach numbers 1.5 to 4.63, for AOAS to 45 deg and control deflections of 0 and 10 deg at M_∞ of 1.5 and 2.0 and 0 to 20 deg at M_∞ of 2.35 to 4.63. The data were taken at a Reynolds number of 2.5×10^6/ft and boundary-layer trips were also used. The model had a hollow camber, and camber axial force measurements were given separately in Ref. 41. These results were added to the forebody axial force measurements to compare with the AP98 and new AOA axial force prediction method presented in this book.

Figure 5.35 shows the comparisons of the AP98 to the data of Ref. 41 for $\varphi = 0$ deg and $\varphi = 45$ deg. Figures 5.35a and 5.35b give C_A, C_N, and C_M for $M_\infty = 1.5$ at $\delta = 0$ and $\delta_W = 10$ deg at $\varphi = 0$ deg. In general, the AP98 gives good comparisons to data. Figures 5.35c and 5.35d give similar results for $M_\infty = 2.87$, and Figs. 5.35e and 5.35f give results for $M_\infty = 4.6$. Overall, for this configuration, at $\varphi = 0$ deg roll, the AP98 gives good comparisons to data.

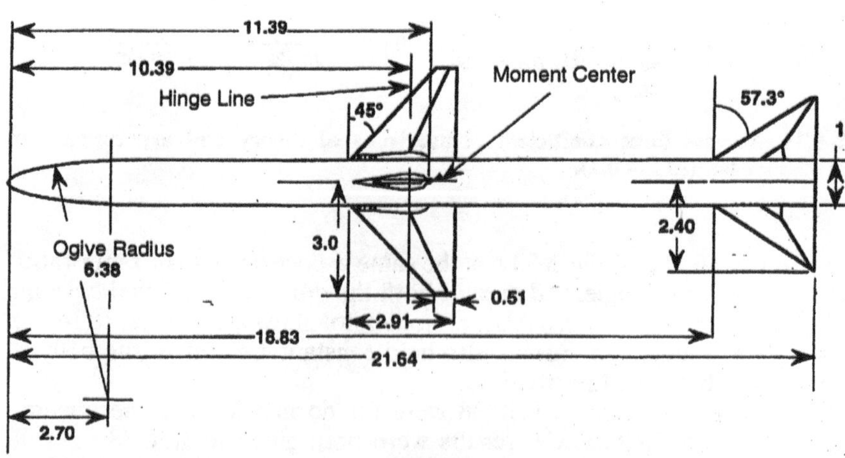

Fig. 5.34 Wing–body–tail configuration used in validation process.[41]

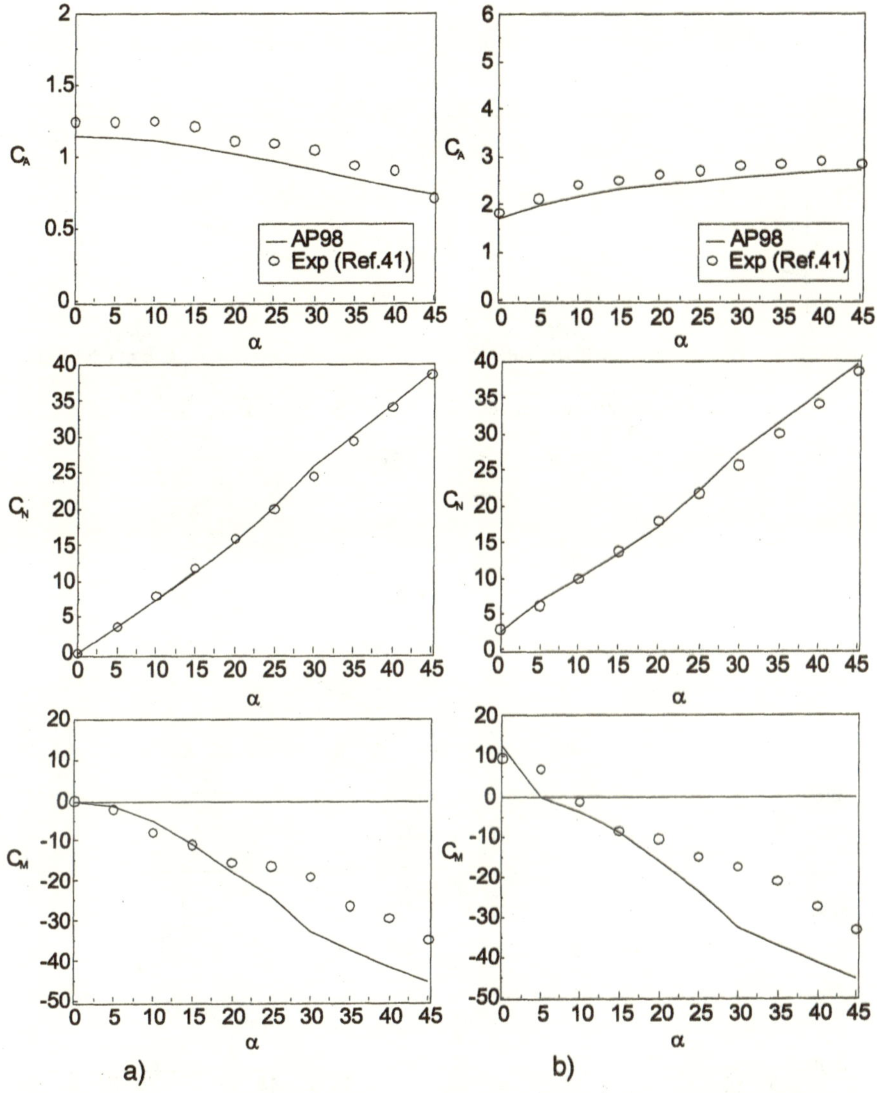

Fig. 5.35 Comparison of experiment and theory for C_A, C_N, and C_M for Fig. 5.34 wing control case. a) $M_\infty = 1.5$, $\delta_W = 0°$, and $\varphi = 0°$, b) $M_\infty = 1.5$, $\delta_W = 10°$, and $\varphi = 0°$, c) $M_\infty = 2.87$, $\delta_W = 0°$, and $\varphi = 0°$, d) $M_\infty = 2.87$, $\delta_W = 20°$, and $\varphi = 0°$, e) $M_\infty = 4.6$, $\delta_W = 0°$, and $\varphi = 0°$, f) $M_\infty = 4.6$, $\delta_W = 20°$, and $\varphi = 0°$, g) $M_\infty = 1.5$, $\delta_W = 0°$, and $\varphi = 45°$, h) $M_\infty = 1.5$, $\delta_W = 10°$, and $\varphi = 45°$, i) $M_\infty = 2.87$, $\delta_W = 0°$, and $\varphi = 45°$, j) $M_\infty = 2.87$, $\delta_W = 20°$, and $\varphi = 45°$, k) $M_\infty = 4.6$, $\delta_W = 0°$, and $\varphi = 45°$, l) $M_\infty = 4.6$, $\delta_W = 20°$, and $\varphi = 45°$

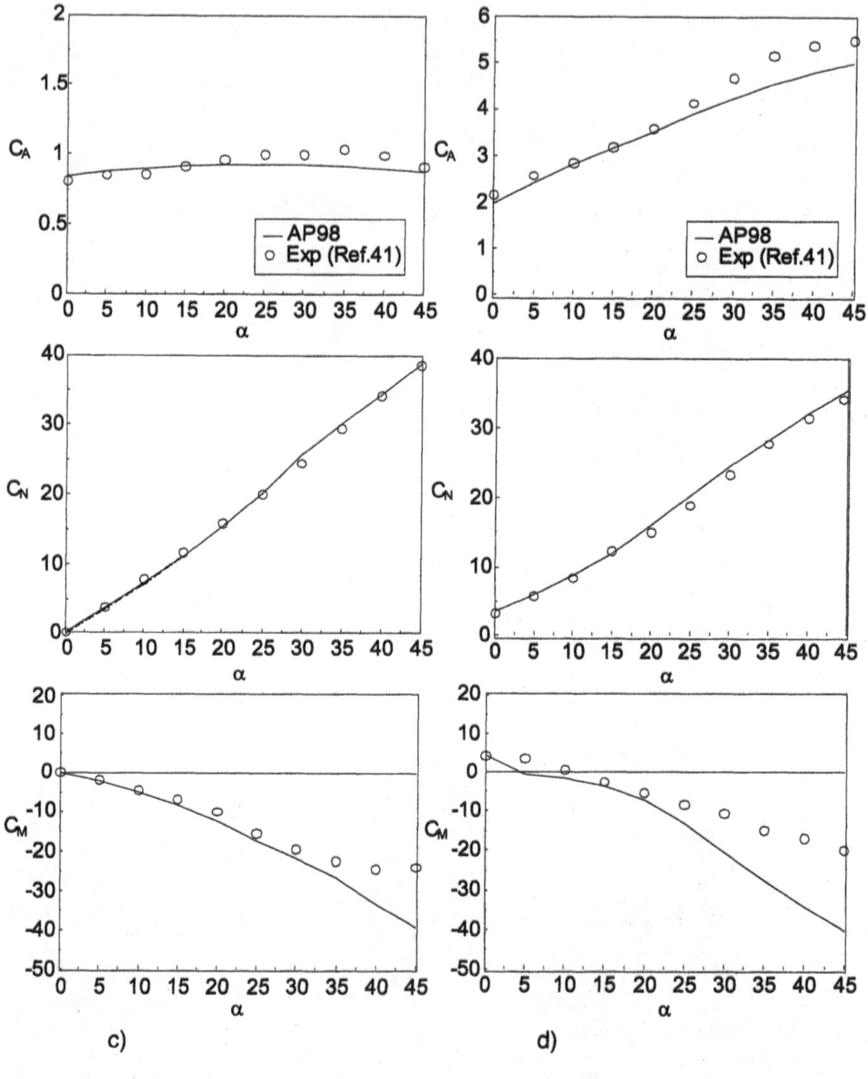

Fig. 5.35 *(Continued)*

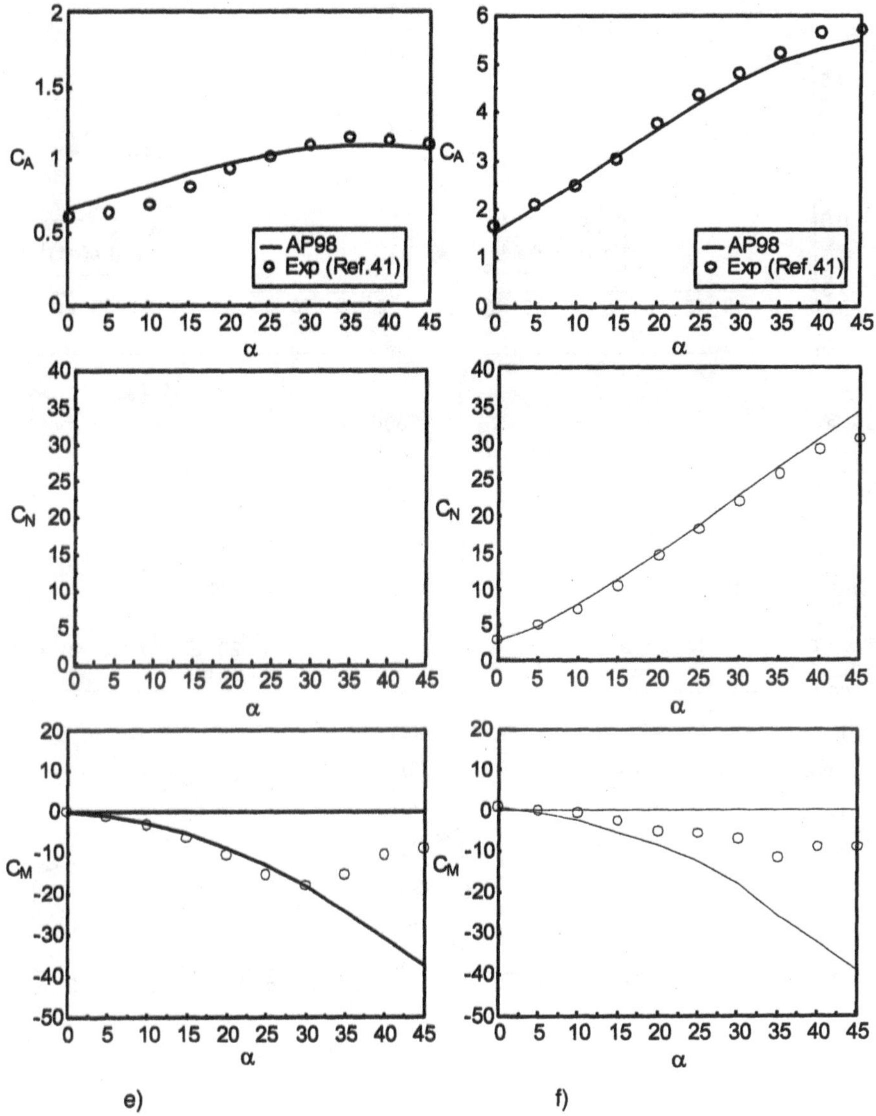

Fig. 5.35 ***(Continued)***

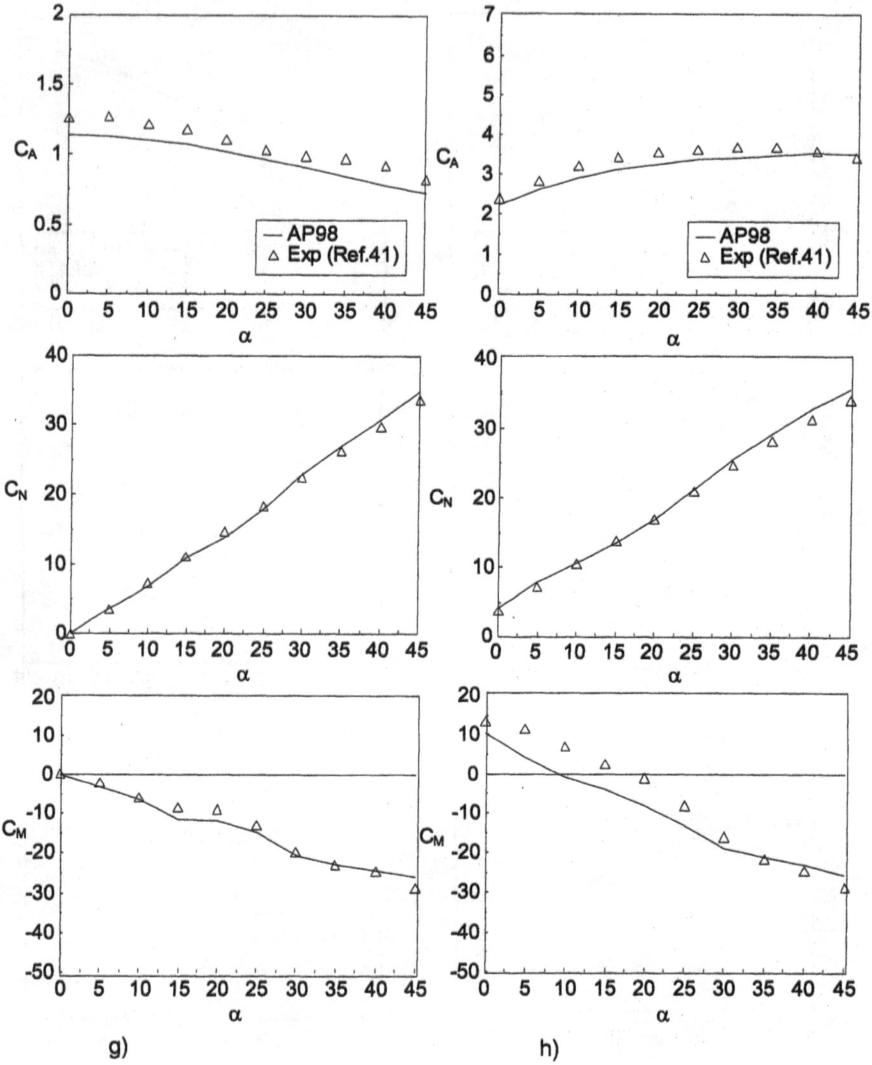

Fig. 5.35 ***(Continued)***

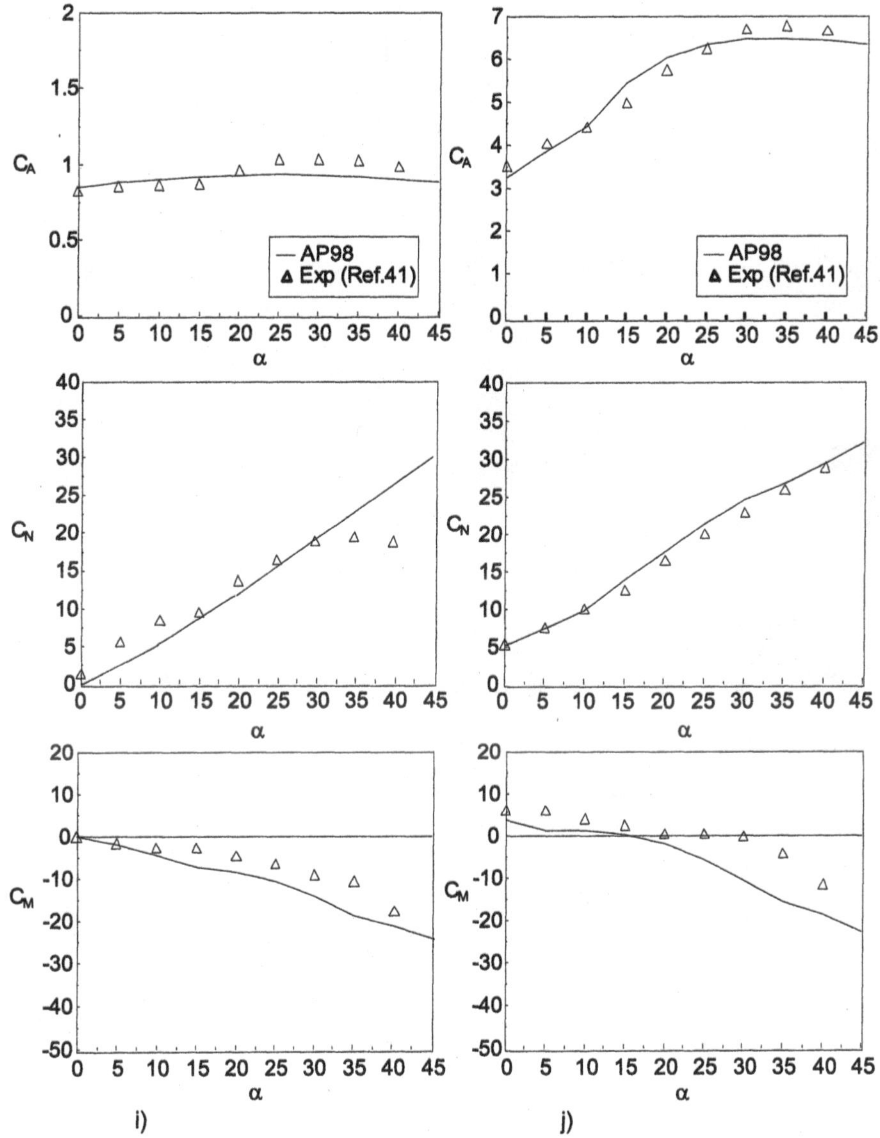

Fig. 5.35 ***(Continued)***

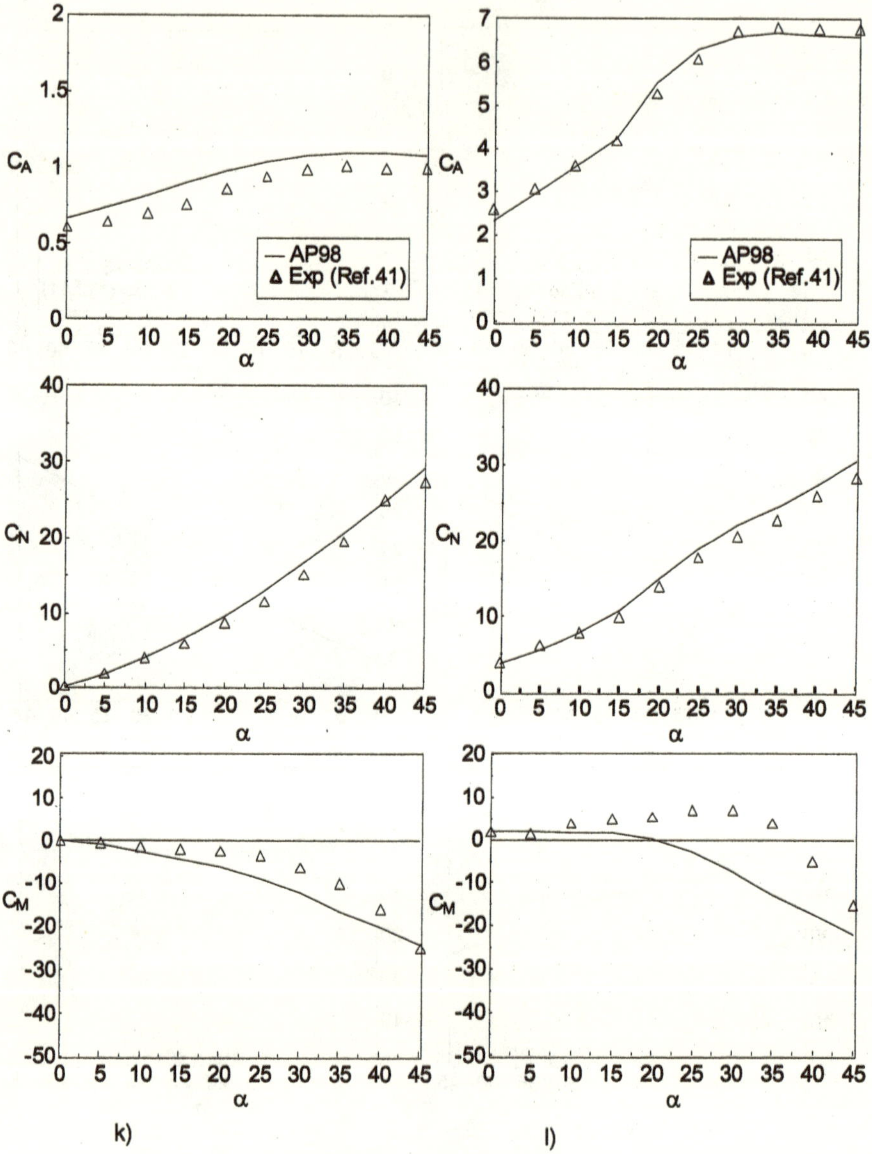

Fig. 5.35 *(Continued)*

The worst case errors are for center of pressure at higher Mach number and AOA, where the bow shock intersects the wing shocks. This nonlinear phenomenon is not modeled in the $\varphi = 0$ deg roll orientation at all. For the $\varphi =$ 45 deg roll, the center of pressure shift, Eq. (24), partially accounts for this phenomenon, but not entirely. Center of pressure errors approach a full caliber or 5% of the body length at $M_\infty = 4.6$ and $\alpha = 40$ deg. However, the average center of pressure errors for several Mach numbers and over a range of AOAs is less than 3%. The average axial force and normal force coefficient errors are on the order of 5%.

Figures 5.35g and 5.35h present $M_\infty = 1.5$ results for $\varphi = 45$ deg roll. These figures give C_A, C_N, and C_M for $\delta = 0$ and 10 deg. Figures 5.35i and 5.35j give similar results for $M_\infty = 2.87$ and $\delta = 0$ and 20 deg, and Figs. 5.35k and 5.351 give results for $M_\infty = 4.6$ and $\delta = 0$ and 20 deg. Note that good agreement is obtained between experimental data and the AP98 for all static aerodynamics at all three Mach numbers and for all control deflections. Here, the worst case center of pressure error is less than 3% of the body length.

Figure 5.36 presents results for the Fig. 5.34 configuration where the tail versus the wing is used for control. Here, only the $\delta_T = -10$ deg and -20 deg results are shown as the $\delta_T = 0$ deg control is basically the same as that in Fig. 5.35 for $\delta_W = 0$ deg. The AP98 shows the improvements in axial force at high AOA methodology of Section 5.7 to be quite useful. Figures 5.36b, d, and f give the tail control, $\varphi = 45$ deg roll results. Note the good agreement of the AP98 with the data. Again, the worst case error on pitching moment results is a center of pressure error of less than 3% of the body length.

In viewing the comparison of theory to experiment in Figs. 5.35 and 5.36 for both the wing and tail control alternatives of the Fig. 5.34 configuration, several conclusions can be drawn. In terms of accuracy, the AP98 gives acceptable predictions for C_A, C_N, and C_M for $\varphi = 0$ and 45 deg roll when α and δ are of the same or opposite signs. Some improvement is needed in the nonlinear methodology for higher supersonic speeds to account for suspected internal shock wave interactions. These interactions appear to have the most pronounced effect on pitching moments, particularly at the $\varphi = 0$ deg roll orientation. Even with this problem, average accuracy goals of $\pm 10\%$ on C_A and C_N and $\pm 4\%$ of the body length for center of pressure are easily met.

To summarize the overall comparisons of the semi-empirical methods of Chapters 3 to 5 in predicting static aerodynamics, the following statements seem to apply. First of all, for planar aerodynamics, semi-empirical methods give quite acceptable accuracy for static aerodynamics. "Acceptable," once again, means $\pm 10\%$ for C_A and C_N and $\pm 4\%$ of body length for x_{CP}. These types of codes and aerodynamics are most useful in preliminary design. If more accuracy in aerodynamics is needed, or a particular issue is of great importance, resort needs to be made to the CFD methods of Chapter 2 or experimental data. Finally, additional work is still needed in semi-empirical nonlinear methods associated with wing–tail interference and internal shock interactions that occur between the bow shock and a lifting surface or between lifting surfaces.

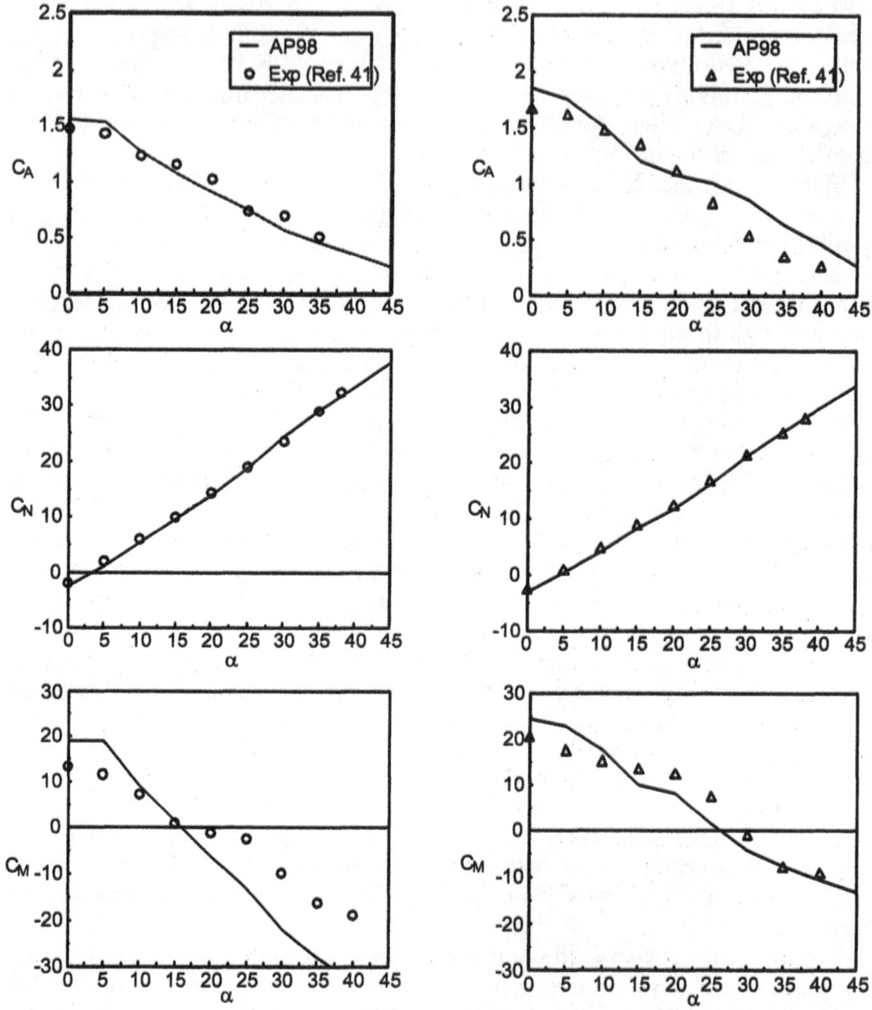

a) b)

Fig. 5.36 Comparison of theory and experiment for C_A, C_N, and C_M for Fig. 5.34 tail control case. a) $M_\infty = 1.5$, $\delta_T = -10°$, and $\varphi = 0°$, b) $M_\infty = 1.5$, $\delta_T = -10°$, and $\varphi = 45°$, c) $M_\infty = 2.87$, $\delta_T = -20°$, and $\varphi = 0°$, d) $M_\infty = 2.87$, $\delta_T = -20°$, and $\varphi = 45°$, e) $M_\infty = 4.6$, $\delta_T = -20°$, and $\varphi = 0°$, f) $M_\infty = 4.6$, $\delta_T = -20°$, and $\varphi = 45°$

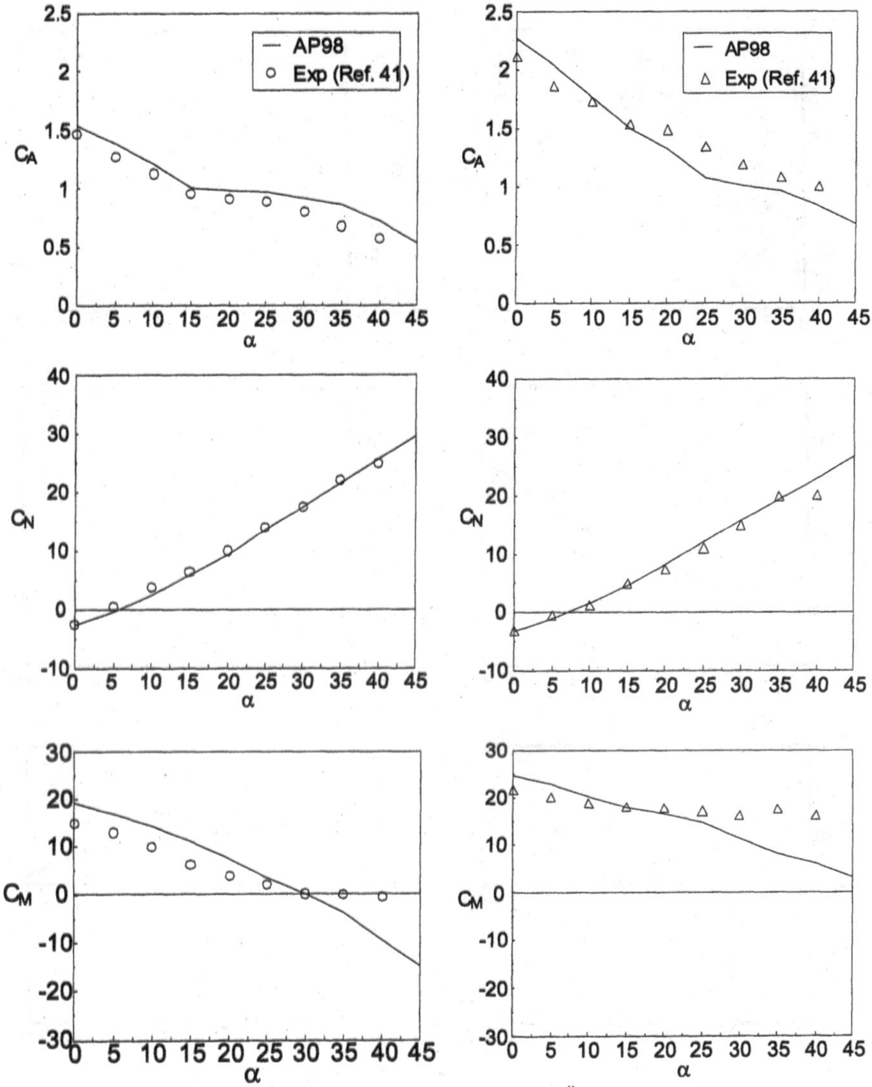

c) d)

Fig. 5.36 ***(Continued)***

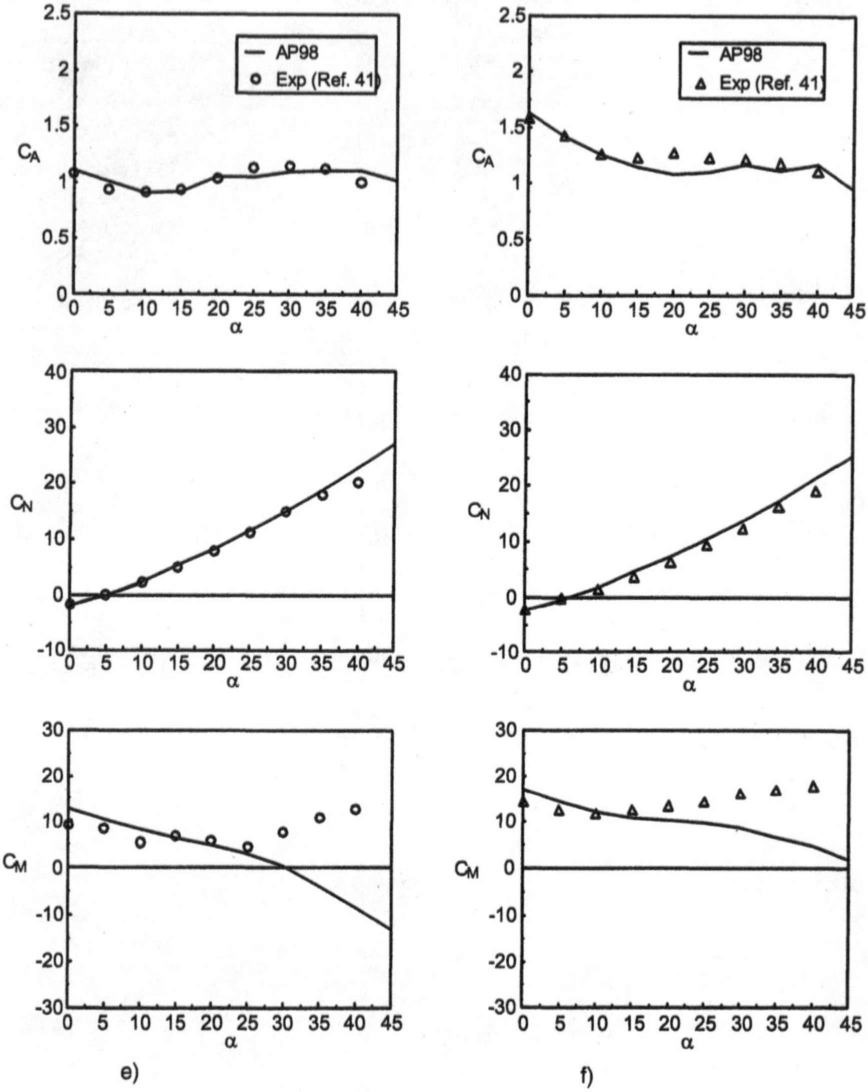

Fig. 5.36 ***(Continued)***

References

[1]Ericsson, L. E., and Reding, J. P., "Asymmetric Vortex Shedding for Bodies of Revolution," *Tactical Missile Aerodynamics,* edited by M. J. Hemsch and J. N. Nielsen, Vol. 104, AIAA Progress in Astronautics and Aeronautics Series, AIAA, New York, 1986, pp. 243–296.

[2]Allen, J. H., and Perkins, E. W., "Characteristics of Flow Over Inclined Bodies of Revolution," NACA RM A50L07, 1965.

[3]Jorgensen, L. H., "Predictions of Static Aerodynamic Characteristics for Slender Bodies Alone and With Lifting Surfaces to Very High Angles of Attack," NASA TR R-474, Sept. 1977.

[4]NASA Langley Research Center Tri-service missile data base, transmitted from NASA/LRC Jerry M. Allen to Naval Surface Warfare Center, Dahlgren Div., Nov. 1991 (Submitted for publication).

[5]Baker, W. B., Jr., "Static Aerodynamic Characteristics of a Series of Generalized Slender Bodies With and Without Fins at Mach Numbers from 0.6 to 3.0 and Angles of Attack from 0 to 180°," Arnold Engineering Development Corp. Rept. AEDC-TR-75-124, Vols. 1 and 2, Tullahoma, TN, May 1976.

[6]Maynes, R. D., and Gebert, G. A., "An Investigation of High Angle of Attack Aerodynamics on a Finned and Finless Body With a Fineness Ratio of 20.0," AIAA Paper No. 94-0724, 32nd Aerospace Sciences Meeting, Jan. 1994.

[7]Moore, F. G., and McInville, R. M., "A New Method for Calculating Wing Alone Aerodynamics to Angle of Attack 180°," Naval Surface Warfare Center, Dahlgren Div. Rept. NSWCDD/TR-94/3, March 1994.

[8]Stallings, R. L., Jr., and Lamb, M., "Wing-Alone Aerodynamic Characteristics for High Angles of Attack at Supersonic Speeds," NASA TP 1889, July 1981.

[9]Nielsen, J. N., Hemsch, M. J., and Smith, C. A., "A Preliminary Method for Calculating the Aerodynamic Characteristics of Cruciform Missiles to High Angles of Attack Including Effects of Roll Angle and Control Deflections," Office Naval Research Rept. ONR-CR215-216,4F, Arlington, VA Nov. 1977.

[10]Moore, F. G., and McInville, R. M., "Extension of the NSWCDD Aeroprediction Code to the Roll Position of 45 Degrees," Naval Surface Warfare Center, Dahlgren Div., Rept. NSWCDD/TR-95/160, Dec. 1995.

[11]Meyer, J., "Effects of the Roll Angle on Cruciform Wing–Body Configurations at High Incidences," *Journal of Spacecraft and Rockets,* Vol. 31, No. 1, Jan.–Feb. 1994, pp. 113–122.

[12]Moore, F. G., McInville, R. M., and Hymer, T., "The 1995 Version of the NSWC Aeroprediction Code: Part 1 — Summary of New Theoretical Methodology," Naval Surface Warfare Center, Dahlgren Div., Rept. NSWCDD/TR-94/379, Feb. 1995.

[13]Stallings, R. L., Jr., Lamb, M., and Watson, C. B., "Effect of Reynolds Number on Stability Characteristics of a Cruciform Wing–Body at Supersonic Speeds, NASA TP 1683, July 1980.

[14]Agnone, A. M., Zakkay, V., Tory, E., and Stallings, R., "Aerodynamics of Slender Finned Bodies at Large Angles of Attack, AIAA Paper 77-0666, 10th Fluid Plasma Dynamics Conference, Albuquerque, NM, June 1977.

[15]Moore, F. G., McInville, R. M., and Hymer, T., "The 1998 Version of the NSWC Aeroprediction Code: Part 1 — Summary of New Theoretical Methodology," Naval Surface Warfare Center, Dahlgren Div., Rept. NSWCDD/TR-98/1, April 1998.

[16]Washington, W. D., and Spring, D. J., "An Experimental Investigation of Wing–Tail Interference for a Typical Supersonic Missile," AIAA Paper 82-1339, AIAA 9th Atmospheric Flight Mechanics Conference, San Diego, CA, Aug. 1982.

[17]Aiello, G. F., and Bateman, M. C., "Aerodynamic Stability Technology for Maneuverable Missiles, Vol 1 Configuration Aerodynamic Characteristics," Air Force Systems Command, AFFDL-TR-76-55, Wright Patterson Air Force Base, OH, March 1979.

[18]Washington, W. D., "Computer Program for Estimating Stability Derivatives of Missile Configurations," U.S. Army Missile Command, RD-76-25, Redstone Arsenal, AL, May 1976.

[19]Hemsch, M. J., and Mullen, J., Jr., "Analytical Extension of the MISSILE1 and MISSILE2 Computer Programs," Nielsen Engineering Research, Rept. NEAR TR-272, March 1982.

[20]Dillenius, M. F. E., and Hemsch, M. J., Sawyer, W. C., Allen, J. M.; and Blair, A. B., Jr., "Comprehensive Missile Aerodynamics Programs for Preliminary Design," AIAA-82-0375. AIAA 20th Aerospace Science Meeting, Jan. 1982.

[21]Pitts, W. C., Nielsen, J. N., and Kaaturi, G. E., "Lift and Center of Pressure of Wing–Body–Tail Combinations at Subsonic, Transonic, and Supersonic Speeds," NACA TR 1307, 1957.

[22]Moore, F. G., and Hymer, T., "An Improved Method for Predicting Axial Force at High Angle of Attack," Naval Surface Warfare Center, Dahlgnen Div., Rept. NSWCDD/TR-96/240, Feb. 1997.

[23]Ingram, C. W., Ball, K., Dinkeloo, C., and Shida, D., "Preliminary Design Aerodynamic Prediction Methodology for Missiles with Nose Bluntness," Air Force Systems Command, AFFDL-TR-78-117, Wright Patterson Air Force Base, OH, Sept. 1978.

[24]Aiello, G. D., and Bateman, M. C., "Aerodynamic Stability Technology for Maneuverable Missiles,[11]Air Force Systems Command, AFFDL-TR-76-55, Vol. 1, Wright Patterson Air Force Base, OH, Dec. 1979.

[25]Graves, E., and Fournier, R., "Stability and Control Characteristics at Mach Numbers of 0.2 to 4.63 of a Cruciform Air-to-Air Missile with Triangular Canard Controls and a Trapezoidal Wing," NASA TM-X 3070, Nov. 1974.

[26]Smith, E. H., Hebbar, S. K., and Platzer, M., "Aerodynamic Characteristics of a Canard-Controlled Missile at High Angles of Attack," AIAA Paper No. 93-0763, Reno, NV, Jan. 1993.

[27]Vukelich, S. R., and Jenkins, J. E., "Missile DATCOM: Aerodynamic Prediction on Conventional Missiles Using Component Build-up Techniques," AIAA Paper 84-0388, Reno, NV, 1984.

[28]Monta, W. J., "Supersonic Aerodynamic Characteristics of a Sparrow III Type Missile Model with Wing Controls and Comparison with Existing Tail-Control Results, NASA TP 1078, Nov. 1977.

[29]Monta, W. J., "Supersonic Aerodynamic Characteristics of an Air-to-Air Missile Configuration with Cruciform Wings and In-Line Tail Controls," NASA TM S-2666, 1972.

[30]Jorgensen, L. H., "Prediction of Static Aerodynamics Characteristics for Slender Bodies Alone and with Lifting Surfaces to Very High Angles of Attack," NASA TR R-474, Sept. 1977.

[31]Gudmundson, S. E., and Torngren, L., "Supersonic and Transonic Wind Tunnel Tests on Slender Ogive–Cylinder Body Single and in Combination with Cruciform Wings and Tails of Different Sizes," Aeronautical Research Inst. Sweden, Aeronautics Dept., TFFA AU-772, Stockholm, Sweden, April 1972.

[32]Robinson, R. B., "Wind Tunnel Investigation at a Mach Number of 2.01 of the Aerodynamic Characteristics in Combined Angles of Attack and Sideslip of Several Hypersonic Missile Configurations with Various Canard Controls," NACA RM L58A21, March 1958.

[33]Smith, E. H.; Hebbar, S. K.; and Platzer, M., "Aerodynamic Characteristics of a Canard Controlled Missile at High Angles of Attack," AIAA Paper No. 93-0763, 31st Aerospace Sciences Meeting, Reno, NV, Jun 1993.

[34]Vukelich, S. R., and Jenkins, J. E., "Missile DATCOM: Aerodynamic Prediction on Conventional Missiles Using Component Build-Up Techniques," AIAA Paper 84-0388, Reno, NV, 1984.

[35]Spearman, L. M., "Aerodynamic Characteristics in Pitch of a Series of Cruciform-Wing Missiles with Canard Controls at a Mach Number of 2.01," NASA TN D839, May 1961.

[36]Blair, A. B., Jr.; Allen, J. M.; and Hernandez, G., "Effect of Tail-Fin Span on Stability and Control Characteristics of a Canard Controlled Missile at Supersonic Mach Numbers," NASA TP 2157, June 1983.

[37]Lesieutre, D. J., Love, J. F., and Dillenius, M. F., "M3HAX Aerodynamic Analysis for Finned Vehicles with Axisymmetric Bodies," NEAR TR 493-D, Mountain View, CA, Feb. 1996.

[38]Howard, R. M., and Dunn, A., "Missile Loads at High Angles of Attack," Engineering Note in *Journal of Spacecraft and Rockets*, Vol. 28, No.1, Jan.–Feb. 1991.

[39]Dietz, W. E., Jr., and Altstatt, M. C., "Experimental Investigation of Support Interference on an Ogive Cylinder at High Incidence," *Journal of Spacecraft and Rockets*, Vol. 16, No. 2, March–April; 1979.

[40]Canning, T. T., and Nielsen, J. N., "Experimental Study of the Influence of Supports on the Aerodynamic Loads on a Ogive Cylinder at High Angles of Attack," AIAA Paper 81-00070, 19th Aerospace Sciences Meeting, St. Louis, MO, Jan. 1981.

[41]Monta, W. J., "Supersonic Aerodynamic Characteristics of a Sparrow III Type Missile Model with Wing Controls and Comparison with Existing Tail-Control Results," NASA TP 1078, Nov. 1977.

Chapter 6

Aerodynamics of Noncircular Body Configurations

Nomenclature

A_C	= cross-sectional area of circular cylinder (ft^2)
A_{eq}	= cross-sectional area of circular cylinder equal to that of body with noncircular cross section
A_i	= apparent mass values from slender body theory (from Ref. 16)
A_{ref}	= reference area (maximum cross-sectional area of body, if a body is present, or planform area of wing if wing alone) (ft^2)
a,b	= semimajor and semiminor axis, respectively, of ellipse
a'	= body shape parameter (see Ref. 36)
$A(x)$	= body cross-sectional area as a function of position along body axis (ft^2)
AR	= wing or tail aspect ratio
Cir	= circumference of body (ft)
C_A	= axial force coefficient
$(C_A)_{\alpha=0}$	= axial force coefficient at 0-deg. AOA
C_{A_B}	= axial force coefficient component due to base pressure
C_{A_f}	= axial force coefficient component due to skin friction
C_{A_W}	= axial force coefficient component due to forebody and boattail or flare pressure
C_D	= drag coefficient
C_d, C_{d_c}	= local and total crossflow drag coefficients
$(C_{d_c})_1$, $(C_{d_c})_2$	= subcritical and supercritical value, respectively, of C_{d_c}
C_M	= pitching moment coefficient
C_{M_L}	= linear component of pitching moment coefficient
$C_{M_{NL}}$	= nonlinear component of pitching moment coefficient
$(C_n/C_{n_O})_{SB}$, $(C_n/C_{n_O})_N$	= ratio of the local normal force coefficient of a body with a noncircular cross section to that with a circular cross section calculated by slender body and Newtonian theory, respectively
C_N	= total normal-force coefficient
C_{N_B}	= normal-force coefficient of the body

C_{N_L} = linear component of the normal-force coefficient
$C_{N_{NL}}$ = nonlinear component of the normal-force coefficient
$(C_{N_\alpha})_W, (C_{N_\alpha})_T$ = normal-force coefficient slope of wing and tail, respectively
$C_{N_{T(V)}}$ = normal-force coefficient on tail due to wing-shed vortices
C_P = pressure coefficient
C_{P_B} = base pressure coefficient
C_{P_i} = local pressure coefficient at orifice i (Table 6.1)
C_{P_O} = stagnation pressure coefficient
c_r = root chord of lifting surface (ft), as measured at the point where the wing joins the body
d, d_B = diameter of body and base diameter of body, respectively
F = ratio of wing–body interference factor of a noncircular cross-sectional configuration to that of a circular cross-sectional configuration
k = parameter used to define corner radius for squares and triangles ($k = r_n/W_M$)
$k_{B(W)}, k_{B(T)}$ = ratio of additional body normal force coefficient due to the presence of a deflected wing or tail to that of the wing or tail alone at $\alpha = 0$ deg
$k_{W(B)}, k_{T(B)}$ = ratio of normal force contribution of a deflected wing or tail in the presence of a body to that of the wing or tail alone at $\alpha = 0$ deg
$K_{B(W)}, K_{B(T)}$ = ratio of additional body normal-force coefficient in the presence of a wing or tail to that of the wing or tail alone at $\delta = 0$ deg
$K_{W(B)}, K_{T(B)}$ = ratio of wing or tail normal-force coefficient in the presence of a body to that of the wing or tail alone at $\delta = 0$ deg
ℓ_{ref} = reference length that is body diameter
$\ell_1, \ell_2, \ell_3, \ell_i$ = individual segments of body length where body has variable noncircular cross section
M = Mach number
M_N = Mach number normal to body = $M \sin \alpha$
M_{N_C} = normal Mach number where flow transitions from subcritical to supercritical conditions
M_{N_O} = value of normal Mach number to body used in Newtonian correction factor calculation
M_∞ = freestream Mach number
NF = Newtonian correction factor
NF_0 = Newtonian correction factor for crossflow Mach number of zero on squares and triangles
NF_1 = Newtonian correction factor for an ellipse at $\alpha \geq 20$ deg
p, p_∞ = pressure and freestream pressure ($\mathrm{lb/ft^2}$)
Q = dynamic pressure = $1/2\ \rho_\infty V_\infty^2$
r = local body radius (ft)

r_{eq}, d_{eq} = radius and diameter, respectively, of a circular cross-sectional body that has same cross-sectional area as that of noncircular cross-sectional body
r_n = corner radius of a rounded corner on square or triangle
R_N = Reynolds number
R_{NC} = Reynolds number where flow transitions from subcritical to supercritical conditions
R_{N_D} = Reynolds number based on body diameter
$R_{N_{eff}}$ = an effective Reynolds number above which the flow transitions from subcritical to supercritical conditions
s = radius of body plus wing or tail semispan
SBTSF = slender body theory scaling factor
V_∞ = freestream velocity
V_N = velocity normal to body
V_{NW} = velocity normal to wing
W = length of one side of a triangle or square
W_m = maximum diameter of a triangle or square as measured normal to the velocity vector
x = distance along body axis (ft)
x_{CP} = center of pressure measured about some reference location
x_m = reference location about which center of pressure is measured
α = AOA (rad or deg)
$\alpha_1, \alpha_2, \alpha_c\ \alpha_D, \alpha_M$ = values of AOA upon which the nonlinear term of the wing–body and body–wing interference factors are based
γ = ratio of specific heats
δ = deflection angle of control surface (rad, deg)
η = parameter used in viscous crossflow theory for nonlinear body normal force (in this context, it is the ratio of the normal force of a circular cylinder of given length-to-diameter ratio to that of a cylinder of infinite length)
λ = taper ratio of fin = c_t/c_r
ρ = density (slugs/ft^3) of air
φ = roll position of missile ($\varphi = 0$ deg corresponds to fins in the plus (+) orientation) and the leeward plane. $\varphi = 45$ deg corresponds to fins rolled to the cross (x) orientation

I. Background and Survey of Nonaxisymmetric Body Methods

THE desire to increase weapon range and manueverability, design weapons that are more optimum from an aircraft total drag and radar signature standpoint, or provide optimum loadout of multiple missiles in a ship's vertical launcher has driven weapons designers to consider nonaxisymmetric body shapes. Some typical shapes of interest in design tradeoffs are shown in Fig. 6.1. Configurations shown in Fig. 6.1 include circular body

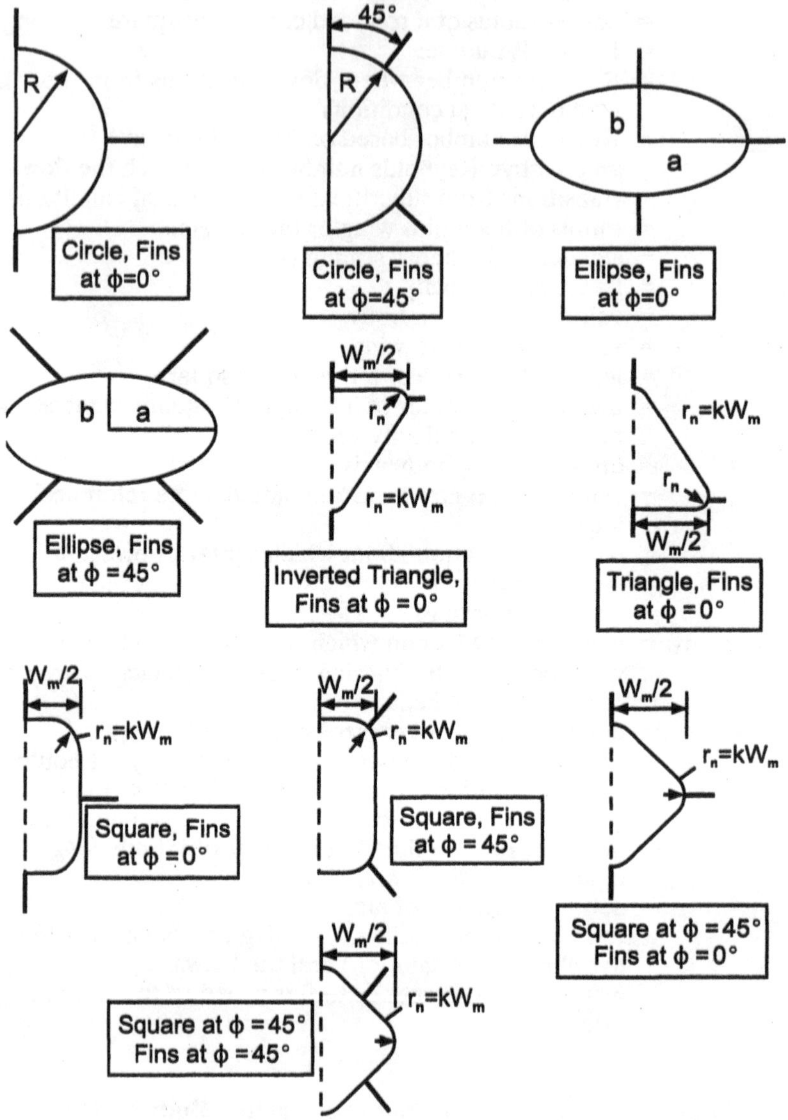

Fig. 6.1 Noncircular cross-section, wing–body configurations of interest for design studies.

cross sections with fins at either $\varphi = 0$ or $\varphi = 45$ deg, elliptical or square cross sections with fins at either $\varphi = 0$ or 45 deg, triangular or inverted triangular cross sections with fins at $\varphi = 0$, and squares rotated 45 deg (diamonds) with fins at $\varphi = 0$ or 45 deg. The squares, diamonds, and triangles can also have rounded corners. While most missiles in the United States and foreign countries in existence today have axisymmetric body configurations, these conceptual design tradeoffs of various configurations other than axisymmetric require engineering estimates of aerodynamics. Current state-of-the-art methods for predicting aerodynamics of nonaxisymmetric body shapes with engineering accuracy are much more limited than for axisymmetric bodies. This is primarily driven by the fact that getting reasonable accuracy of the aerodynamics requires an accurate description of the body geometry. To describe the geometry of a complex body shape accurately can take days or weeks depending on the requirements of the aerodynamics code being used.

Present approaches to computing aerodynamics fall into two basic classes. The first class requires the description of the body geometry in some detail for aerodynamic computations. Methods that fall into this class may use local slope approaches (for example, tangent-wedge, tangent-cone, Newtonian impact theory,) to estimate the surface pressure. These techniques basically need a freestream Mach number and angle between the freestream velocity vector and the tangent to a local point on the body to compute a local pressure coefficient (and other thermodynamic properties if desired). No influence of surrounding elements is felt by the particular body element of interest. A typical computer program today that uses this approach is the Supersonic–Hypersonic–Arbitrary Body–Programs.[1] Chapter 4 gives the theoretical basis for many simple local slope methods.

Another approach that requires detailed geometry input is the so-called paneling method. Here the configuration is described by a distribution of sources, sinks, and doublets so that the body boundary condition of flow tangency is met. The configuration is once again described by panels, but here the flow from one panel can affect another panel, as opposed to local slope approaches. A state-of-the-art program that uses this approach is the PANAIR code.[2]

A third approach that falls into the class of aerodynamic codes requiring detailed geometry inputs is use of the so-called numerical codes (see Chapter 2). These codes typically solve the full inviscid[3] set of equations coupled with boundary-layer methods or various versions of the viscous Navier–Stokes equations.[4] Geometry must be input even more accurately here due to the requirement of accurate second-order, partial derivatives in fluid dynamics equations. Weeks or months of staff time can easily be required to describe a complex configuration for these types of codes.

The second class of codes calculating the aerodynamics of a nonaxisymmetric body is based on an equivalent axisymmetric body. The beauty of this approach is the fact that an existing axisymmetric body code can be used to calculate nonaxisymmetric body aerodynamics if the area distribution of the body is known. The area distribution requires a lot less time to define than a detailed configuration geometry. On the other hand, one should anticipate possibly larger errors in predicting aerodynamics than an approach where detailed geometry is required. The ease of use, fast turnaround, and lower cost may be worth the reduced accuracy, however. This type of approach was pioneered by the methods of Whitcomb[5] and Jorgensen.[6–10]

The Missile DATCOM[11] code uses the Jorgensen method to estimate aerodynamics of ellipitic cross-sectional shapes to any angle of attack (AOA) and other cross-sectional shapes at small AOA. However, no such code is available for other than elliptical cross-sectional shapes at high AOA. There are several areas where the Jorgensen approach needs improvement. First, the method was derived on the basis of slender body theory for low AOA and modified Newtonian theory at higher AOA. The modified Newtonian theory, strictly speaking, is accurate only for higher Mach numbers and, therefore, needs to be corrected for lower Mach numbers. Second, most of the data available for making these empirical corrections to modified Newtonian theory are at low speed, and thus additional data or computational fluid dynamics (CFD) efforts are needed for higher Mach number. Third, Jorgensen correlation factors based on slender body theory and modified Newtonian theory work reasonably well in predicting normal forces for elliptical configurations but not so well for triangular and square cross-sectional bodies. Fourth, the Jorgensen approach uses slender body theory for interference between the wings and bodies with no nonlinear corrections for AOA, Mach number, or wing shape. The Missile DATCOM[11] improves upon this problem by the use of the equivalent AOA method.[12] In this way, nonlinear corrections of wing–body lift are included and the AOA boundary is extended to 20 to 25 deg for interference effects. Finally, no corrections are included in the Jorgensen method for the axial force wave component of drag for the noncircular bodies compared to circular bodies.

While the Jorgensen method has its shortcomings, it is still the method most preferable for a semi-empirical aerodynamic prediction method. The methods of Chapters 3 to 5 were all derived on the basis of an axisymmetric body, and it would be much easier to integrate the Jorgensen type of technology to an existing axisymmetric body code,[13] as opposed to starting from scratch to create a new code. The goal of this chapter is, therefore, to modify the work of Jorgensen to improve upon the shortcomings stated previously. If successful, a more robust, more accurate, yet simpler method for computing aerodynamics of nonaxisymmetric body missile configurations will be available.

II. Review of Jorgensen Method

Jorgensen's method[6–10] for static aerodynamics of a body alone is defined by

$$C_A = (C_{A\alpha=0}) \cos^2 \alpha \tag{1}$$

$$C_N = \frac{\sin(2\alpha)\cos(\alpha/2)}{A_{\text{ref}}} \int_0^{\ell} \left(\frac{C_n}{C_{n0}}\right)_{SB} \left(\frac{\mathrm{d}A}{\mathrm{d}x}\right) dx + \frac{2\eta C_{d_c} \sin^2 \alpha}{A_{\text{ref}}} \int_0^{\ell} \left(\frac{C_n}{C_{n0}}\right)_N r\,dx \tag{2}$$

$$C_M = \frac{\sin(2a)\cos(a/2)}{A_{\text{ref}}\ell_{\text{ref}}}\int_0^{\ell}\left(\frac{C_n}{C_{no}}\right)_{SB}\frac{dA}{dx}(x_m - x)dx + \frac{2\eta C_{d_c}\sin^2 a}{A_{\text{ref}}\ell_{\text{ref}}}\int_0^{\ell}\left(\frac{C_n}{C_{no}}\right)_N r(x_m - x)dx \tag{3}$$

$$\frac{x_{CP}}{\ell_{\text{ref}}} = -\frac{C_M}{C_N} + \frac{x_m}{\ell_{\text{ref}}} \tag{4}$$

Equations (2) and (3) allow for continually varying cross sections along the body. Equation (1) is an assumed approximation for correlating axial force with AOA. It also requires a calculation of, or that experimental data be available for, axial force coefficient at zero AOA. Equations (2) and (3) both include the factors $(C_n/C_{no})_{SB}$ and $(C_n/C_{no})_N$. These factors represent the slender body and Newtonian approximations to the local normal force coefficient per unit length of the desired cross-sectional shape (C_n) to the similar coefficient for the equivalent circular cross-sectional shape (C_{no}). The radius of the equivalent cross-sectional shape is determined by

$$r_{eq} = \sqrt{\frac{A(x)}{\pi}} \tag{5}$$

where $A(x)$ represents the area of the nonaxisymmetric body as it varies along the body x axis. The first terms of both Eq. (2) and (3) are the terms due to potential flow, and the second terms are those due to the viscous crossflow. Because slender body theory is applicable only for small AOAs, the coefficient $(C_n/C_{no})_{SB}$ can only be used for the first term of Eqs. (2) and (3). On the other hand, Newtonian impact theory is applicable at any AOA and can, therefore, be used for the second term of Eqs. (2) and (3). The term C_{d_c} is the crossflow drag coefficient for the equivalent body of revolution. Finally, both Eq. (3) and (4) are defined about some reference point x_m. $(C_n/C_{no})_{SB}$ and $(C_n/C_{no})_N$ are given in Ref. 6 for several ellipses. $(C_n/C_{no})_N$ is also given for some squares. Vukelich and Jenkins[11] gives the slender body theory values of $(C_n/C_{no})_{SB}$ for several configurations.

To apply the methodology of Eqs. (1) through (4) to wing–body–tail configurations, Jorgensen[6] suggests replacing the first term of Eqs. (2) and (3) with that computed by Ref. 14. Using this approach, Eqs. (2) and (3) become

$$C_N = C_{N_L}\frac{\sin 2a}{2a} + \frac{2\eta C_{dC}\sin^2 a}{A_{\text{ref}}}\int_0^{\ell}\left(\frac{C_n}{C_{no}}\right)_N r\,dx \tag{6}$$

$$C_M = C_{M_L} \frac{\sin 2a}{2a} + \frac{2\eta C_{dc} \sin^2 a}{A_{\text{ref}} \ell_{\text{ref}}} \int_0^{\ell} \left(\frac{C_n}{C_{no}} \right)_N r(x_m - x) dx \qquad (7)$$

C_{N_L} and C_{M_L} are the potential normal force and pitching moments computed by linear theory and slender body theory as defined by Chapters 3 and 4.

Reviewing Eqs. (1) through (7), it is appropriate to specifically point out the weak points suggested in Section 6.1. First of all, Eq. (1) assumes $(C_A)_{a=0}$ of the equivalent axisymmetric body is the same as that for the noncircular body and that C_A varies as $\cos^2 a$ with AOA. These assumptions need further investigation. Second, $(C_n/C_{no})_N$ is accurate primiarily at high Mach number. A correction for lower crossflow Mach numbers ($M_N < 2.0$) is probably required for accurate prediction of static aerodynamics at all AOAs and M_∞'s. Third, additional analytical equations need to be derived for $(C_n/C_{no})_N$ for other than elliptical cross-sectional shapes. Additional equations are required to expand the approach of Ref. 6 to a broader range of cross sections. Fourth, while Jorgensen discusses the effect of crossflow drag coefficient as a function of Reynolds number, he does not offer any methodology to correct for the sudden decrease in the crossflow drag coefficient for Reynolds numbers above the critical value as a function of body cross-sectional shape. Finally, configuration aerodynamics need to be defined in such a way as to allow the interference terms between the wing and body to include nonlinearities. Moore et al.[13] and Chapter 5 defined these nonlinearities for circular bodies. This methodology has been validated to high AOA and a large range of Mach numbers and configurations. However, this methodology also needs to be adjusted for noncircular bodies being considered.

Most of the remainder of this chapter will take each of the problems discussed and define modifications or new methods necessary to attempt to achieve a more accurate and robust way of treating noncircular bodies than is available in either Ref. 6 or 11.

III. Body-Alone Axial Force Approach

The axial force coefficient has three components that arise from wave drag generated by the flow as it is compressed on the body surface, friction of the air as it passes over the body, and pressure on the afterbody or base caused by the separation of the flow from the body surface. Mathematically, the axial force coefficient is defined by

$$C_A = C_{A_W} + C_{A_f} + C_{A_B} \qquad (8)$$

Jorgensen[6] alludes to the use of equivalent axisymmetric axial force for use in noncircular body axial force calculations. It is reasonable to assume that the base drag methodology of Section 4.8.5 could be extended to a noncircular body by use of the equivalent diameter. That is,

$$C_{A_B} = -\left(C_{P_B}\right)_{3-D} \left(\frac{d_B}{d_{\text{ref}}} \right)^3 ; \begin{pmatrix} \text{squares, triangles} \\ \text{ellipse, } a/b \le 1.5 \end{pmatrix} \qquad (9a)$$

Here, $(d_B)_{eq}$ and $(d_{\text{ref}})_{eq}$ of Eq. (9a) are the base and reference diameters of the equivalent body of revolution computed from Eq. (5), and C_{P_B} is the negative base pressure coefficient that occurs on a circular cylinder with no boattail as a function of Mach number and AOA (see Fig. 4.28).

Equation (9a) applies to configurations which, while noncircular in shape, are not very flat. For elliptical cross-sectional bodies with large values of semimajor to semiminor axes, the relation in Eq. (9a) does not hold. In such cases, the base pressure coefficient is more like a two- versus three-dimensional value and the boattail effect should be squared rather than cubed. Hence, for ellipses with $a/b \geq 5$,

$$C_{A_B} = -\left(C_{P_B}\right)_{2\text{-}D}; \begin{pmatrix}\text{ellipses with}\\ a/b \geq 5.0\end{pmatrix} \tag{9b}$$

For ellipses, where $1.5 \leq a/b < 5$, a linear interpolation between Eqs. (9a) and (9b) is assumed. That is,

$$C_{A_B} = -\left(C_{P_B}\right)_{3\text{-}D}\left(\frac{d_B}{d_{\text{ref}}}\right)_{eq} - \left(\frac{d_B}{d_{\text{ref}}}\right)_{eq}^2\left[\left(C_{P_B}\right)_{2\text{-}D} - \left(C_{P_B}\right)_{3\text{-}D}\left(\frac{d_B}{d_{\text{ref}}}\right)_{eq}\right]\left(\frac{a/b - 1.5}{3.5}\right); \begin{pmatrix}\text{ellipse with}\\ 1.5 < a/b < 5.0\end{pmatrix} \tag{9c}$$

Values of $(C_{P_B})_{2-D}$ and $(C_{P_B})_{3-D}$ both come from Fig. 4.28.

Intuitively, the skin-friction axial force of a noncircular body should be directly proportional to the wetted area of the air on the body surface. As a first approximation, this can be estimated by the ratio of the circumference of the noncircular body to that of the circular body. Then,

$$\left(C_{A_f}\right)_{NC} = \left(\frac{(\text{Cir})_{NC}}{2\pi r_{eq}}\right)\left(C_{A_f}\right)_{eq} \tag{10}$$

The term $(C_{A_f})_{eq}$ is the skin-friction drag of the equivalent circular body of radius r_{eq}.

The wave drag term of Eq. (8) is more complicated than the other two components due to the fact it is a function of the slope along the body surface as well as the area distribution. Hence, to compute the wave drag on an equivalent axisymmetric body would require the product $(r dr/dx)_{eq}$ to be the same as for the noncircular body. The best way to analyze this term would be through numerical computations using codes such as full Euler[3] or Navier–Stokes.[4] However, until such time as these computations are available or experimental results become available, the present approach is to assume the wave drag term of the noncircular cross-sectional body is the same as that of the equivalent axisymmetric cross-sectional body.

The method of Chapter 5 will be assumed for AOA changes in C_A, and Chapter 3 methods will be used for fin values of C_A. Hence, to compute C_A, we first determine the equivalent axisymmetric body. The axial force at low AOA is computed as currently done in Chapters 3 and 4 for this equivalent axisymmetric body with or without fins. The body-alone value of C_A is then adjusted using Eqs. (9) and (10) for the appropriate noncircular shape, and the fin values of C_A are held constant. C_A at AOA is then adjusted according to the Section 5.7 methodology.

IV. Newtonian and Slender Body Theory Factors

When applying Eqs. (3) to (7) to noncircular bodies, values of the parameters $(C_n/C_{no})_{SB}$ and $(C_n/C_{no})_N$ are required for the particular noncircular shape of interest. C_{d_c} of these equations remains the crossflow drag coefficient of a circular cylinder with the radius defined by Eq. (5). Values of these parameters for ellipses using both slender body and modified Newtonian theory are given in Ref. 6 and shown here in Fig. 6.2 for convenience.

Values of $(C_n/C_{no})_N$ for squares and triangular shapes of various corner radii are defined in Ref. 15. Vukelich et al.[16] contained approximate values of the slender body apparent mass parameters that allowed computation of $(C_n/C_{no})_{SB}$ for the body configurations of Fig. 6.1 with no corner radius. These values were then allowed to go to one in a linear fashion when k goes to 0.5. Values of $(C_n/C_{no})_N$ and $(C_n/C_{no})_{SB}$ are given in Figs. 6.3 and 6.4 for

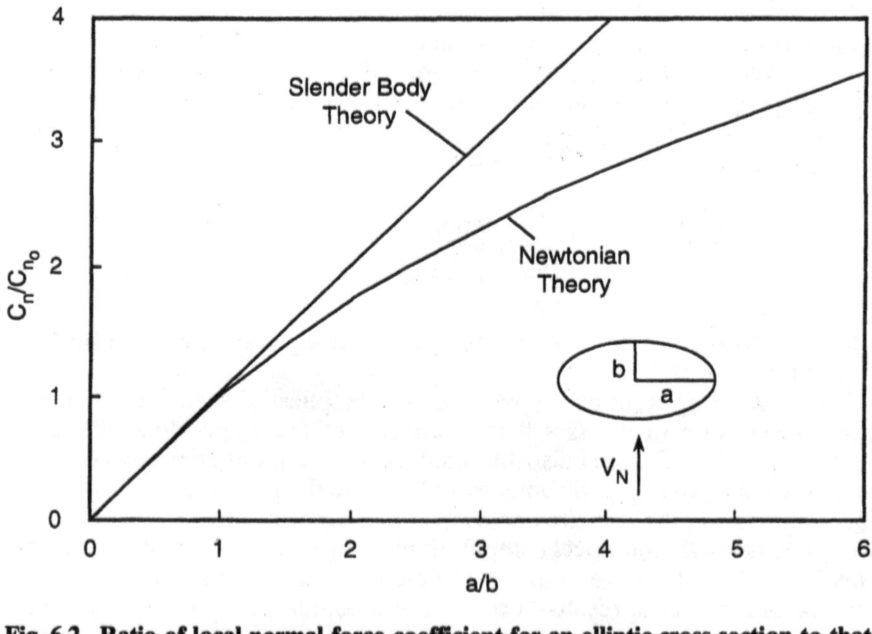

Fig. 6.2 Ratio of local normal-force coefficient for an elliptic cross section to that for the equivalent circular cross section.

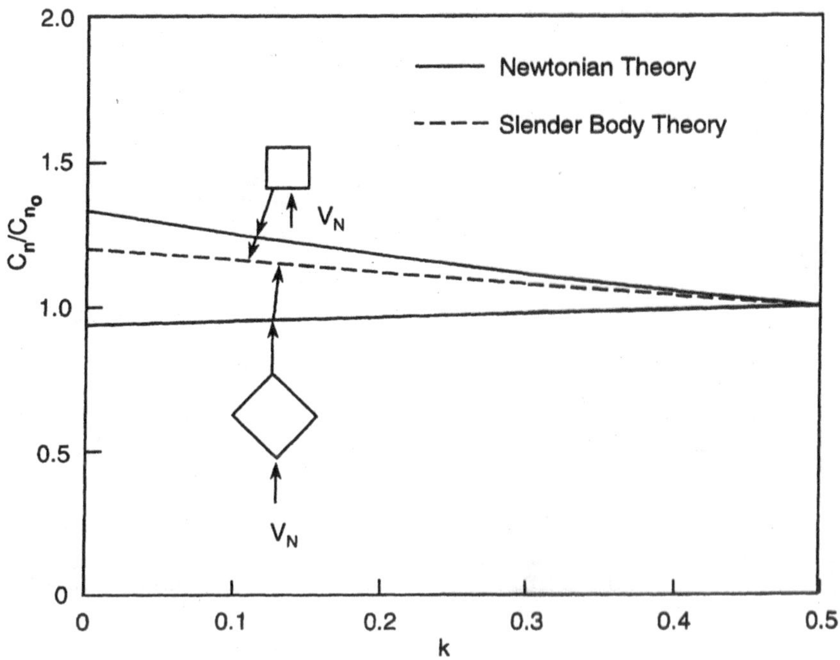

Fig. 6.3 Ratio of local normal-force coefficient for a square cross section to that for the equivalent circular cross section.

the square and triangular cross sections, respectively, as a function of the corner radius parameter k. Values of $(C_n/C_{no})_{SB}$ at $k = 0$ were computed by

$$\left(\frac{C_n}{C_{no}}\right)_{SB} = \frac{A_i}{A_{eq}} \tag{11}$$

where A_i is the "apparent mass" value given by Ref. 16 and A_{eq} is the equivalent circular cross-sectional area for each of the noncircular shapes of interest. (The "apparent mass" or method of inertial coefficients is a method used to compute stability derivatives of slender configurations.)

It is interesting to note from Figs. 6.3 and 6.4 that slender body theory gives constant values of $(C_n/C_{no})_{SB}$ for both the triangle and square[16] independent of their orientation. On the other hand, Newtonian theory values of $(C_n/C_{no})_N$ vary depending on the orientation of the triangle or square. Newtonian impact theory is a function of the sine of the angle between the velocity vector and a tangent to the body surface to the second power. As a result, the configuration that has the base of the triangle normal to the flow has a fairly high value of the parameter $(C_n/C_{no})_N$, whereas the value when the triangle is inverted is about 1/4 of the larger value when the corner is

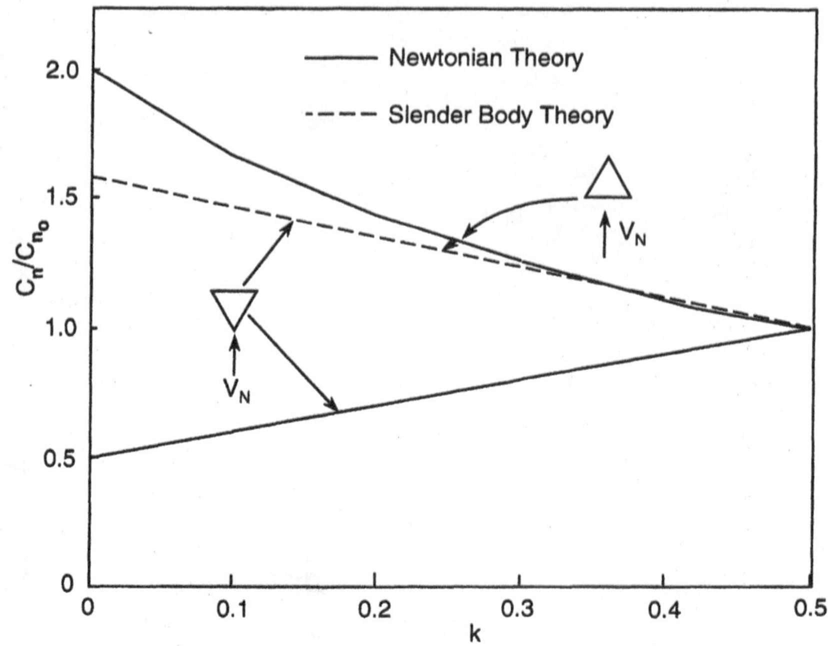

Fig. 6.4 Ratio of local normal-force coefficient for a triangular cross section to that for the equivalent circular cross section.

sharp ($k = 0$). On the other hand, Vukelich et al.[16] shows the apparent mass to be independent of orientation of the triangle and square, so a value of slender body theory is obtained for the parameter $(C_n/C_{no})_{SB}$ that is independent of orientation and is between the two values computed by Newtonian theory for both the square and triangle.

Figures 6.3 and 6.4 address one of the weak areas in the Refs. 6 and 11 methodology discussed previously. A second problem that needs to be addressed is to provide a correction as a function of Mach number for the Newtonian theory curves of Figs. 6.2 through 6.4 Because the slender body curves are used at low AOA and values of the linear term of normal force are small, it is not as critical that these curves be corrected for nonslender bodies. On the other hand, the Newtonian theory curves are used at high AOA and could potentially be erroneous at lower Mach numbers.

Equations (6) and (7) for the body-alone normal force and pitching moment will, therefore, be rewritten to include the factor NF, which represents the correction in the modified Newtonian values of $(C_n/C_{no})_N$ because the crossflow Mach number is not infinite. Thus,

$$C_N = C_{N_L}\left(\frac{C_n}{C_{no}}\right)_{SB} + C_{N_{NL}}\left(\frac{C_n}{C_{no}}\right)_N \text{NF} \tag{12}$$

$$C_M = C_{M_L}\left(\frac{C_n}{C_{no}}\right)_{SB} + C_{M_{NL}}\left(\frac{C_n}{C_{no}}\right)_N \mathrm{NF} \tag{13}$$

The subscript L and NL in Eqs. (12) and (13) represent the linear and nonlinear normal force and pitching moment components, respectively. Both Eqs (12) and (13) are based on body aerodynamics of a circle of equivalent cross-sectional area to that of the noncircular body cross section.

To compute NF, use will be made of available experimental data to compare $(C_n/C_{no})_N$ of Figs. 6.2 through 6.4. If values of these parameters are not available directly, then the parameter NF can be approximated numerically by comparing total force and moment data from tests to that predicted by Ref. 13 when NF is 1. NF can then be computed to bring the theory more in line with experimental data. References used for this phase of the work were 6, 7, and 17 through 24.

A brief explanation of why the factor NF is needed is in order. Basically, Newtonian impact theory assumes that the component of the momentum normal to the surface of an air particle is lost upon direct impact on a body surface. The theory is derived based on very high Mach numbers, and past comparisons with experiment have shown that the simple theory does indeed give quite accurate pressure information on simple body shapes as Mach number becomes large and the particle impact angle is near normal to the surface. As a consequence of the normal momentum of each air particle being deposited on a surface upon impact, the pressure coefficient on all areas that the flow does not see directly ("shadowed regions"), or leeward plane areas, is assumed to be zero. This Newtonian assumption becomes increasingly erroneous as Mach number decreases. While the ratio $(C_n/C_{no})_N$ tends to mitigate the error somewhat, because the error occurs on both the circular and noncircular configurations, it is logical to assume that the error is configuration-dependent. If Eqs. (12) and (13) were applied only for crossflow Mach numbers greater than about 2 to 5, one could probably neglect the factor NF and still achieve reasonable accuracy for engineering codes. However, because most tactical weapons fly in the range of freestream Mach numbers less than 6, it is very important to have the factor NF defined.

The first noncircular configuration investigated is the elliptical cross section of Fig. 6.2. This configuration has the most experimental data available and, therefore, allows the easiest definition of the factor NF. Results for elliptical configurations with $a/b = 0.5$, 2.0, and 3.0 are given in Fig. 6.5. Figure 6.5 was derived primarily from the data of Jorgensen[6,7] and Shereda et al.[24] in conjunction with the 1998 Aeroprediction Code, or AP98 (Moore et al.[13]) As a result of the combined usage of data and a code, some of the factor NF in Fig. 6.5 could be from errors in the AP98 itself. However, as seen in Fig. 6.5, if the errors are from the AP98, they are fairly consistent in terms of a general trend as a function of a/b.

Figure 6.5 was derived for AOAs greater than 20 deg because this is the region where one obtains the largest amount of separation in the leeward

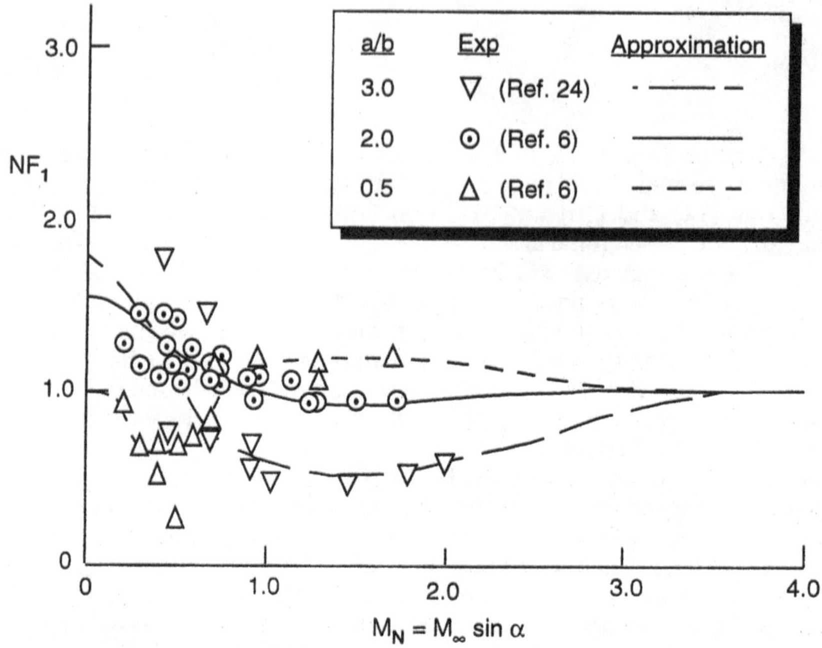

Fig. 6.5 Newtonian correction factor for an elliptical cross section ($a \geq 20°$).

plane of the vehicle and, therefore, the region where Newtonian theory is least accurate. To blend in the factor $(NF)_1$ with AOA, a linear variation is assumed between AOA of 0 and 20 deg. That is,

$$\begin{aligned} NF &= 1 + \left[(NF)_1 - 1\right]\frac{a}{20} \quad ; \quad a \leq 20 \\ NF &= (NF)_1 \quad ; \quad a > 20 \end{aligned} \tag{14a}$$

Several points are worthy of note in Fig. 6.5. First, for all values of a/b NF_1 is close to 1 at crossflow Mach numbers of 3.0 and higher. Second, at low crossflow Mach numbers, a/b = 2.0 and 3.0 configurations generate a large amount of nonlinear normal force compared to a circle, which is partly explained by Ref. 25. Lijewski et al.[25] shows that the drag coefficients of various flat shapes coalesce at high Mach number but vary widely at low Mach numbers. The ellipse approaches a flat surface as a/b gets large. Third, the configuration for a/b = 0.5 appears to indicate a region of supercritical crossflow Reynolds number around M_N = 0.5 where the factor NF_1 decreases below 1. However, for a/b = 0.5, the value of NF more closely resembles Newtonian theory (NF is closer to 1) than the larger values of a/b when M_N is less than about 0.8. Finally, to estimate the effects of elliptical shapes other than a/b = 0.5, 2.0, or 3.0, a linear assumption is made that the factor varies between 1.0 for $a/b = 1.0$ and its value in the Fig. 6.5 for a/b other than 1. Mathematically, this is expressed as

$$
\begin{aligned}
\mathrm{NF} &= (\mathrm{NF})_{a/b=3} && ; \quad a/b > 3.0 \\
\mathrm{NF} &= \left\{(\mathrm{NF})_{a/b=3} - (\mathrm{NF})_{a/b=2}\right\}(a/b - 2) + (\mathrm{NF})_{a/b=2} && ; \; 2 < a/b \le 3.0 \\
\mathrm{NF} &= \left\{(\mathrm{NF})_{a/b=2} - 1\right\}(a/b - 1) + 1 && ; \; 1 \le a/b < 2.0 \\
\mathrm{NF} &= \left\{(\mathrm{NF})_{a/b=0.5} - 1\right\}2(1 - a/b) + 1 && ; \; 0.5 \le a/b < 1.0 \\
\mathrm{NF} &= (\mathrm{NF})_{a/b=0.5} && ; \quad a/b \le 0.5
\end{aligned}
\tag{14b}
$$

The second configuration of interest is the square cross section (see Figs. 6.1 and 6.3). Unfortunately, there are not as many data available for the square and triangular cross sections as needed, particularly at Mach numbers between about 1 and 5. However, lower Mach number data is available for crossflow drag coefficient (see Refs. 17, 18, 20, 21, 22, and 25 through 31) and some higher Mach number data in Ref. 7. The data of Refs. 20 and 31 in particular were useful in defining the Newtonian correction factor at low crossflow Mach numbers. Both these tests had the model mounted in a two-dimensional flow sense and normal to the walls of the wind tunnel. Thus,

$$M_N = M_\infty \sin 90 = M_\infty$$

The data from both Ref. 20 and Ref. 31 were at a subcritical Reynolds number up to supercritical. The values of crossflow drag coefficient were taken at subcritical Reynolds numbers for application here and at a $M_N \approx 0.4$.

To compute a Newtonian correction factor from the data of Refs. 20 and 31 basically involves computing the value of $(C_n/C_{no})_N$ experimentally and comparing it with the analytical curve of Fig. 6.3. That is,

$$\mathrm{NF} = \frac{(C_d/C_{dc})(W_m/d_{eq})}{(C_n/C_{no})_N} \tag{15a}$$

The values of C_d and C_{d_c} come from Refs. 20 and 31 and $(C_n/C_{no})_N$ from Fig. 6.3. To compute W_M/d_{eq} of Eq. (15a) for the squares and triangles, we must compute this term as a function of the corner radius (see Fig. 6.1) at $\varphi = 0$ and 45 deg for the square and for the triangle and inverted triangle of Fig. 6.1. Here, W_M is the maximum diameter the flow sees in the normal direction of the body and d_{eq} is the cross-sectional area of the body. Moore et al.[15] derived these terms for configurations with square and triangular cross sections, and they are given as follows:

Square at $\varphi = 0$ deg roll

$$\frac{W_M}{d_{eq}} = \frac{1}{2\sqrt{\dfrac{1-(4-\pi)k^2}{\pi}}} \tag{15b}$$

Square at $\varphi = 45$ deg roll

$$\frac{W_M}{d_{eq}} = \frac{\sqrt{2}}{2\left[1+2k\left(\sqrt{2}-1\right)\right]\sqrt{\dfrac{1-(4-\pi)k^2}{\pi}}} \tag{15c}$$

Triangle and inverted triangle

$$d_{eq} = 2W_M \frac{\left[1+2k\left(\sqrt{3}-1\right)\right]}{\sqrt{\pi}} \left\{ \frac{\sqrt{3}}{4} - \frac{k^2\left(3\sqrt{3}-\pi\right)}{\left[1+2k\left(\sqrt{3}-1\right)\right]^2} \right\}^{1/2} \tag{15d}$$

It should be pointed out that C_{d_c} measured in Ref. 20 was the traditional value of 1.2, whereas that measured in Ref. 31 was 1.0. It was noted in Ref. 31 that flow velocity was lost through holes in the mount of the wind tunnel wall at each end of the model, which accounted for this loss. Hence, some of the data of Ref. 31 could have been affected by this loss, but in comparing them to Ref. 20 data, they appeared to be consistent for configurations other than the circular cylinder values.

Results of Eq. (15) calculations at $M_N \cong 0.40$, based on the Ref. 31 and 30 data, respectively, are shown in Fig. 6.6a. Figure 6.6a shows the factor is much higher for the square rotated in the $\varphi = 45$ deg roll position than in the $\varphi =$ 0 roll. This makes sense from the standpoint that it is well-known that Newtonian theory gives a reasonable approximation to the pressure near the region of a blunt tip, even though the Mach number is low. On the other hand, when one gets away from the blunt tip, the pressures deviate substantially from experiment. Also, for the squares in both the $\varphi = 0$ and $\varphi = 45$ deg roll orientations, Newtonian theory assumes $C_P = 0$ in the rear of the configuration, whereas experimental data show a fairly large negative pressure coefficient. The combination of these two factors is why the correction factor is needed.

Figure 6.6a gives values of NF near $M_N = 0.4$. To correct for Mach number, Ref. 7 data were used in conjunction with the AP98 to back out the Newtonian correction factor for crossflow Mach numbers other than 0.4. Jorgensen[7] had data for Mach numbers of 1.98 and 3.88 for AOAs to 20 deg and 14 deg, respectively, on squares, diamonds, triangles, and inverted triangles with sharp corner radius. Hence, crossflow Mach numbers from 0 to about 1.0 can be obtained for $k = 0$. Figure 6.6b shows the results from Ref.

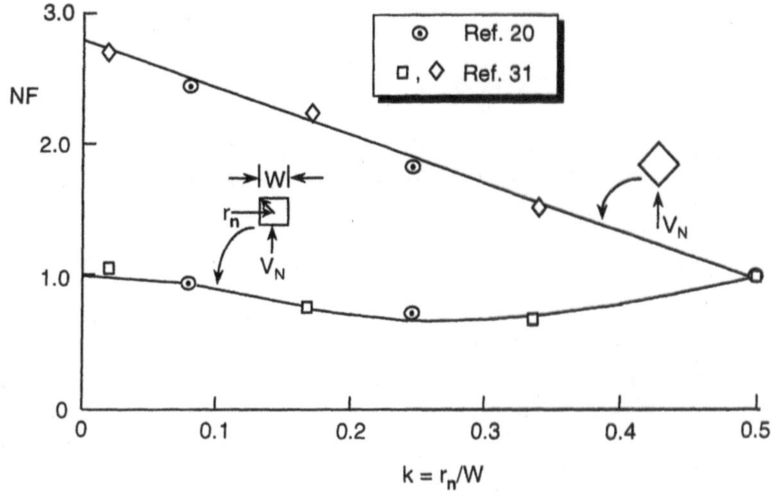

Fig. 6.6a Newtonian correction factor for squares as a function of corner radius at $M_N \cong 0.4$.

7 data for crossflow Mach numbers less than about 0.95 for squares at both the 0 and 45 deg roll orientations. These data are then extrapolated to a value of NF = 1.0 at $M_N = 2.6$. For squares at $\varphi = 0$ deg, a value of 1.0 for NF is obtained at $M_N = 0.4$; whereas, for diamonds, NF = 1.0 is obtained at $M_N = 2.6$. Note that Ref. 7 data are in agreement with Ref. 21 and 30 data at $M_N = 0.4$. That is, the value of NF from Fig. 6.6a is identical to that in 6.6b at $M_N = 0.4$.

Now to derive a value of NF_0 (NF at $M_N = 0$) as a function of corner radius, Figs. 6.6a and 6.6b have been combined. The value of NF in Fig. 6.6a at $k = 0$ has been adjusted upward to correspond to the value at $M_N = 0$ in Fig. 6.6b. Then the value of NF_0 follows the trends in Fig. 6.6a except for the initial point at $k = 0$ being adjusted upward. The adjusted values of NF_0 for the squares and diamonds are given in Fig. 6.6c.

To compute Newtonian correction factor for squares, one first picks off the value of NF_0 from Fig. 6.6c as a function of k and roll orientation. Then Fig. 6.6b is used to correct the value of NF_0 as a function of crossflow Mach number. The mathematics of this process are as follows:

Square

$$NF = NF_0 \quad ; \quad M_N \leq M_{N_O}$$

$$NF = NF_0 - \frac{M_N - M_{N_O}}{1 - M_{N_O}}(NF_0 - 1) \quad ; \quad M_{N_O} < M_N \leq 1.0$$

$$NF = 1.0 \quad ; \quad M_N > 1.0 \tag{16a}$$

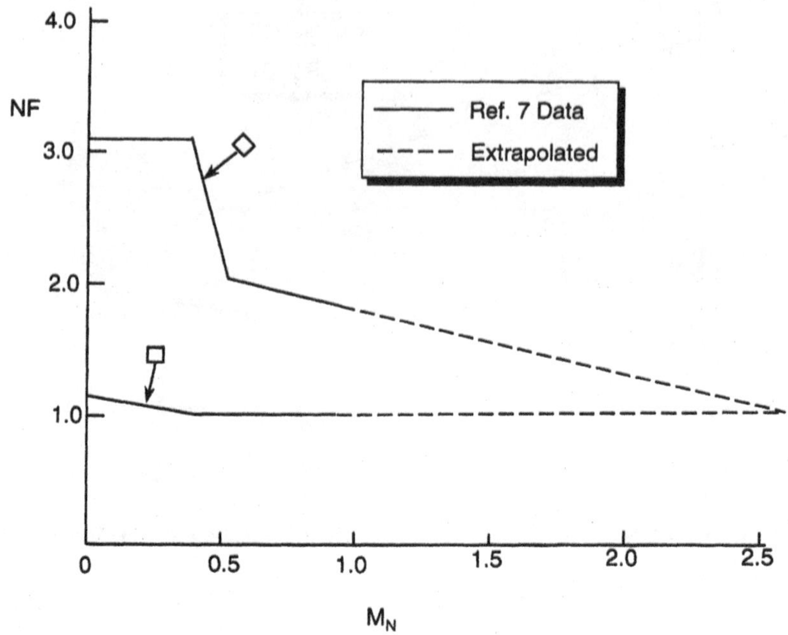

Fig. 6.6b Newtonian correction factor for squares as a function of crossflow Mach number ($k = 0$).

Diamond

$$
\begin{aligned}
\mathrm{NF} &= \mathrm{NF}_0 && ; M_N \le M_{N_O} \\
\mathrm{NF} &= \mathrm{NF}_0 - \left(\frac{M_N - M_{N_O}}{0.55 - M_{N_O}}\right)[0.55(\mathrm{NF}_0 - 1)] && ; M_{N_O} < M_N \le 0.55 \\
\mathrm{NF} &= (\mathrm{NF})_{M_N=0.55} - \frac{M_N - 0.55}{2.05}\left[(\mathrm{NF})_{M_N=0.55} - 1\right] && ; 0.55 < M_N \le 2.6 \\
\mathrm{NF} &= 1.0 && ; M_N > 2.6
\end{aligned}
\tag{16b}
$$

where

$$
\begin{aligned}
M_{N_O} &= 0.4 - 2.67k \; ; \; k \le 0.15 \\
M_{N_O} &= 0 \; ; \; k > 0.15
\end{aligned}
\tag{16c}
$$

Equation (16c) tends to delay the decrease in NF with increasing values of M_N somewhat if the square has fairly sharp or sharp corners (similar to the

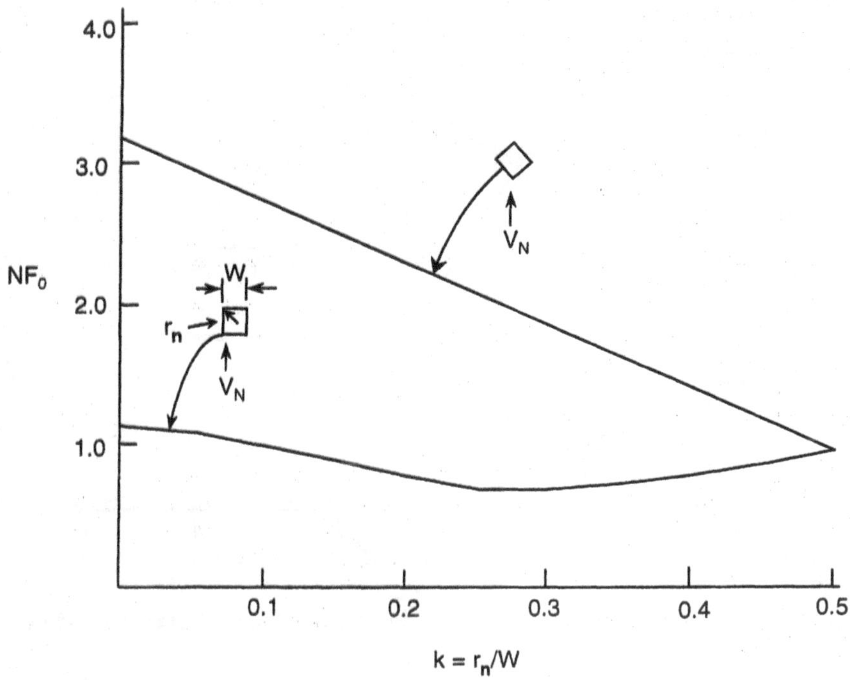

Fig. 6.6c Newtonian correction factor for squares as a function of corner radius at $M_N = 0$.

large *a*/*b* elliptic case). However, both the sharp and rounded results of NF computed by Eq. (16) approach 1.0 as M_N approaches 1.

Data for triangular cross sections are more meager than for squares. Delany and Sorensen[31] has the best data, with some data available in Refs. 19 and 20. Using the data from these references, all of which are for low Mach numbers, a process similar to that for the square cross section can be used to derive a Newtonian correction factor for triangular shapes. This factor is shown in Fig. 6.7a. The only data that were available were for $k =$ 0.02, 0.1, and 0.4. Hence, between these points, a straight line was drawn. Once again it is interesting to note that, when the flow is to the flat side of the triangle, the Newtonian correction factor is close to 1. However, when the flow hits the angled side of the triangle, the correction factor is much higher. Once again, the process used for the square cross section that corrects for crossflow Mach numbers other than 0.4, will be used for the triangular cross sectional shapes.

Figure 6.7b thus gives the value of NF as a function of crossflow Mach number using a combination of Ref. 7 and the AP98, as done for squares. Then Fig. 6.7c combines Figs. 6.7a and 6.7b to correct for the corner radius based on a crossflow Mach number of 0. The mathematics that define the

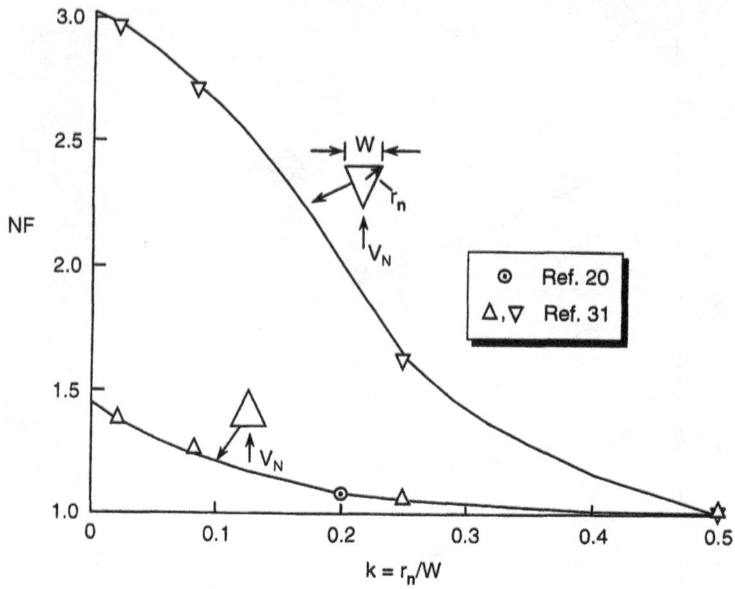

Fig. 6.7a Newtonian correction factor for triangles as a function of corner radius at $M_N = 0.4$.

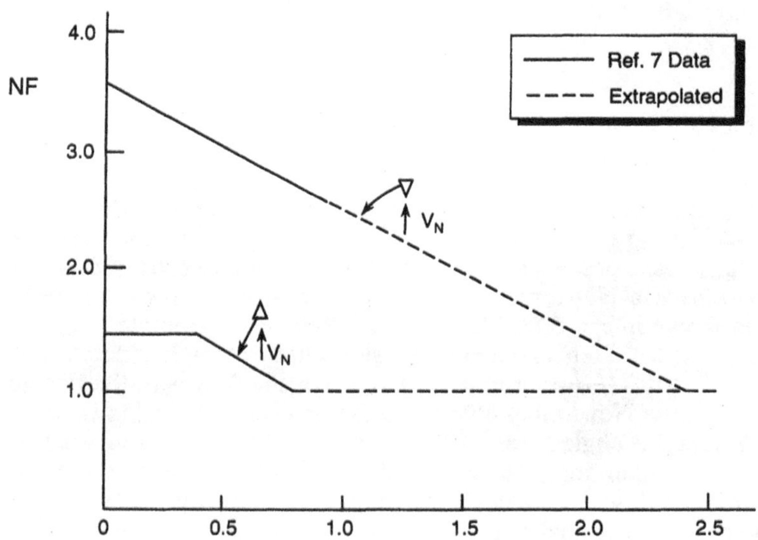

Fig. 6.7b Newtonian correction factor for triangles as a function of crossflow Mach number ($k = 0$).

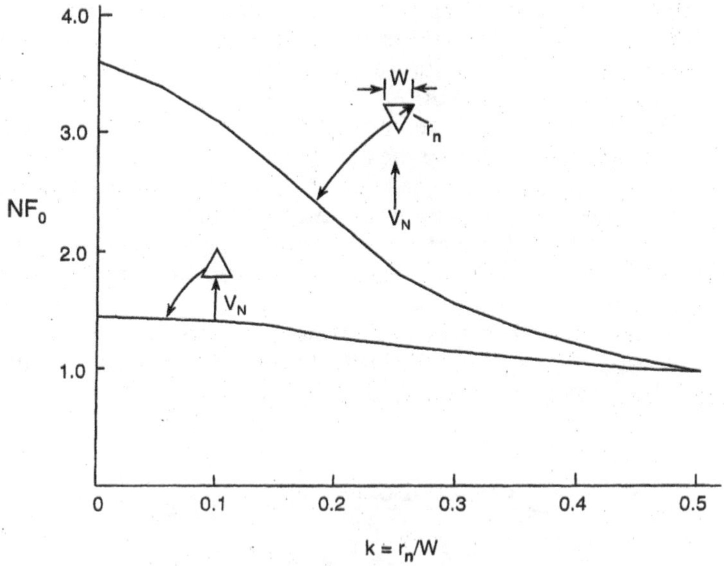

Fig. 6.7c Newtonian correction factor for triangles as a function of corner radius at $M_N = 0$.

calculation of NF is given by Eq. (16d) for the triangle and Eq. (16e) for the inverted triangle.

Triangle

$$
\begin{aligned}
NF &= NF_0 &&; \quad M_N \leq M_{N_O} \\
NF &= NF_0 - \left(\frac{M_N - M_{N_O}}{0.80 - M_{N_O}}\right)(NF_0 - 1) &&; \quad M_{N_O} < M_N \leq 0.80 \\
NF &= 1.0 &&; \quad M_N > 0.80
\end{aligned}
\qquad (16d)
$$

Inverted triangle

$$
\begin{aligned}
NF &= NF_0 - \left(\frac{M_N - M_{N_O}}{2.4 - M_{N_O}}\right)(NF_0 - 1) &&; \quad M_N \leq 2.4 \\
NF &= 1.0 &&; \quad M_N > 2.4
\end{aligned}
\qquad (16e)
$$

M_{N_O} of Eqs. (16d) and (16e) is defined by Eq. (16c), just as for the square cross-sectional cases.

V. Reynolds Number Effect on Crossflow Drag Coefficient

Section 2 of Chapter 5 presented the methodology used to calculate the crossflow drag coefficient of a circular cross-sectional shape. It was shown that C_{d_c} can have two values for low values of M_N, depending on the critical crossflow Reynolds number and critical crossflow Mach number. An approach to determine which value of C_{d_c} to use was also given for circular cross-sectional bodies of revolution.

Because the methodology for computing aerodynamics of nonaxisymmetric bodies is based on computing the aerodynamics of an equivalent axisymmetric body, the question arises as to whether a relationship can be derived for critical crossflow Reynolds numbers as a function of body cross-sectional shape. Polhamus[20] and Stoy and Vukelich[26] both give values of C_{d_c} as functions of crossflow Reynolds number and body cross-sectional shape for M_N values near zero.

Polhamus[20] correlated crossflow drag coefficient to a parameter ($R_N\,k^{1.3}$). For squares at $\varphi = 0$, it appears the critical value of this parameter based on Ref. 20 can be defined as a function of k. Thus for square cross sections at $\varphi = 0$ deg,

$$\frac{R_{NC}}{\left(R_{NC}\right)_{k=0.5}} = \frac{0.083 + 1.36k - 1.44k^2}{k^{1.31}}; \; 0.02 \le k \le 0.5 \qquad (17)$$

For the value of $k \le 0.02$, a value of 0.02 could be used in Eq. (17). If the square is rotated to $\varphi = 45$ deg (diamond), then a better approximation to the data of Ref. 20 is

$$\frac{R_{NC}}{\left(R_{NC}\right)_{k=0.5}} = \frac{6.33k - 24.1k^2 + 26.1k^3}{k^{1.31}}; \; 0.02 \le k \le 0.5 \qquad (18)$$

For ellipses, an approximation to the data of Ref. 18 is

$$\frac{R_{NC}}{\left(R_{NC}\right)_{a/b=1.0}} = \left(a/b\right)^{1.5} \qquad (19)$$

Finally, for triangular cross-sectional shapes with the base normal to the flow, the correlation equation for the square at $\varphi = 0$ deg, Eq. (17) appears to correlate the data of Ref. 31 reasonably well. Likewise, when the flow is in the direction of the triangular tip or inverted triangle where the base is to the rear, the equation for the square rotated 45 deg [Eq. (18)] appears acceptable. Figure 6.8a gives the values of the functions for the squares and triangles defined by Eqs. (17) and (18), whereas Fig. 6.8b gives the values of the function for ellipses defined by Eq. (19).

Physically, what Eqs. (17) through (19) are saying and Fig. 6.8 shows, is that when the configuration has corners or approaches a flat plate in the direction normal to the velocity vector, the critical value of crossflow Reynolds number increases. The amount of this increase is proportional to the sharpness of the corners or to the elongation of the body (ellipse). Therefore, the

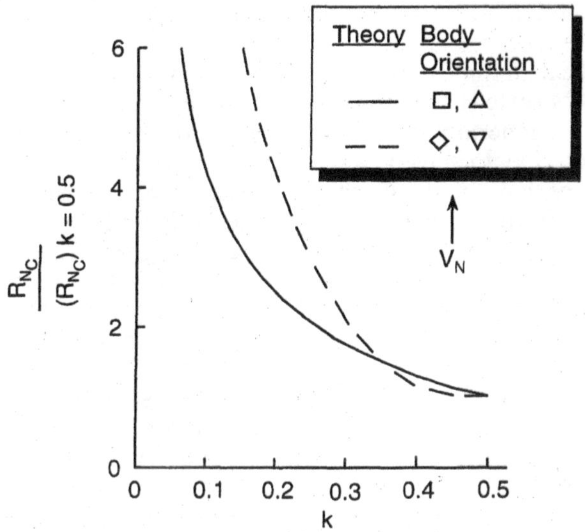

Fig. 6.8a Critical Reynolds number for squares and triangles.

mechanism that causes the "drag bucket" for circular cylinders is harder to establish itself for most noncircular cross sections.

VI. Scaling Considerations Based on Slender Body Theory

Sections 6.2 through 6.5 discussed the methods used to calculate the aerodynamics of a body with a noncircular cross section. When the fins are attached to the body, as shown in Fig. 6.1, and one is interested in calculating the aerodynamics of the wing–body configuration based on some reference area, scaling considerations must be considered. Hence, before we discuss

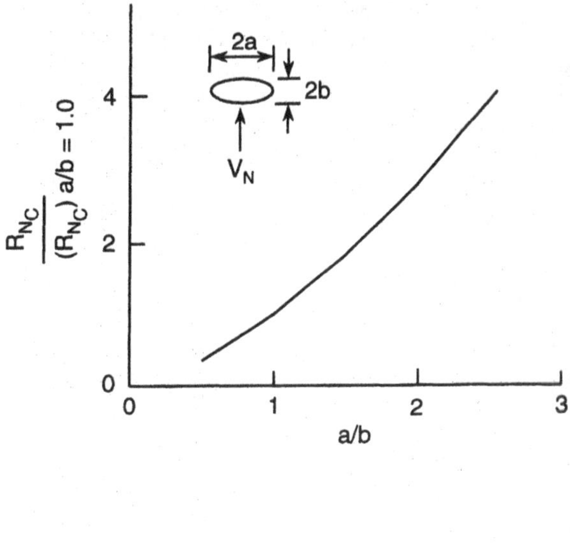

Fig. 6.8b Critical Reynolds number of ellipses.

the wing–body aerodynamics of nonaxisymmetric bodies, some fundamental scaling principles will be considered.

In the normal course of calculating aerodynamics of a given configuration, understanding the effects of scale is important. For example, when one tests a model in the wind tunnel, generally that model is geometrically scaled in terms of its configuration shape, so aerodynamics will be approximately the same. However, experience has taught us that boundary-layer transition naturally occurs on a smooth wind tunnel model at a much higher Reynolds number than on a rougher flight vehicle. This knowledge, along with tunnel test conditions, allows us to predict aerodynamics on a scaled wind tunnel model and indicate the approximate difference in these aerodynamics to a flight vehicle model. The major difference is generally a slightly different value of axial force coefficient.

The matter of scaling is quite important in development of a semi-empirical aerodynamic prediction method for nonaxisymmetric bodies. The reason is that we are trying to accomplish this task based on an equivalent axisymmetric body. Second, the scaling issue is important because, in some of the literature, tests are conducted on noncircular cross-sectional bodies with a constant dimension for the body diameter, and the results are compared to those on a circular cylinder with a diameter equal to that same dimension. Fin shape is generally the same for these cases. Other tests have been conducted on noncircular bodies where the body cross-sectional area is held constant and equal to that of a circular body. Again, fin shape is generally constant. The question arises as to whether one can use the same overall approach to calculate aerodynamics of these two cases and if so, are there scaling factors that need to be considered to compare one case to another?

To investigate first-order effects of scale on linear normal force of bodies and wing–body combinations, slender body theory is an excellent method to consider. Ashley and Landahl[32] and Nielsen[33] were the primary materials used for this analysis. Slender body theory basically assumes that the configuration is fairly smooth and slender with no discontinuities along the body surface. This type of contour will then minimize the perturbation in the freestream due to the presence of the body. It allows the equations of motion to be linearized as a result. Generally, the flow is assumed to be isentropic due to only weak shock waves being created by this slender, smooth body. While these assumptions seem rather strong, some good qualitative insight in terms of aerodynamic trends can be obtained, although the magnitudes of the numbers can be in substantial error. Therefore, slender body theory is quite useful for the purpose of investigating the issues addressed here concerning scaling.

The normal force of a given wing–body configuration is given by

$$N = N_B + N_{W(B)} + N_{B(W)} \tag{20}$$

where N_B is the normal force of the body alone, $N_{W(B)}$ is the normal force on the wing in presence of the body, and $N_{B(W)}$ is the additional normal force on the body caused by the presence of the wing. Figure 6.9 is a sketch of a typical wing–body showing important configuration dimensions.

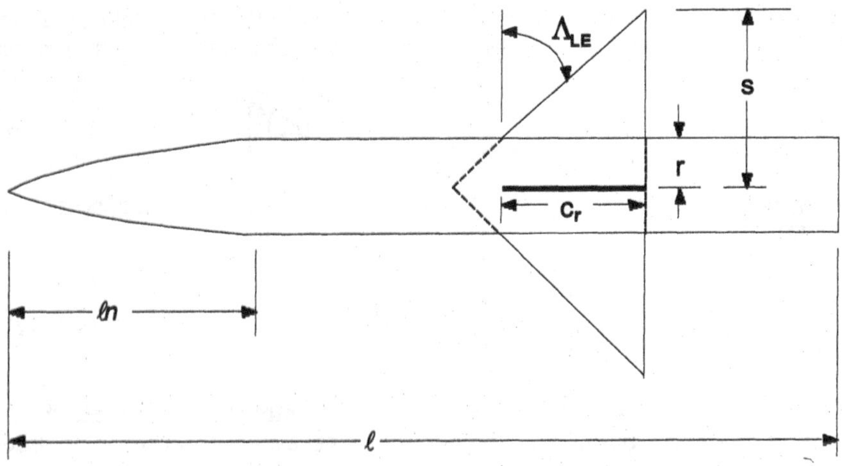

Fig. 6.9 Circular wing–body geometrical parameters needed for slender body theory considerations.

Slender body theory defines the body-alone lift as simply that due to the nose with no contribution from the afterbody (assuming the afterbody has a constant radius). This value is

$$N_B = 2aQ\pi r_b^2 \tag{21}$$

Here, r_b is the radius of the base of the configuration. Slender body theory also assumes the wings extend to the midbody. The normal force of the last two terms of Eq. (20) is

$$N_{W(B)} + N_{B(W)} = [K_{W(B)} + K_{B(W)}]\, N_W \tag{22}$$

Also,

$$K_{W(B)} + K_{B(W)} = \left(1 + \frac{r}{s}\right)^2 \tag{23}$$

and

$$N_W = 2\pi a\, Q\, (s - r)^2 \tag{24}$$

Substituting Eqs. (21), (22), (23), and (24) into Eq. (20) and dividing through by QA_{ref} to put Eq. (20) in the form of a coefficient, one obtains

$$C_N = \frac{2a\pi}{A_{\text{ref}}}\left\{\underbrace{r_b^2}_{\substack{\text{body}\\\text{term}}} + \underbrace{s^2\left[1 - 2(r/s)^2 + (r/s)^4\right]}_{\substack{\text{wing-body}\\\text{term}}}\right\} \tag{25}$$

If one combines the two terms in Eq. (25), one obtains the usual expression given in Refs. 32 and 33 for total configuration normal force coefficient as

$$C_N = \frac{2\pi a s^2}{A_{ref}}\left[1 - (r/s)^2 + (r/s)^4\right] \tag{26}$$

using body cross-sectional area as A_{ref}. If $r = r_b$, Eq. (26) can be further written as

$$C_N = 2a\left[\left(\frac{s}{r}\right)^2 + (r/s)^2 - 1\right] \tag{27}$$

Equation (27) gives us some very simple and important information concerning scaling. It basically says that if you take two different wing–body configurations that both obey slenderness requirements, the normal force coefficient of the two configurations will be the same at any Mach number for a given AOA if r/s of the two configurations is equal.

The question we now need to address is if one can use the information given by Eq. (25) to help answer some of the scaling issues associated with creating an axisymmetric body to calculate aerodynamics of a noncircular cross-sectional body. To address this question, several examples will be considered. The first example is illustrated in Fig. 6.10. This figure maintains a body circular diameter that is equal to one side of the square. Thus, r/s for the two cases is the same. Referring back to Eq. (25), where body and wing–body contributions have been separated, the implication of this analogy is that the wing–body term will remain constant for the two cases in Fig. 6.10 because r/s is the same. However, the body cross-sectional area of the noncircular case is different from the circular case. Thus, the first term of Eq.

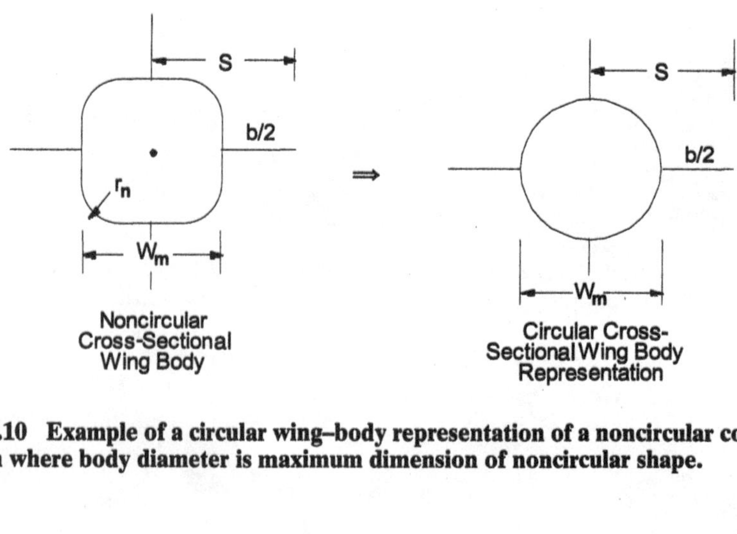

Fig. 6.10 Example of a circular wing–body representation of a noncircular configuration where body diameter is maximum dimension of noncircular shape.

(25) must be multiplied by this ratio. In other words, the aerodynamics of the noncircular cross-sectional configuration of Fig. 6.9 can be estimated using the circular cross-sectional case of Fig. 6.9 and Eq. (25) as

$$C_N = \frac{2\alpha\pi}{A_C}\left\{\left(\frac{A_{NC}}{A_C}\right) r^2 + s^2\left[1 - 2(r/s)^2 + (r/s)^4\right]\right\} \tag{28}$$

Using this information for the computational process, it says we should multiply the body term by (A_{NC}/A_C) if we represented it by an equivalent circular body of diameter W_M. This process applies also to comparing predictions to wind tunnel data that were taken where the diameter was held constant and equal to W_M.

A second example of estimating aerodynamics of a noncircular body by a circular body is illustrated in Fig. 6.11. Here the circular body diameter is equal to the equivalent diameter of a circular cylinder of the same cross-sectional area of the noncircular shape. For this case, the body contribution of Eq. (25) has a factor of 1 because cross-sectional areas of the circular and noncircular shapes are the same. However, r/s values of these two cases are different. This means that the wing–body contribution of Eq. (25) for the squares or triangles must be multiplied by the factor

$$(\text{SBTSF})_{WB} = \frac{\left(\frac{s}{r}\right)_{eq}^2\left[1 - 2\left(\frac{r}{s}\right)_{eq}^2 + \left(\frac{r}{s}\right)_{eq}^4\right]}{\left(\frac{s}{r}\right)_1^2\left[1 - 2\left(\frac{r}{s}\right)_1^2 + \left(\frac{r}{s}\right)_1^4\right]} \tag{29}$$

Here, the subscript 1 refers to the original noncircular body and the subscript *eq* refers to the body with a cross-sectional area equal to that of body one.

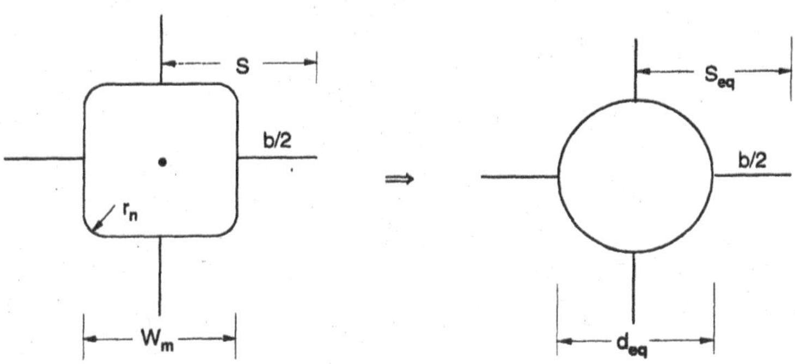

Fig. 6.11 Example of a circular wing–body representation of a noncircular configuration where body diameter is equivalent diameter.

The above analysis has shown that, even if aerodynamics are estimated based on some reference diameter, they can be compared to wind tunnel data based on a different diameter by using the slender body theory scaling factors (SBTSF) of A_{NC}/A_C for the body, or Eq. (29) for the wing–body.

As an example application of Eq. (29) scaling factors, consider the configuration of Fig. 6.12. Equations (15b), (15c), and (15d) define $d_{eq}(=2r_{eq})$ and $W_m(=2r_1)$ for the triangle as a function of W and the corner rounding parameter k. For simplicity, let us assume the corner is sharp so $k = 0$ and $W_m = W$. For this case, $d_{eq} = 0.743W$. Also, for simplicity, assume $W = b$. Then, applying Eq. (29) to Fig. 6.12 (because here we are keeping the area of the circle the same as the area of the triangle), one obtains for the SBTSF

$$\text{SBTSF} = \frac{\left(\dfrac{0.377W + W/2}{0.743W}\right)^2 \left[1 - 2\left(\dfrac{0.377W}{0.377W + W/2}\right)^2 + \left(\dfrac{0.377W}{0.377W + W/2}\right)^4\right]}{\left(\dfrac{W/2 + W/2}{W}\right)^2 \left[1 - 2\left(\dfrac{W/2}{W/2 + W/2}\right)^2 + \left(\dfrac{W/2}{W/2 + W/2}\right)^4\right]}$$

$$= \frac{1.39}{1}\frac{[1 - 0.37 + 0.034]}{[1 - 0.5 + 0.0625]} + 1.64$$

Notice that had we represented the triangular shape by a circular shape of diameter W, the wing–body SBTSF is 1, because the r/s of these two shapes is the same. However, the body term must be multiplied by $\left(\dfrac{0.743W}{W}\right)^2$ or 0.522. Also, the value of SBTSF from Eq. (29) is a function of $b/2$, k, and configuration shape as well as $(r/s)_1$ and $(r/s)_{eq}$. A single case was chosen where values of these parameters were selected for illustration purposes.

To summarize, if we define $(\text{SBTSF})_1$ and $(\text{SBTSF})_2$ as those associated with the body and wing–body normal force components, respectively, then one can write

$$C_N = C_{N_B}(\text{SBTSF})_1 + \left[C_{N_{W(B)}} + C_{N_{B(W)}}\right](\text{SBTSF})_2 \tag{30}$$

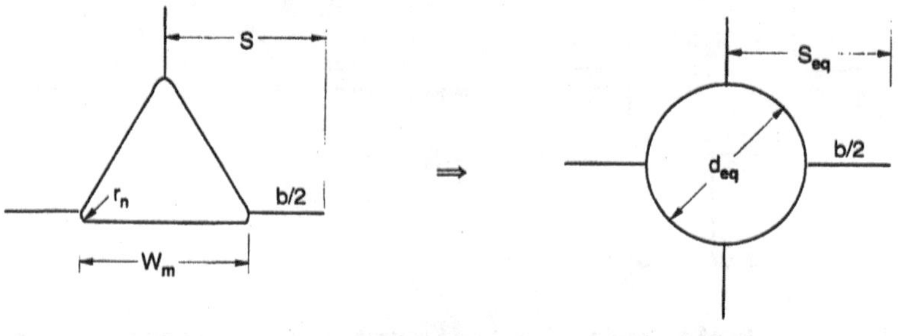

Fig. 6.12. Example application for SBTSF of Eq. (29).

If the aerodynamics are based on $\frac{\pi W^2}{4}$ as a reference area and wind tunnel data were taken that way, then

$$\begin{aligned}(\text{SBTSF})_1 &= \left(\frac{d_{eq}}{W}\right)^2 \\ (\text{SBTSF})_2 &= 1\end{aligned} \tag{31}$$

On the other hand, if the aerodynamics are based on $\frac{\pi d_{eq}^2}{4}$ as a reference area and wind tunnel data were taken that way, then

$$\begin{aligned}(\text{SBTSF})_1 &= 1 \\ (\text{SBTSF})_2 &= \frac{\left(\frac{s}{r}\right)_{eq}^2 \left[1 - 2\left(\frac{r}{s}\right)_{eq}^2 + \left(\frac{r}{s}\right)_{eq}^4\right]}{\left(\frac{s}{r}\right)_1^2 \left[1 - 2\left(\frac{r}{s}\right)_1^2 + \left(\frac{r}{s}\right)_1^4\right]}\end{aligned} \tag{32}$$

Finally, if the aerodynamics are based on $\frac{\pi d_{eq}^2}{4}$ and we wish to compare these coefficients to those that are based on a reference area of $\frac{\pi W^2}{4}$ (where wind tunnel data were also taken based on $\frac{\pi W^2}{4}$), then

$$\begin{aligned}(\text{SBTSF})_1 &= \left(\frac{d_{eq}}{W}\right)^2 \\ (\text{SBTSF})_2 &= \frac{\left(\frac{s}{r}\right)_{eq}^2 \left[1 - 2\left(\frac{r}{s}\right)_{eq}^2 + \left(\frac{r}{s}\right)_{eq}^4\right]}{\left(\frac{s}{r}\right)_1^2 \left[1 - 2\left(\frac{r}{s}\right)_1^2 = \left(\frac{r}{s}\right)_1^4\right]}\end{aligned} \tag{33}$$

VII. Wing–Body Configurations with Noncircular Cross Sections

Jorgensen[6] used a combination of two approaches for wing–body aerodynamics of configurations with noncircular cross sections. The first approach was to simply use Newtonian theory to approximate $(C_n/C_{no})_N$ of Eqs. (2) and (3) for wing–body configurations where the body was noncircular and the wing was a simple extension of this. He then used a modified version of Ref. 14 to compute the slender body or linear theory term of Eqs. (2) and

(3). As shown in Ref. 6, this approach significantly overpredicted the normal force while giving reasonable results for center of pressure. The primary reason for this overprediction of normal force was failure to account for the nonlinearities that occur in the wing–body interference factor as AOA increases.

Missile DATCOM[16] improved upon the Jorgensen[6] approach by using the equivalent AOA method[12] to incorporate nonlinearities in the wing–body interference factor. This increased the AOA capability to 20 to 30 deg. On the other hand, Ref. 16 did not include triangular or square cross-sectional shapes for high AOA due to the fact that Newtonian theory was not developed for those shapes. Also, the Newtonian correction factor for lower Mach number was not available. Vukelich et al.[16] also did not include the Newtonian correction factor for all Mach numbers for ellipses.

Moore et al. [13] and Moore and McInville[34] (summarized in Chapter 5) developed a new approach for incorporating nonlinearities in wing–body and body–wing interference factors. In this approach, each interference term was divided into a linear and nonlinear component. The linear term was estimated by linear theory or slender body theory and the nonlinear term was estimated directly through use of large missile component wind tunnel databases. For regions where data were not available, the method was extrapolated based on engineering judgment and comparison with aerodynamics on various other missile aerodynamic databases. The net result of this approach was average accuracy levels at roll positions of 0 and 45 deg of $\pm 10\%$ on normal and axial force and $\pm 4\%$ of body length on center of pressure to AOAs of 90 deg. Exceptions to this accuracy were at low Mach number and high AOA where wind tunnel data or critical crossflow Reynolds number were in question and at high Mach number and AOA for a configuration with two sets of lifting surfaces, where internal shock interactions became important. As a result, the improved approach for incorporating nonlinearities to wing–body interference factors, presented in Chapter 5, will be the method used here for noncircular bodies.

Figure 6.1 gives a qualitative pictorial view of configurations for which aerodynamics are desired. It is believed that this set of configurations can be made broad enough to encompass most tactical weapons of interest to the community at large. It includes circles of various diameter, ellipses of various eccentricity, and triangles and squares of various orientations as well as corner radii. To complement body geometry, wings have been included at both $\varphi = 0$ and 45 deg roll orientations on all configurations except the triangular shape (where wings were limited to the $\varphi = 0$ deg roll orientation only). The question that now must be addressed is how the Chapter 5 methodology will vary for noncircular wing–body configurations of Fig. 6.1.

Fortunately, Nelson,[35] Est and Nelson,[36] and Sigal[37] have investigated $K_{W(B)}$, $K_{B(W)}$, and $k_{W(B)}$ for low AOAs for noncircular configurations. Nelson[35] and Est and Nelson[36] defined low AOA values of $K_{W(B)}$ for elliptical, square, and triangular cross-sectional shapes at moderate supersonic Mach numbers using a Euler code in conjunction with low Mach number experimental data. Because the equivalent AOA method used slender body theory for $K_{B(W)}$, no equivalent data for noncircular cross sections were given in Refs. 35 and 36 for this parameter. Figure 6.13 gives the wing–body interference term for elliptical cross-sectional shapes after being divided by that of the circular cross section near $\alpha = 0$ deg. This figure was derived from the data of Ref. 35 in conjunction with the circular body results of Ref. 14. The Ref. 35 results were given as a function of a'/s_1, so they had to be translated to equivalent circular values through the relationship

$$\frac{r_{eq}}{s} = \frac{a'}{s_1}\left[\frac{1}{a'/s_1\left(1-\sqrt{C_1}\right)+\sqrt{C_1}}\right] \tag{34}$$

where $C_1 = a'/b'$.

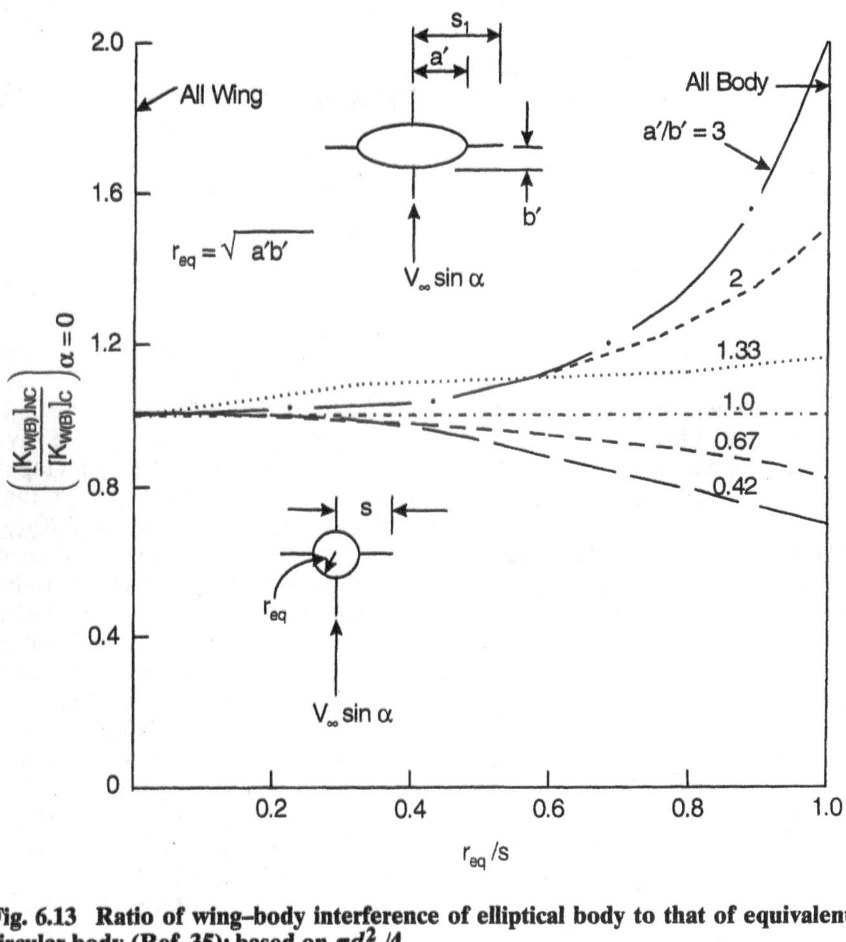

Fig. 6.13 Ratio of wing–body interference of elliptical body to that of equivalent circular body (Ref. 35): based on $\pi d_{eq}^2/4$.

It is interesting to note from Fig. 6.13 that, for most practical missile configurations where r_{eq}/s typically varies from about 0.1 to 0.7 and a/b is generally greater than 0.5 but less than 2.0, the noncircular $k_{W(B)}$ deviates from the circular value by, at most, 18% and, for most cases, is less than that.

Est and Nelson[36] also showed that, for the conditions investigated, $K_{W(B)}$ varied only slightly from slender body theory for any of the cross-sectional shapes. As a result, current nonlinear models of Chapter 5 will be used directly for this term.

As a first approximation, the body–wing interference term will be assumed to vary from the circular cylinder values in a proportional manner to the wing–body carryover. That is, if one defines

$$F = \frac{\left[K_{W(B)}\right]_{NC,a}}{\left[K_{W(B)}\right]_{C,a}} \tag{35}$$

then

$$\left[K_{B(W)}\right]_{NC,a} = F\left[K_{B(W)}\right]_{C,a} \tag{36}$$

$$\left[k_{B(W)}\right]_{NC,a} = F\left[k_{B(W)}\right]_{C,a} \tag{37}$$

Also, as already mentioned,

$$\left[k_{W(B)}\right]_{NC,a} = \left[k_{W(B)}\right]_{C,a} \tag{38}$$

It should be emphasized that Fig. 6.13 is the low AOA value of $K_{W(B)}$ and $K_{B(W)}$ and says nothing about how $K_{W(B)}$ and $K_{B(W)}$ will vary with AOA. As a result, some assumptions will be made. Chapter 5 defined $K_{W(B)}$ and $K_{B(W)}$ as shown in Figs. 5.14a and 5.14b. As seen in Fig. 5.14a, $K_{W(B)}$ has five parameters, which are defined in tables as functions of wing aspect and taper ratios for various values of r/s, AOA, and M_∞. It will be assumed here that the Fig. 6.13 results will be applied to the slender body theory value of $K_{W(B)}$ and $K_{B(W)}$ [that is, the first term of Eq. (28) in Chapter 5]. Then if a_c is held constant, $dK_{W(B)}/da$ will change. However, a_D and a_M will be assumed to be the same as the circular body shape. In other words, at low to moderate AOA, the body cross section is allowed to change the interference factors from their circular-body values; whereas, at high AOA, it is not. Also, as a first approximation, the values of $\Delta K_{w(B)}$ and $\Delta K_{B(W)}$ available in Tables 5.4 and 5.9 will be assumed to be independent of cross-sectional shape at AOA.

Figure 6.13 was based on fins located at roll position of zero. Chapter 5 treats nonlinear fin carryover interference for roll positions of both 0 and 45 deg. As a result, the slender body theory results of Fig. 6.13 will be applied to the body with the fins in both the $\varphi = 0$ and 45 deg orientations. Of course, onlinearities with Mach number, AOA, and fin shape are different for these roll orientations and will be used as presently done in Chapter 5. As already stated, Eqs. (35) through (38) are used to compute $K_{B(W)}$, $k_{W(B)}$, and $k_{B(W)}$

for noncircular cross sections when the methods of Refs. 35 and 36 are used to compute $K_{W(B)}$ for the ellipse.

Est and Nelson[36] also presented results for $[K_{W(B)}]_a = 0$ for the triangular and square shapes. Two methods were used in that reference. For k values between 0.125 and 0.5 and r/s between 0.167 and 0.667, a numerical code was used. To obtain the value of $K_{W(B)}$ for $r/s = 1.0$, incompressible data of Ref. 18 was used in conjunction with the incompressible form of the Bernouli equation.[36] That is,

$$p + \frac{\rho V^2}{2} = ¢ = p_\infty + \frac{\rho_\infty V_\infty^2}{2} \tag{39}$$

Because $C_P = \dfrac{p - p_\infty}{\frac{1}{2}\rho_\infty V_\infty^2}$, Eq. (9) can be written as

$$C_P = 1 - \left(\frac{V}{V_\infty}\right)^2 \tag{40}$$

Values of C_P are available from Ref. 18 at low Mach number and for the squares with $k = 0.2$ and triangles with $k = 0.175$. Following Ref. 36,

$$K_{W(B)} = \frac{V_{NW}}{V_\infty} = \frac{V}{V_\infty}\cos\theta \tag{41}$$

Then, using Eq. (40) in (41), one obtains

$$K_{W(B)} = \cos\theta\sqrt{1 - C_p} \tag{42}$$

The V_{NW} term of Eq. (41) is the velocity normal to the wing, and θ is the angle between the normal to the wing and a tangent to the body surface. If the wing is perpendicular to the body surface, then $\theta = 0$ deg, and Eq. (34) becomes

$$K_{W(B)} = \sqrt{1 - C_p} \tag{43}$$

Results from Ref. 18 were used for C_P at the various locations of the wings in Fig. 6.1. Several values of C_P were available in Ref. 18 as a function of Reynolds number. Table 6.1 gives the pressure tap number where values of C_P were taken from Ref. 18 as a function of the body/wing orientations of Fig. 6.1. Also given in the table are the values of $K_{W(B)}$ computed from Eq. (43) where the C_P values of Ref. 18 were averaged for both subcritical and supercritical Reynolds numbers. While there was significant variation in C_P for various Reynolds numbers, as seen in Table 6.1, when values were averaged, there was not a great deal of difference in $K_{W(B)}$ between the subcritical and supercritical case.

Values from Table 6.1 are nondimensionalized by the slender body theory value of $K_{W(B)}$ for a circle at $r/s = 1$ (which is 2.0) and plotted in Fig. 6.14. A straight line interpolation is assumed between the value of $k = 0.175$ or 0.2 to 0.5. For values of $[K_{W(B)}]_{NC}$ at $k = 0$ and 0.1, Ref. 29 data were used. These data, taken at $M_\infty = 0.31$, are also at a low enough Mach number to be considered incompressible, so it complements the data of Ref. 18 quite

Table 6.1 Estimated values of $K_{W(B)}$* for r/s = 1.0 at low Mach number for various body cross sections and fin locations (data from Ref. 18)

Configuration shape and fin position	 $k = 0.2$	 $k = 0.2$	 $k = 0.2$	 $k = 0.2$	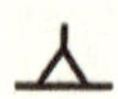 $k = 0.175$	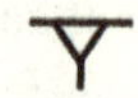 $k = 0.175$
Pressure tap orifice number from Ref. 18	11, 23	5, 20	5, 29	1, 11, 17, 23	8, 24	8, 24
$K_{w(B)}$						
Subcritical R_N	1.52	1.6	1.59	1.34	1.48	N/A
Supercritical R_N	1.40	1.56	1.76	1.32	1.70	1.54

$$* \ K_{W(B)} = \sqrt{1 - \sum_{i=1}^{N} C_{P_i}/N}$$

nicely. However, only two cases of Fig. 6.1 were available in Reg. 29. These were the cases for the square body at roll = 0 and fins at roll = 45 deg and the square body at roll = 45 deg and fins with roll = 0 deg. The other cases shown in Fig. 6.14 were then extrapolated from their values at k = 0.2 to k = 0 based on data for the two cases from Ref. 29 as guides. Also, note that Fig. 6.14 was derived based on calculations of a wing–body with body circular cross-sectional diameter of W.

In analyzing Fig. 6.14, it is seen that all cross sections are less efficient than the circle in producing wing–body carryover normal force near $M_\infty = 0$ and at $r/s = 1.0$. For $k = 0$, the value of $K_{W(B)}$ for the square varies between 1.14 and 1.5 depending on its orientation and the location of the fin, whereas the triangle and inverted triangle are approximately 1.5. These values compare to the circular cylinder value of $[K_{W(B)}]_{r/s=1} = 2.0$. Compare these values to those of Fig. 6.13, which are greater than one for values of $a/b \geq 1.0$. Also note that Fig. 6.13 is based on a reference area of $\frac{\pi d_{eq}^2}{4}$ versus $\frac{\pi W^2}{4}$ for Fig. 6.14. This brings us to a dilemma as to why the interference factors appear so different. The answer lies in the scaling.

Referring back to Section 6.6, the values of Fig. 6.14 must be multiplied by the slender body theory scaling factor, Eq. (29). On the other hand, Fig. 6.13 was derived based on $\frac{\pi d_{eq}^2}{4}$ and the scalar factor has already been incorporated to that figure. Using the example application of Section 6.6 for a triangle, if $k = 0$, the SBTSF is 1.64. Multiplying this value by the value of Fig. 6.13 of 0.75, one obtains a value of $[K_{W(B)}]_{NC}/[K_{W(B)}]_C = 1.23$. This says that, when we take a configuration such as the triangular shape of Fig. 6.12, represent it by a circular shape of diameter d_{eq}, keep the fins constant, the value of the circular $K_{W(B)}$ must be increased by 23%. Notice that had we represented the triangular shape by a circular shape of diameter W, the

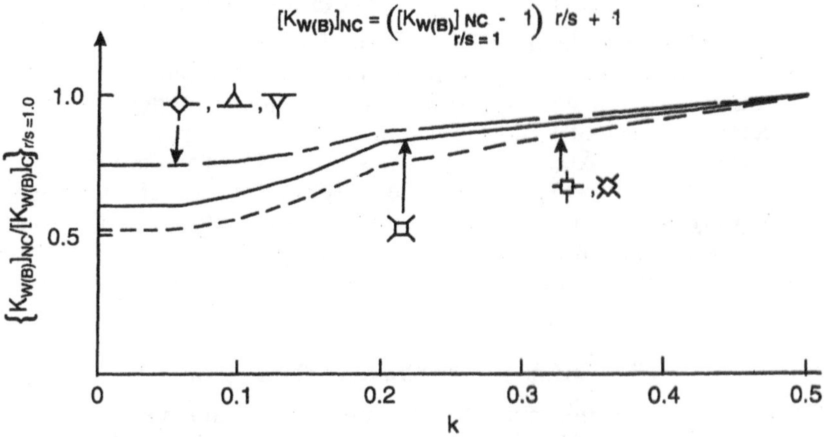

Fig. 6.14 Effect square and triangle body cross sections on low Mach number values of $K_{W(B)}$ (data from Ref. 18 and 29 and based on $A_{ref} = \pi W^2/4$).

wing–body SBTSF is 1 because the r/s of these two shapes is the same. However, the body term must be multiplied by $\left(\frac{0.743W}{W}\right)^2$ or 0.522. Also, $K_{W(B)}$ of Fig. 6.13 is applied directly. In summary, Eqs. (32) and (33) are needed for square and triangular cross-sectional shapes because $K_{W(B)}$ (Fig. 6.14) was derived on the basis of a body with diameter W. Equations (32) and (33) are not needed for elliptical shapes because $K_{W(B)}$ (Fig. 6.13) was derived on the basis of a body with diameter d_{eq}. If one were interested in deriving aerodynamics of an ellipse based on $\frac{\pi W^2}{4}$ versus $\frac{\pi d_{eq}^2}{4}$, then new SBTSFs would be needed for this case in analogy to those for square and triangular shapes.

Figure 6.14 gives values of $K_{W(B)}$ for squares and triangles at the limiting value of $r/s = 1$. Thus, to relate the value of $K_{W(B)}$ at some value of r/s other than 1 to Fig. 6.14, a linear assumption is used, similar to what occurs in slender body theory.[14] Hence,

$$\left[K_{W(B)}\right]_{NC} = \left(\left[K_{W(B)}\right]_{\substack{NC\\ r/s=1}} - 1\right) r/s + 1 \tag{44}$$

Using the value of $[K_{W(B)}]_{NC}$ computed from Eq. (44) for a given r/s, nonlinear models of Chapter 5 are once again used to relate the wing–body aerodynamics as functions of AOA and Mach number. Equations (35) through (38) are used for the other interference terms, in analogy to the elliptical cross-sectional case.

VIII. Wing–Body–Tail Configurations

Wings and tails will be assumed to be in line at roll positions of either φ = 0 or 45 deg. This will allow the new wing–tail interference model developed and discussed in Section 5.6 to be used. While this model will probably be affected by noncircular shapes, it is believed that it is probably less of an effect than on either $K_{W(B)}$ or $K_{B(W)}$. As a result of this assumption, the tail can be analyzed just as the wing, for the body cross section of interest at the tail. The only difference will be the addition of the downwash effect on the tail produced by the wing. The nonlinear wing–tail interference model of Section 5.6 is, therefore, used directly for both axisymmetric and nonaxisymmetric bodies.

IX. Variable Body Cross-Sectional Shapes

The discussion to this point has assumed a constant noncircular body cross-sectional shape. That is, the nose, afterbody, and boattail or flare all have the same cross-sectional shape. These shapes could be circular, elliptical, square, or triangular and oriented as shown in Fig. 6.1. In principle, the methodology for noncircular shaped missile configurations discussed in Sections 6.2 through 6.8 for constant cross-sectional shape can be applied to configurations that have variable cross sections. Different values of the parameters $(C_n/C_{n_O})_{SB}, (C_n/C_{n_O})_N$ NF, and interference factors for each of the different cross-sectional shapes will be required. From a practical standpoint, only two different cross-sectional shapes will be allowed with a transition region between. For example, if the front of the missile were circular and the back elliptical, a transition region between circular and elliptical shape is necessary from a practical standpoint. Figure 6.15 illustrates a variable cross-sectional elliptical shape taken from Jorgensen.[6] The front or nose of the missile consists of an ellipse of a/b = 0.5 and the rear an ellipse

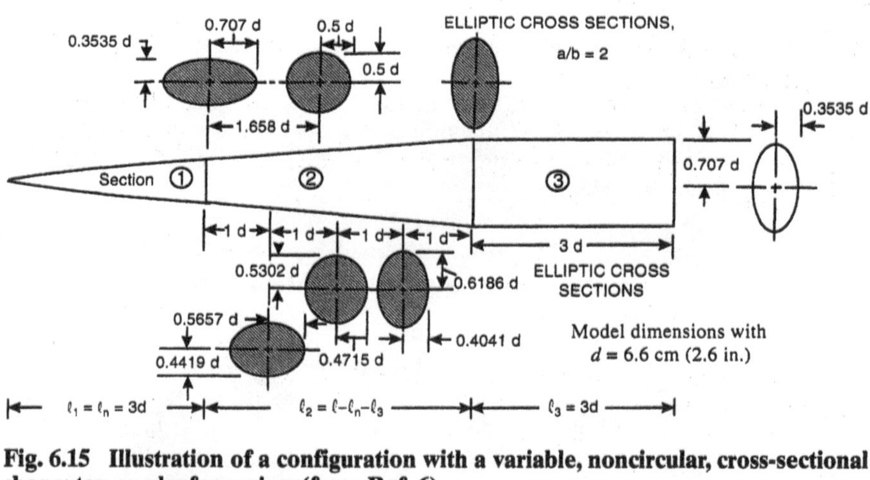

Fig. 6.15 Illustration of a configuration with a variable, noncircular, cross-sectional shape: top or planform view (from Ref. 6).

of $a/b = 2.0$ when viewed from the top, or planform view. It is necessary, therefore, to have a region that smoothly contours the nose ellipse to that of the afterbody. The cross-sectional area of the Fig. 6.15 configuration remains constant from the end of the nose to the end of the body. This also means the equivalent diameter of a circular cross section also remains constant. However, the noncircular slender body and Newtonian factors are significantly different, which means the normal force, pitching moment, and center of pressure will change substantially.

Assuming the nose of Fig. 6.15 has a constant cross-sectional shape and the afterbody has a length of constant cross-sectional shape of length ℓ_3 (which is equal to the boattail or flare length if a boattail or flare is present), then the normal force, pitching moment, and center of pressure for the representative body-alone configuration of Fig. 6.15 can be written as

$$C_N = \sum_{i=1}^{3} \left\{ \left[\left(\frac{C_n}{C_{no}} \right)_{SB} C_{N_L} \right]_i + \left[\left(\frac{C_n}{C_{no}} \right)_N (\text{NF})\, C_{N_{NL}} \right]_i \right\} \tag{45}$$

$$C_M = \sum_{i=1}^{3} \left\{ \left[\left(\frac{C_n}{C_{no}} \right)_{SB} C_{N_L} (x_m - x_{CP}) \right]_i + \left[\left(\frac{C_n}{C_{no}} \right)_N (\text{NF})\, C_{N_{NL}} (x_m - x_{CP}) \right]_i \right\} \tag{46}$$

$$x_{CP} = -\frac{C_M}{C_N} \tag{47}$$

In Eqs. (45) and (46), each segment of the body is assumed to have its individual linear and nonlinear component of normal force and, likewise, to have its individual center of pressure relative to some desired reference location, x_m. The terms x_{CP} and x_m of Eq. (46) are in dimensions of body diameters or calibers. The local values of $(C_n/C_{no})_{SB}$ and $(C_n/C_{no})_N$ for segments 1 and 3 are known from Figs. 6.2 through 6.4, and NF is known from Figs. 6.5 through 6.7. The values of these parameters for segment 2 are assumed to be the average of those for segments 1 and 3 for this case. Equations (45) through (47) thus allow some approximate accounting for the variation of body cross-sectional shape along its length.

X. Summary of Computational Procedure for Aerodynamics of Nonaxisymmetric Body Configurations

The method chosen to compute aerodynamics on configurations of noncircular cross-sectional shapes extends the method of Jorgensen[6] for the body alone, uses the approach of Nelson[35] and Est and Nelson[36] to correct

$K_{W(B)}$ at low AOA, and uses the methods of Chapter 5 to include nonlinearities in all of the interference terms. No changes in the parameters $k_{W(B)}$ and $C_{NT(V)}$ computed by Chapter 5 methods for circular shapes have been made for noncircular shapes. $K_{B(W)}$ and $k_{B(W)}$ are assumed to vary in a similar fashion to $K_{W(B)}$ with respect to body cross-sectional shape.

To compute the aerodynamics on a noncircular shape, the procedure outlined below is recommended.

1) Obtain the cross-sectional area distribution as a function of length. This may be available from a drawing if not given directly.
2) Approximate the body by one that is consistent with the logic of the AP98 (or some other code). The AP98 code requires a nose, afterbody, and boattail/flare. The afterbody and boattail/flare length can be close to zero if the configuration is only a nose shape. If the configuration is a nose-afterbody with no boattail/flare, then the boattail/flare length can be zero.
3) Determine if the noncircular cross-sectional shape is the same or approximately the same. If it is, then pick an option from Fig. 6.1 that most closely represents the cross section and roll position of interest.
4) Based on this cross-sectional shape and the cross-sectional area distribution from instruction 1 above, a circular body of diameter d_{eq} is formed of given nose, afterbody, and boattail/flare length. Factors $(C_n/C_{n_O})_{SB}$, $(C_n/C_{n_O})_N$ and NF are computed for the shape of interest. R_{NC} and M_{NC} are also defined for a given circular shape.
5) If fins are present, interference factors are computed for the cross section and fin orientation of interest at $a = 0$, based on a body of diameter d_{eq}. These aerodynamics are combined with the equivalent body aerodynamics by multiplying the wing–body contributions by a slender body theory scaling factor.
6) Total nonlinear aerodynamics for the equivalent axisymmetric body are computed based on methods of Chapter 5 or the AP98.
7) Total configuration aerodynamics are then computed using the circular body computations from instruction 6 above in conjunction with the noncircular body information from 4 and 5.
8) If the configuration has a variable noncircular cross-sectional shape, then the length ℓ_3 must be defined because it is assumed $\ell_1 = \ell_n$. Also, if a boattail/flare is present, ℓ_3 is assumed to be the boattail or flare length. Choose two cross-sectional shapes from Fig. 6.1 to represent ℓ_1 and ℓ_3. Instructions 4 through 7 are then repeated for the variable cross-sectional shape.

XI. Comparison of Method to Experiment

The performance of the nonaxisymmetric body methodology was evaluated by applying it to an extensive array of configurations for which wind tunnel data were found.[15] Several of those examples will be given here to illustrate the capability and robustness of the methodology. In all calculations, optimum values of M_{NC} and R_{NC} were used. These values were determined by trial and error compared to wind tunnel data from each wind tunnel test where circular body data were available. On the other hand, if one is computing aerodynamics on a noncircular cross-sectional shape for which no experimental data exist, it is recommended that one of two alternatives be exercised. The first alternative is to use the average default values of R_{NC} and M_{NC}, which are 180,000 and 0.1, respectively. The second alternative is to modify these values to more closely reflect the performance of a given wind tunnel if previous experience allows this to be done. At low Mach numbers, correct selection of these parameters is quite important, whereas, at higher Mach numbers, it is not so critical.

The first set of validation runs was performed on the large experimental data set of Ref. 7. The geometric configurations contained within this set are shown in Fig. 6.16. Included are two circular bodies, 1.4 in. in diameter, with 3.0-caliber tangent ogive noses, and either a 7.0-caliber or 3.0-caliber cylindrical afterbody, giving l/d (length/diameter) ratios of 10 and 6. The noncircular bodies in the data set are squares, diamonds (squares at a 45-deg roll position), triangles, inverted triangles (triangles at a 60-deg roll position), and ellipses with axis ratios of 2:1, 1.5:1, 0.67:1 (1.5:1 at 90-deg roll), and 0.5:1 (2:1 at 90-deg roll). All bodies have the same cross-sectional area as the circle, and the distribution of that area along the longitudinal axis is the same as for the circle. The squares, diamonds, triangles, and inverted triangles have very small corner radii and were, therefore, assumed to have a value of k equal to 0. All noncircular bodies are identical in length to the $l/d = 10$ circular cylinder, except for a 2:1 ellipse and a 0.5:1 ellipse, which have the same length as the $l/d = 6$ cylinder. Mach numbers were 1.98 and 3.88 in the Ref. 7 tests. Jorgensen and Nelson[38] later tested the elliptical shape body of Fig. 6.16 for $M_\infty = 0.6$ to 2.0 and AOAs up to 56 deg.

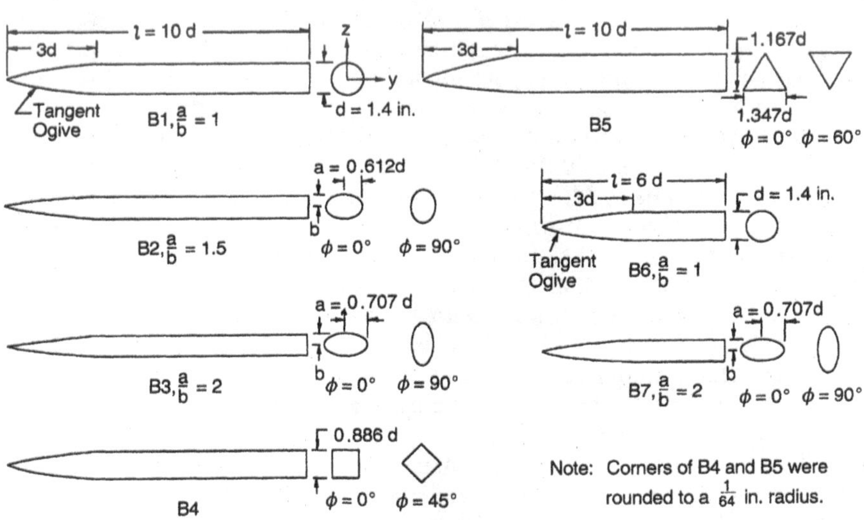

Fig. 6.16 Body alone configurations[2] with elliptical, square, diamond, triangular, and inverted triangular shapes.

The case shown here will thus be the elliptical 10-caliber-long body tests of Ref. 32, which go to 56-deg AOA, and the square and triangular tests of Ref. 7, which go only to a 24-deg AOA. Not all results will be shown, as the Refs. 7 and 38 databases were fairly extensive. Most of these results are shown in Ref. 15, however.

Figures 6.17 to 6.19 give the elliptical body results for Mach numbers of 0.6, 1.2, and 2.0. Results shown are for ellipticity values of 0.5, 1.0, and 2.0 and are given in terms of normal and axial force coefficients and center of pressure. Also, the axial force coefficient does not include a base drag component. The AP98 results in the figures correspond to the methodology presented in this chapter. In examining the comparisons of theory and experiment in Figs. 6.17 to 6.19, it is seen that the theory does a fairly good job of predicting most of the aerodynamics on the $a/b = 2.0$ elliptical case at all Mach numbers. The normal force and center of pressure predictions are quite encouraging as they are well within the average accuracy levels of ± 10 and $\pm 4\%$ of body length, respectively. The axial force prediction comparisons are not as good as desired. However, the poorer comparison of theoretical axial force to data could be due to measurement accuracy where it is difficult to get accurate axial force measurements with a sting designed for measuring normal force at high AOAs. The $a/b = 0.5$ results are not quite as good as the $a/b = 2.0$ results, particularly at high AOA and low M_∞. This could be due to the critical value of crossflow Mach number prediction, which was assumed to have a value of 0.1. The data appear to support supercritical flow up to an AOA of 56 deg, whereas the theory indicates the flow transitions to subcritical conditions around an AOA of 36–40 deg. Also, the center of pressure prediction for the $a/b = 0.5$ case at low AOA appears to indicate a center of pressure much farther toward the nose tip at $M_\infty = 0.6$ than the theory predicts. It is not clear what mechanism causes this. Because the nose length is 3 calibers and there is no boattail present, intuition would lead one to expect the center of pressure to lie somewhere between 1.5 and 3.0 calibers from the nose tip near $\alpha = 0$ deg.

Theoretical and experimental results for the squares and triangles of Fig. 6.16 are given in Figs. 6.20 and 6.21, respectively. Only the 10-caliber-long configuration results at $M_\infty = 1.98$ are shown. Here the results are given in terms of lift coefficient, lift-to-drag ratio, and center of pressure. In general, comparison of theory and experiment for the squares and diamonds is quite encouraging, although not as good as the circular cross-sectional shapes. The triangular-shaped body predictions for lift coefficient tend to be somewhat low as AOA increases. However, this is to be expected because the values for the circular cylinder results are also low. Lift-to-drag ratio predictions are quite good, with the peak values being reasonably well predicted. Center of pressure prediction for the triangular shape is well within the $\pm 4\%$ of body length used as a criterion for axisymmetric bodies. However, the inverted triangle center of pressure predictions slightly exceed this value.

The second case considered is a 10-caliber wing–body and wing–body–tail case shown in Fig. 6.22 (Ref. 6). The body cross section is an ellipse with an $a/b = 2.0$, and the nose length is 3.0 calibers. Figures 6.23 and 6.24 present

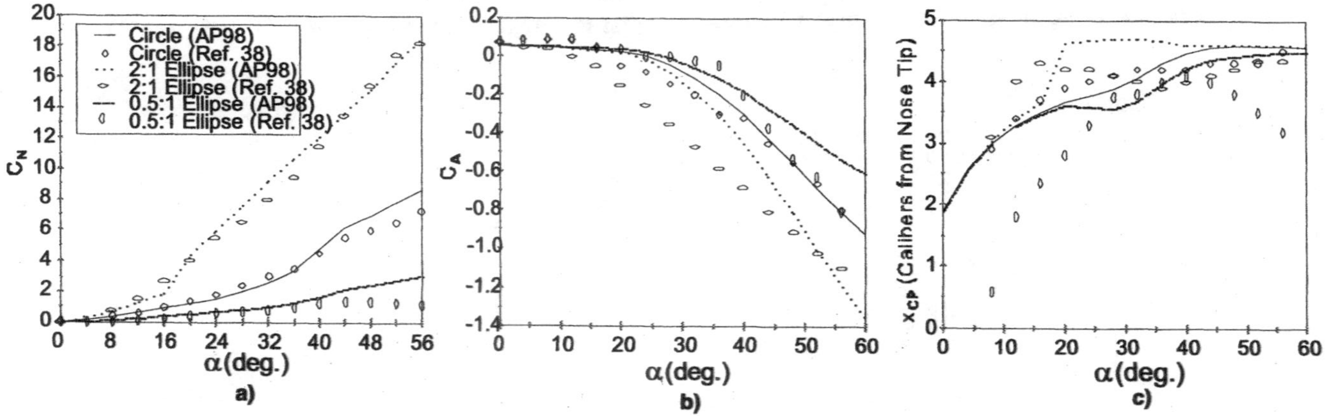

Fig. 6.17 Aerodynamic data for 2:1 and 0.5:1 ellipses of Fig. 6.16 compared with circular body at M_∞ = 0.6: a) normal force coefficient, b)axial force coefficient, and c) center of pressure.

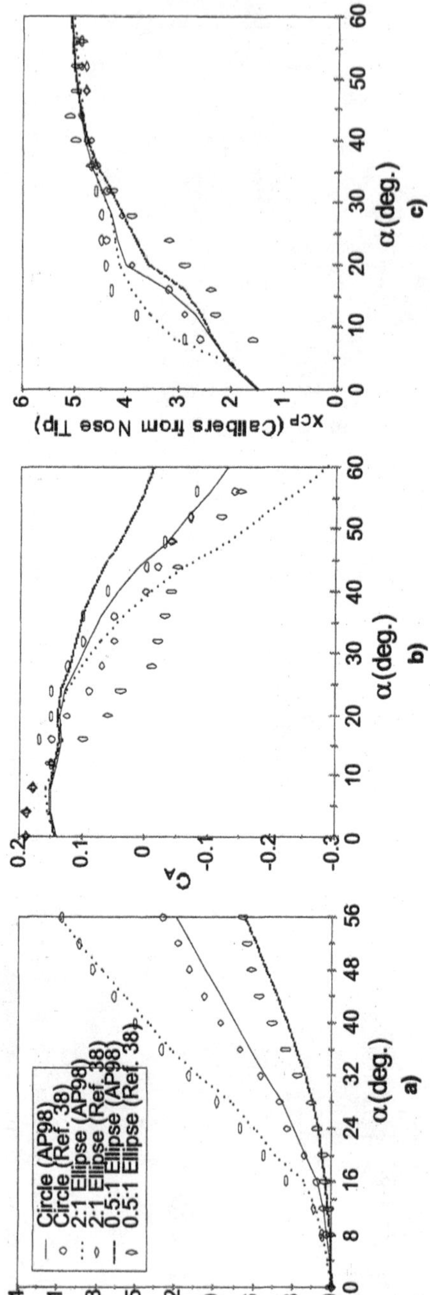

Fig. 6.18 Aerodynamic data for 2:1 and 0.5:1 ellipses of Fig. 6.16 compared with circular body at $M_\infty = 1.2$: a) normal force coefficient, b) axial force coefficient, and c) center of pressure.

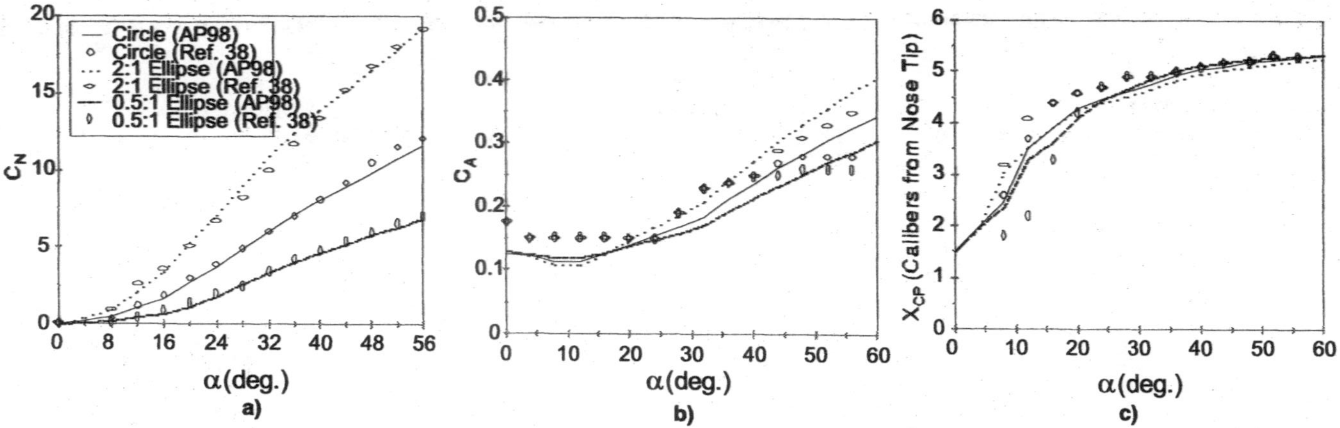

Fig. 6.19 Aerodynamic data for 2:1 and 0.5:1 ellipses of Fig. 6.16 compared with circular body at M_∞ = 2.0: a) normal force coefficient, b) axial force coefficient, and c) center of pressure.

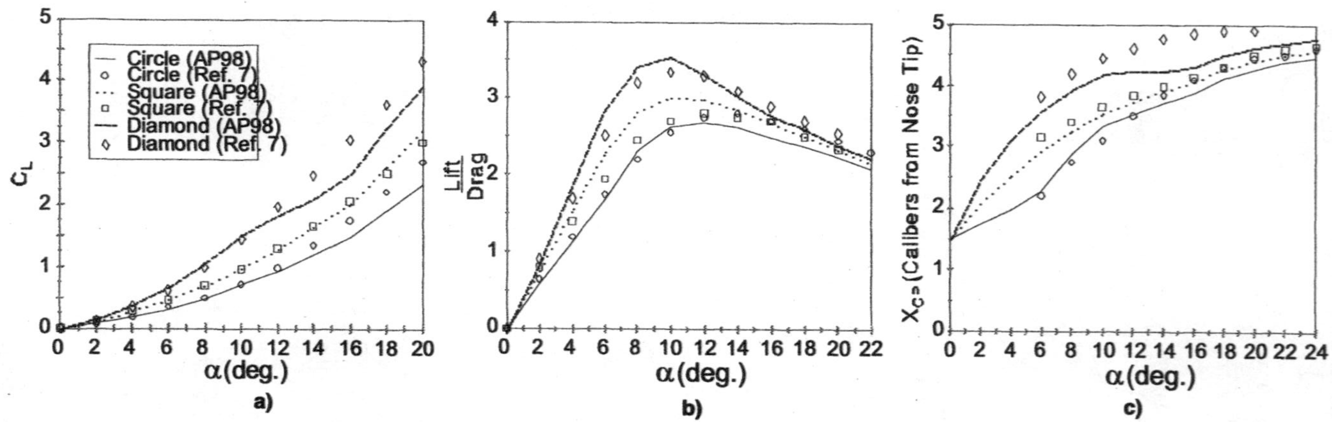

Fig. 6.20 Aerodynamic data for squares ($k = 0.0$) and diamonds ($k = 0.0$) of Fig. 6.16 compared with circular body at $M_\infty = 1.98$ ($\ell/d = 10$): a) lift coefficient, b) lift-to-drag ratio, and c) center of pressure.

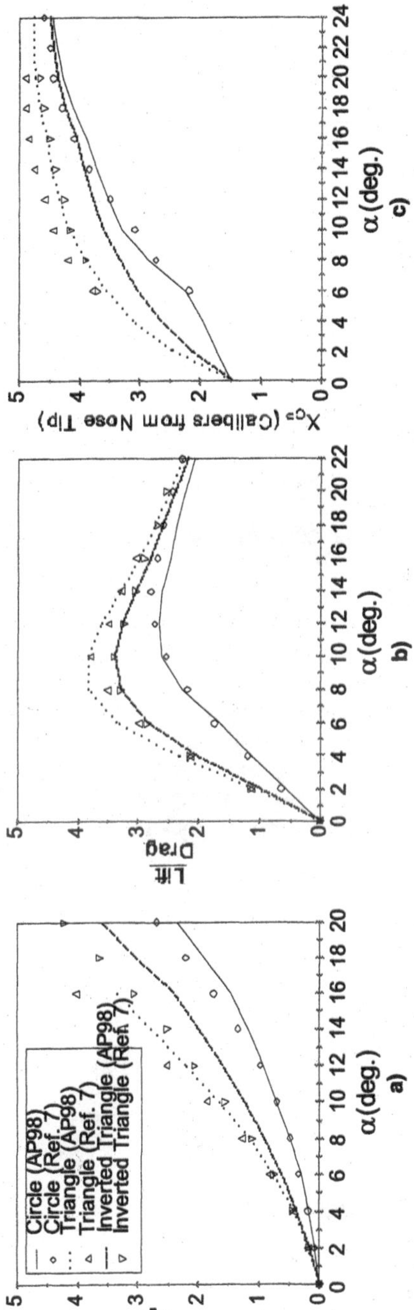

Fig. 6.21 Aerodynamic data for triangles ($k = 0.0$) and inverted triangles ($k = 0.0$) of Fig. 6.16 compared with circular body at $M_\infty = 1.98$ ($\ell/d = 10$): a) lift coefficient, b) lift-to-drag ratio, and c) center of pressure.

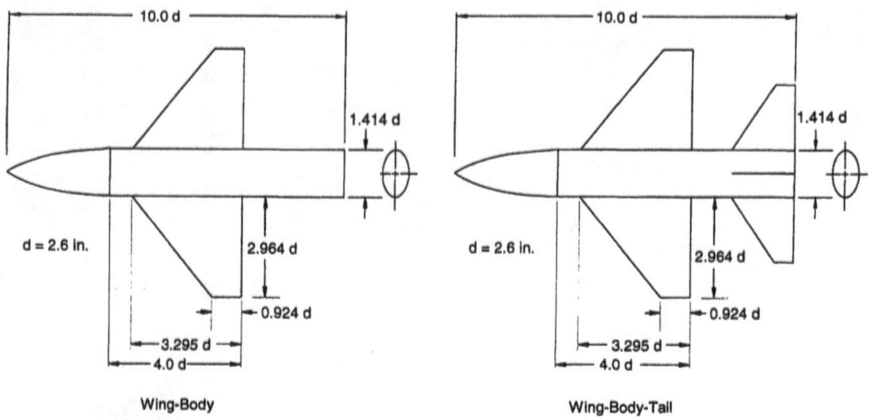

Fig. 6.22 Geometry of the wing–body and wing–body–tail configuration with 2:1 elliptical bodies.[6]

the normal force and center of pressure comparisons of experiment and theory at $M_\infty = 0.6$ and 2.0 for the wing–body and wing–body–tail cases, respectively. Data for the wing–body–tail case at $M_\infty = 2.0$ were available only to an AOA of 34 deg, whereas all other cases have data to an AOA of 60 deg. As seen in Figs. 6.23 and 6.24, theoretical predictions are quite good for center of pressure and fair to good for normal force coefficient.

The third configuration shown (see Fig. 6.25) is a waverider configuration taken from Ref. 39. Lift, drag, and pitching moment comparisons of theory and experiment at $M_\infty = 14.0$ and to an AOA of 25 deg are shown in Fig. 6.26. Results are quite encouraging, even though this configuration does not quite fit the triangular shape, which has 60-deg angles in all corners. Note

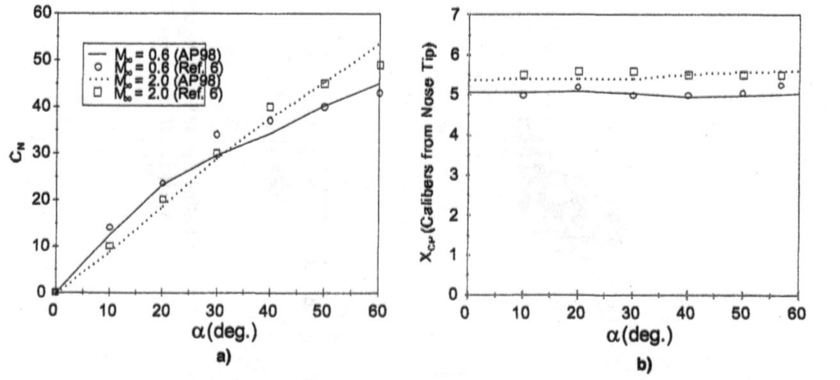

Fig. 6.23 Aerodynamic data for the wing–body configuration of Fig. 6.22 with a 2:1 elliptical cross-sectional body: a) normal force coefficient and b) center of pressure.

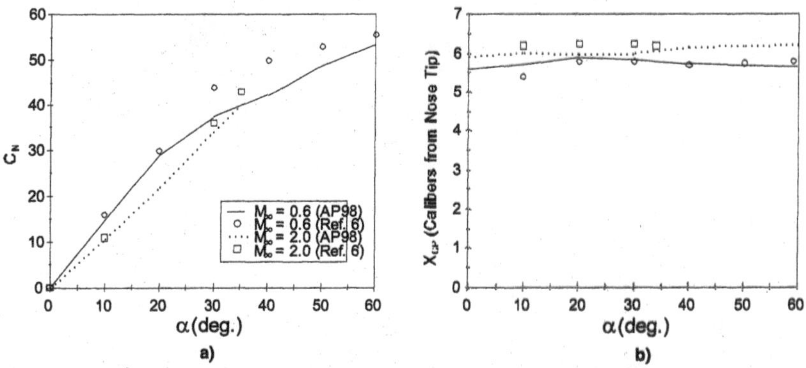

Fig. 6.24 Aerodynamic data for the wing–body–tail configuration of Fig. 6.22 with a 2:1 elliptical cross-sectional body: a) normal force coefficient and b) center of pressure.

that these results are based on a 375-in.^{2} planform area and a reference length of 10.8 in. It should be noted that C_L at $a = 0$ is nonzero, indicating that this configuration has a slight asymmetry or camber in the pitch plane. One could actually try to adjust theory to experiment by adjusting the AOA in the theory that closely responds to the value of C_L at $a = 0$. However, no attempt was made to do this. Even though the Fig. 6.25 configuration does not exactly fit one of the Fig. 6.1 options, the theory can still be used to give approximations to static aerodynamics, as seen in Fig. 6.26.

The final configuration considered, taken from Ref. 40, is shown in Fig. 6.27. It consists of a lifting body with variable sweep wing panels. In this case, we have a body that appears at first glance to be an inverted triangle, but the lower angle is only 45 deg and not 60 deg. In this sense, it resembles the lower one-half of a diamond. For comparison purposes, it was decided to consider this example as both an inverted triangle and a diamond, both with sharp corners, because the top corners were thought to be more influential on the flowfield. The wing panels were modeled at a 40-deg leading-edge sweep. Computations were preformed for Mach numbers of 0.4, 0.6, and 0.8 with corresponding Reynolds numbers based on the maximum body chord of 5.04×10^6, 7.2×10^6, and 9.36×10^6, respectively. The lift coefficients for the wind-tunnel tests and AP98 computations for both body shapes are shown for the three Mach numbers in Fig. 6.28. The coefficient values shown are based on the planform area of the body (96.2 in.^{2}). On balance, both approaches give acceptable results, with the diamond being somewhat better in most cases. The diamond may give these slightly better results because it more closely represents the true angle on the body's lower surface, and it is windward side aerodynamics that dominate the flowfield.

To summarize this chapter, an improved method has been developed to compute aerodynamics of noncircular cross-sectional shapes in an approximate sense. The improved method is based on the Jorgensen approach of

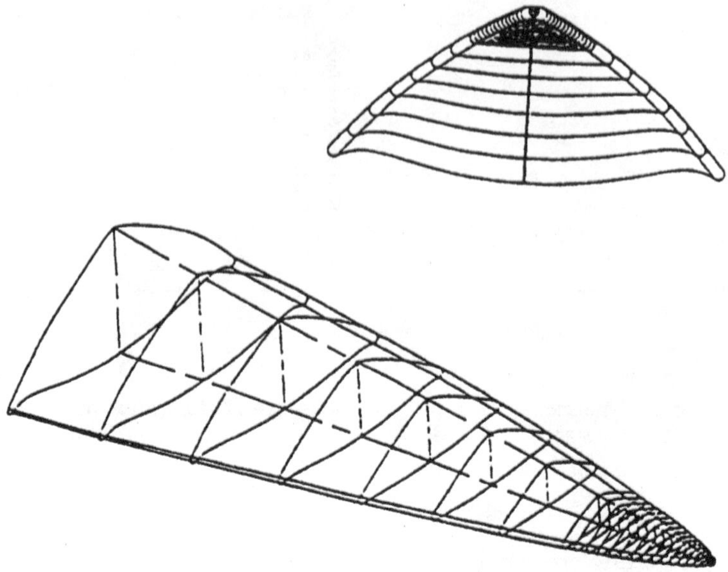

Fig. 6.25 Wire frame geometry of the waverider (adapted from Ref. 39).

computing aerodynamics on a noncircular body using circular body methods, the Nelson approach for noncircular wing–body interference corrections at low AOA, and the method of Chapter 5 for including nonlinearities in wing–body interference aerodynamics. The new method extends these approaches in several significant ways to make the method more general and applicable to most configurations for noncircular cross section.

Specific extensions include the following: derivation of the Newtonian approximation to the ratio of local normal force coefficient per unit length of a noncircular shape to the similar coefficient of a circular shape for squares, diamonds, triangles, and inverted triangles; an empirical correction to these Newtonian factors to account for finite versus very high Mach numbers; derivation of an empirical estimate of critical crossflow Reynolds and Mach numbers as functions of noncircular geometry shape; a method to treat wing–body interference factor corrections as a function of body geometry and freestream parameters; derivation of slender body theory scaling factors to allow aerodynamics to be computed using wind-tunnel data; and an approximate way of treating configurations that have variable, noncircular cross-sectional shapes.

The method presented in this chapter was applied to all of the noncircular configurations found in the literature for which data were available. These configurations included the following: elliptical cross-sectional shapes with a/b from 0.5 to 3.0, Mach numbers from 0.6 to 3.88, AOAS as high as 60 deg and some cases with wings; squares and triangular cross-sectional shapes with sharp and rounded edges at mostly lower Mach numbers, but some data

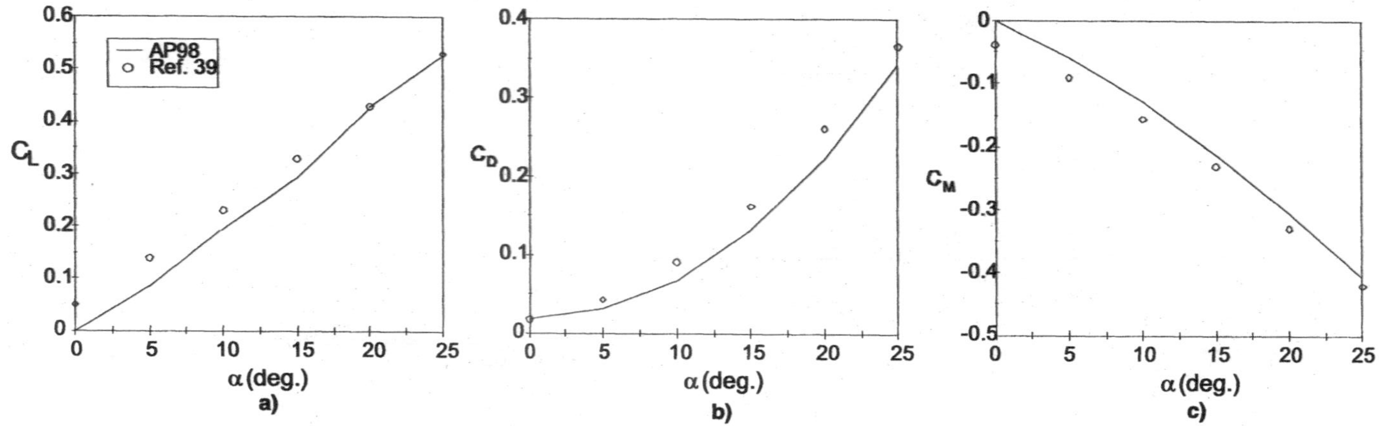

Fig. 6.26 Aerodynamic data for the Mach 14 waverider of Fig. 6.25: a) lift coefficient, b) axial force coefficient, and c) moment coefficient.

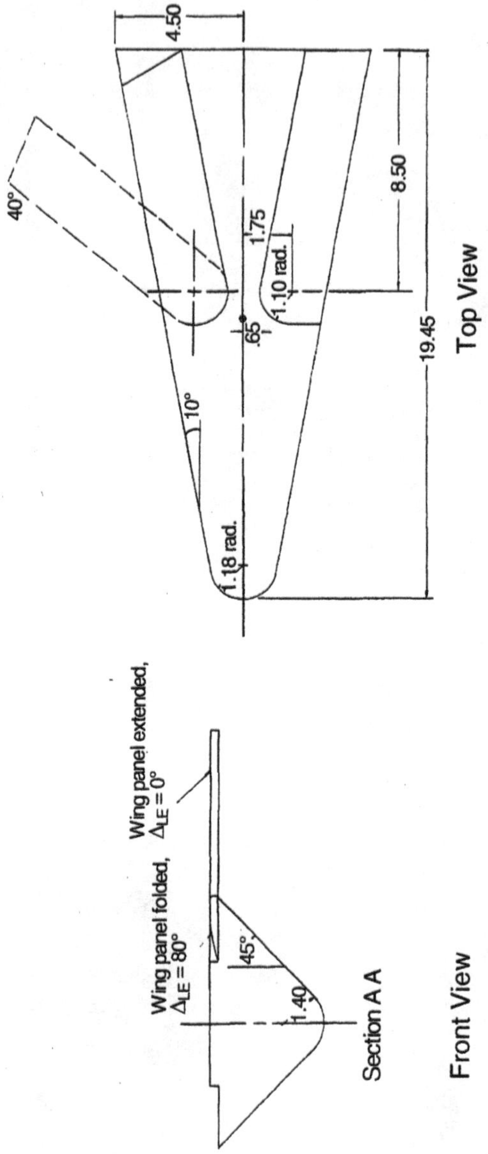

Fig. 6.27 Geometry of the variable wing sweep lifting body (from Ref. 40).

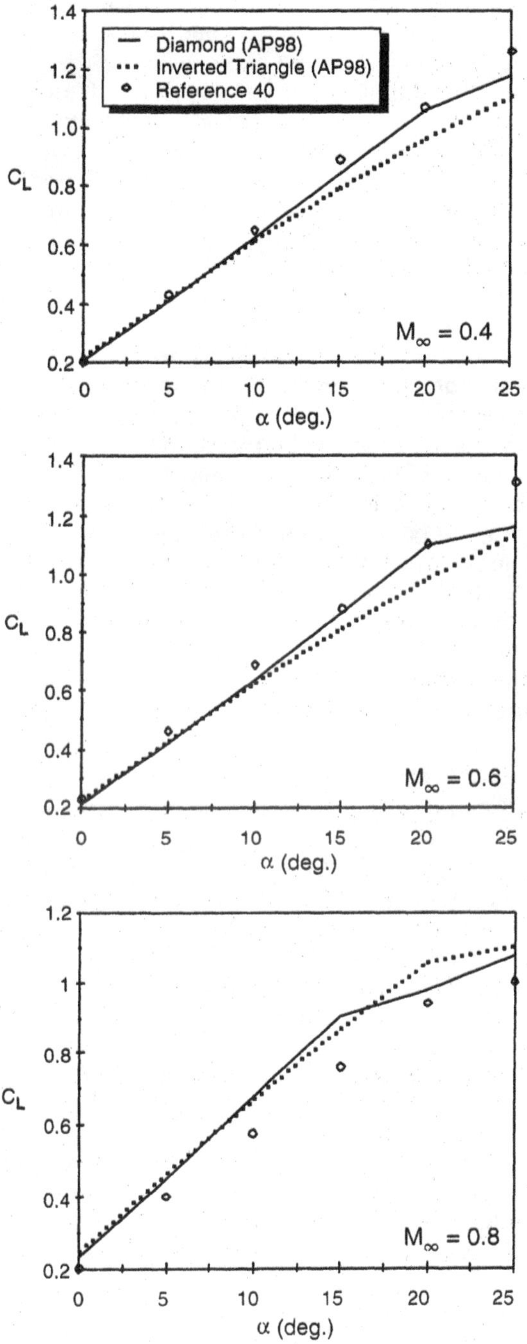

Fig. 6.28 Lift coefficients for the body–wing configuration of Fig. 6.27 computed by two methods and compared with experiment at three Mach numbers.

at Mach numbers as high as 4, AOAs as high as 58 deg, and some cases with wings; a single configuration with a variable cross-sectional shape at Mach numbers from 0.6 to 2.0 and AOAs to 60 deg; and two configurations that were quite complex and did not fit within the exact requirements of the geometry options. Results for planar aerodynamics using a semi-empirical code were, in general, very good. Although it is too early to state the overall accuracy of the nonaxisymmetric body aerodynamic prediction method, it appears that the normal force is almost as good as the circular body predictions. To date, we have not found a circular body configuration where average accuracy on C_A and C_N exceeds $\pm 10\%$ and x_{CP} exceeds $\pm 4\%$ of the body length. Here "average" means enough Mach numbers and AOAs to get a good statistical sample.

Although additions to the state-of-the-art in computing approximate aerodynamics are significant, several challenges remain. The first and foremost challenge is to compare the present predictions of static aerodynamics to more data. Because many databases shown in the literature focus on normal force, additional data or computational fluid dynamics computations may be needed to validate and/or modify the axial force and center of pressure prediction techniques. It is hoped that ballistics range tests ongoing will partially address this problem for axial force and low AOA center of pressure prediction. Second, it is suspected that, once more zero AOA axial force data are available, the assumption of wave drag on the nonaxisymmetric body being equal to that of the axisymmetric body may need to be adjusted. Finally, it is suspected that the Newtonian correction factor for triangles and squares could be improved upon with more data or CFD results.

References

[1]Gentry, A. E., Smyth, D. N., and Oliver, W. R., "The Mark IV Supersonic—Hypersonic Arbitrary Body Program," Air Force Flight Dynamics Lab. AFFDL TR-73-159, Wright Patterson Air Force Base, Dayton, OH, 1973.

[2]Sidwell, K. W., Baruah, P. K., and Bussoletti, J. E., "A Computer Program for Predicting Subsonic or Supersonic Linear Potential Flows About Arbitrary Configurations Using a Higher Order Panel Method," Vol. 2, NASA CR 3252, 1980.

[3]Wardlaw, A. B., and Davis, S., "A Second-Order-Gudonov Method for Supersonic Tactical Missiles," Naval Surface Warfare Center, Dahlgren Div. NSWC TR-86-506, Dahlgren, VA, 1986.

[4]Walters, R. W., Slack, D. C., Cimmella, P., Applebaum, M. P., and Frost, C., "A Users Guide for GASP," Virginia Polytechnic Inst. State Univ., Dept. Aerospace Ocean Engineering, Blacksburg, VA, Nov. 1990.

[5]Whitcomb, R. T., "A Study of the Zero-Lift Drag Rise Characteristics of Wing–Body Combinations Near the Speed of Sound," NACA Rept. 1273, 1956.

[6]Jorgensen, L. H., "Prediction of Static Aerodynamic Characteristics for Slender Bodies Alone and with Lifting Surfaces to Very High Angles of Attack," NASA TR R-474, Sept. 1977.

[7]Jorgensen, L. H., "Inclined Bodies of Various Cross Sections at Supersonic Speeds," NASA Memo 10-3-58A, 1958.

[8]Jorgensen, L. H., "Prediction of Static Aerodynamic Characteristics for Space-

Shuttle-Like and Other Bodies at Angles of Attack from 0° to 180°," NASA TN D-6996, 1973.

[9]Jorgensen, L. H., "Estimation of Aerodynamics for Slender Bodies Alone and with Lifting Surfaces at α's from 0° to 90°," *AIAA Journal,* Vol. 11, No. 3, March 1973, pp. 409–412.

[10]Jorgensen, L. H., "A Method for Estimating Static Aerodynamic Characteristics for Slender Bodies of Circular and Noncircular Cross Section Alone and with Lifting Surfaces at Angles of Attack from 0° to 90°," NASA TN D-7228, 1973.

[11]Vukelich, S. R., and Jenkins, J. E., "Missile DATCOM: Aerodynamic Prediction on Conventional Missiles Using Component Build-Up Techniques," AIAA Paper no. 84-0388, 1984.

[12]Hemsch, M. J., and Nielsen, J. N., "Equivalent Angle-of-Attack Method for Estimating Nonlinear Aerodynamics of Missile Fins," *Journal of Spacecraft and Rockets,* Vol. 20, No. 4, July–Aug. 1983.

[13]Moore, F. G., McInville, R. M., and Hymer, T., "The 1998 Version of the NSWC Aeroprediction Code: Part 1—Summary of New Theoretical Methodology," NSWCDD/TR-98/1, Dahlgren, VA, April 1998.

[14]Pitts, W. C., Nielsen, J. N., and Kaattari, G. E., "Lift and Center of Pressure of Wing–Body–Tail Combinations at Subsonic, Transonic, and Supersonic Speeds," NACA TR 1307, 1957.

[15]Moore, F. G., McInville, R. M., and Hymer, T. C., " An Improved Semiempirical Method for Calculating Aerodynamics of Missiles with Noncircular Bodies," NSWCDD/TR-97/20, Sept. 1997.

[16]Vukelich, S. R., Stoy, S. L., Burns, K. A, Castillo, J. A., and Moore, M. E., "Missile DATCOM—Volume 1—Final Report," Wright Patterson Air Force Base, AFWAL-TR-86-3091, Dayton, OH, Dec. 1988.

[17]Jones, N. H., and Zollars, G. J., "Investigation of the Aerodynamics of Square Cross Section Missiles," U.S. Air Force Academy Rept. USAFA-TR-83-15, Colorado Springs, CO, Sept. 1983.

[18]Polhamus, E.C., Geller, B. W., and Greenwald, K.J., "Pressure and Force Characteristics of Noncircular Cylinders as Affected by Reynolds Number with a Method Included for Determining the Potential Flow About Arbitary Shapes," NASA TR R-46, 1959.

[19]Lockwood, V. E.,"Effects of Reynolds Number and Flow Incidence on the Force Characteristics of a Family of Flat-Front Cylinders," NASA TN D-3932, May 1967.

[20]Polhamus, E. C., "Effect of Flow Incidence and Reynolds Number on Low-Speed Aerodynamic Characteristics of Several Noncircular Cylinders with Applications to Directional Stability and Spinning," NACA TN 4176, Jan. 1958.

[21]Daniel, D. C., Yechout, T. R., and Zollars, G. J., "Experimental Aerodynamic Characteristics of Missiles with Square Cross Sections," *Journal of Spacecraft and Rockets,* Vol. 19, March–April 1982, pp. 167–172.

[23]Schneider, W., "Experimental Investigation of Bodies with Noncircular Cross Section in Compressible Flow," AGARD-CP-336, Symposium on Missile Aerodynamics, Trondheim, Norway, Sept. 1982, pp. 19-1 to 19-15.

[23]Sellers, M. E., and Siler, L. G., "Pressure and Static Force Test of Three Elliptic Missile Body Configurations at Mach Numbers 1.5 to 5.0," Arnold Engineering Development Corporation Rept. AEDC-TSR-83-V44, 1983.

[24]Shereda, D. E., Amidon, P. F., Dahlem, V., III, and Brown-Edwards, E., "Pressure Test of Three Elliptical Missile Body Configurations at Mach Numbers 1.5 to 5.0" Wright-Patterson Air Force Base AFWAL AL TM 84-236-FIMG Dayton OH, Dec. 1984.

[25]Lijewski, L. E., Zollars, G. J., Yechout, T. R., and Haupt, B. F., "Experimental Flowfield Measurements of Missiles with Square Cross Sections," AIAA-82-0055, AIAA 20th Aerospace Sciences meeting, Orlando, FL, Jan. 1982.

[26]Stoy, S. L., and Vukelich, S. R., "Prediction of Aerodynamic Characteristics of Unconventional Missile Configurations using Component Buildup Techniques," AIAA-86-0489, AIAA 24th Aerospace Sciences Meeting, Reno, NV, Jan. 1986.

[27]Amidon, P. F., "Supersonic Aerodynamic Characteristics of Elliptic Cross Section Bodies," AIAA 85-1607, AIAA 18th Fluid Dynamics Plasmadynamics Lasers Conference, Cincinnati, OH, July 1985.

[28]Vukelich, S. R., "Aerodynamic Prediction of Elliptically-Shaped Missile Configurations Using Component Build-up Methodology," AIAA-85-0271, AIAA 23rd Aerospace Sciences Meeting, Reno, NV, Jan. 1985.

[29]Zollars, G. J., Yechout, T. R., Daniel, D. C., and Lijewski, L. E., "Experimental Aerodynamic Characteristics of Missiles with Square Cross Sections," U.S. Air Force Academy, USAFA-TN-83-8, Colorado Springs, CO, May 1983.

[30]Hoerner, S. F., *Fluid Dynamic Drag,* Hoerner Fluid Dynamics, Brick Town, NJ, pp. (1965) 16-15, 16-16.

[31]Delany, N. K., and Sorensen, N. E., "Low Speed Drag of Cylinder of Various Shapes," NACA TN 3038, Nov. 1953.

[32]Ashley, H., and Landahl, M., *Aerodynamics of Wings and Bodies,* Addison Wesley Longman Reading, MA, 1965.

[33]Nielsen, J. N., *Missile Aerodynamics,* NEAR, Mountain View, CA, 1988.

[34]Moore, F. G., and McInville, R., "Extension of the NSWCDD Aeroprediction Code to the Roll Position of 45 Degrees," NSWCDD/TR-95/160, Dahlgren, VA, Dec. 1995.

[35]Nelson, H. F., "Wing–Body Interference Lift for Supersonic Missiles with Elliptical Cross-Section Fuselages," *Journal of Spacecraft and Rockets,* Vol. 26, No. 5, Sept.–Oct. 1989, pp. 322–329.

[36]Est, B. E., and Nelson, H. F., "Wing–Body Carryover and Fin Center of Pressure for Missiles with Noncircular Fuselage Cross Sections," AIAA 91-2856, Atmospheric Flight Mechanics Conference, New Orleans, LA, Aug. 1991.

[37]Sigal, A., *Tactical Missile Aerodynamics,* edited by J. Nielsen, Progress in Aeronautics and Astronautics Series, AIAA, Washington, DC, 1991 Chapter 5, pp. 171–217

[38]Jorgensen, L. H., and Nelson, E. R., "Experimental Aerodynamic Characteristics for Bodies of Elliptic Cross Section at Angles of Attack from 0° to 58° and Mach Numbers from 0.6 to 2.0," NASA TM X-3129, 1975.

[39]Gillum, M. J., and Lewis, M. J., "Analysis of Experimental Results on a Mach 14 Waverider with Blunt Leading Edges," AIAA 96-0812, 34th Aerospace Sciences Meeting Exhibit, Reno, NV, Jan. 1996.

[40]Spencer, B., "Longitudinal Aerodynamic Characteristics at Mach Numbers from 0.40 to 1.10 of a Blunted Right-Triangular Pyramidal Lifting Reentry Configuration Employing Variable-Sweep Wing Panels," NASA TN D-1518, 1963.

Chapter 7

Aerodynamic Heating at Hypersonic Mach Numbers, Including Real Gas Effects

Nomenclature

a = speed of sound (ft/s)
C_P = pressure coefficient
D = diameter (ft)
e = internal energy (ft^2/s^2)
H = temperature-based heat transfer coefficient (ft-lb/ft^2-s-°R)
H_o = total enthalpy (ft^2/s^2)
H_1 = enthalpy-based heat transfer coefficient (slug/ft^2-s)
h = specific enthalpy (ft^2/s^2) and altitude (ft)
h_{aw} = adiabatic wall enthalpy
K = temperature (°K)
M = Mach number
N = Mangler transformation factor
Pr = Prandtl number
p = pressure (lb/ft^2)
$\dot{q}$ = heat transfer (ft-lb/ft^2-s or Btu/ft^2-s)
R = gas constant [for air R = 1716 (ft-lb)/(slug-°R)]
r_C = shock radius of curvature at vertex (ft)
R_N = Reynolds number
r = body or shock radius (ft)
r_c = recovery factor
r_n = nose radius of curvature (ft)
S = entropy (ft-lb)/(slug-°R)
s = boundary-layer running length (ft)
T = temperature (° K or R)
T_P = temperature of a perfect gas
T_R = temperature of a real gas
t = time
u = velocity in streamwise direction (ft/s)
V = total velocity
V_1, V_2 = velocity (ft/s)
V_r, V_θ = velocity along and normal to a conical ray

V_{max} = maximum velocity
x = streamwise coordinate (ft)
x, r, φ = cylindrical coordinate system with $\varphi = 0$ = leeward plane
x, y, z = rectangular coordinate system
Z = compressibility factor (= 1 for a perfect gas)
α = AOA (deg)
β = $\sqrt{M^2 - 1}$
γ = ratio of specific heats (c_p/c_v)
δ = boundary-layer thickness (ft)
δ^* = boundary-layer displacement thickness (ft)
δ_{eq} = angle used in Newtonian theory (= angle between velocity vector and local body slope)
η, η_1 = exponent used in second-order shock expansion theory
θ = local body slope
θ_c, θ_s = cone half-angle and shock angle, respectively
Λ_{LE} = control surface leading edge sweep (deg)
λ = $\rho V^2/\beta$
μ = viscosity coefficient (slug/ft-s)
ν = Prandtl–Meyer angle (deg)
ρ = density (slug/ft^3)
σ = shock wave angle (deg)
ω = shock standoff distance (ft)

Subscripts
aw = adiabatic wall
b = body
c = cone
e = boundary-layer edge
f = frozen flow
L = local conditions
l = laminar and freestream value
m = match point
N = normal direction
o = total and reference values
p = parallel
s = based on boundary-layer running length
t = turbulent and tangential velocity component
w = wall or surface
o = stagnation conditions
2 = downstream of shock or expansion
∞ = freestream conditions

Superscripts
$'$ = inviscid surface conditions behind parallel shock or wing leading edge
$*$ = eckert reference conditions

I. Introduction

THE accelerating pace of technological advances in recent years has given rise to a new generation of potential military threats for which appropriate countermeasures must be developed. In the case of surface-launched tactical missiles, new targets that must be intercepted are faster, more maneuverable, and more difficult to detect and track. Many design and operational factors must be considered if successful engagements are to be possible under these conditions. One of the more critical concerns lies in the severe aerodynamic environment to which the missile is exposed as a result of high velocities needed to deal with sophisticated targets that may be approaching quite rapidly and detected only at relatively close range. Depending on the given situation, effective defensive measures could call for a missile capable of upper supersonic or even hypersonic flight.

As the performance envelope is extended to the hypersonic region above Mach 6, two important aerodynamic phenomena become increasingly important. The first of these relates to the manner in which the air molecules in the vicinity of the missile store energy at elevated temperatures created by the strong bow shock. At temperatures up to about 1400 °R, heat dissipated by the shock is stored in the translational and rotational energy modes of gas molecules. The fluid behavior is adequately described by the perfect gas law, and specific heats remain constant. As temperature rises, vibrational energy modes of molecules become excited. The perfect gas law is no longer valid, and specific heats become functions of temperature, as opposed to being constant when the perfect gas law is obeyed. Above about 4500 °R, air molecules begin to dissociate and chemical reactions occur, leading to even greater complexities. This departure from perfect gas behavior is often referred to generally as real-gas effects. (Note that "real gas effects" as defined here refer to traditional usage by the aerodynamics literature as opposed to the definition from a chemical standpoint.[1]) It tends to become significant above approximately Mach 6 to 8. The overall result is to lower the gas temperatures below those that would be predicted by a perfect gas analysis. The increase in modes of internal energy storage means that less energy goes into the translational mode, which is directly measured as temperature.

One may rightfully ask, why be concerned about real gas effects? A good illustration of the importance was given by Anderson[1] and extended by Moore et al.[2] These references showed the results of static temperature behind a normal shock where the perfect gas assumption was made, and then these results were compared to a real gas. For a perfect gas, the static temperature ratio across a normal shock is

$$\frac{T_2}{T_\infty} = 1 + \frac{2(\gamma - 1)}{(\gamma + 1)^2}\frac{(\gamma M_\infty^2 + 1)}{M_\infty^2}(M_\infty^2 - 1) \tag{1}$$

where T_2 is temperature behind the shock and subscript $_\infty$ represents conditions in the freestream ahead of the shock. Figure 7.1 plots these results as a function of Mach number for an altitude of 170,000 ft. At this altitude, the

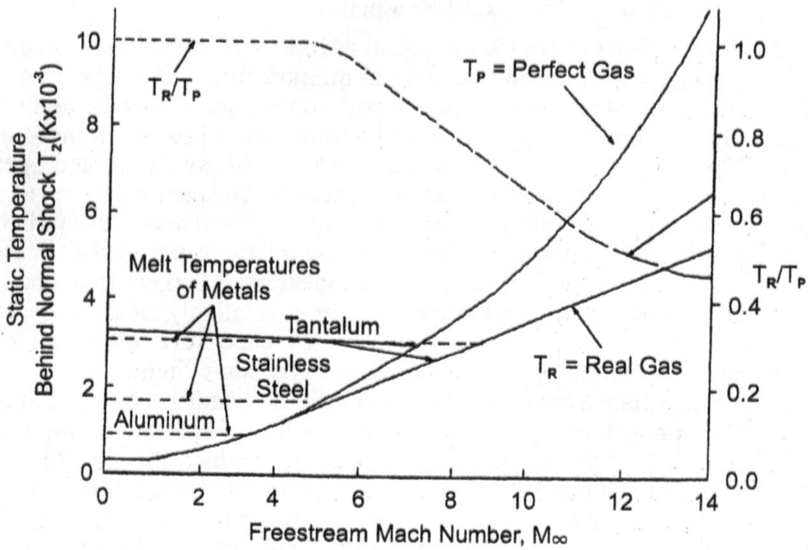

Fig. 7.1 Temperature behind a normal shock as a function of freestream Mach number (H = 170,000 ft).

speed of sound is approximately 1100 ft/s and the freestream air temperature is approximately 283°K. The normal shock would occur in the vicinity immediately ahead of the blunted portion of a seeker that may be mounted on a missile nose to detect an incoming target. Note that temperatures of interest to tactical weapons aerodynamicists can be very high for high Mach number conditions assuming a perfect gas. Also shown on the figure are real gas results taken from Ref. 1. Note, in particular, the plot of T_R/T_P, the ratio of the real gas to perfect gas temperature. For Mach numbers of 6 or less, this ratio is unity or near unity. For this reason, aerodynamic computations below $M_\infty = 6$ could neglect real gas effects with little error. However, as one goes significantly above $M_\infty = 6$, errors in temperature using the real gas assumption can be quite large. This is of particular importance to materials and structures engineers who must design the system to withstand these temperatures. Also shown in Fig. 7.1 is the melting point of typical structural materials used in present-day missile design. The actual-use temperature is less than the melting-point temperature. For missiles that fly at any appreciable time above the maximum-use temperature of a given material, some form of active cooling or insulation would be required, meaning additional dead weight and, hence, less performance for the missile. It is, therefore, obvious that a reasonably accurate estimate of temperature is essential for the design of the missile guidance and control unit and the structure of the weapon.

Before we finish the discussion on the importance of real gas effects, it should be noted that pressure is also affected, but to a much lesser extent than temperature. One is led to the conclusion that if temperature is not of concern and aerodynamic forces and moments are of major interest, the assumption of a perfect gas can be made with reasonable validity for Mach numbers much higher than 6. The reason is that pressure is mainly governed by the kinetic energy of the flow and not the potential energy. On the other hand, the temperature is a function of both and, therefore, varies considerably due to the real gas effects. In fact, many of the hypersonic flow assumptions are based on the assumption of a thin shock layer and $M_\infty >> 1$, and reasonable results for aerodynamic forces can be obtained.

The temperatures shown in Fig. 7.1 are so-called inviscid temperatures. To obtain the temperatures that are needed to perform heat transfer analysis requires the local temperature at the wall. This means that the temperatures at the outer edge of the boundary layer must be used to compute the local surface temperature.

This chapter will describe the process to compute the temperature at an outer edge of the boundary layer where the Mach number is high enough for real gas effects to become important. The chapter will then describe some simple engineering approximations to carry this inviscid temperature to the wall so that the convective portion of the heat transfer information can be made available for those interested in performing approximate heat transfer analysis.

II. Real Gas Computational Procedure

The computational procedures discussed in this section will be limited to frozen or equilibrium chemically reacting flows. If one defines the characteristic length of the flowfield to be the body length, then the time it takes for a fluid element to traverse the shock wave and full body length is

$$t_1 \simeq l_B/V$$

Defining the time it takes for a chemical reaction to come into equilibrium as t_2, then if

$$t_1 >> t_2$$

the flow is truly in chemical equilibrium for all practical purposes. On the other hand, if

$$t_2 >> t_1$$

we may assume that the flow is "frozen" at the chemistry that exists just behind the shock. That is, we can compute the flow as if it were a perfect gas except we have a different value of $\gamma = \gamma_f$ behind the shock. Finally, if

$$t_1 \approx t_2$$

we have non-equilibrium flow, which means that the chemical processes are occurring and changing significantly along the body. This results in an order-of-magnitude more difficulty in making real gas computations. What can be done, however, from a more practical standpoint is to compute the local conditions assuming "frozen" and "equilibrium" flows and assume that the non-equilibrium case falls in between these two. That is, generally the equilibrium and frozen flow cases present outer boundaries for flow properties and the non-equilibrium case falls in between these outer boundaries.

Because the focus of the present work is to incorporate real gas effects to engineering methods, only equilibrium and frozen flows will be considered. The computations involved with non-equilibrium flow are inconsistent with an approximate mathematical model of the flowfield and are more consistent with solution of the full Euler or Navier–Stokes equations, discussed in Chapter 2.

The methods described in Chapter 4 for high Mach number, perfect gas computations were second-order shock expansion theory (SOSET) in combination with modified Newtonian theory (MNT), when the body had a blunt or truncated nose tip. The focus in that chapter was on obtaining accurate pressure distributions so that accurate forces and moments could be obtained for low angle of attack (AOA). Because that is the preferred method for a perfect gas, the question arises as to what modifications are necessary to allow real gas computations to be made for local surface properties. This includes temperature, in particular, in addition to pressure.

The major change in the computational procedure for real versus perfect gases has to do with the equation of state, Eq. (15a) from Chapter 2. For a perfect gas, the internal energy can be defined as the product of a constant times the single thermodynamic property, temperature [Eq. (15b) of Chapter 2]. However, for a real gas, the internal energy or any other thermodynamic property must be defined as a function of two other thermodynamic properties. If one limits his or her interest to air, the number of chemical reactions involved as the air is heated is quite limited. This is because air contains about 80% nitrogen, 20% oxygen, and some trace elements. Hence, as air is heated and the internal energy modes of vibration, dissociation, and ionization are excited, a maximum of 22 chemical reactions occur.

One approach to solve for the internal energy or any other thermodynamic property is to solve a set of 28 nonlinear algebraic equations at each point in the computational domain. This is a rather lengthy process. Tannehill and Mugge,[3] with later extension by Srinivasan et al.,[4] reduced this process to a series of curve fits of various thermodynamic variables through use of an effective $\bar{\gamma}$. Thus, knowing p and ρ at some point allows one to obtain the following properties:

$$\begin{aligned} \bar{\gamma} &= \bar{\gamma}(p, \rho) \\ h &= h(p, \rho, \bar{\gamma}) \\ e &= e(p, \rho, \bar{\gamma}) \\ T &= T(p, \rho) \\ a &= a(e, p) \end{aligned} \tag{2a}$$

Srinivasan et al.[4] gave additional curve fits of the properties

$$
\begin{aligned}
\rho &= \rho(h, p) \\
\rho &= \rho(p, T) \\
S &= S(e, \rho) \\
\rho &= \rho(p, S) \\
a &= a(p, S) \\
e &= e(p, S)
\end{aligned}
\tag{2b}
$$

Thus, through use of the curve-fit equations, knowing any two flow properties, the remaining properties can be determined in the flow. Knowing the properties given by Eq. (2), the only remaining properties are local velocity and compressibility factor. These can be determined from the constancy of total enthalpy relation for an equilibrium or frozen flow where

$$
H_O = \text{¢} = h_\infty + \frac{V_\infty^2}{2} = h_2 + \frac{V_2^2}{2} \tag{3}
$$

and

$$
Z = \frac{p}{\rho R T} \tag{4}
$$

An example of the use of curve-fit equations is given in Fig. 7.2. In this figure, local temperature is shown as a function of density and pressure ratios, all referenced to some set of standard conditions that could be the sea-level standard atmosphere, for example.

To compute pressure, density, and other thermodynamic properties for a real gas requires that properties be known across shock waves where high-temperature effects are present. An example of these high-temperature effects is shown in Fig. 7.1. Hence, the next several sections of this chapter will discuss how to obtain the properties across normal and oblique shock waves, expansion waves, and conical flows where the temperature is high enough for real gas effects to occur. All of these real gas properties will be needed for extending SOSET and MNT to real gases, which will be discussed later in the chapter.

III. Normal and Oblique Shock Waves in Real Gas Environments

To compute the flow over two- or three-dimensional configurations using the shock expansion theory requires an initial starting solution. This solution is provided by computing the flowfield variables across a normal (in the case of a blunt-nose body or blunt-leading-edge wing) or oblique shock for sharp-nosed bodies or wing leading edges. If the flow is low enough in Mach

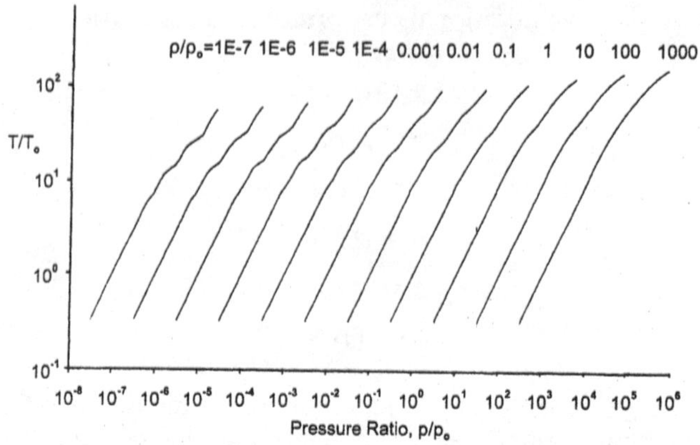

Fig. 7.2 Example of curve-fit equations for real gas equation of state[3,4]; $T = T\ (P,\ \rho)$.

number so that real gas effects are negligible or if one is only interested in approximate prediction of forces and moments, a perfect gas can be assumed, and the shock wave relations of Ref. 5 can be used directly. This last situation results from the fact that real gas effects have a fairly small effect on pressure (usually less than 10%) but can have a large effect on density and temperature. Because we are interested in temperature profiles along the body as inputs for structural and heat transfer analysis, we must consider both normal and oblique shock wave computations in real gases.

A. Normal Shock Waves

The conservation of mass, momentum, and energy—Eqs. (12), (13), and (14) of Chapter 2—for steady ($\partial/\partial t = 0$), adiabatic ($q = 0$) flow with no body forces ($f = 0$), reduce to the following for flow across a normal shock wave (see Fig. 7.3):

$$\rho_1 V_1 = \rho_2 V_2 \tag{5}$$

$$p_1 + \rho_1 V_1^2 = p_2 + \rho_2 V_2^2 \tag{6}$$

$$h_1 + \frac{V_1^2}{2} = h_2 + \frac{V_2^2}{2} \tag{7}$$

Solving Eqs. (6) and (7) for p_2 and h_2, while using Eq. (5) for V_2, there is obtained

$$p_2 = p_1 + \rho_1 V_1^2 (1 - \rho_1/\rho_2) \tag{8}$$

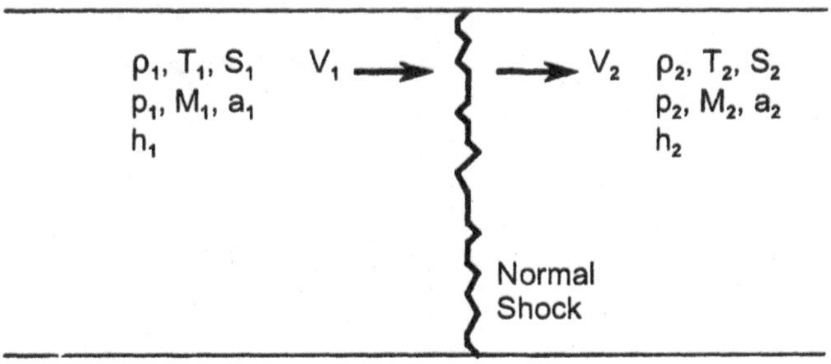

Fig. 7.3 Normal shock wave flow.

$$h_2 = h_1 + V_1^2/2\left[1 - (\rho_1/\rho_2)^2\right] \tag{9}$$

Note that all terms on the right-hand side of Eqs. (8) and (9) are known except for ρ_2. To solve Eqs. (8) and (9), one guesses a value of ρ_1/ρ_2. A value of 0 or 0.01 is adequate. Equations (8) and (9) are then solved for p_2 and h_2. Knowing p_2 and h_2, a new value of ρ_2 is computed from the equilibrium thermodynamic and chemistry model discussed in Section 7.2. New values of p_2 and h_2 are then computed and, once again, new values of ρ_2 computed from the equilibrium chemistry model. This process is repeated until the change in ρ_2 is small and within the desired accuracy for convergence. This then defines the correct values of p_2, h_2, and ρ_2, behind the shock. T_2, a_2, and e_2 can then be computed from Section 7.2 using the curve-fit procedure from Eq. (2a). The compressibility factor Z is defined as

$$Z = \frac{p}{\rho RT} \tag{10}$$

For a perfect or thermally perfect gas, Z is 1 but, for a real gas that undergoes chemical reactions, Z represents the level of dissociation that takes place. Because R is the gas constant for air, and p, ρ, and T are computed by Eqs. (8) and (9) and Section 7.2, Z can be calculated by Eq. (10). Also

$$V_2 = (\rho_1/\rho_2)\, V_1 \tag{11}$$

$$\gamma_2 = a_2^2\, \rho_2/p_2 \tag{12}$$

$$M_2 = V_2/a_2 \tag{13}$$

The normal shock solution for equilibrium air is given in Refs. 6 and 7. Figure 7.4 gives T_2/T_1 as a function of freestream pressure and velocity. Note the large difference between the perfect and real gas temperatures as temperature increases and as atmospheric pressures decrease (higher altitudes).

For the flowfield over a blunt body, properties vary between shock and

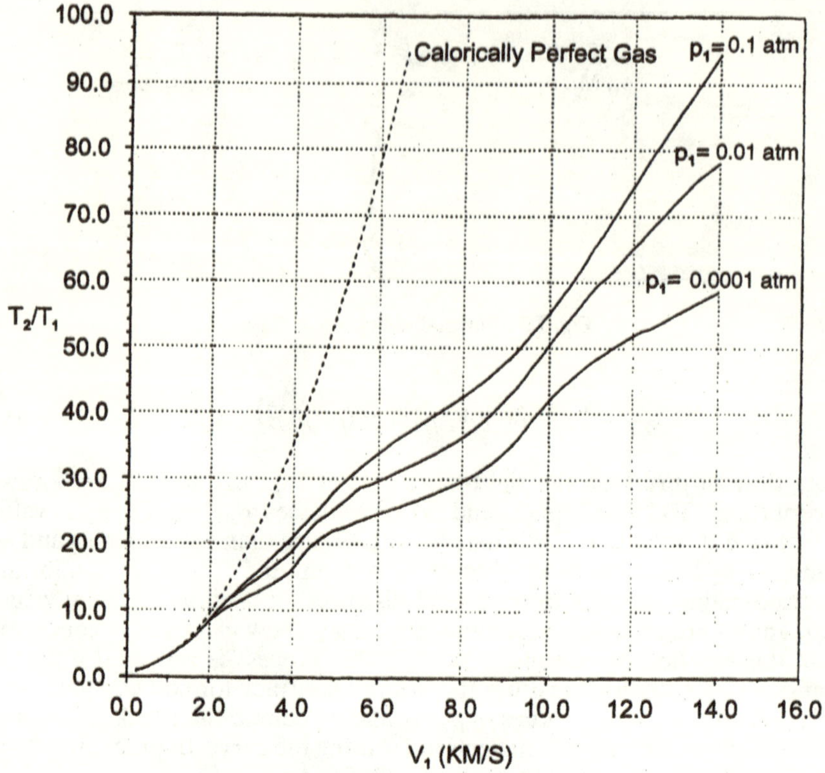

Fig. 7.4 Influence of freestream pressure and velocity on normal shock temperature ratio for equilibrium air (T_1 = 225°K).

body. Hence, frozen flow would consist of computing equilibrium air properties immediately behind the shock and then treating the flowfield as a perfect gas thereafter, with constant values of $\gamma = \gamma_2$ and compressibility factor $Z = Z_2$. Equilibrium chemically reacting flow would allow different values of these parameters as a function of distance between the body and shock. However, if one is only interested in properties along the body surface from a method such as MNT combined with SOSET, only the stagnation streamline that wets the body surface needs to be examined. Wittliff and Curtis[6] showed that, along the stagnation streamline, γ and Z change very little. Hence, effectively, the flow is frozen at values that exist behind the shock (within 1 or 2%) between the shock and body stagnation point. This assumption will be made here as well.

B. Oblique Shock Waves: Two-Dimensional or Wedge Flows

Examining Fig. 7.5, it soon becomes clear that one can use the previous normal shock relations of Eqs. (5) through (9) to solve for flow parameters

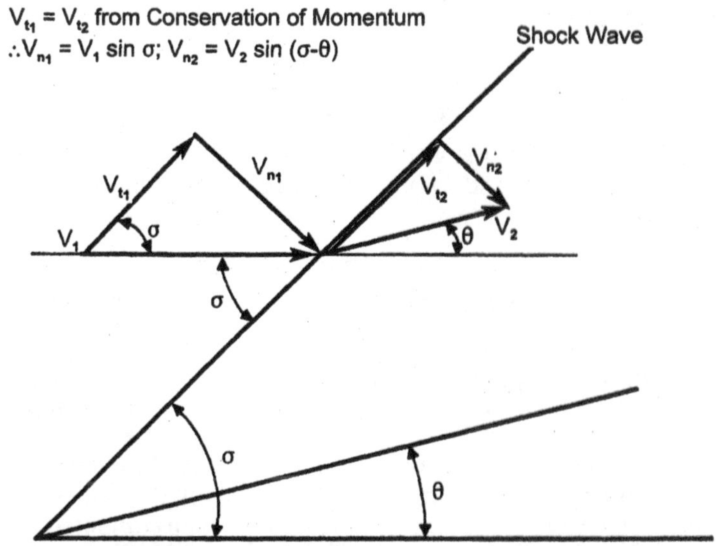

Fig. 7.5 Oblique shock geometry for wedge flow with attached shock.

behind an oblique shock if V_1 and V_2 components of flow velocity normal to the shock wave are replaced with their equivalent components normal to the oblique shock wave. That is,

$$V_{n_1} = V_1 \sin \sigma \tag{14}$$

$$V_{n_2} = V_2 \sin(\sigma - \theta) \tag{15}$$

Here, θ is the flow direction, which, for attached flow on a wedge, is constant and equal to the wedge angle. Equations (5), (8), and (9) now become, with these substitutions,

$$\rho_1 V_1 \sin \sigma = \rho_2 V_2 \sin (\sigma - \theta) \tag{16}$$

$$p_2 = p_1 + \rho_1 V_1^2 \sin^2 \sigma (1 - \rho_1/\rho_2) \tag{17}$$

$$h_2 = h_1 + \frac{V_1^2 \sin^2 \sigma}{2}\left[1 - (\rho_1/\rho_2)^2\right] \tag{18}$$

Also, because $V_{t_1} = V_{t_2}$ from the conservation of momentum, then

$$V_1 \cos \sigma = V_2 \cos (\sigma - \theta)$$

or

$$\frac{V_1}{V_2} = \frac{\cos(\sigma - \theta)}{\cos(\sigma)} \tag{19}$$

Combining Eq. (19) with Eq. (16), there is obtained

$$\frac{\rho_1}{\rho_2} = \frac{\tan(\sigma - \theta)}{\tan \sigma} \tag{20}$$

Expanding tan $(\sigma - \theta)$ using a trigonometric identity, going through some algebraic manipulation, and taking the weak shock solution, one obtains from Eq. (20)

$$\tan \sigma = (\rho_2/\rho_1 - 1)/(2 \tan \theta) - \sqrt{\left(1 - \frac{\rho_2}{\rho_1}\right)^2 \frac{1}{4 \tan^2 \theta} - \frac{\rho_2}{\rho_1}} \tag{21}$$

To solve for the flowfield properties, a value of ρ_1/ρ_2 is once again guessed. The shock angle is computed from Eq. (21). With this value of σ and ρ_1/ρ_2 known, p_2 and h_2 are computed from Eqs. (17) and (18). A new value of ρ_2 is computed from the process of Section 7.2 and an iteration takes place until successive changes in ρ_2 are within some error bound. In this way, the values of σ, p_2, ρ_2, and h_2 behind the shock are defined. T_2 and a_2 are then computed from Section 7.2, and Z, V_2 from Eqs. (10) and (19). Local Mach number and the isentropic exponent, γ, are then computed from Eqs. (12) and (13).

It should be pointed out that the isentropic exponent, defined as

$$\gamma = \left[\frac{\partial(\ell n\, p)}{\partial(\ell n\, \rho)}\right]_s$$

is the same as the gamma of Eq. (12). This is different from the "equivalent" gamma defined in Ref. 3 as

$$\bar{\gamma} = \frac{h}{e}$$

Here, the word "equivalent" in quotes is used because $\gamma = \bar{\gamma}$ for a perfect gas. However, for a real gas where vibration and or dissociation are present, the isentropic exponent used in true speed of sound computations is different from the "equivalent" gamma.

Because the body is a wedge, flow properties behind the straight shock are constant throughout the flowfield. Hence, pressure p_2 is also the pressure on the surface, which can be used to compute forces and moments. Also, frozen and equilibrium chemically reacting flows aft of the shock wave give the same results because all properties are constant between the body and shock. Figure 7.6 gives the results of the shock angle versus wedge half-angle at an altitude of 100,000 ft for two velocities for a perfect and equilibrium chemically reacting gas. Note the small change between the perfect gas and real gas at small flow deflections and Mach numbers. However, as flow deflection and Mach number increase, the difference between the real and perfect gas shock angle increases.

Figures 7.7 through 7.9 show the temperature, density, and pressure for the same 100,000 altitude condition at two velocities as a function of wedge angle.

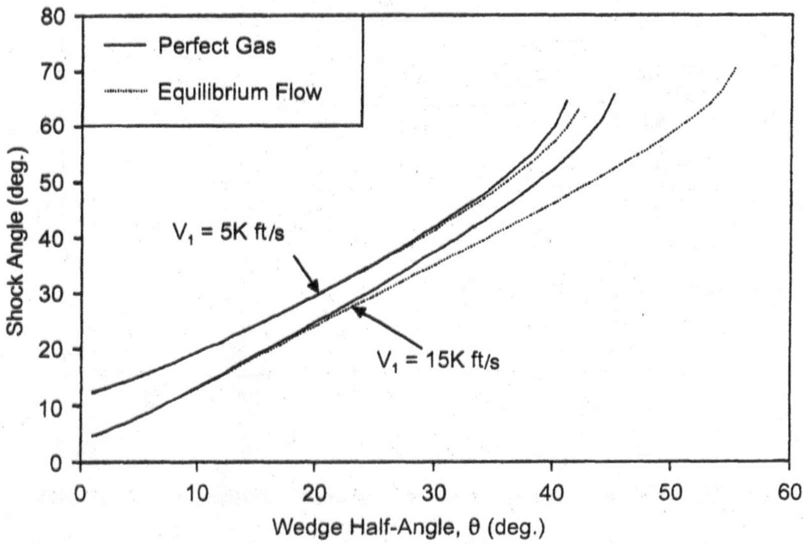

Fig. 7.6 Deflection angle/wave angle velocity diagram for oblique shocks (altitude = 100,000 ft).

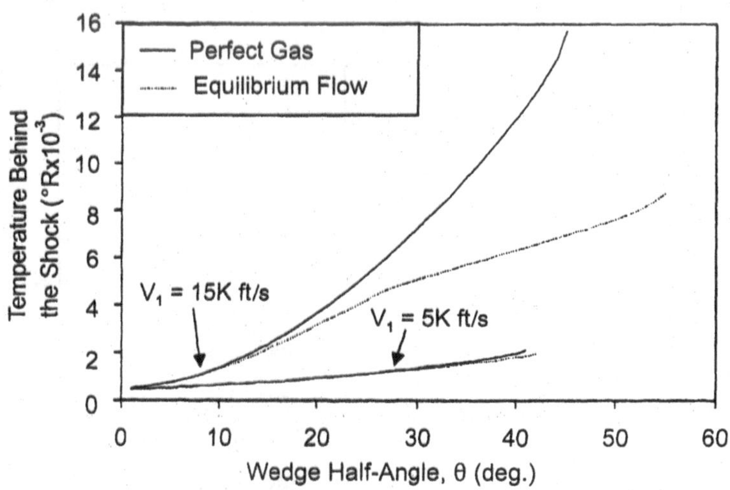

Fig. 7.7 Temperature in the shock layer for a wedge (altitude = 100,000 ft).

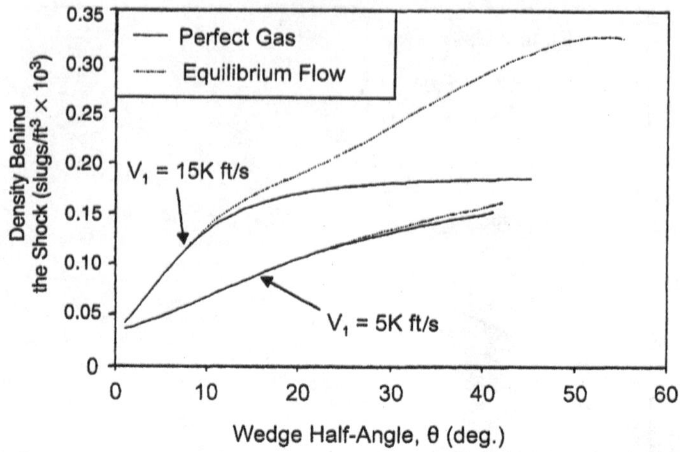

Fig. 7.8 Density in the shock layer for a wedge (altitude = 100,000 ft).

Note that the major effect of real gases is in the temperature and density at higher velocities and large wedge angles. The pressure is only affected slightly.

C. Oblique Shock Waves: Axisymmetric Conical Flows

The flowfield between the shock and conical surface is not constant here as was the case with the wedge. However, flow properties are constant along rays emanating from the conical tip for a perfect, frozen, or equilibrium chemically reacting gas. Furthermore, entropy is constant throughout the flowfield between the shock and body for these flow types.

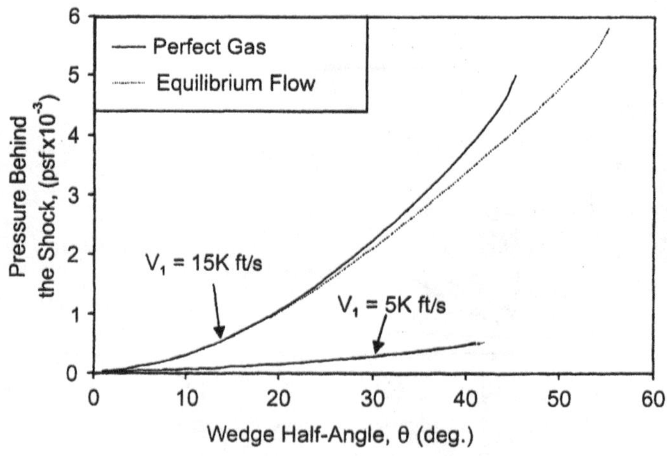

Fig. 7.9 Pressure in the shock layer for a wedge (altitude = 100,000 ft).

Continuity and momentum equations in spherical coordinates for this type of flow become[8]

$$\frac{dp}{d\theta} = -\rho V_\theta \left(\frac{a^2}{V_\theta^2 - a^2} \right) (V_r + V_\theta \cot \theta) \tag{22}$$

$$\frac{dV_\theta}{d\theta} = V_\theta \tag{23}$$

$$\frac{dV_\theta}{d\theta} = \left(\frac{a^2}{V_\theta^2 - a^2} \right) \left(2V_r + V_\theta \cot \theta - \frac{V_\theta^2 V_r}{a^2} \right) \tag{24}$$

The boundary condition on the body surface requires $V_\theta = 0$ there. Figure 7.10 shows the geometry and nomenclature used for conical flow. These equations represent a set of three first-order ordinary differential equations for five unknowns. The remaining two equations come from the high-temperature curve-fit process of Section 7.2:

$$\rho = \rho\,(p,S) \tag{25}$$

$$a = a\,(p,S) \tag{26}$$

1. Frozen Flow

For frozen flow, the equilibrium chemically reacting flow properties are computed across the shock instantaneously, and the values of γ and Z are set

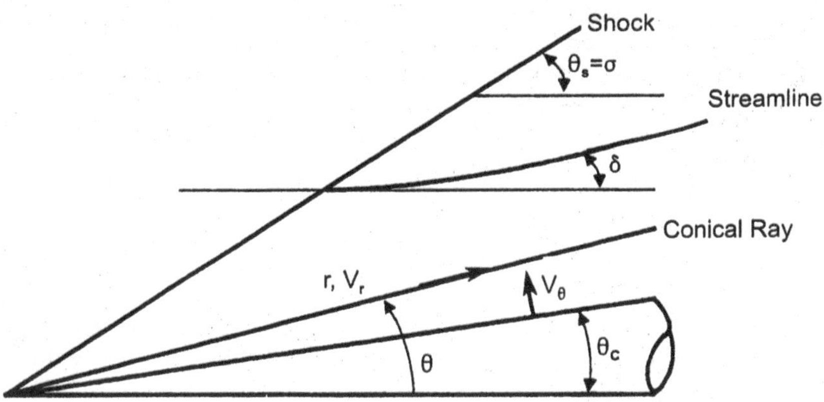

Fig. 7.10 Oblique shock geometry for a cone (spherical coordinates).

equal to their values just downstream of the shock. Also, because the flow angle θ behind the shock is not constant and not equal to the cone angle, as was the case for a wedge, Eq. (21) is not the most appropriate equation to use in the solution process because this involves three unknowns. Using Eqs. (17) and (18) instead requires only a double iteration on σ and ρ_1/ρ_2 to solve Eqs. (22) through (24) from the body to the shock. The solution proceeds as follows:

1) Guess values of σ and ρ_1/ρ_2. Perfect gas values are adequate.
2) Solve Eqs. (17) and (18) for p_2, h_2.
3) Solve the appropriate form of Eq. (2) for a new value of ρ_2 [i.e., $\rho_2 = \rho_2\ (p_2, h_2)$ or $\rho_2 = \rho_2\ (p_2, S)$ if S is given from blunt body flow].
4) Repeat steps 2 and 3 until successive changes in ρ_2 are within some acceptable tolerance. This defines the values of ρ_2, p_2, and h_2 behind the shock for the guessed angle σ.
5) Calculate values of a_2 and T_2 behind the shock from appropriate thermodynamic curve-fit equations and a value of $Z = Z_f$ and $\gamma = \gamma_f$ from

$$Z_f = \frac{p_2}{\rho_2 R T_2}$$

and

$$\gamma_f = a_2^2 \frac{\rho_2}{p_2}$$

The values of γ_f and Z_f are fixed at these values for the solution of Eqs. (22) through (24).

(Recall from the wedge flow computations that the values of $\bar{\gamma}$ and γ are different. The question arises as to which to use for frozen flow. The answer is that either can be used, or for that matter, other options for frozen flow could be used. However, to be consistent with the use of total temperature, pressure, and density relationships along the body where the flow is in fact isentropic and the isentropic exponent γ is used, the isentropic exponent γ is the one that will be used for frozen flow computations.)
6) Define initial values of V_r and V_θ behind the shock from the momentum and energy equations. That is,

$$(V_r)_2 = V_1 \cos\theta \tag{27}$$

$$(V_\theta)_2 = -\left[V_{\max}^2 - 2h_2 - (V_r)_2^2\right]^{1/2} \tag{28}$$

7) Numerically integrate Eqs. (22) through (24) from the shock, where $\theta = \sigma$, to the body, where $\theta = \theta_c$, using the fourth-order Runge–Kutta integration method. During the integration, new values of p, V_r, and V_θ are computed at each $\Delta\theta$ interval (note that θ is decreasing). At each interval, new values of a^2 and ρ must be computed based on the values of p, V_θ, and V_r that are

obtained from the solution of the Eqs. (22) through (24). These new values are

$$a^2 = \left(\frac{\gamma_f - 1}{2}\right)(V_{max}^2 - V_r^2 - V_\theta^2) \tag{29}$$

$$\rho = \frac{\gamma_f p}{a^2} \tag{30}$$

8) At the body, check the value of V_θ. If it is not zero, guess a new value of σ.
9) Repeat steps 1 through 8 until V_θ is zero to some acceptable tolerance. Once this has occurred, the entire flowfield between the shock and body is defined. Only the values of properties at the body surface are saved as they are all that will be used in the shock expansion theory.
10) Calculate the entropy behind the shock from the appropriate thermofit Eq. (2b). This value of entropy will be used later in the shock expansion theory. If entropy values are known from the blunt body stagnation solution, this step can be omitted.

2. *Equilibrium Chemically Reacting Flow*

For equilibrium chemically reacting flows, the computation process illustrated for frozen flows is used with two exceptions:

1) The values γ and Z are allowed to vary between the body and shock.
2) The values of "a" and "ρ" at each point in the computation process are computed via thermofit Eq. (2) based on the latest computed value of "p" and the value of "S,". which is constant behind the shock. That is,

$$a = a(p, S)$$

$$\rho = \rho(p, S) \tag{31}$$

Both of these relations are given in Ref. 4 and they replace the equations for "ρ" and "a" in the frozen flow computations discussed previously.

Knowing the values of V_θ, V_r, S, p, a, and ρ during the solution process of Eqs. (22) through (24), the remaining thermodynamic properties can be computed as follows:

$$h = (V_{max}^2 - V_\theta^2 - V_r^2)/2 \tag{32}$$

$$e = h - p / \rho \tag{33}$$

$$M = \frac{\left[V_\theta^2 + V_r^2\right]^{1/2}}{a} \tag{34}$$

$$T = T(p, S) \tag{35}$$

$$Z = \frac{p}{\rho RT} \tag{36}$$

$$\bar{\gamma} = h/e\,;\ \gamma = \frac{a^2\rho}{p} \tag{37}$$

To illustrate the effects of real gas computations on conical shapes, Figs. 7.11 through 7.14 have been prepared. These figures duplicate the results of Ref. 8 within numerical accuracy. Results here are similar to those of the wedge case presented previously for the chemical equilibrium cases. The trends of decreased shock angle, higher temperatures, lower densities, and near constant pressures of real gases compared to perfect gases holds true for the cone also. In particular, note the almost constant pressure of Fig. 7.12 between real and perfect gases. This is what allows one to use perfect gas solutions to approximate forces and moments on configurations that are in a real gas environment. These trends for conical flow will not all be true for Prandtl–Meyer expansion flows considered in the next section of this chapter.

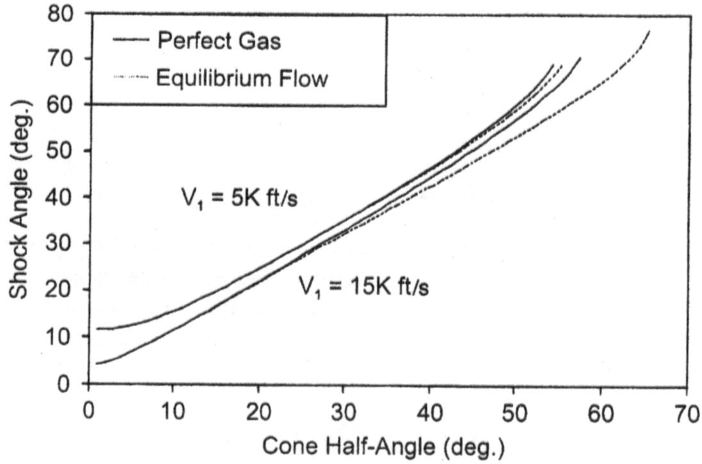

Fig. 7.11 Effect of freestream velocity and cone angle on cone shock wave angle (H = 100,000 ft).

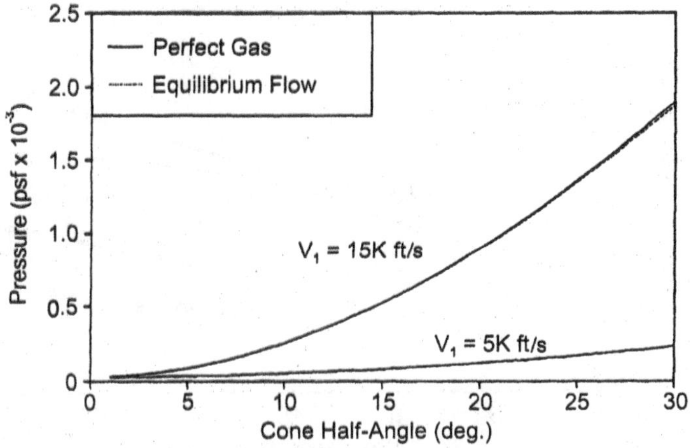

Fig. 7.12 Effect of freestream velocity and cone half-angle on cone surface pressure (H = 100,000 ft).

IV. Computation of Properties Across Expansion Waves in Real Gas Environments

Prandtl–Meyer expansion (PME) flow is standard in the aerodynamics literature (see, for example, Refs. 9 through 12). For convenience, a brief summary of the theory and results is presented in this section.

The PME consists of an isentropic turning of a streamline from its initial conditions at state 1 to its final conditions at state 2 (see Fig. 7.15). Note that, while isentropic (for equilibrium air), this process applies from streamline to streamline, and each streamline can have different conditions. Referring to

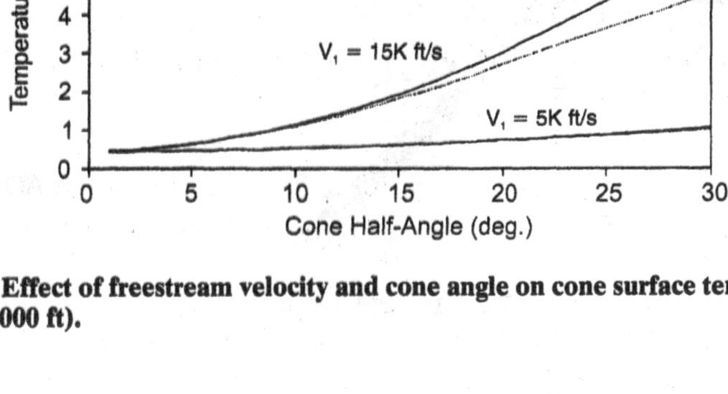

Fig. 7.13 Effect of freestream velocity and cone angle on cone surface temperature (H = 100,000 ft).

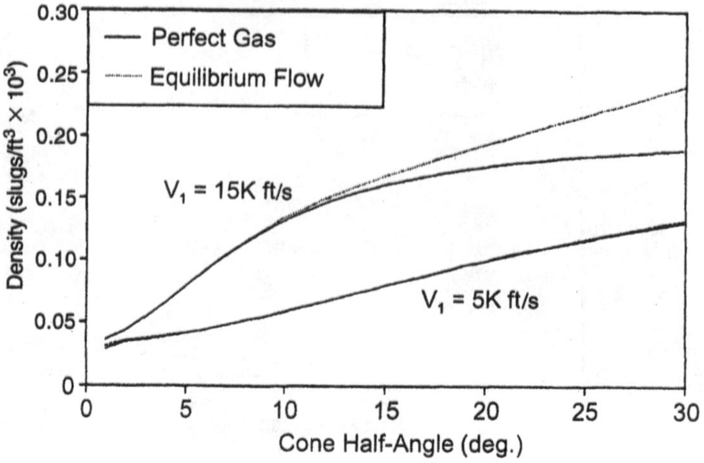

Fig. 7.14 Effect of freestream velocity and cone angle on cone surface density (H = 100,000 ft).

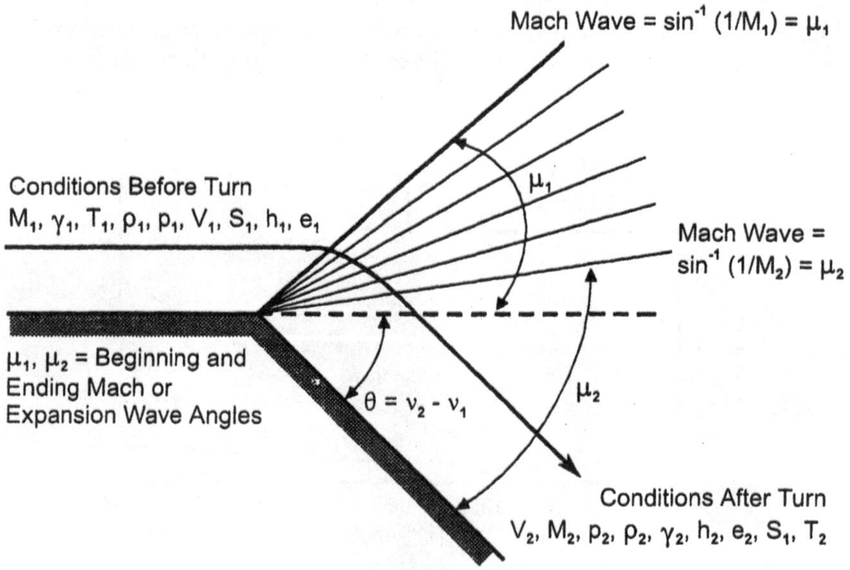

Fig. 7.15 PME process.

Fig. 7.15, this means that we do not necessarily have constant conditions in the entire flowfield leading up to the turn. Indeed, in general, conditions are not constant. However, on the body surface, because it is considered a streamline, the flow is isentropic for our flowfield. The Mach waves are actually characteristic lines along which no flow information is transmitted upstream. Of course, as conditions from streamline to streamline change, the Mach lines or characteristics will curve to accommodate these changes. If conditions ahead of the turn are constant throughout the flowfield, the characteristics will be straight.

Another important feature of PME flows is that the velocity parallel or tangential to the Mach line will be constant on either side. This comes from the conservation of mass and momentum equations used in the derivation of the equations for flow across shock waves. This result is independent of type of flow (i.e., shock or expansion wave). Looking at an incremental change in the velocity as the flow turns around a corner, we can derive the equations for PME flows. Referring to the bottom sketch of Fig. 7.16, we can write the following from geometry:

$$\mathrm{d}V_n \cos\mu = (V + \mathrm{d}V)(-\mathrm{d}\theta) \simeq -V\mathrm{d}\theta \tag{38}$$

$$\sin\mu = \frac{\mathrm{d}V}{\mathrm{d}V_n} \tag{39}$$

The figure shows μ_1, but the process applies to any portion of the expansion, so the subscripts have been dropped in the analysis. Substituting Eq. (39) into (7.38), we get

$$\mathrm{d}V\frac{\cos\mu}{\sin\mu} = -V\mathrm{d}\theta$$

or

$$-\mathrm{d}\theta = \cot\mu\,\frac{\mathrm{d}V}{V} \tag{40}$$

But, because

$$\sin\mu = \frac{1}{M},$$

then

$$\cot\mu = \sqrt{M^2 - 1}$$

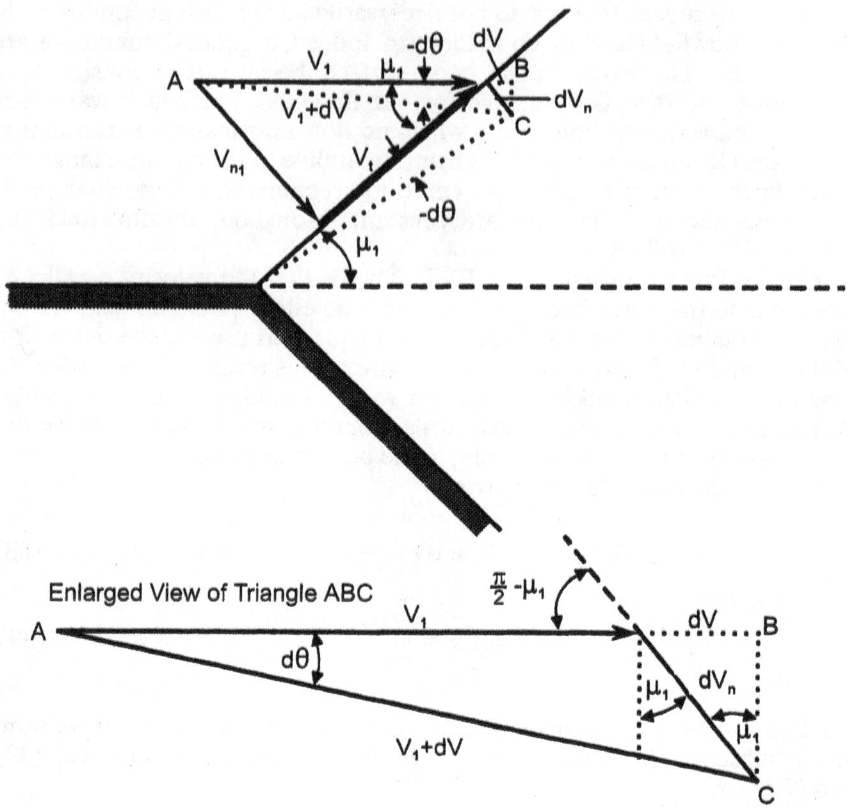

Fig. 7.16 Geometry used in derivation of Prandtl–Meyer equation.

Equation (40) becomes

$$-\mathrm{d}\theta = \sqrt{M^2 - 1}\,\frac{\mathrm{d}V}{V} \tag{41}$$

From the adiabatic energy equation, one can write

$$H_o = \not{C} = h + \frac{V^2}{2} \tag{42}$$

Differentiating Eq. (42), solving for d V/V, we obtain

$$\frac{\mathrm{d}V}{V} = \frac{\mathrm{d}(H_o - h)}{2(H_o - h)} \tag{43}$$

Also,

$$M = \frac{V}{a} = \frac{\sqrt{2(H_o - h)}}{a} \tag{44}$$

Substituting Eqs. (43) and (44) into (41), the general equation for PME flows is obtained

$$-d\theta = \frac{1}{2(H_o - h)} \left[\frac{2(H_o - h)}{a^2} - 1 \right]^{1/2} d(H_o - h) \tag{45}$$

This equation is quite general, as the only assumption is isentropic flow along a streamline. It, therefore, could apply to an isentropic expansion or compression and to perfect, frozen, or equilibrium chemically reacting flows.

For a perfect gas, γ is constant and equal to 1.4 for air. Equation (45) integrates to that defined by Eqs. (17) and (18) of Chapter 4. For frozen flow, the process is quite similar to a perfect gas except γ is frozen at some constant value. The initial values for γ_f and Z_f come from the conditions behind a normal shock if the body is blunt or behind an oblique shock if the body has a sharp tip. Once values of γ_f and Z_f are known behind the shock, then the rest of the flow can be basically solved as a perfect gas. That is, Eqs. (17) and (18) from Chapter 4 are used where $\gamma = \gamma_f$. Knowing the value of Mach number M_2 behind a turn from Eqs. (17) and (18) from Chapter 4, remaining flow properties can be computed from the following thermodynamic relationships:

$$T_2 = T_1\left(1 + \frac{\gamma_f - 1}{2} M_1^2\right) \Big/ \left(1 + \frac{\gamma_f - 1}{2} M_2^2\right) \tag{46}$$

$$p_2 = p_1\left(1 + \frac{\gamma_f - 1}{2} M_1^2\right)^{\frac{\gamma_f}{\gamma_f - 1}} \Big/ \left(1 + \frac{\gamma_f - 1}{2} M_2^2\right)^{\frac{\gamma_f}{\gamma_f - 1}} \tag{47}$$

$$\rho_2 = p_2 /(RZT_2) \tag{48}$$

$$a_2 = \sqrt{\frac{\gamma_f p_2}{\rho_2}} \tag{49}$$

$$V_2 = M_2 a_2 \tag{50}$$

$$h_2 = \frac{V_{\max}^2 - V_2^2}{2} \tag{51}$$

$$e_2 = h_2 - p_2/\rho_2 \tag{52}$$

For the case of equilibrium chemically reacting flows where the ratio of specific heats and compressibility factor can change as flow expands around the surface, one can no longer assume a perfect gas and integrate Eq. (45) in closed form. Equation (45) must be solved through numerical integration. In differential equation form, Eq. (45) becomes

$$\frac{d(H_o - h)}{d\theta} + \frac{2(H_o - h)}{\left[\frac{2(H_o - h)}{a^2} - 1\right]^{1/2}} = 0 \tag{53}$$

Once again, fourth-order Runge–Kutta, or other appropriate numerical integration schemes, can be used. The $\Delta\theta$ interval $(\theta_2 - \theta_1)$ is divided into a given number of equally spaced intervals. The initial values of H_o, h, θ_1, θ_2 (defined), a, M, p, ρ, and S are known. These all come from the conditions behind a normal shock or oblique shock, of course, the conditions behind the shock must be taken to the surface of the body and, in the case of a blunt body or leading edge, taken to the point where the PME process begins. For a blunt body, this process will be discussed in the MNT and wedge and cone sections in this chapter.

One positive aspect about using the fourth-order Runge–Kutta integration routine (as opposed to Simpson's rule or some other integration process where h_2 is required as an initial guess to start the integration process) is the fact that no initial guess of h_2 is necessary, nor any follow-on iteration to correct this guess based on the boundary conditions. Unfortunately, while Eq. (53) is given in a very straightforward manner for integration, $a(h, S)$ is not available from either Ref. 3 or 4. As a result, either a resort to tables[13] must be used for $a(h, S)$, or Eq. (53) must be put in a form compatible with the thermofit equations of Refs. 3 and 4. Because the thermofit equation process is much faster than referring to tables, and at only a small sacrifice in accuracy, another alternative was investigated for solving Eq. (3).

This approach uses the Prandtl–Meyer equation (41) in conjunction with the one-dimensional momentum equation:

$$dV = -\frac{dp}{\rho V} \tag{54}$$

Substituting Eq. (54) into (41), one obtains

$$\frac{dp}{d\theta} = \frac{\rho V^2}{\sqrt{\frac{V^2}{a^2} - 1}} \tag{55}$$

To solve Eq. (55), one knows initial values of p, θ, $\Delta\theta$, ρ, a, and V. Equation (55) is integrated using, again, fourth-order Runge–Kutta to obtain a value of p at the next integration interval. Knowing p and the fact that entropy is constant, updated values of

$$\left.\begin{aligned} \rho &= \rho(p, S) \\ a &= a(p, S) \\ e &= e(p, S) \end{aligned}\right\} \tag{56a}$$

can be computed from the thermofit equations of Ref. 4. Then,

$$h = e + p/\rho \tag{56b}$$

and

$$V = \sqrt{2(H_o - h)} \tag{56c}$$

Equations (56a), (56b), and (56c) are then used with Eq. (55) in the integration.

Figures 7.17 through 7.20 give pressure, temperature, density, and Mach number for perfect, equilibrium, and frozen flow cases. The initial conditions for the expansion were $T_1 = 6140$ deg, $M_1 = 1$, and $p_1 = 1.2$ atm. It is quite interesting to note that, while pressure is not affected to any appreciable extent in a compression process, compared to perfect gas computations, pressure in an expansion process is higher for equilibrium versus perfect gas computations (see Fig. 7.17). Also, the equilibrium temperature is higher in an expansion as opposed to lower in a compression compared to perfect gas computations (see Fig. 7.18). Note that density of equilibrium flows is nearly

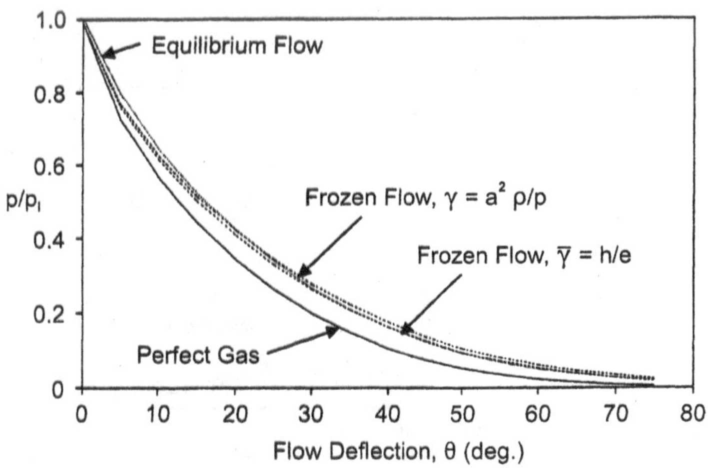

Fig. 7.17 PME flow: pressure versus flow deflection angle ($M_1 = 1.0$, $T_1 = 6140$°K, and $p_1 = 1.2$ atm).

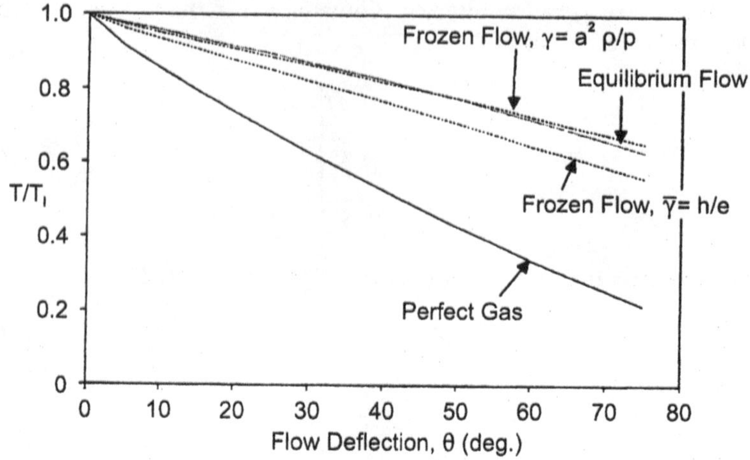

Fig. 7.18 PME flow: temperature versus flow deflection angle (M_1 = 1.0, T_1 = 6140°K, and p_1 = 1.2 atm).

the same as the perfect gas solution for an expansion (Fig. 7.19), whereas this density for an equilibrium compression is higher compared to a perfect gas solution. Finally, Mach number is lower for an equilibrium expansion compared to a perfect gas (Fig. 7.20). Frozen flow cases are also shown in Figs. 7.17 through 7.20 where the gas conditions that exist at the initial condition state are held constant throughout the expansion process. Two values of the initial state are held fixed corresponding to the isentropic exponent ($\gamma = a^2\rho \ / \ p$) and the so-called "equivalent" gamma ($\bar{\gamma} = h/e$). Note that, for the expansion process, there is very little difference between the two, which

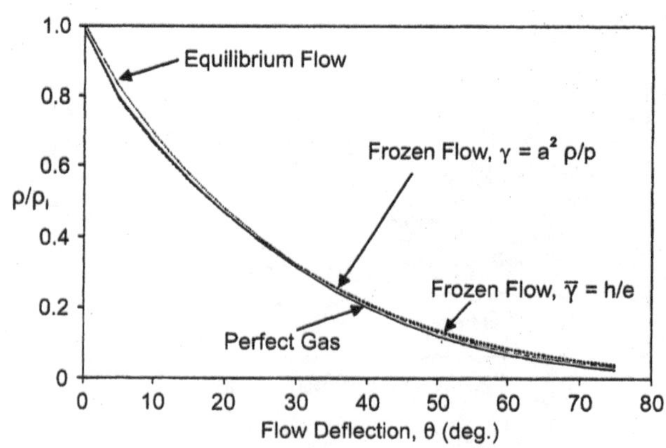

Fig. 7.19 PME flow: density versus flow deflection angle (M_1 = 1.0, T_1 = 6140°K, and p_1 = 1.2 atm).

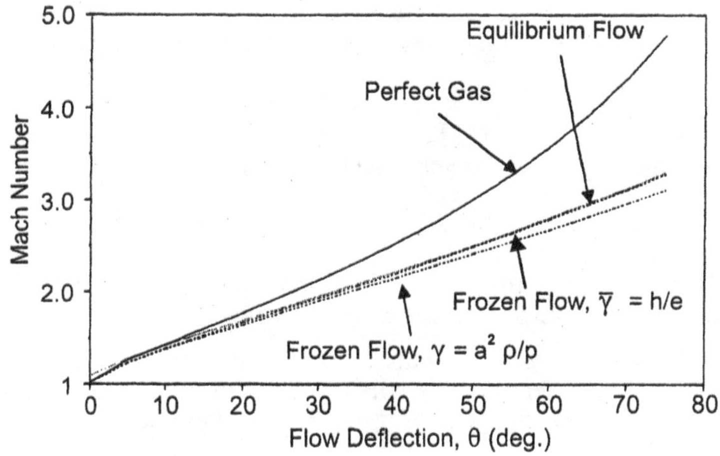

Fig. 7.20 PME flow: Mach number versus flow deflection angle (M_1 = 1.0, T_1 = 6140°K, and p_1 = 1.2 atm).

indicates that, for this set of initial conditions, results for both values of gamma are about the same. This may not necessarily be the case, however, had the frozen flow values of $\bar{\gamma}$ and γ been those that existed behind a normal shock on a blunt body.

V. Modified Newtonian Theory for Real Gases

Newtonian theory for perfect gases was discussed in Section 4.4. The pertinent equations given there will be repeated here for convenience. The pressure coefficient at any point on the blunt nose portion of the body is

$$C_P = C_{P_o} \sin^2 \delta_{eq} \tag{57}$$

where

$$\delta_{eq} = \sin^{-1} [\sin \theta \cos a - \sin a \cos \varphi \cos \theta] \tag{58}$$

$$C_{P_o} = \frac{p_{O_2} - p_\infty}{\frac{1}{2} \rho_\infty V_\infty^2} \tag{59}$$

See Fig. 4.2 for the angle δ, θ, and φ. For perfect gases, perfect gas normal shock wave relations yielded Eq. (32) from Chapter 4 for C_{P_o}. However, Eq. (32) from Chapter 4 no longer holds for real gas computations across normal or oblique shock waves. The values of thermodynamic properties computed by the Section 7.3 process are the appropriate values to use. We will thus assume we know the defining conditions behind the shock, that is,

$$\gamma_2, p_2, \rho_2, h_2, a_2, V_2, M_2, T_2, Z_2, S_2$$

from the process described in Section 7.3.

The question that must be addressed is twofold:

1) How do these parameters vary along the stagnation streamline between the body and shock?
2) How do these properties vary from the stagnation point around the body to the point where shock expansion theory is applied?

We will address these questions for both frozen flow and equilibrium chemically reacting flows where MNT is used on the blunt portion of a configuration for pressure, temperature, force, and moment computations.

A. Frozen Flow

For frozen flow, it is assumed that the γ and the compressibility factor are fixed at the values they have just behind the shock (i.e., $\gamma = \gamma_f, Z = Z_f$). It is also known that, along the stagnation streamline, the flow is isentropic for equilibrium flow conditions. Furthermore, it is known from many previous works (see, for example, Refs. 5, 13, and 14) that the Mach number behind a normal shock is much less than 1 for high freestream Mach numbers. Also, this Mach number must go to 0 at the stagnation point, which is only a slight distance away from the shock. This has led to approximate theories in the stagnation region based on constant density, temperature, and pressure, which, in effect, is the Newtonian assumption in the shock layer. That is, the shock is assumed to lie on the body and the conditions behind the shock are the same as those on the body surface. The physics of the flow between the shock and body is what allows one to assume γ_f and Z_f are in fact approximately constant for both equilibrium and frozen flow between the shock and body and along the stagnation streamline. Witcliff and Curtis[14] indicate a maximum 2.3% error in making this assumption and, in most cases, the error was much less.

A second assumption will also be made. This assumption will allow the total temperature, pressure, and density relationships to be used in a local sense, along the body where $\gamma = \gamma_f$ and M_∞ is the local value. Strictly speaking, these equations relating total to static conditions are derived based on perfect gas assumptions and with constant $\gamma = 1.4$. The accuracy of this assumption (combined with others) can be assessed in comparison with exact calculations over a blunt-nosed body by solution of full Euler equations. Such an assessment will be done later in this chapter.

Returning now to Eq. (57) and recognizing the fact that all the properties behind the shock are known from the real gas computations of the flow across a normal shock, one can write

$$C_{P_o} = \frac{p_{O_2} - p_\infty}{\frac{1}{2}\rho_\infty V_\infty^2} = \frac{2}{\gamma M_\infty^2}\left(\frac{p_{O_2}}{p_\infty} - 1\right) = \not{C} \tag{60}$$

where

$$p_{O_2} = p_2\left(1 + \frac{\gamma_f - 1}{2} M_2^2\right)^{\frac{\gamma_f}{\gamma_f - 1}} = \not{C} \tag{61}$$

along the stagnation streamline. Here p_2, ρ_2, h_2, V_2, a_2, M_2, γ_f, Z_f, are the values calculated immediately behind the shock from Section 7.3. Also,

$$H_o = h_2 + \frac{V_2^2}{2} \tag{62}$$

$$T_{O_2} = T_2\left(1 + \frac{\gamma_f - 1}{2} M_2^2\right) \tag{63}$$

$$\rho_{O_2} = \rho_2\left(1 + \frac{\gamma_f - 1}{2} M_2^2\right)^{\frac{1}{\gamma_f - 1}} \tag{64}$$

Now, because at the stagnation point local velocity and Mach number are zero, the static pressure, temperature, density, and enthalpy at the stagnation point are simply the local stagnation values behind the shock given by Eqs. (61) through (64). Thus, at the stagnation point, for frozen flow

$$\left.\begin{aligned}
\rho_S &= \rho_{O_2} \\
T_S &= T_{O_2} \\
h_S &= H_o \\
p_S &= p_{O_2} \\
M_S &= V_S = 0 \\
\gamma &= \gamma_f = \not{C} \\
\overline{C}_P &= \overline{C}_{pf} = \not{C} \\
\overline{C}_v &= \overline{C}_{vf} = \not{C} \\
Z_S &= Z_f = \not{C} \\
S &= S_2 = \not{C} \text{ along body}
\end{aligned}\right\} \tag{65}$$

To continue the computations around the blunt surface, an isentropic expansion is assumed and Eq. (57) is applied at each point on the surface where δ_{eq} is given by Eq. (58) and C_{P_o} is constant at the value given by Eq. (60). Knowing δ_{eq}, C_{P_o}, and C_P allows one to calculate the local pressure at a given point. That is,

$$p_L = p_\infty\left[1 + \frac{\gamma_\infty M_\infty^2}{2} C_P\right] \tag{66}$$

Because the flow is frozen, we can treat it just like a perfect gas except that the values of $\gamma = \gamma_f$ and $R_f = RZ_f$. Hence, knowing local pressure from Eq. (66), total pressure, temperature, density, and other flow properties behind the normal shock [Eqs. (61) to (64)], the fact that total conditions are constant along a streamline, the local conditions at any point on the blunt body can be calculated by

$$M_L = \left\{\left[\left(\frac{p_{O_2}}{p_L}\right)^{\frac{\gamma_f - 1}{\gamma_f}} - 1\right]\left(\frac{2}{\gamma_f - 1}\right)\right\}^{1/2} \tag{67}$$

$$T_L = T_{O_2} \Big/ \left(1 + \frac{\gamma_f - 1}{2} M_L^2\right) \tag{68}$$

$$\rho_L = \rho_{O_2} \Big/ \left(1 + \frac{\gamma_f - 1}{2} M_L^2\right)^{\frac{1}{\gamma_f - 1}} \tag{69}$$

$$a_L = \left(\frac{\gamma_f p_L}{\rho_L}\right)^{1/2} \tag{70}$$

$$V_L = M_L a_L \tag{71}$$

$$h_L = H_o - \frac{V_L^2}{2} \text{ and } e_L = h_L - p_L/\rho_L \tag{72}$$

B. Equilibrium Flow

Equilibrium chemically reacting flow properties at the stagnation point are computed just like the frozen flow case. This statement is based on the rationale discussed previously of only slight changes in the specific heats and compressibility factor between the body and shock. However, for isentropic expansion around the body, γ and Z are allowed to vary from the values at the stagnation point in contrast to the frozen flow case where these parameters are frozen at values behind the shock. Hence, for equilibrium chemically reacting flows, Eq. (65) defines the stagnation point conditions. Also, Eq. (66) defines the pressure at the next point around the body. To get the remaining properties, recourse is once again made to the real gas curve-fits of Section 7.2. To obtain ρ, we know local pressure and entropy so that

$$\rho_L = \rho(p,S) \tag{73}$$

Then, knowing pressure and density

$$h_L = h(p,\rho), T_L = T(p,\rho), a_L = a(p,\rho) \tag{74}$$

The remaining quantities can be computed from thermodynamic relations as follows:

$$V_L = \sqrt{2\,(H_o - h_L} \tag{75}$$

$$M_L = V_L/a_L \tag{76}$$

$$\gamma_L = a_L^2 \rho_L / p_L \tag{77}$$

$$e_L = h_L - p_L/\rho_L \tag{78}$$

$$Z_L = p_L / (\rho_L\, RT_L) \tag{79}$$

This process is continued around the body surface to the point where the shock expansion method begins. Again, Eqs. (66), (67), and (73) through (79) provide initial conditions for the shock expansion process.

VI. Second-Order Shock Expansion Theory for Real Gases

Recall that SOSET was defined in Chapter 4 by Eqs. (19) and (20), repeated here for convenience:

$$p = p_C - (p_C - p_2)\, e^{-\eta}$$

$$\eta = \frac{\left(\frac{\partial p}{\partial s}\right)_2 (s - s_2)}{p_C - p_2}$$

Referring back to Fig. 4.7, p_C is the tangent cone pressure on a conical segment of the body; p_2 is the pressure just behind an expansion corner; and s and s_2 are distances along the body surface and to the point 2, respectively.

We desire to develop the process to allow computations to proceed around the body using modified versions of Eqs. (19) and (20) from Chapter 4 for real gas effects. To do this requires the calculation of pressures on wedges and cones based on real gases (see Section 7.3), and pressures behind an expansion corner computed by the PME for real gases (see Section 7.4). If the configuration has a blunt nose tip, then the Section 7.5 process is used for the starting solution up to the match point where SOSET is applied. The match point is defined exactly as in Section 4.6 for perfect gases where $\theta_M = 25.95$ deg.

In essence, the SOSET for real gases is the same for perfect gases (see Chapter 4), except that thermodynamic properties are determined using the processes of Chapter 7, as opposed to those of Chapter 4, for perfect gases. There are two exceptions. First, Hudgins[8] found an approximate formula to estimate pressures on cones in high-temperature environments at zero AOA. This equation, in pressure coefficient form, is

$$\left(C_P\right)_{\alpha=0} = \frac{2}{\gamma_\infty M_\infty^2}\left[1.4932\ (M_\infty \sin\theta_C)^{1.9854} + 0.3017\right] \tag{80}$$

Hudgins[8] indicates accuracies of better than 1.5% for Eq. (80) over a wide range of M_∞ and θ_c. It should be pointed out that for values of $M_\infty \sin\theta_c \leq 1.5$, real gas effects on pressure are negligible and Eq. (16c) from Chapter 4 can be used instead of Eq. (80). For values of $M_\infty \sin\theta_c > 1.5$, Eq. (80) replaces Eq. (16c) from Chapter 4. For $\alpha > 0$, Eq. (16a) from Chapter 4 still applies where $(C_P)_{\alpha=0}$ is defined by either Eq. (80) or (16c) from Chapter 4, depending on the value of $M_\infty \sin\theta_c$.

The second exception for real versus perfect gases involves the pressure gradient $(\partial p/\partial s)_2$. Moore et al.[2] went through an extensive derivation process to develop a formula for the pressure gradient behind a corner where the perfect gas assumption was not made. The detailed derivation will not be repeated here, but the steps in the process of Ref. 2 will be listed. These steps consisted of

1) Conversion of Euler equations from rectangular to streamline coordinates and derivation of characteristic equations.
2) Derivation of the pressure change along a left running characteristic (or Mach line) because this is the mechanism for differential pressure change along the surface.
3) For AOA computations, a method for accurately computing Δp and ΔT due to α.
4) An algebraic approximation as a solution to an ordinary differential equation.

The result of this process was Eq. (24) from Chapter 4. This equation can be used for perfect or real gas computations. Equation (24) from Chapter 4 is a function of properties upstream of the Prandtl–Meyer turn and properties downstream of the turn, which must be defined by the method of Section 7.4.

To apply SOSET, two cases will be considered and the process will be summarized for each. These two cases are pointed and blunt bodies at AOA.

For a pointed body at $\alpha > 0$,

1) Use Eq. (80) to compute $(C_P)_{\alpha=0}$, if $M_\infty \sin\theta_c > 1.5$, or Eq. (16c) from Chapter 4, if $M_\infty \sin\theta_c \leq 1.5$.
2) Use Eq. (16a) from Chapter 4 to compute $C_P(\alpha, \theta, \varphi, M)$ at the nose tip.
3) Use Eq. (80) to solve for a new value of $\theta_C = \theta_{eff}$ based on the value of C_P at $\varphi = 180$ deg (windward plane) computed from step 2.

4) Use this value of cone half-angle ($\theta_C = \theta_{eff}$) to solve for the real gas cone solution at $a = 0$ from Section 7.3.3. This gives the value of entropy in the windward plane, along with temperature and other thermodynamic properties in the windward plane. Entropy remains constant all around the body at this value determined in the windward plane.
5) Knowing entropy, S, and C_P (a, θ, φ, M) at any point around the body from Eq. (16a) from Chapter 4 and Eq. (7.80), other thermodynamic properties around the body can be determined from the thermofit curve-fits of Refs. 3 and 4 and described in Section 7.2. These properties are

$$\left.\begin{aligned} T(\theta, \varphi, a) &= T\left[p(\theta, \varphi, a), S\right] \\ \rho(\theta, \varphi, a) &= \rho\left[p(\theta, \varphi, a), S\right] \\ e(\theta, \varphi, a) &= e\left[p(\theta, \varphi, a), S\right] \\ h &= e + p/\rho \\ V &= \left[2(H - h)\right]^{1/2} \\ M &= V/a \\ \gamma &= \frac{a^2 \rho}{p} \\ Z &= p/(\rho RT) \end{aligned}\right\} \tag{81}$$

6) The properties from step 5 are advanced along the body by some amount $\Delta\theta$ through the PME of Section 7.4. The PME is used at each point around the body (i.e., every φ plane). This defines all the properties at point 2 shown in Fig. 4.7.
7) Compute pressures along the conical segment of the body to second-order accuracy by Eqs. (19), (20), and (24) from Chapter 4 if Eq. (20) from Chapter 4 is positive. If Eq. (20) from Chapter 4 is negative, SOSET reverts back to either first-order shock expansion theory or tangent cone theory. Equation (25) from Chapter 4 can then be used with the value η_1 specified by the user to determine whether shock expansion or tangent cone theory or a blend of the two is used.

For a blunt body at $a > 0$,

1) Use Section 7.5 to determine the values of the thermodynamic properties all along and around the nose tip up to the match point, which is defined as $\theta_M = 25.95$ deg. This defines conditions at point I of Fig. 4.7.
2) Repeat steps 6 and 7 of the pointed body solution for $a > 0$, except in the leeward plane area ($\varphi \leq 30$ deg) where Eq. (16d) replaces (16a), both from Chapter 4.

As examples of the application of the SOSET and MNT extended to real gases, a couple of cases will be considered. The first case is a 20% blunt cone (Fig. 7.21a) with a cone half-angle of 15 deg. Figures 7.21b through 7.21e

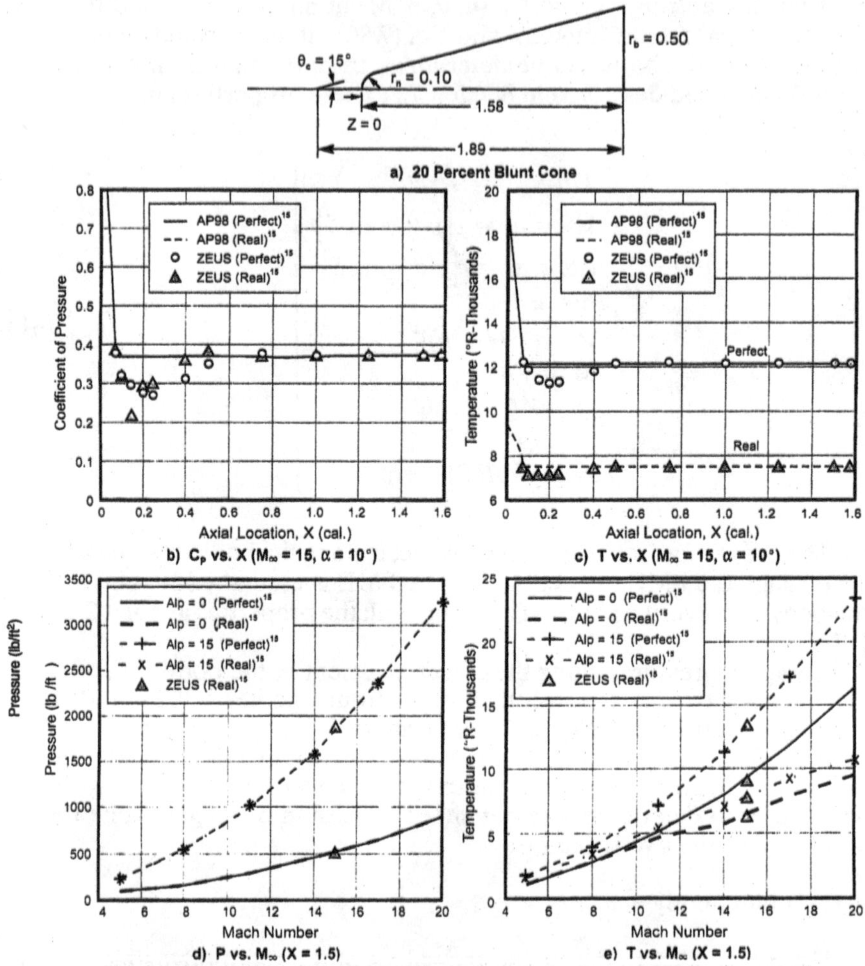

Fig. 7.21 Comparison of real and perfect gas results for temperature and pressure on a 20% blunt cone (h = 100,000 ft, and φ = 180°). a) 20% blunt cone, b) C_p vs X (M_∞ = 15 and α = 10°), c) T vs X (M_∞ = 15 and α = 10°), d) P vs M_∞ (X = 1.5), e) T vs M_∞ (X = 1.5)

illustrate the results of the current method (shown as Ref.[16]) compared to that of an exact Euler solution.[15] Results shown are perfect and real gas pressure coefficients and temperatures in the windward plane at $M_\infty = 15$ and $\alpha = 10$ deg (see Figs. 7.2b and 7.2c). Notice that, except for the overexpansion region where the blunt nose joins the cone, the approximate methods of this chapter agree quite well with the exact Euler solutions of Ref. 15. Also notice that the pressure coefficients of real and perfect gas are almost identical, which will lead to forces and moments being nearly identical. Also note that the real and perfect gas temperatures in Fig. 7.21c differ by almost a factor of two. This can have significant effect on aeroheating, which will be discussed in Section 7.7 of this chapter.

Figure 7.21d and 7.21e illustrate the comparison of real and perfect gas computations using the approximate methods of this chapter at an x station of 1.5. This value of x is located near the base of the blunt cone. Also shown are Euler[15] results for $M_\infty = 15$. Figure 7.21d illustrates the fact that pressure is little affected by the real gas process, all the way to $M_\infty = 20$. Figure 7.21e illustrates that real gas temperature effects begin around $M_\infty = 6$ and increase as Mach number increases. All of Fig. 7.21 was at the altitude of 100,000 ft.

A second example is a cone–cylinder–flare case shown in Fig. 7.22a. It has a 10-deg cone half angle and a 7-deg flare half angle. The pressure as a function of position around the body is given in Fig. 7.22b for three x locations: $x = 2$ is on the cone, $x = 4$ is on the cylinder, and $x = 10$ is at the base of the flare. Both perfect and equilibrium gas pressures are shown, where calculations are performed based on methods presented in this chapter. Very little difference between the two is seen. Figure 7.22c gives the temperature in the windward plane all along the body for both perfect and real gas computations. Note that, for pointed cone surfaces, real gas temperatures differ from perfect gas results by only 10 to 15% for this set of conditions versus a factor of 2 for similar conditions when the nose was blunt. Figures 7.22d and 7.22e give temperature around the cone at $x = 2$ and the cylinder at $x = 4$ for both perfect and real gas conditions. As seen in the figure, the largest deviation between the real and perfect gas is in the windward plane.

To summarize this section, SOSET from Ref. 2 has been modified for inclusion of real gas effects. It was seen that pressures are little affected by real gas effects, whereas temperatures are substantially affected, particularly if the body has a blunt tip. Hence, if one is interested in only forces and moments from approximate methods such as presented in this chapter, perfect gas computations are probably satisfactory. However, if one is interested in performing aeroheating analysis, real gas effects must be considered as Mach number increases above about 6.

VII. Aerodynamic Heating at Hypersonic Mach Numbers

The first six sections of this chapter have been laying the foundation to allow reasonably accurate estimates of the inviscid surface temperature so that approximate estimates of aeroheating information could be determined all along and around a body using MNT in conjunction with SOSET. The discussion in this section will be primarily taken from Ref. 17, which utilized

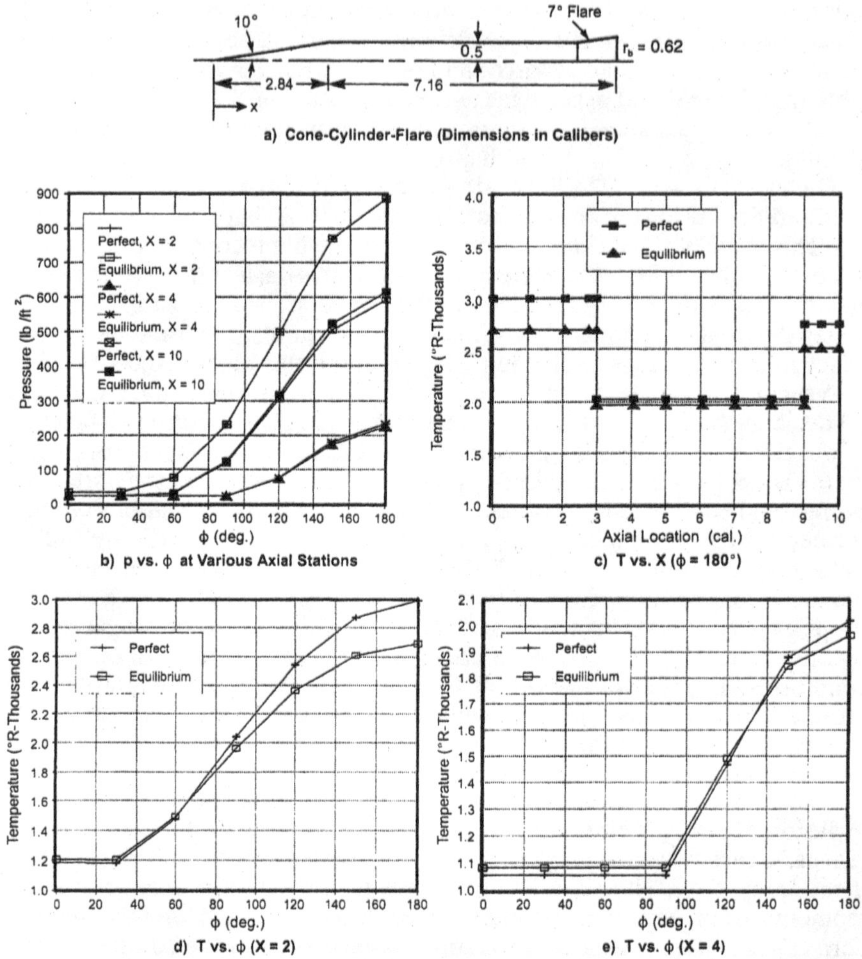

Fig. 7.22 Comparison of real and perfect gas results for temperature and pressure on a cone–cylinder–flare (h = 100,000 ft, M_∞ = 15, and α = 10°) from 1998 Aeroprediction Code.[16] a) Cone–cylinder–flare (dimensions in calibers), b) p vs φ at various axial stations, c) T vs X (φ = 180°), d) T vs φ (X = 2), e) T vs φ (X = 4)

inviscid surface temperatures at the outer edge of the boundary layer to determine temperatures at the surface. Several issues will be discussed here. These include the entropy layer effects that are dominant on a blunt body at high Mach numbers, engineering approximations to compute heat transfer coefficients on both the body and lifting surfaces, and comparison of theory to experiment.

A. Entropy Layer Effects

As Mach number increases, shock layer decreases; the percent of shock layer that is viscous (boundary layer) increases; and, if the nose of the body is blunt, a strong entropy gradient is created near the body surface. As the streamlines approach the body surface downstream, the boundary layer will eventually "swallow" the entropy layer. As a result, to assume the entropy value at the body surface is the same as that at the outer edge of the boundary layer becomes increasingly erroneous as flow proceeds downstream. Thus, to counter this and attempt to calculate a more accurate value of entropy at the edge of the boundary layer, a mass balance technique may be applied. Referring to Fig. 7.23, the mass-balance technique basically says that the flow within an element of area that goes through the shock (area 1 of Fig. 7.23) at the nose tip remains constant downstream (area 2 of Fig. 7.23). That is,

$$\begin{aligned} \dot{m}_1 &= \dot{m}_2 \text{ or} \\ \rho_\infty A_\infty V_\infty &= \rho_e A_e V_e \end{aligned} \tag{82}$$

The subscript e in Eq. (82) represents boundary-layer edge conditions. Substituting areas A_1 and A_2 of Fig. 7.23 for A_∞ and A_e, respectively, one obtains the radial location on the shock of the streamline that intersects the boundary layer at the local point of interest. That is,

$$r_s^2 = \frac{\rho_e V_e}{\rho_\infty V_\infty}(2r_b + \delta + \delta^*)(\delta - \delta^*) \tag{83}$$

The determination of r_s is an iterative process because the local boundary-layer thickness is dependent on local entropy. A value of entropy must be assumed and, in conjunction with the local pressure value that is known and is not affected by the boundary layer (to a first approximation), the other local thermodynamic properties can be determined. A boundary-layer thickness is then computed. At the present time, this computation is done by basic incompressible relations with the use of Eckert's[18] reference enthalpy properties to account for compressibility across the boundary layer. For laminar flow,

$$\delta_\ell = \frac{5.0\ s}{\sqrt{R_{N_s}}} \tag{84}$$

$$\delta_\ell^* \frac{1.729\ s}{\sqrt{R_{N_s}}} \tag{85}$$

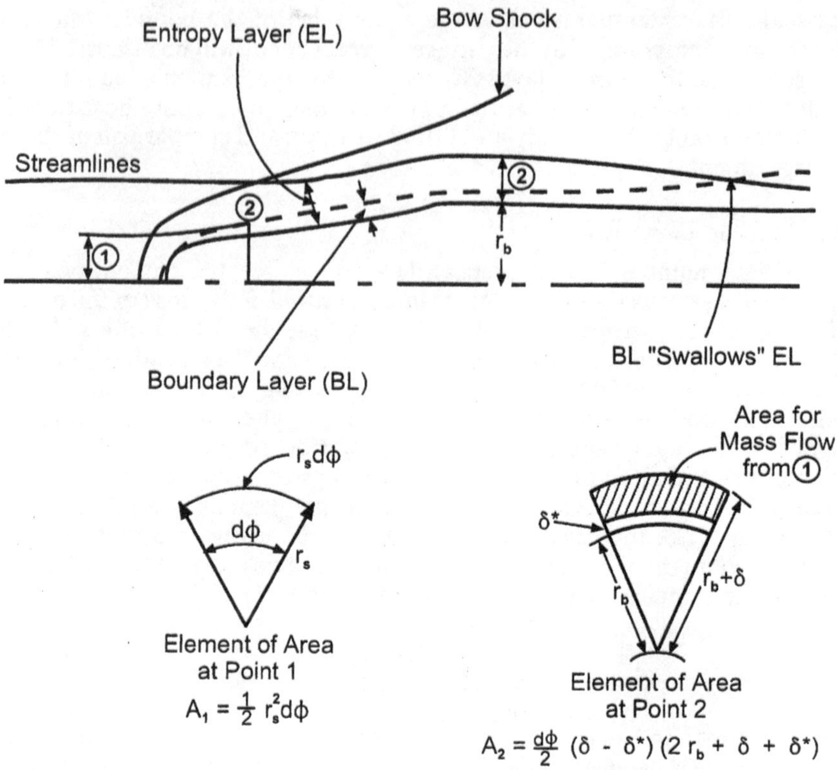

Fig. 7.23 Determination of entropy at outer edge of boundary layer. [16]

where s is the local boundary-layer coordinate and R_{N_s} is the Reynolds number based on s and the local reference enthalpy properties. For turbulent flow,

$$\delta_t = \frac{0.37\ s}{\left(R_{N_s}\right)^{0.2}} \tag{86}$$

$$\delta_t^* = \frac{0.046\ s}{\left(R_{N_s}\right)^{0.2}} \tag{87}$$

For local Reynolds numbers below some lower value R_{N_L}, laminar values are used. For local Reynolds numbers above an upper limit, R_{N_U}, the turbulent value is used. For transitional cases with $R_{N_L} < R_{N_s} < R_{N_U}$, the boundary-layer thickness is given by

$$\delta = \delta_\ell + \frac{R_{N_s} - R_{N_L}}{R_{N_U} - R_{N_L}} (\delta_t - \delta_\ell) \tag{88}$$

The Reynolds number adjusted value of δ^* is determined in a similar fashion. Knowing δ and δ^* for the assumed value of entropy, r_s can be determined. The shock slope at this location is determined by first assuming an analytical expression for the shock shape.[1]

$$x = r_N + \omega - r_c \cot^2 \beta \left[\left(1 + \frac{r_s^2 \tan^2 \beta}{r_c^2} \right)^{1/2} - 1 \right] \tag{89}$$

r_N is the nose radius and ω is the shock standoff distance given by

$$\frac{\omega}{r_N} = 0.143 \exp \left(\frac{3.24}{M_\infty^2} \right) \tag{90}$$

r_c is the radius of curvature of the shock at its vertex (assuming a hyperbolic shock shape), which is given by

$$\frac{r_c}{r_N} = 1.143 \exp \left(\frac{0.54}{(M_\infty - 1)^{1/2}} \right) \tag{91}$$

β is the limiting shock wave angle downstream from the nose curvature region. By differentiating Eq. (89) to get dr_s/dx and evaluating it at r_s, the shock slope at this point can be found. Then, the change in entropy across the shock can easily be determined. If it matches the assumed value at the body point, then the original assumption was correct and the solution can proceed to the next body station. If it does not match, a second value of entropy is assumed and the process is repeated. If this value also fails to match, then a variable secant iteration process can be started, which usually converges quite rapidly to the correct local entropy.

If the body under consideration is a blunted cone, the local entropy value will lie somewhere between that found behind a normal shock at the freestream Mach number and that found behind the attached oblique shock associated with the solution for a sharp cone of the same half-angle. Tables 7.1 and 7.2 show examples of normal shock and sharp-cone entropy values for several cone angles. As shown, they can be substantially different with the difference increasing as the Mach number goes higher. Pressures, because of the boundary-layer assumption, are not affected, but temperatures and densities at the boundary-layer edge can vary significantly.[17]

Figure 7.24 gives an indication of how entropy varies along the body surface as a function of distance downstream from the stagnation point using the present approach. As can be seen in Fig. 7.24, at a distance of 60 nose radii downstream, entropy values have largely stabilized to the sharp-cone value.

Table 7.1 Normal shock and cone entropies, $M_\infty = 8$ and altitude = 150,000 ft

Cone half-angle, deg	Normal shock entropy, ft^2/s^2-°R	Cone entropy, ft^2/s^2-°R
5	60631	51914
10	60631	52206
15	60631	52924
20	60631	53917
25	60631	54963
30	60631	55962

B. Engineering Approximations for Aeroheating

The primary focus of interest for aeroheating is to obtain reasonably accurate engineering estimates of heat transfer coefficient and adiabatic wall temperature (also known as recovery temperature) at flight conditions that are the most pressing and points on the body and fins that are the most critical. In most situations, this means conditions where dynamic pressure is the highest on the body or wing surface. Some local hot spots are not adequately treated by approximate methods such as will be described here, but must be treated by a Navier–Stokes code such as discussed in Chapter 2. An example of this type of problem area is a bow shock intersecting a wing.

The heat transfer coefficient and adiabatic wall temperature are related through[17]

$$H = \frac{\dot{q}_w}{(T_{aw} - T_w)} \tag{92}$$

Table 7.2 Normal shock and cone entropies, $M_\infty = 15$ and altitude = 150,000 ft

Cone half-angle, deg	Normal shock entropy, ft^2/s^2-°R	Cone entropy, ft^2/s^2-°R
5	69548	52139
10	69548	53706
15	69548	55766
20	69548	57632
25	69548	59374
30	69548	61391

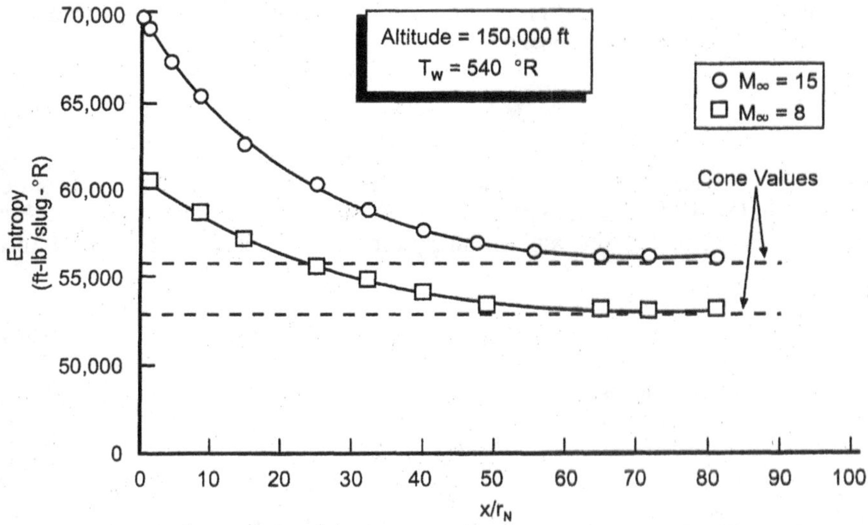

Fig. 7.24 Variation of entropy along windward symmetry plane of 1.5-in. nose radius, 15-deg half-angle cone.

where H is the heat transfer coefficient, $\dot{q}_w$ is the surface heat transfer rate, T_{aw} is the adiabatic wall temperature, and T_w is the wall temperature. Adiabatic wall temperature is defined as the wall temperature for no heat transfer at the wall and is

$$T_{aw} = T_e + r_c (T_0 - T_e) \tag{93}$$

where T_e is the boundary-layer edge temperature (inviscid solution surface temperature of Sections 7.5 and 7.6) and T_0 is the stagnation temperature. The recovery factor r_c is defined as

$$r_c = \frac{T_{aw} - T_e}{T_0 - T_e} \tag{94}$$

and is a measure of the degree of efficiency involved in converting the kinetic energy of external flow into heat within the boundary layer. If this conversion were 100% efficient, adiabatic wall temperature and stagnation temperature would be the same. T_{aw} is the highest possible temperature that can be reached within the boundary layer. The length of time required for the wall temperature to reach this upper limit is dictated by the rate of heat transfer $\dot{q}_w$. If the surface actually reaches T_{aw}, it can be seen from Eq. (92) that the heat transfer rate would go to zero, thus the term adiabatic wall temperature. If the wall were heated at a value greater than T_{aw}, it would be termed a hot wall and would have a negative heat transfer to the boundary layer.

The heat transfer coefficient for high-temperature flows is often expressed in terms of enthalpies as

$$H_1 = \frac{\dot{q}_w}{h_{aw} - h_e} \tag{95}$$

where H_1 is the enthalpy-based heat transfer coefficient, h_{aw} is the adiabatic wall enthalpy, and h_w is the enthalpy of the fluid at the body surface. In this case, h_{aw} is defined as the wall enthalpy for no heat transfer at the wall and is

$$h_{aw} = h_e + r_c\,(H_0 - h_e) \tag{96}$$

and the recovery factor, r_c is

$$r_c = \frac{h_{aw} - h_e}{h_0 - h_e} \tag{97}$$

At temperatures above about 1500 °R, the enthalpy formulation is the more rigorously correct of the two. Below that point, they are equivalent, because the specific heats of the gas are constant and $h = C_pT$. In either case, the recovery factor can be related approximately to the Prandtl number of the fluid by

$$r_c = \sqrt{Pr} \text{ (laminar)} \tag{98}$$

and

$$r_c = (Pr)^{1/3} \text{(turbulent)} \tag{99}$$

The heat transfer is normalized in the form given by Eq. (92) because the coefficient H remains fairly constant over a wide range of wall temperatures. The heat transfer itself varies from conditions where the wall is cold ($T_W < T_{aw}$), to conditions where the wall is hot ($T_w > T_{aw}$), to conditions where $T_w = T_{aw}$ and no heat transfer occurs. Given H and T_{aw} for a designated Mach number, AOA, and altitude combination, it is possible to compute the actual heat transfer rate for any wall temperature. Then, given the composition and structure of the wall, its rate of temperature change can be determined. By performing these computations sequentially at a selected number of points along a missile's trajectory, a time history of the surface temperature over the body can be constructed.

For regions of flow near a stagnation point, Section 7.5 is used to determine the properties just behind the shock and at the stagnation point. The nose tip stagnation point heating rate can then be determined by[19]

$$\dot{q}_w = 0.763\ Pr^{-0.6}\sqrt{\rho_0\mu_0}\sqrt{\frac{du_e}{dx}}\,(h_{aw} - h_w) \tag{100}$$

where the ρ_0 and μ_0 are the stagnation conditions determined from the Section 7.5 process. The Prandtl number Pr is also found at the stagnation point by the curve fit process discussed in Section 7.2 where

$$Pr = Pr\,(T_0, \rho_0) \tag{101}$$

The stagnation point streamwise velocity gradient, du_e/dx of Eq. (100), can be determined from Newtonian theory as

$$\frac{du_e}{dx} = \frac{1}{r_N}\sqrt{\frac{2(p_0 - p_\infty)}{\rho_0}} \tag{102}$$

where a spherical nose tip of radius r_N is assumed. If the body has a sharp nose, r_N is set to 0.01 in. The adiabatic wall enthalpy h_{aw} is determined from Eq. (96) using Eq. (98) for r_c, and h_w is the enthalpy of the fluid at the wall temperature. At the stagnation point, the flow will always be laminar.

If the stagnation line is on the leading edge of a wing surface, then for the wing in the plus "+" roll orientation, the boundary-layer heating is computed at the midspan of the upper and lower horizontal panels. If the wings are swept back, the freestream velocity must be resolved into components normal and parallel to the leading edge. That is.

$$\begin{aligned} V_N &= V_\infty \cos \Lambda_{LE} \\ V_P &= \sqrt{V_\infty^2 - V_N^2} \end{aligned} \tag{103}$$

The normal velocity component is passed through a normal shock and conditions behind the normal shock are obtained similarly as for the body. V_P is not affected by the shock and V_N goes to zero at the stagnation point as for the body.

The method of Beckwith and Gallagher[20] is used for swept cylinder heating on the wing leading edges. Here, laminar and turbulent boundary-layer approximations are given. For laminar flow, the heating rate is similar to Eq. (100) except for a cos Λ_{LE} term and a different constant coefficient at the beginning of the equation. That is,

$$\dot{q}_{w,\ell} = 0.57\; Pr^{-0.6}\sqrt{\rho_0\mu_0}\;\sqrt{\frac{du_e}{dx}}\;(h_{aw} - h_w)\,(\cos \Lambda_{LE})^{1.1} \tag{104}$$

All of the conditions of Eq. (104) are determined based on the normal velocity V_N of Eq. (103) taken across a normal shock instead of V_∞

Likewise, for turbulent flow,

$$\dot{q}_{w,t} = 1.04\ Pr^{-0.6} \frac{\left(\rho^* \mu^*\right)^{0.8}}{\left(\mu_0'\right)^{0.6}} (V_P \sin \Lambda_{LE})^{0.6} \left(\frac{du_e}{dx}\right)_{SL}^{0.2} (h_{aw} - h_w) \tag{105}$$

The * superscript denotes evaluation of the Eckert reference enthalpy,[18] which is given by

$$h^* = 0.5\,(h_w + h_e) + 0.22\,(h_{aw} - h_e) \tag{106}$$

where h_e is enthalpy at the outer edge of the boundary layer. Also, the μ_0' represents the coefficient of viscosity evaluated at a modified value of T_0 and ρ_0 based on conditions determined across a normal shock with upstream velocity V_N from Eq. (103) plus a component at the stagnation point of V_P from Eq. (103). That is, at the leading edge (or stagnation line)

$$\left.\begin{aligned} p_0' &= p_0 + \frac{1}{2}\rho_0 V_P^2 \\ H_0' &= H_0 \\ h' &= h + \frac{1}{2} V_P^2 \\ \rho_0' &= \rho_0\left(p_0', H_0'\right) \\ T_0' &= T_0\left(p_0', \rho_0'\right) \\ \mu_0' &= \mu_0\left(T_0', \rho_0'\right) \end{aligned}\right\} \tag{107}$$

The streamwise velocity gradient of Eq. (105) at the stagnation line is defined by MNT as

$$\left(\frac{du_e}{dx}\right)_{SL} = \frac{1}{r_{LE}} \sqrt{\frac{2\left(p_0' - p_\infty\right)}{\rho_0'}} \tag{108}$$

where r_{LE} is the leading edge radius of the control surface (which is assumed to be cylindrical). If a sharp leading edge is specified, r_{LE} is set equal to 0.01 in. for heating computations.

The laminar or turbulent status of the flow is determined by computing the Reynolds number based on the leading edge diameter and the freestream conditions.

$$R_{N_D} = \frac{\rho_\infty V_\infty D_{LE}}{\mu_\infty} \tag{109}$$

If R_{N_D} is below some user-specified lower limit, R_{N_L}, the laminar heat transfer is used. If R_{N_D} is greater than a specified upper limit, R_{N_U}, the turbulent value is used. For transitional values of R_{N_D} between R_{N_L} and R_{N_U}, the heat transfer is given by

$$\dot{q}_w = \dot{q}_{w,\ell} + \left(\frac{R_{N_D} - R_{N_L}}{R_{N_U} - R_{N_L}} \right) \dot{q}_{w,t} \tag{110}$$

Once the stagnation point conditions are known on the body nose or fin leading edge regions, one can march downstream to determine values of the heat transfer coefficient and adiabatic wall temperature based on some specified value of T_w. To do this, conditions at the outer edge of the boundary layer are required. These values are the inviscid surface temperature and pressure values of Sections 7.5 and 7.6. In the case of pressure, the value at the outer edge of the boundary layer is approximately constant across the boundary layer. Entropy at the outer edge of the boundary layer is computed by the mass-balance technique of Section 7.6 for a blunt surface. If the surface is sharp with an attached shock, entropy is assumed to be that computed behind the shock from Section 7.3. Also, because the inviscid flow is adiabatic, the total enthalpy, H_O, is constant throughout the flowfield for equilibrium flow.

Knowing the value of entropy at the body boundary-layer edge from either the oblique shock or mass-balance technique and the value of inviscid temperature at the outer edge of the boundary layer, then all remaining thermodynamic quantities at any body or wing surface point can be obtained from the curve-fit relations and thermodynamic relations. That is, if p_b and S_b are the known values of pressure and entropy at the body point (which are taken to be boundary-layer edge conditions), then

$$\left.\begin{array}{llll}
\rho_b = \rho_b(b_b, S_b) & ; \; h_b = h_b(p_b, \rho_b) & ; \; a_b = a_b(p_b, S_b) \\
T_b = T_b(p_b, \rho_b) & ; \; e_b = h_b - p_b/\rho_b & ; \; V_b = \sqrt{2(H_o - h_b)} \\
M_b = V_b/a_b & ; \; \gamma_b = h_b/e_b & ; \; Z_b = \dfrac{p_b}{\rho_b R T_b} \\
(P_r)_b = (P_r)_b(T_b, \rho_b) & ; \; \mu_b = \mu_b(T_b, \rho_b) &
\end{array}\right\} \tag{111}$$

Given appropriate fluid properties, laminar and turbulent heat transfer rates are computed as[21]

$$\dot{q}_{w,\ell} = 0.332\,(Pr^*)^{-0.667} \frac{\rho^* V_b}{\sqrt{\dfrac{R^*_{N_\ell}}{N_\ell}}} (h_{aw} - h_w) \tag{112}$$

and

$$\dot{q}_{w,t} = 0.185\,(Pr^*)^{-0.667}\,\frac{\rho^* V_b}{\left(\ell n\left(\dfrac{R^*_{N_e}}{N_t}\right)\right)^{2.584}}\,(h_{aw} - h_w) \tag{113}$$

The * conditions are evaluated at the reference enthalpy h * given by Eq. (106).

The e subscript of Eqs. (112) and (113) once again denotes the conditions at the edge of the boundary layer and, in this case, corresponds to the b subscript of Eq. (111). N_ℓ, and N_t are the laminar and turbulent Mangler transformation factors that are equal to 3.0 and 2.0, respectively. These factors provide an approximation to three-dimensional results even though the basic heat transfer relations were originally derived for a flat plate. The degree of turbulence is determined by first computing a local Reynolds number based on the boundary-layer running length s and local reference conditions.

$$R^*_{N_e} = \frac{\rho^* V_b s}{\mu^*} \tag{114}$$

As before, if $R^*_{N_e}$ is less than a user-specified lower limit R_{N_L}, the flow is assumed to be fully laminar. For $R^*_{N_e}$ values greater than a specified upper limit R_{N_U}, the flow is assumed to be fully turbulent. For intermediate transitional values of $R^*_{N_e}$ between R_{N_L} and R_{N_U}, the heat transfer is given by

$$\dot{q}_w = \dot{q}_{w,\ell} + \left(\frac{R^*_{N_e} - R_{N_L}}{R_{N_U} - R_{N_L}}\right)\dot{q}_{w,t} \tag{115}$$

The boundary-layer running length s of Eq. (114) is assumed to be distance along a meridian line of the body for simplicity.

Now knowing $\dot{q}_w$ from either Eq. (100), (104), (105), (112), or (113), the heat transfer coefficient can be computed from Eq. (95). As already mentioned, the adiabatic wall or recovery temperature is computed from Eq. (93) and the adiabatic wall enthalpy is computed from Eq. (96), using Eq. (98) or (99) for the recovery factor. These three pieces of information, given a specified value of wall temperature at launch, are generally all that is needed for boundary-layer heating information. This information is generally combined with details of a flight structure in terms of materials and structural design to perform heat transfer during flight where radiation, conduction, and convection are all considered simultaneously. Information provided here is for only the convective portion of the heat transfer.

C. Example Application of Approximate Methods for Boundary-Layer Heating

Good heat transfer information at conditions where real gas effects are important is not as readily available as desired. Two simple cases are shown here where data are available.

The first configuration considered consisted of a 15-deg half-angle cone with a nose radius of 0.375 in. Freestream pressure and temperature were 2.66 lb/ft^2 and 89.971 °R. The Mach number was 10.6 at an AOA of 20 deg. The wall temperature was 540 °R. Heat transfer rates along the windward symmetry plane are shown as a function of distance along the body in Fig. 7.25. The results of Chapter 7 are shown as the 1998 Aeroprediction Code (AP98).[16] For comparison purposes, numerical results are shown from Ref. 22 and corresponding experimental results from Ref. 23. The methodology used in Ref. 22 was much more involved than the approach used in the Chapter 7 methods because it included complex streamline tracking methods and the axisymmetric analogue technique to model streamline spreading effects. AP98 and another approximate code, MINIVER,[24] tend to underpredict experimental data and the computational results of Ref. 22 by 15 to 20%, a performance that is quite creditable considering the simplified solution approach. At the nose tip stagnation point, AP98 gives a heat transfer rate of 33.99 Btu/ft^2-s, while Ref. 22 predicts 35.94 Btu/ft^2-s, which again is good agreement.

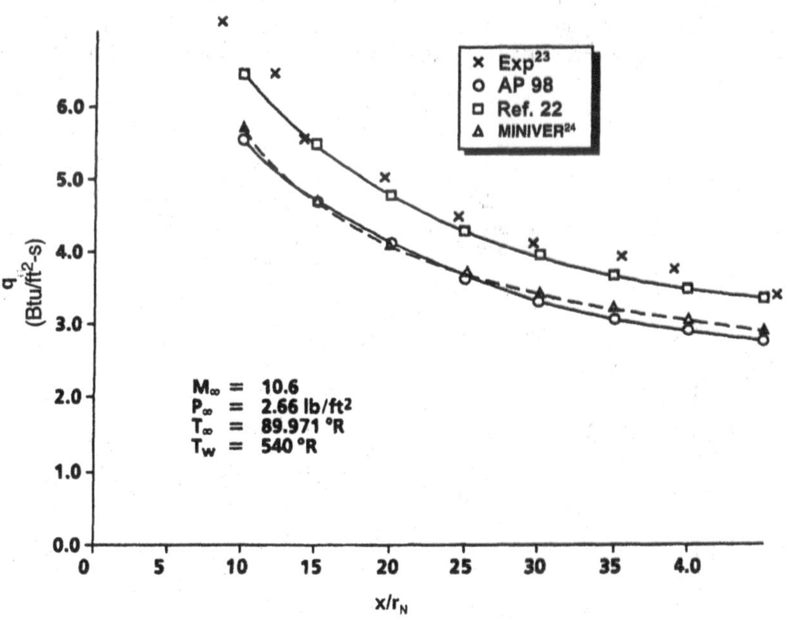

Fig. 7.25 Heat transfer rates for 0.375-in. nose radius, 15-deg half-angle cone at $\alpha = 20°$ and $\varphi = 180°$.

The second configuration considered was a spherically blunted, 15-deg half-angle cone with a nose radius of 1.1 in. Freestream pressure and temperature were 2.66 lb/ft^2 and 89.971 °R, respectively. The Mach number was 10.6 and AOAs of 5 and 10 deg were considered. The wall temperature was a constant 540 °R. Figures 7.26 and 7.27 show heating rate predictions on the windward symmetry plane as a function of position along the body for the 5- and 10-deg AOA cases, respectively. Also shown are results from AEROHEAT, an engineering-type code that used streamline tracking and axisymmetric analogue methodology and, from a second approximate code that used more sophisticated streamline tracking as well as a more exact boundary-layer formulation.[25] MINIVER results are also plotted with experimental data.[23] AP98 does well beyond about 5 nose radii downstream but tends to overpredict heating rates near the nose tip. The region of overprediction is somewhat less at the higher AOA. This discrepancy is believed to be caused by the use of heating rate expressions that are based on flat plate relations and thus are incapable of modeling the effects of the rapidly changing boundary-layer edge conditions in this region. The aeroprediction results are slightly better than those from MINIVER because of the inclusion of variable entropy effects. Figures 7.28 through 7.31 show the computed variation of heating rates in the circumferential direction around the body. The φ = 180 deg location corresponds to the windward symmetry plane. Figures 7.28 and 7.29 illustrate results for a 5-deg AOA at points 4.86 and 10.13 nose radii downstream of the nose tip, respectively. At 4.86 nose radii, the AP98 results are uniformly high, a result that could be anticipated because of the tendency to overpredict heating rates in the nose region. At 10.13 nose radii, agreement with both numerical and experimental results is good, with the greater differences occurring on

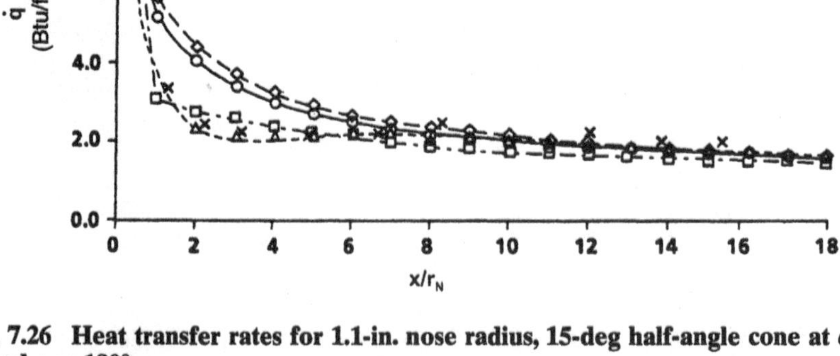

Fig. 7.26 Heat transfer rates for 1.1-in. nose radius, 15-deg half-angle cone at α = 5° and φ = 180°.

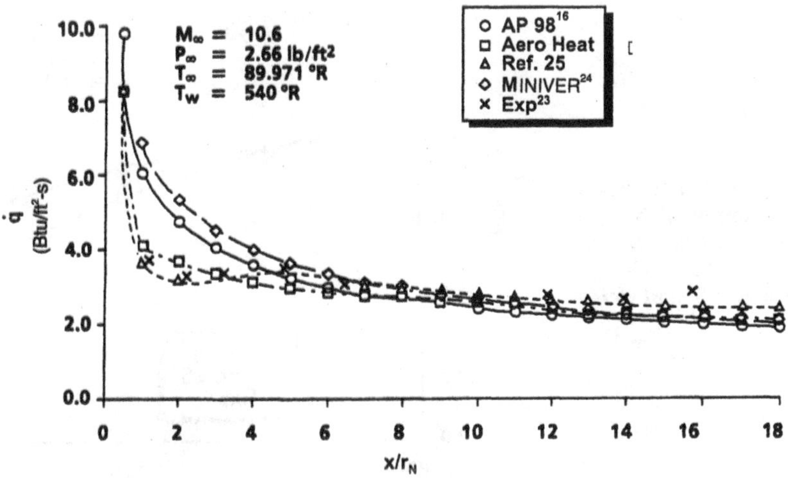

Fig. 7.27 Heat transfer rates for 1.1-in. nose radius, 15-deg half-angle cone at $\alpha = 10°$ and $\varphi = 180°$.

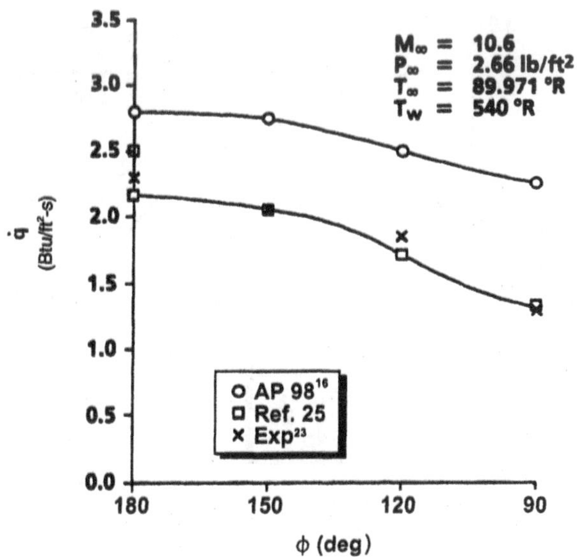

Fig. 7.28 Circumferential variation of heating rates for 1.1-in nose radius, 15-deg half-angle cone at $\alpha = 5°$ and $x/r_N = 4.86$.

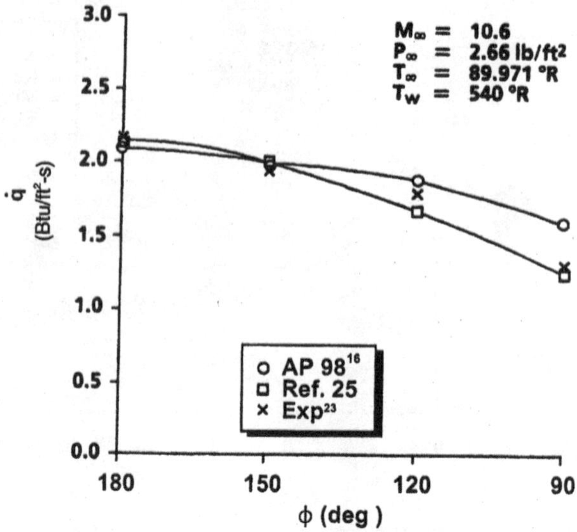

Fig. 7.29 Circumferential variation of heating rates for 1.1-in. nose radius, 15-deg half-angle cone at $\alpha = 5°$ and $x/r_N = 10.13$.

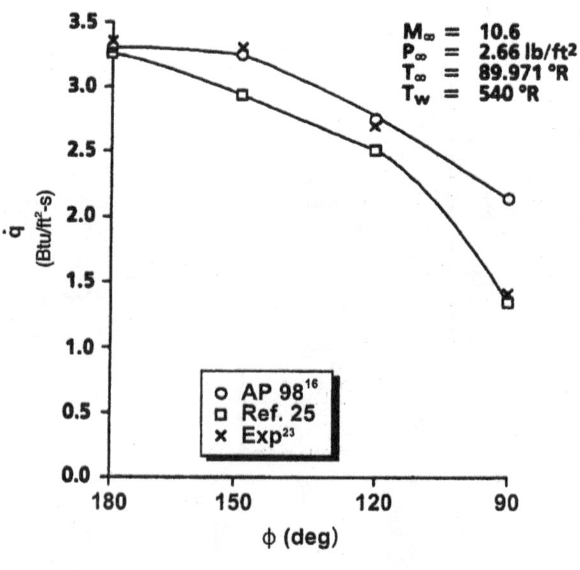

Fig. 7.30 Circumferential variation of heating rates for 1.1-in. nose radius, 15-deg half-angle cone at $\alpha = 10°$ and $x/r_N = 4.86$.

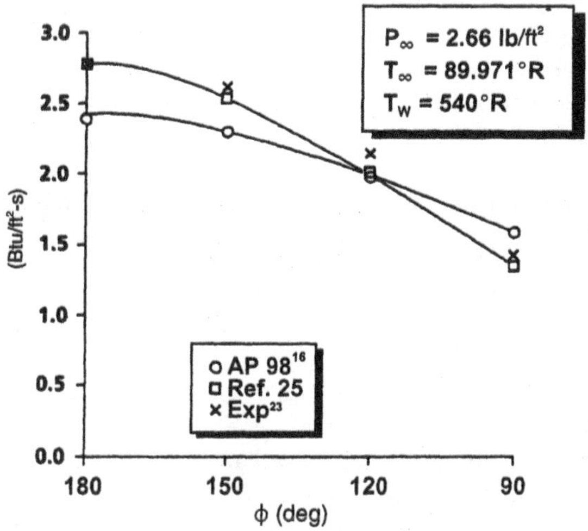

Fig. 7.31 Circumferential variation of heating rates for 1.1-in. nose radius, 15-deg half-angle cone at $\alpha = 10°$ and $x/r_N = 10.13$.

the leeward side of the body. Figures 7.30 and 7.31 show the 10-deg AOA results for the same streamwise locations. At the 4.86 nose radii station, except for the leeward plane, the aeroprediction results are actually more in agreement with the experiment than are predictions of the more sophisticated numerical methodology of Ref. 25. At 10.13 nose radii, the agreement is not as good but is still within 15% of both experimental data and other computational predictions. No results are shown from MINIVER for these circumferentially varying cases because it does not explicitly provide a capability to handle three-dimensional problems.

References

[1]Anderson, J. D., *Modern Compressible Flow,* 2nd ed., McGraw–Hill, New York, 1990.

[2]Moore, F. G., Armistead, M. A., Rowles, S. H., and DeJarnette, F. R., "Second-Order Shock-Expansion Theory Extended to Include Real Gas Effects," Naval Surface Warfare Center, Dahlgren Div. Rept. NAVSWC TR 90-683, Feb. 1992.

[3]Tannehill, J. C., and Mugge, P. H., "Improved Curve Fits for the Thermodynamic Properties of Equilibrium Air Suitable for Numerical Computation Using Time-Dependent or Shock-Capturing Methods," NASA CR-2470, 1974.

[4]Srinivasan, S., Tannehill, J., and Weilmuenster, K., "Simplified Curve Fits for the Thermodynamic Properties of Equilibrium Air," Engineering Research Ins., Iowa State Univ. Rep. ISU-ERI-AMES 86401, Ames, IA, June 1986.

[5]Ames Research Staff, "Equations, Tables and Charts for Compressible Flow," NACA TR 1135, 1953.

[6]Wittliff, C. E., and Curtis, J. T., "Normal Shock Wave Parameters in Equilibrium Air," Cornell Aero Lab. Rep. CAL-III, Buffalo, NY, Nov. 1961.

[7]Marrone, P. V., "Normal Shock Waves in Air: Equilibrium Composition and Flow

Parameters for Velocities from 26,000 to 50,000 ft/sec," Cornell Aero Lab. Rep. No. AG-1729-A-2, Buffalo, NY. Aug. 1962.

[8]Hudgins, H. E., Jr., "Supersonic Flow About Right Circular Cones at Zero Yaw in Air at Chemical Equilibrium, Part 1 — Correlation of Flow Properties," Picatinny Arsenal, TM 1493, Aug. 1965.

[9]Truitt, R. W., *Hypersonic Aerodynamics,* Ronald, New York, 1959.

[10]Hayes, W. D., and Probstein, R. F., *Hypersonic Flow Theory,* Academic International Press, New York, 1966.

[11]Owczarek, J. A., *Fundamentals of Gas Dynamics,* International Textbook, Scranton, PA, 1964.

[12]Heims, S., "Prandtl–Meyer Expansion of Chemically Reacting Gases in Local Chemical and Thermodynamic Equilibrium," NACA TN 4230, March 1958.

[13]Hayes, W. D., and Probstein, R. F., *Hypersonic Flow Theory,* Academic International Press, New York, 1966.

[14]Witcliff, C. E., and Curtis, J. T., "Normal Shock Wave Parameters in Equilibrium Air," Cornell Aero Lab. Rep. CAL-III, Buffalo, NY, Nov. 1961.

[15]Priolo, F. J., and Wardlaw, A. B., Jr., "High-Temperature Effects for Missile-Type Bodies Using the Euler Solver, ZEUS," AIAA Paper No. 91-3259, AIAA Applied Aerodynamics Conference, Baltimore, MD, Sept. 1991.

[16]Moore, F. G., McInville, R. M., and Hymer, T. C., "The 1998 Version of the NSWC Aeroprediction Code: Part 1—Summary of New Theoritical Methodology," NSWCDD/TR-98/1, April 1998.

[17]McInville, R. M., and Moore, F. G., "Incorporation of Boundary Layer Heating Predictive Methodology into NSWCDD Aeroprediction Code," Naval Surface Warfare Center, Dahlgren Div. Rept. NSWCDD/TR-93/29, Dahlgren, VA, Feb. 1993.

[18]Eckert, E. R. G., "Engineering Relations for Heat Transfer and Friction in High-Velocity Laminar and Turbulent Boundary Layer Flow over Surfaces with Constant Pressure and Temperature," *Transcript of the American Society of Mechanical Engineers,* Vol. 78, No. 6, Aug. 1956.

[19]Anderson, J. D., *Hypersonic and High Temperature Gasdynamics,* McGraw–Hill, 1989.

[20]Beckwith, I. E., and Gallagher, J. J., "Local Heat Transfer and Recovery Temperatures on a Yawed Cylinder at Mach Numbers of 4.15 and High Reynolds Numbers," NACA TRR-104, 1961.

[21]Eckert, E. R. G., "Survey of Boundary Layer Heat Transfer at High Velocities and High Temperatures," WADC TR 59-624, April 1960.

[22]DeJarnette, F. R., and Hamilton, H. H., "Aerodynamic Heating on 3-D Bodies Including The Effects of Entropy Layer Swallowing," *Journal of Spacecraft and Rockets,* Vol. 12, No. 1, 1975 pp. 5–12.

[23]Cleary, J. W., "Effects of Angle of Attack and Bluntness on Laminar Heating Rate Distribution of a 15° Cone at a Mach Number of 10.6," NASA TN D-5450, 1969.

[24]Hender, D. R., "A Miniature Version of the JA70 Aerodynamic Heating Computer Program, H800(MINIVER)," McDonnell–Douglas Astronautics Co., Rep. MDC G0462, June 1970.

[25]Riley, C. J., and DeJarnette, F. R., "Engineering Aerodynamic Heating Method for Hypersonic Flow," *Journal of Spacecraft and Rockets, Vol. 29, No. 3, May–June 1992.*

Chapter 8

Applications of Aerodynamics

Nomenclature

AR	= aspect ratio = b^2/A_W
A_{ref}	= reference area (maximum cross-sectional area of body, if a body is present, or planform area of wing if wing alone) (ft^2)
A_W	= wing area
b_T, b_W	= tail or wing span (not including body) (ft)
$\text{BM}(x)_B$	= body bending moment as a function of streamwise coordinate (lb-ft)
$\text{BM}(y)_{W/T}$	= wing/tail bending moment as a function of spanwise coordinate (lb/ft)
C_A	= axial force coefficient
$C_{A\text{BODY}}$	= axial force coefficient of the body alone
$C_{A\text{SF}}$	= axial force coefficient of a single fin
C_D	= drag coefficient
C_L	= lift coefficient
$C_{\ell P}$	= roll damping moment coefficient
C_M	= pitching moment coefficient about some reference point
C_{M_α}	= pitching moment coefficient derivative
$C_{M\delta}$	= pitching moment coefficient derivative with respect to control deflection
$C_{M_q} + C_{M\dot{\alpha}}$	= pitch damping moment coefficient
$c(y)_T, c(y)_W$	= local chord of tail or wing as a function of spanwise coordinate (ft)
$C_{nB}, (C_{nB})_L, (C_{nB})_{\text{NL}}$	= local, total, linear, and nonlinear normal-force coefficient of the body
$(C_{nB})_{\text{NL,ADJ}}$	= nonlinear component of body local normal-force coefficient adjusted for decreased nose loads
$(C_{nB})_{T(V)}$	= component of body local normal-force coefficient due to wing–tail interference
$[(c_n(y))_L]_{T,W}$	= linear component of local normal-force coefficient on tail or wing as a function of spanwise coordinate

$[c_n(y)]_{\text{LEEWARD, WINDWARD}}$	= portion of total local normal-force coefficient assigned to leeward or windward lifting surfaces
$[c_n(y)]_{T,W}$	= local normal-force coefficient on tail or wing as a function of spanwise coordinate
$[c_n(y)]_{T,V}$	= portion of local normal-force coefficient on lifting surfaces attributed to wing–tail interference
$\Delta C_{n_{B(T)}}, \Delta C_{n_{B(W)}}$	= local normal-force coefficient load induced on the body due to presence of tail and wing, respectively
C_N	= total normal-force coefficient
C_{N_B}	= normal-force coefficient of the body
$\Delta C_{N_{B(T)}}, \Delta C_{N_{B(W)}}$	= increment to normal-force coefficient on body due to presence of tail or wing
$C_{N_{T(B)}}, C_{N_{W(B)}}$	= normal-force coefficient of tail or wing in presence of body
$C_{N_{T(V)}}$	= component of normal-force coefficient in tail region due to interference from wing
$[C_{N_{T(V)}}]_{\text{BODY, TAIL}}$	= portion of total wing–tail interference term assigned to body or tail
$[C_{N_{T(V)}}]_{\text{LEEWARD, WINDWARD}}$	= portion of tail component of wing–tail interference term assigned to leeward or windward tail surfaces
C_{N_a}	= normal-force coefficient derivative
$(C_{N_a})_W$	= change of wing normal-force coefficient with AOA (rad^{-1})
C_P	= pressure coefficient
C_{P_B}	= base pressure coefficient
$(c_r)_{T,W}$	= root chord of tail or wing lifting surface (ft)
d_B	= diameter of body at base
d_r	= reference diameter of body (ft)
i_1, i_4	= interference factors in wing–tail interference calculations for windward and leeward plane wings, respectively
k, $k_{W(B)}$	= ratio of normal-force contribution of a deflected lifting surface in the presence of a body to that of the lifting surface alone at $a = 0$ deg
$K_{W(B)}$	= ratio of lifting surface normal-force coefficient in the presence of a body to that of the lifting surface alone at $\delta = 0$ deg
$K_{B(W)}$	= ratio of additional body normal-force coefficient in the presence of a wing to that of the wing alone at $\delta = 0$ deg
$\left([K_{B(W)}]_{\min}\right)_{\varphi=0^\circ,\varphi=45^\circ}$	= minimum value of $K_{B(W)}$ at the $\varphi = 0$ deg or 45 deg roll positions

ℓ = length of body
M = Mach number
m = mass (lb–s^2/ft)
N = normal force (lb)
NF = number of fins
ρ_∞ = freestream density
Q = dynamic pressure (lb/ft^2)
r = local body radius (ft)
r_n = nose radius
r_{VORT} = vortex radius (ft)
s = wing or tail semispan plus the body radius in wing–body lift methodology
$VC(y)$ = vortex strength-local chord parameter
$V(x)_B$ = body shear as a function of streamwise coordinate (lb)
$V(y)_{W/T}$ = wing or tail shear as a function of spanwise coordinate (lb)
V_∞ = freestream velocity
W = weight
$w(x)_B$ = local body load per unit length as a function of streamwise coordinate (lb/ft)
$w(y)_T$ = local tail load per unit length as a function of spanwise coordinate (lb/ft)
$w(y)_W$ = local wing load per unit length as a function of spanwise coordinate (lb/ft)
x = streamwise coordinate (ft)
x_{CP} = center of pressure (in feet or calibers from some reference point that can be specified) in the x direction
y = spanwise lifting surface coordinate (ft)
y_{DIST} = spanwise distance from path of vortex shed from wing to local point on tail (ft)
y_{VORT} = spanwise distance from centerline of body to path of vortex shed from wing
z = coordinate in vertical direction
α = AOA (rad, deg)
δ = deflection angle of control surface (rad, deg)
Λ_{LE} = leading edge sweep angle of wing or tail
λ = taper ratio = c_t/c_r
φ = roll position of missile ($\varphi = 0$ deg corresponds to fins in the plus (+) orientation and $\varphi = 45$ deg corresponds to fins in cross (×) orientation)

Subscript

TR = corresponds to conditions at trim

I. Introduction

MOST of this book has discussed ways to calculate aerodynamics, with little emphasis placed on the uses of these aerodynamics once computed. An exception to this was in Chapter 7, which was devoted to calculating aerodynamic heating at hypersonic Mach numbers. This chapter will focus on several applications of aerodynamics. These include defining structural loads that are needed in preliminary structural design, minimum drag design, multifin design tradeoffs, and trajectory analysis. The focus here is to remind the aerodynamicist of the purpose for what he or she does, that is, the application.

II. Structural Loads

A. Background

There are two primary classes of structural analysis. The first and most sophisticated is a finite element analysis of the entire structure. For this type of application, the structural designer is interested in having accurate nonlinear pressure loads as a function of position along and around the body and as a function of position along the chord and out the span of a wing. These loads must be known as a function of Mach number, angle of attack (AOA) and altitude. Ultimately, what will drive the design is the set of "worst case" conditions in terms of loads. If this set of worst case conditions has too severe an implication on the weight of the vehicle, this set of conditions can be reduced by limiting flight to a less severe environment. To calculate aerodynamics compatible with a finite element structural analysis will generally require a numerical solution of the Navier–Stokes or Euler equations discussed in Chapter 2.

The second type of structural analysis, which is more compatible with the semi-empirical aerodynamic methods of Chapters 3 to 7, is a beam-type analysis. In this type of analysis, the structural analysis involves knowledge of the running loads along the body axis and out the span of the wings, but not around the body or along the chord of the wings. Again, this information is needed as a function of Mach number, AOA, and altitude so that the "worst case" (highest structural load) condition can be computed.

In evaluating semi-empirical aerodynamic methods presented in Chapters 3 to 7, it is clear that the linear component of loads can be easily put in the form needed for a beam-type structural analysis. This is because the pressure distribution is known along and around the body and along the chord and out the span of the wings. However, interference-induced pressures on the wings and body are not computed as part of the normal component buildup approach. Also, nonlinear loads are known only in terms of total force and moment components, with no information available as to how to distribute these loads over the body and lifting surfaces. As a result, this chapter will focus on taking interference loads and all nonlinear loads and distributing them over the body and lifting surfaces. This distribution will be done to be compatible with overall force and moment computations. These distributed loads can be added to the linear load distribution so a total load distribution over the body and lifting surfaces can be made available for structural analysis.

B. Approach to Distribute Loads

The approach to accomplish the task of developing a total load distribution based on semi-empirical aerodynamic methods of Chapters 3 to 7 is taken from Ref. 1 and is

1) Pick a missile configuration that has two sets of lifting surfaces and a good set of force and moment wind tunnel data over a fairly broad range of freestream conditions. The configuration chosen is shown in Fig. 8.1, and the wind tunnel data set is given in Ref. 2. The configuration in Fig. 8.1 is the same case considered in Section 5.8 and shown as Fig. 5.34.
2) Use a Navier–Stokes code and refine the mesh until the total force and moment results do not change to an appreciable extent. The code selected was the thin layer Navier–Stokes code referred to as OVERFLOW.[3]
3) Compare the computational fluid dynamics (CFD) predicted force and moment and center of pressure data to experimental data of Ref. 2. If the comparisons agree reasonably well over a broad range of conditions, we will assume the code is accounting for all of the physics reasonably well and that local loads on the body and lifting surfaces are also correct.
4) Derive semi-empirical methods for distributing the nonlinear component of aerodynamic loads over the body and lifting surfaces for use in structural analysis, using CFD loads for comparison.

Table 8.1 presents the results of CFD calculations compared to experiment (the reader is referred to Ref. 4 for the detailed discussion of Navier–Stokes computations and theory discussed in this section).

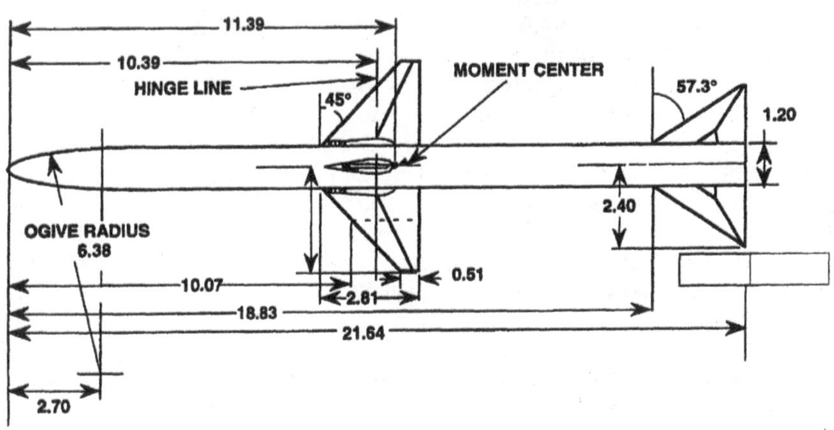

Fig. 8.1 Wing–body–tail missile configuration.

Table 8.1 Comparison of CFD calculations[1] to experiment[2]

Mach number	AOA, a	Roll, φ	CL % difference, exp vs CFD	CD % difference, exp vs CFD	x_{CP} difference % body length, exp vs CFD
1.50	10	0	1.8	6.1	0.78
1.50	25	0	1.1	2.6	0.57
1.50	40	0	−0.7	0.4	0.07
2.87	10	0	2.1	5.9	1.44
2.87	25	0	−1.5	−0.7	0.07
2.87	40	0	−1.4	−2.9	0.38
4.60	10	0	0	−8.3	1.13
4.60	25	0	1.2	4.3	0.35
4.60	40	0	−6.1	−7.4	2.42
1.50	10	45	2.4	15.0	0.85
1.50	25	45	0.4	0.6	0.44
1.50	40	45	−1.5	−0.7	−0.09
2.87	10	45	−1.6	12.1	1.22
2.87	25	45	−3.0	3.1	0.77
2.87	40	45	−0.9	−0.8	0.56
4.60	10	45	−5.6	6.7	1.58
4.60	25	45	−0.6	3.6	1.21
4.60	40	45	0.5	1.8	0.82

Errors in Table 8.1 are defined as difference between experiment and CFD results divided by experimental results. As seen in the table, comparisons are quite good, with a couple of exceptions. These are at low AOAs (a = 10 deg), where the error in reading data points from the graph can be a significant portion of the data (an estimated 5%), and the case for $a = 40$ deg, $\varphi = 0$ deg, and $M_\infty = 4.6$. This case shows the CFD results to be off by 6 to 7% on both lift and drag and off by more than 2% of body length on center of pressure. It is suspected that strong internal shock interactions that occur at this AOA and Mach number require a full Navier–Stokes solution versus a thin layer code to account for streamwise and circumferential gradients of stress terms. The other possibility is that a finer grid is needed than that used for this case.

Overall, results of Table 8.1 were considered quite acceptable for use in deciding how to distribute nonlinear aerodynamic loads for use in semi-empirical aerodynamic codes. The average accuracy levels of the lift coefficient was better than 2%, and the center of pressure, about 1% of the body length. These are the main force and moment contributors to local load, shear, and bending moment for use in structural analysis. The remainder of Section 8.2 will discuss the approach of how each nonlinear aerodynamic component is treated to accomplish the task of distributing these loads in a semi-empirical code.

The total normal force for a wing–body–tail configuration can be defined by

$$C_N = C_{N_B} + C_{N_{W(B)}} + \Delta C_{N_{B(W)}} + C_{N_{T(B)}} + \Delta C_{N_{B(T)}} + C_{N_{T(V)}} \qquad (1)$$

The first term in Eq. (1) is the normal force coefficient of the body alone. The second term is the normal force coefficient of the wing in the presence of the body and the third represents the additional component of normal force on the body due to the presence of the wing. The fourth term is the normal force coefficient of the tail in the presence of the body and the fifth represents the incremental normal force on the body as a result of the tail. The final term is the normal force on the tail, usually negative, caused by the vortices shed from the wing.

Each of the terms in Eq. (1) has a linear and nonlinear component. Either linear theory or slender body theory is used to determine the linear portion of each term and nonlinear contributions are computed directly from wind tunnel databases with engineering judgment and extrapolation being used where database information is limited. Overall average accuracy is maintained at ±10% for the total normal and axial force coefficients and ±4% of total body length for the center of pressure.

In the following discussion, each component of Eq. (1) will be considered individually and the methodology that has been developed to distribute its nonlinear elements over the body surface will be briefly discussed. First, the $\varphi = 0$ deg roll position will be addressed and then any changes or additions that are necessary to extend the methodology to the $\varphi = 45$ deg roll position will be presented. Those interested in more details of theoretical discussions in this section are once again referred to Ref. 4.

C. Roll Position of $\varphi = 0$ deg

1. Body Loads at the $\varphi = 0$ deg Roll Position

Four of the terms in Eq. (1) contribute to body loading. These are C_{N_B}, $\Delta C_{N_{B(W)}}$, $\Delta C_{N_{B(T)}}$, and $C_{N_{T(V)}}$. Only the first three of these will be discussed in this section. The wing–tail interference will be reserved for the section on lifting surface loads because it is connected directly to them and is shared with tail surfaces. Also, as will be shown later, a portion of the $C_{N_{T(V)}}$ term has been distributed onto the body, based on CFD results.

2. Body-Alone Loads

The body-alone component, C_{N_B} is composed of a linear and a nonlinear load. The linear load is computed analytically above $M_\infty = 1.2$ by either Van Dyke's hybrid theory combined with modified Newtonian theory or second-order shock expansion theory. Pressure coefficients are computed as a function of position along and around the body and then integrated along the body to obtain local linear normal force. Both of these theories were discussed in Chapter 4.

The nonlinear body normal force coefficient contribution is computed by a modification of the Allen–Perkins viscous crossflow theory of Chapter 5 and distributed uniformly along the body surface.

The total local normal force coefficient at any point along the body surface is the sum of the linear and nonlinear components. That is,

$$C_{nB} = (C_{nB})_L + (C_{nB})_{NL} \tag{2}$$

Comparison of Eq. (2) with results from CFD computations[4] indicated that the predicted body load in the nose region was too high at low AOAs. It is believed that this discrepancy occurs because, at low AOAs, the point at which body vortices begin to separate is toward the afterbody of the missile. Then, as AOA is increased, the separation point moves forward and the full effect of the body-shed vortices is apparent. To account for this effect, the nonlinear load contribution of Eq. (2) was adjusted at low AOAs.

On the nose itself, the nonlinear load is reduced at an AOA below $\alpha = 30$ deg by a factor that varies linearly with AOA from 1 at $\alpha = 30$ deg to 0 at $\alpha = 0$ deg. No adjustment is made at an AOA above 30 deg. There is a transition region, which extends for a distance of two calibers from the end of the nose, within which the nonlinear load component is allowed to recover gradually from its reduced value at the end of the nose to its normally computed level at that location. Because the total normal force computed by semi-empirical methods of Chapter 5 agrees well with experimental results, the portion of the nonlinear loads that are removed from the nose and transition regions must be added back into the loads on the remainder of the body, consistent with the total loads and center of pressure of the Chapter 5 results. Thus, for points on the body more than two calibers downstream of the nose, the normally computed local nonlinear loads are increased by terms that represent the total load reductions in the nose and transition regions, respectively, spread evenly over the total remaining body length.

3. Body–Wing and Body–Tail Interference Loads

For the discussion to follow, forward lifting surfaces will be referred to as wings, and aft lifting surfaces will be referred to as the tail. In the vicinity of any lifting surface, the body is subjected to increased loads caused by interference effects. These contributions, represented in Eq. (1) by $\Delta C_{N_{B(W)}}$ for the wing and $\Delta C_{N_{B(T)}}$ for the tail, are also computed as separate linear and nonlinear components with afterbody effects being considered for rear lifting surfaces. Therefore, the region of influence of the lifting surfaces on the body is consistent with the method currently used for moment and center of pressure calculations in Chapter 3. This method basically assumes the region of carryover lift onto the body is based on the region influenced by local Mach lines from the wings or tails as they sweep across the body surface. The region of load on the body consists of a trapezoidal shape with total height defined by the magnitude of the aerodynamic interference terms $\Delta C_{N_{B(W)}}$ and $\Delta C_{N_{B(T)}}$, Mach number, body diameter, and root chord lengths of the wing or tail.[4]

In situations where the wing or tail is located near the aft end of the body, it is possible for Mach lines to extend beyond the body, and some of the interference loading will be lost unless further adjustments are made. For these cases, the cross-sectional area of the body enclosed by Mach lines and the aft end of the body is determined and compared to the total cross-sectional area that would be enclosed by Mach lines if the body were not cut off. Maximum interference load values are then adjusted to compensate for the lost area and maintain the correct total of interference normal force as computed by the method of Chapter 5.

4. Lifting Surface Loads at the $\varphi = 0$ deg Roll Position

Total contributions to the normal force of the wing and tail are represented in Eq. (1) by $C_{N_{W(B)}}$ and $C_{N_{T(B)}}$, respectively. Using $C_{N_{W(B)}}$ as an example, these two terms each have the following form:

$$C_{N_{W(B)}} = \left[K_{W(B)} \sin a + k_{W(B)} \sin \delta\right] \left(C_{N_a}\right)_W \qquad (3)$$

Nonlinearities are introduced to $C_{N_{W(B)}}$ by defining both a linear and a nonlinear contribution for the normal force of the isolated wing, represented by $(C_{N_a})_W$ and for each of the interference factors $K_{W(B)}$ and $k_{W(B)}$, which are introduced to account for the effects of the presence of a body on the normal force of the lifting surface.

The linear part of the local normal force coefficient on a wing or tail surface at a spanwise location y is computed by either lifting surface theory for subsonic flows or three-dimensional thin-wing theory for supersonic flows.

The nonlinear part of the wing-alone lift is determined using a fourth-order analytical method in conjunction with wing-alone databases (see Chapter 5). Only the total values of lift and moment are available from the databases with no surfaces distribution information included. This distribution will be determined in conjunction with that of the interference factors.

From the combined linear and nonlinear wing-alone normal force, C_{N_a} is determined based on a secant slope of the total AOA, $|a + \delta|$, experienced by the lifting surface. This slope is then multiplied by sin a and sin δ terms, which contain the interference factors K and k, respectively, to account for the interference effects of the body on the wing. (All subscripts have been dropped from K and k for simplicity.) The linear contributions to K and k are found from slender body theory. These linear values are then adjusted for nonlinear effects based on information from wind tunnel databases.[5,6,7] As for the wing-alone case, these databases provide only total force and moment measurements and, therefore, a method must be developed to distribute the nonlinear component of K and k along the span.

As mentioned already, the Chapter 3 methods compute the spanwise distribution of the linear normal force of an isolated lifting surface. It will be assumed that nonlinear contributions to the wing-alone normal force and interference effects represented by the K and k factors will be distributed in the same manner. Thus, at any y location, local load $c_n(y)$ for the wing and tail will be

$$[c_n(y)]_W = \left([c_n(y)]_L\right)_W \left(\frac{C_{N_{W(B)}}}{\sum\limits_{\text{SPAN}} \left([c_n(y)]_L\right)_W} \right) \tag{4}$$

and

$$[c_n(y)]_T = \left([c_n(y)]_L\right)_T \left(\frac{C_{N_{T(B)}}}{\sum\limits_{\text{SPAN}} \left([c_n(y)]_L\right)_T} \right) \tag{5}$$

5. Wing–Tail Interference

Chapters 3 and 5 methods give only the total value of the wing–tail interference term [$C_{N_{T(V)}}$ in Eq. (1)]. In this instance, however, not only is a method of spanwise distribution on the tail needed, but it is also apparent that a portion of the downwash effects should be carried over to the body. The division of $C_{N_{T(V)}}$ between the tail and body will be considered first. The method in which this objective is accomplished is not rigorous but does have some intuitive foundation and is probably as well as can be done with the limited amount of CFD data available.

The assumption is made, from slender body theory, that the vortices from the forward lifting surfaces are shed from a spanwise location, y_{VORT}, of

$$y_{\text{VORT}} = \left(\frac{\pi}{4}\right)\left(\frac{b_w}{2}\right) + \frac{D_B}{2} \tag{6}$$

and travel straight back parallel to the velocity vector from this point. A relative value for the effect of this shed vortex on the tail surface at any chordwise point is given by

$$VC(y) = \left(\frac{1}{r_{\text{VORT}} + y_{\text{DIST}}}\right)^2 c(y)_T \tag{7}$$

where

$$y_{\text{DIST}} = |y_{\text{VORT}} - y| \tag{8}$$

and r_{VORT} is a "vortex radius" taken to be 0.04. This value of r_{VORT} could be subject to adjustment in the future as additional CFD or experimental data become available. An inspection of Eq. (7) shows that, at any point y, it consists of the product of a factor that is proportional to the local vortex strength through the inverse square distance term and a second factor, represented by the local chord, which is related to the local area affected by the vortex. The use of an absolute value for y_{DIST} ensures that the relation will be valid regardless of where the vortex location falls on the tail span, even if it lies outboard of the tip. The function of r_{VORT} is to prevent singularities if this distance becomes zero. On the body, a similar relation is used.

$$VC(y) = \left(\frac{1}{r_{\text{VORT}} + y_{\text{DIST}}}\right)^2 (c_r)_T \tag{9}$$

Note that the use of the root chord for the area factor assumes that only the body area directly between the tail surfaces is affected. This assumption is not strictly correct, but it offers a starting point until further refinements can be made.

Using these relations, the apportionment of the wing–tail interference losses can be accomplished by summing the VC terms for both the body and tail and then taking the ratios of the summations to their combined total.

$$\left[C_{N_{T(V)}}\right]_{\text{BODY}} = C_{N_{T(V)}}\left(\frac{\sum\limits_{\text{BODY}} VC(y)}{\sum\limits_{\text{BODY}} VC(y) + \sum\limits_{\text{TAIL}} VC(y)}\right) \tag{10}$$

$$\left[C_{N_{T(V)}}\right]_{\text{TAIL}} = C_{N_{T(V)}}\left(\frac{\sum\limits_{\text{TAIL}} VC(y)}{\sum\limits_{\text{BODY}} VC(y) + \sum\limits_{\text{TAIL}} VC(y)}\right) \tag{11}$$

It will be assumed that the portion of the wing–tail interference load that is allocated to the body by Eq. (10) is distributed evenly along the body directly between the tail surfaces. This additional contribution is added to the total local body load at streamwise locations between the leading and trailing edges of the tail surfaces.

The portion of the wing–tail interference loads assigned to the tail surfaces by Eq. (11) is distributed across the span of the tail by assuming that its magnitude at any y location is proportional to the $VC(y)$ value [from Eq. (7)] at that location.

$$\left[c_n(y)\right]_{T(V)} = \left[C_{N_{T(V)}}\right]_{\text{TAIL}}\left(\frac{VC(y)}{\sum\limits_{\text{SPAN}} VC(y)}\right) \tag{12}$$

These loads are added to those computed in Eq. (5) to get the total load at any spanwise location.

D. Changes for the $\varphi = 45$ deg Roll Position

For the $\varphi = 45$ deg roll position, several changes must be made to the above methodology to produce a more accurate representation of surface loads. It is obvious that there must be some way of accounting for the different aerodynamic environments experienced by windward and leeward plane fins. In addition, distributions of the body carryover loads induced by lifting surfaces behave in a somewhat different fashion than for the $\varphi = 0$ deg case. The following discussion describes modifications made to the methodology presented to accommodate these differences.

1. *Body–Wing and Body–Tail Interference Load Changes*

The body-alone normal force is considered to be independent of roll position. However, the body–wing and body–tail interference effects will change with the roll configuration. The values of these interference factors were chosen to give good agreement with experiment for the total loads and center of pressure. Unfortunately, when compared to CFD data, it appears that they do not produce an accurate distribution of those loads along the body.

On the surface of the body between the lifting surfaces, there is a region of high dynamic pressure. Immediately behind the lifting surfaces, however, if there is an afterbody present, there is an area where the dynamic pressure is much lower. The current nonlinear methodology of Chapter 5 at $\varphi = 45$ deg averages this difference to get the right total normal force and adds a center of pressure shift that accounts for inconsistencies in distribution. If distribution of the body carryover loads is to be modeled correctly, some consideration must be given to the physics of the problem. What appears to be happening, based on experimental data,[1] is that, in the region of the lifting surfaces, there is an enhancement to the body load at both the $\varphi = 0$ deg and $\varphi =$ deg roll positions. At $\varphi = 0$ deg, this enhancement can extend several body diameters downstream into the traditional Mach line influence region. The method used to model this effect was discussed previously. At $\varphi = 45$ deg, the body loads are also strongly influenced in the immediate vicinity of lifting surfaces. Downstream of the fins, however, the high dynamic pressure appears to decrease rapidly and produce a region of overexpansion where body carryover actually becomes negative and total loads fall below those of the body alone. Thus, with an afterbody present, the body carryover load is lower than with no afterbody. (Note that the reverse is true at the $\varphi = 0$ deg roll orientation.)

Chapter 5 models this effect with a lower minimum value of $K_{B(W)}$ at $\varphi = 45$ deg versus $\varphi = 0$ deg and a forward shift of the center of pressure. This approach does not, however, produce proper distribution of interference loads. The following technique was used to partially accomplish the goal of accurate interference load distribution at $\varphi = 45$ deg. It is first assumed that, for the purpose of computing the body carryover load distribution, the minimum value of $K_{B(W)}$ is the same at $\varphi = 45$ deg as it is at $\varphi = 0$ deg. This value of $[K_{B(W)}]_{\min}$ is distributed in the vicinity of the lifting surfaces for both roll orientations as described for $\varphi = 0$ deg roll. However, for the $\varphi = 45$ deg roll position, an amount equal to

$$\Delta K_{B(W)} = \left(\left[K_{B(W)}\right]_{\min}\right)_{\varphi=0^\circ} - \left(\left[K_{B(W)}\right]_{\min}\right)_{\varphi=45^\circ} \tag{13}$$

is subtracted from $K_{B(W)}$ in the region downstream of the fins. This approach appears to represent reasonably well the interference load distribution at $\varphi = 45$ deg while maintaining correct values of forces and moments.

2. *Lifting Surface Load Changes*

When a missile is rolled into the "x" ($\varphi = 45$ deg) position, the lower (windward plane) and upper (leeward plane) lifting surfaces will experience different aerodynamic environments. The normal force computed by the Chapters 3 and 5 methods is the total for all fins, both windward and leeward, and it must be divided between the two planes. To accomplish this division, the approach previously used in Chapter 5 is used. In agreement with slender body theory, it is assumed that, at an AOA of zero, the normal force is evenly distributed between the windward and leeward planes. As the AOA is increased, the load on the windward surfaces is increased linearly up to 90% of the total at $a = 65$ deg. This percentage remains constant up to $a = 90$ deg. On the other hand, the load on the leeward plane decreases in a corresponding fashion so that the total of the two is not changed. Thus, we have

$a \leq 65°$:

$$\left[c_n(y)\right]_{\text{WINDWARD}} = \left[c_n(y)\right]\left(0.5 + 0.4\frac{a}{65}\right) \tag{14}$$

$$\left[c_n(y)\right]_{\text{LEEWARD}} = \left[c_n(y)\right]\left(0.5 - 0.4\frac{a}{65}\right) \tag{15}$$

$a > 65°$:

$$\left[c_n(y)\right]_{\text{WINDWARD}} = 0.9\left[c_n(y)\right] \tag{16}$$

$$\left[c_n(y)\right]_{\text{LEEWARD}} = 0.1\left[c_n(y)\right] \tag{17}$$

A further point to consider is that the normal forces that are computed are normal only to horizontal lifting surfaces in the $\varphi = 0$ deg roll orientation. In the $\varphi = 45$ deg case, the loads given by Eqs. (14) to (17) must be divided by $\cos\varphi$ on the windward side and $\sin\varphi$ on the leeward side to relate them to structural loads that are normal to the surfaces.

The portion of the wing–tail interference $C_{N_{T(V)}}$ that is allocated to the tail surfaces must also be divided between the windward and leeward planes. Once again, relying on methodology developed in Chapter 5, this division is accomplished using the wing–tail interference factors defined during the modified slender body derivation of this term. For the present purpose, if i_1 is the interference factor associated with the windward plane fin and i_4 the one with the leeward plane fin, then

$$\left[C_{N_{T(V)}}\right]_{\text{WINDWARD}} = \left[C_{N_{T(V)}}\right]_{\text{TAIL}} \left(\frac{i_1}{i_1 + i_4}\right) \tag{18}$$

and

$$\left[C_{N_{T(V)}}\right]_{\text{LEEWARD}} = \left[C_{N_{T(V)}}\right]_{\text{TAIL}} \left(\frac{i_4}{i_1 + i_4}\right) \tag{19}$$

This brings us to the second physical phenomenon, which is not explicitly modeled in the approach outlined above. In the $\varphi = 45$ deg roll position, there is a "shadowing" effect on the leeward plane lifting surfaces for high Mach number flows. A "shadowed region" exists near the root chord where Newtonian theory assumes a pressure coefficient of zero. As a result, predicted values of loads on leeward plane lifting surfaces to be presented later will be high but, due to the complexities involved in trying to model this effect, no attempt has been made to do so at this time. The higher the Mach number, the more accurate becomes the Newtonian prediction of zero pressure coefficient in the shadowed region. This means that the current leeward plane loads near the body will become less accurate as Mach number increases.

The final phenomenon that is believed to affect load distributions, but is not yet fully modeled, involves the stronger influence of body vortices on leeward plane fins as their location is moved farther aft on the body. In the development of the semi-empirical approach of Chapter 5, body vortex effects were inherently included in the body carryover load. Hence, the body carryover load was lower for the tail fins than it should have been to offset the fact that vortices actually cause the leeward plane tail fins to lose more normal force the farther aft they are located. To account for this fact, negative contributions from the adjusted body-lifting surface interference loads, which cannot be distributed on the afterbody due to lack of sufficient length, will be carried over to the leeward plane tail fins. This additional interference load will be distributed along the span in the same manner as linear and nonlinear lifting surface loads.

E. Loads, Shear, and Bending Moments

The primary purpose of computing load distributions over the components of a missile body is to allow the determination of shear and bending moment within the structural members. The loads that have been determined so far are nondimensional in the sense that they represent a normal force coefficient per unit length. Multiplying them by the dynamic pressure and the reference area will give local loads in pounds per foot. These loads can be integrated to get the local shear in pounds, and the shear can, in turn, be integrated to get the local moment in pound-feet. For the body, utilizing the results of the theoretical discussions, we have for the dimensionalized local load per unit length, $w(x)$:

$$w(x)_B = \left[(c_{nB})_L + (c_{nB})_{NL,\mathrm{ADJ}} + \Delta c_{nB(W)} + \Delta c_{bB(T)} + (c_{nB})_{T(V)}\right] QA_{\mathrm{ref}} \tag{20}$$

We can then integrate to get the shear, $V(x)_B$, and the bending moment, $BM(x)_B$:

$$V(x)_B = \int_0^x w(x)_B \, \mathrm{d}x \tag{21}$$

$$BM(x)_B = \int_0^x V(x)_B \, \mathrm{d}x \tag{22}$$

The procedure for the lifting surfaces is the same except for the spanwise, rather than the streamwise, variation of the loads. For the wing, using the nondimensional load from Eq. (4), we get

$$w(y)_w = [c_n(y)]_w \, QA_{\mathrm{ref}} \tag{23}$$

For the tail, there may be a contribution from the wing–tail interference in addition to the nondimensional load from Eq. (5). That is,

$$w(y)_T = [c_n(y)]_T + \left[c_n(y)_{T(V)}\right] QA_{\mathrm{ref}} \tag{24}$$

Equations (23) and (24), as presented above, are valid for the $\varphi = 0$ deg roll position. For the $\varphi = 45$ deg case, there will be separate loads for the windward and leeward sets of lifting surfaces. These separate loads can be computed simply by substituting the windward and leeward values of the $[c_n(y)]_w$, $[c_n(y)]_T$, and $[c_n(y)]_{T(V)}$ contributions to Eqs. (23) and (24). The integration for shear and bending moment will be the same for both lifting surfaces and for both roll orientations. With the W/T subscript indicating the appropriate value for either the wing or tail, we have

$$V(y)_{W(T)} = \int_0^y w(y)_{W/T}\,dy \tag{25}$$

$$BM(y)_{W/T} = \int_0^y V(y)_{W/T}\,dy \tag{26}$$

All of the above integrations can be performed with acceptable accuracy using the trapezoidal rule.

F. Method Application

For purposes of validation, results computed by the new methodology are compared to TLNS[3] computer calculations. Distributions of aerodynamic loads, shear, and bending moment were generated[4] by both methods for the wing–body–tail missile configuration of Fig. 8.1 at Mach numbers of 1.50, 2.87, and 4.60 and AOAs of 10, 25, and 40 deg. The configuration shown in Fig. 8.1, with its two sets of lifting surfaces, provides an opportunity to exercise all of the new load distribution methodology. Results of these computations are presented and discussed below. The general format followed is to handle each component (body, wing, tail) separately. Each figure has the local load, shear, and bending moment for each missile component. In all cases, CFD results are plotted together with the methods described in Section 8.2 of this chapter for comparison purposes. The Section 8.2 results are indicated by the 1998 Aeroprediction Code (AP98) on the figures. Due to the number of figures given in Ref. 4, only the highest AOA case (40 deg) will be presented here. This case illustrates some of the more extreme shock interaction results and is the worst case for comparison of semi-empirical load results with CFD in terms of accuracy of loads. Other AOA and additional cases are given in Ref. 4.

Figures 8.2 to 8.4 give the $\varphi = 0$ deg roll results for the body, wing, and tail, respectively. Figures 8.2a, 8.3a, and 8.4a give the local load; Figs. 8.2b, 8.3b, and 8.4b give the shear; and Figs. 8.2c, 8.3c, and 8.4c give the bending moments. In general, the approximate method (in the form of the AP98) comparisons to the CFD results are very good for the body, wing, and tail. Areas of disagreement are primarily on the body load, in the nose region, and in the vicinity of the wing where the AP98 results are slightly high in the nose region (Fig. 8.2a), and the interference results tend to be "smeared" out in the vicinity of the wing. On average, when the loads are integrated, excellent results are obtained for shear and bending moment, however. The only area of disagreement on the wing and tail is the load near the wing root at the higher Mach number. Once again, when the loads are integrated, shear and bending moment comparisons to the CFD results are very good in general.

Figures 8.5 to 8.9 give the $\varphi = 45$ deg roll results. Here, the figures follow the same pattern as for Figs. 8.2 to 8.4, except both the windward and

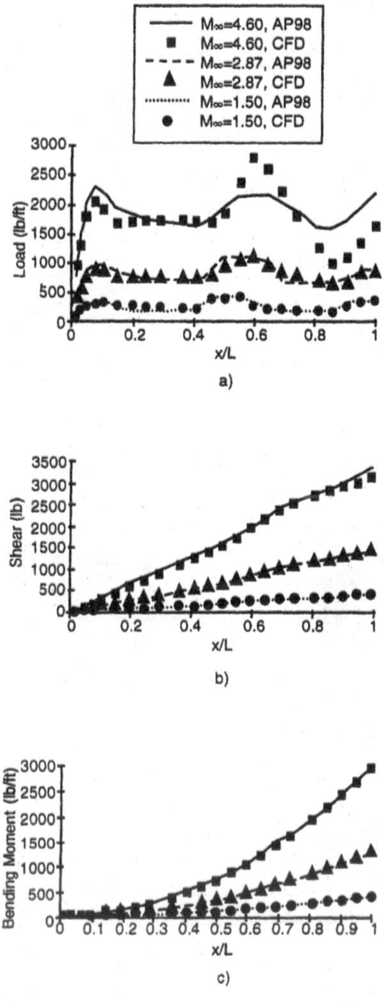

Fig. 8.2 Body load, shear, and bending moment ($\varphi = 0°$ and $\alpha = 40°$) (a, b, and c, respectively).

leeward plane wing and tail results are shown. Once again, the reader is reminded that these results are normal to the lifting surfaces or body. This means that, to compare shear and bending moments to the aerodynamic normal force and pitching moments, results for the windward and leeward fins must be added and multiplied by 0.707.

Note that, for the $\varphi = 45$ deg body load, CFD predictions show a sharp rise for body loads in the vicinity of the wings followed by a rapid decrease below body-alone values behind the wings. This behavior is only partially captured in the AP98 results that show a smoother rise with a lower peak and a lesser decline behind the wings. As discussed previously, modifications were made to the body carryover methodology for the $\varphi = 45$ deg roll

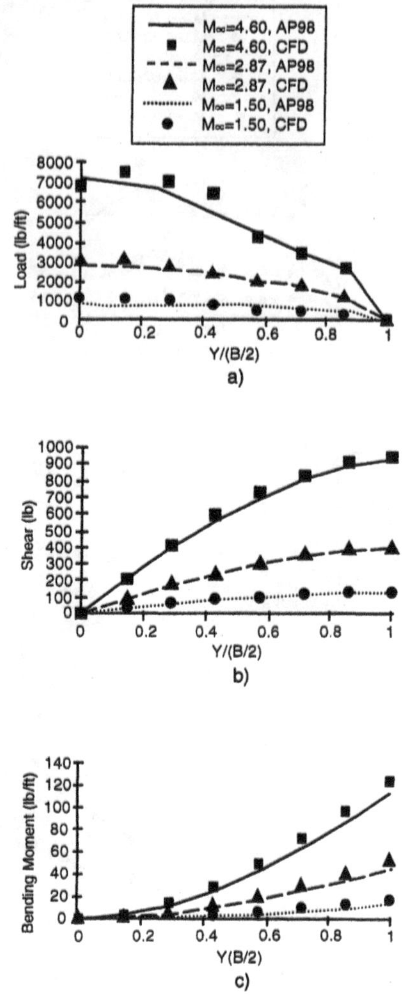

Fig. 8.3 Wing load, shear, and bending moment ($\varphi = 0°$ and $\alpha = 40°$) (a, b, and c, respectively).

position to improve the modeling of this phenomenon. These changes resulted in an improvement in the code's performance in this region, but the sharp surface pressure variations are still not completely modeled. The physical mechanism that gives rise to the pressure spike and following overexpansion is not completely understood at the present time and, because it is most likely associated with strong shock interactions on the surface, it is considered beyond the scope of the present work to try and develop a more rigorous model. Figures 8.5b and 8.5c show the results for the shear force and bending moment distributions. Agreement with CFD results is quite good, especially considering body load variations.

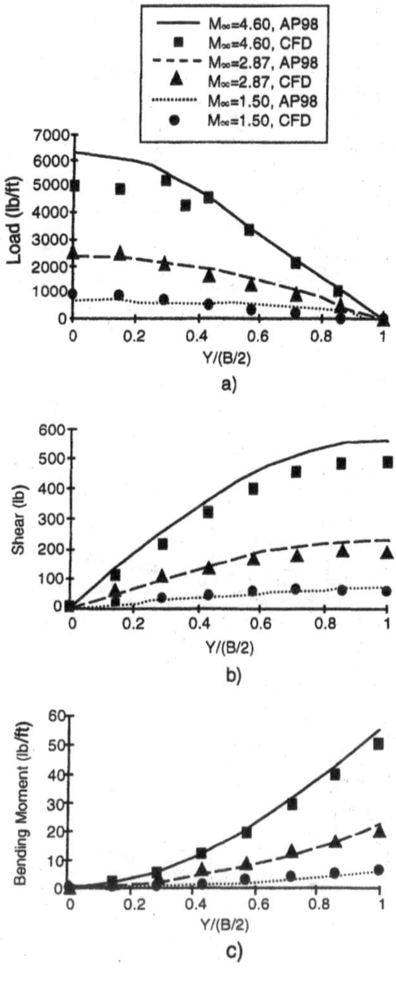

Fig. 8.4 Tail load, shear, and bending moment (φ = 0° and α = 40°) (a, b, and c, respectively).

The spanwise load distributions for the windward plane wings are plotted in Fig. 8.6a. The agreement between the AP98 and CFD results is very good, with only some minor deviations at $M_\infty = 4.60$. The shear loads are shown in Fig. 8.6b. Again, the agreement is quite good between the two methods. The bending moments, presented in Fig. 8.6c, show a general tendency to be slightly low relative to the CFD results, but overall agreement is good.

The load distributions for the leeward wings are shown in Fig. 8.7a. It can be seen that the "shadowing" effect discussed previously is quite evident for the $M_\infty = 2.87$ and $M_\infty = 4.60$ cases. The CFD results predict a large pressure loss on the wings in the shadowed region near the body, which is not modeled in the AP98. Otherwise, agreement is very good. The shear distri-

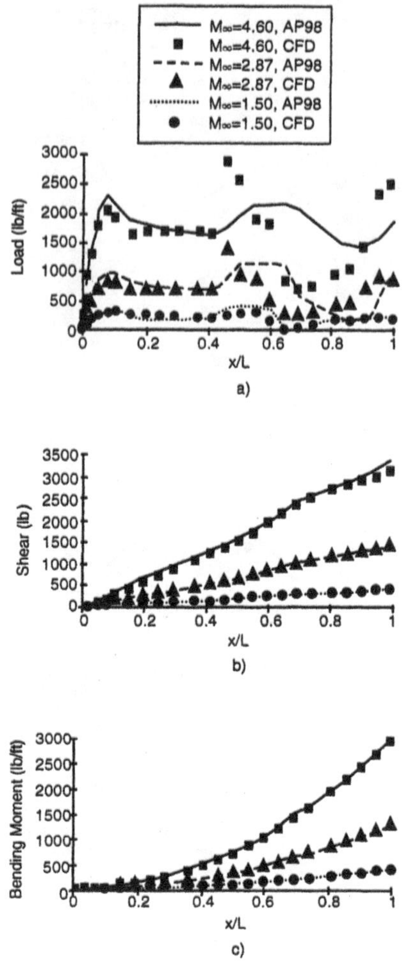

Fig. 8.5 Body load, shear, and bending moment ($\varphi = 45°$ and $\alpha = 40°$) (a, b, and c, respectively).

butions are shown in Fig. 8.7b and, as might be expected, there is significant overprediction at the higher Mach numbers relative to CFD results, but the results are still surprisingly good. The same comments apply to the bending moments, which are presented in Fig. 8.7c.

Figure 8.8 shows the windward tail loads, shear, and bending moment. Agreement between the two approaches is quite good.

The leeward tail loads are plotted in Fig. 8.9a. For both the leeward wings and tails, the "shadowing" effect is very obvious at the two higher Mach numbers. Shear loads, shown in Fig. 8.9b, and bending moments, shown in Fig. 8.9c, are overpredicted relative to CFD results because of this effect. It should be noted that, by modifying the body carryover prediction methodology for the $\varphi = 45$ deg case and applying any excess

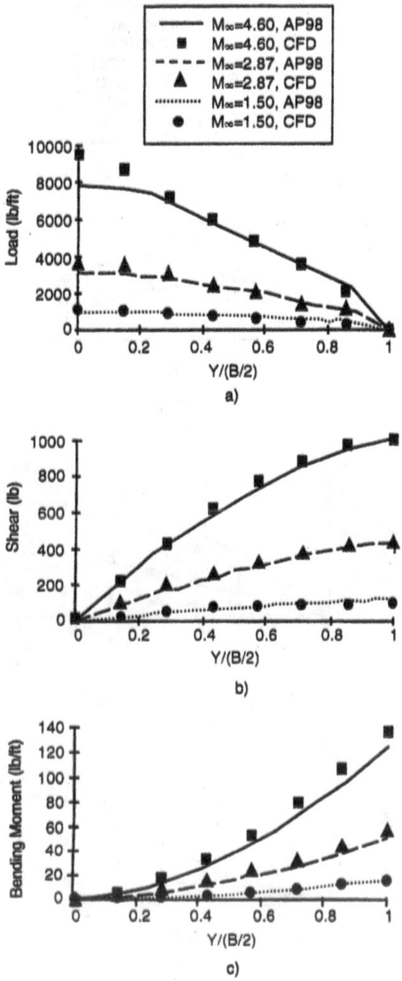

Fig. 8.6 Windward wing load, shear, and bending moment (φ = 45° and α = 40°) (a, b, and c, respectively).

negative contributions to the leeward tail, the results for the leeward tail have been improved substantially over their original values. However, because of the complex nature of leeside flowfields at high AOAs, it is extremely difficult to develop empirical models that will give truly accurate results in these regions.

In summary, methodology has been developed to compute both the linear and nonlinear contributions to surface loads and to distribute these loads over body and lifting surfaces. The new methods include models to simulate lifting surface interference effects on the body for both the φ = 0 deg and φ = 45 deg roll positions. At φ = 45 deg, load distributions are computed for both windward and leeward lifting surfaces. Wing–tail in-

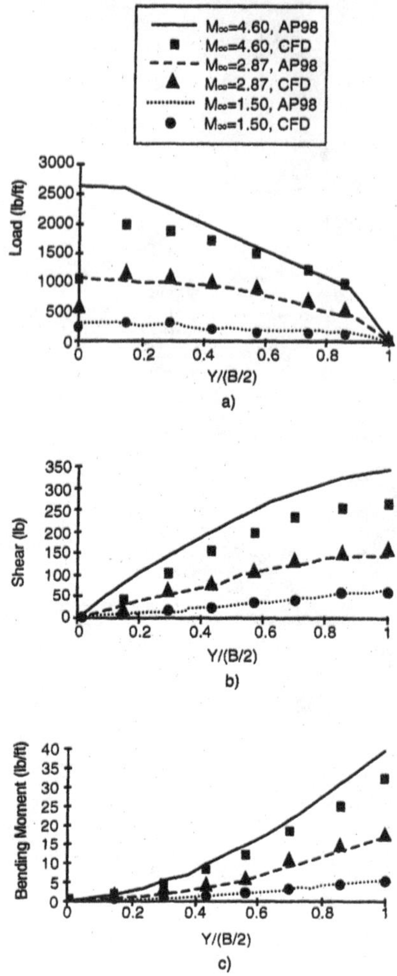

Fig. 8.7 Leeward wing load, shear, and bending moment (φ = 45° and α = 40°) (a, b, and c, respectively).

terference loads are now separated into separate components that are distributed along the body in the vicinity of the tail as well as across the tail surfaces themselves. All loads per unit length are integrated to give shear and bending moment distributions that may be used by structural engineers in the design process.

With these new additions, semi-empirical methods should prove to be much more useful to the structural engineer who is interested in performing preliminary structural analyses of several competing missile designs. To the author's knowledge, it is the only semi-empirical method available to distribute nonlinear aerodynamic loads.

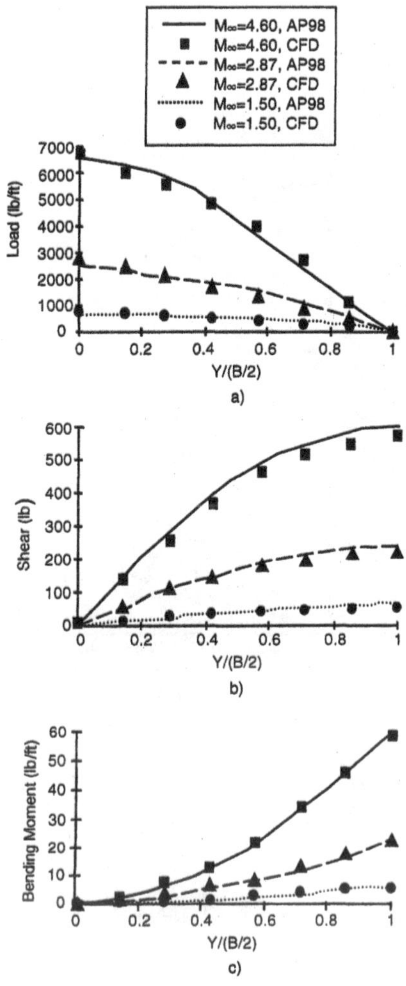

Fig. 8.8 Windward tail load, shear, and bending moment (φ = 45° and α = 40°) (a, b, and c, respectively).

III. Minimum Drag Shapes

There have been numerous attempts at minimizing drag by investigators over the last 60 years. Many of the earliest studies were aimed at deriving minimum drag bodies based on wave drag or wave and skin-friction drag, but not base drag. For example, Von Karman[9] derived a nose shape defined by

$$r = \frac{r_{\text{ref}}}{\sqrt{\pi}} \sqrt{\varphi - \frac{1}{2} \sin 2\varphi} \tag{27}$$

where $\varphi = \cos^{-1} [1 - 2\,x/\ell]$

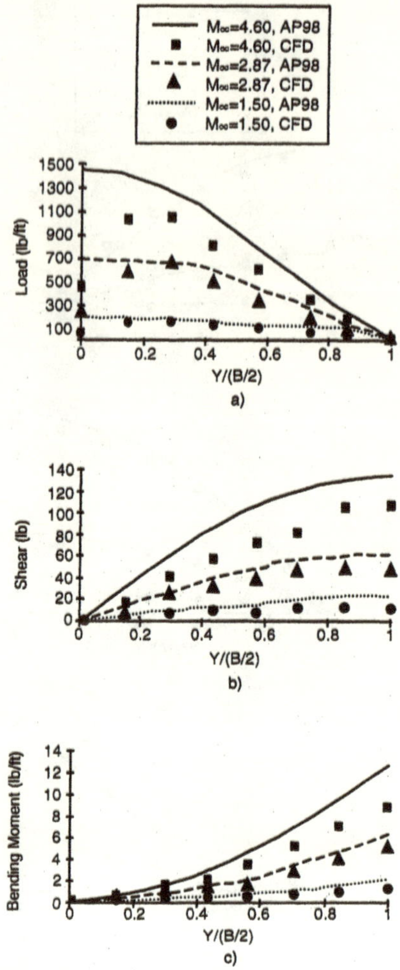

Fig. 8.9 Leeward tail load, shear, and bending moment (φ = 45° and α = 40°) (a, b, and c, respectively).

This shape was derived from linearized theory using the calculus of variations where the boundary conditions chosen for drag minimization were constant length and diameter. Haack[10] generalized Eq. (27) to

$$r = \frac{r_{\text{ref}}}{\sqrt{\pi}} \sqrt{\varphi - \frac{1}{2} \sin 2\varphi + C \sin^3 \varphi} \tag{28}$$

where $\varphi = \cos^{-1} [1 - 2\,x/\ell]$.

C of Eq. (28) is zero for an optimum shape with length and diameter as the boundary conditions; C is 1/3 for length and volume as the boundary conditions. The latter condition is commonly referred to as the Haack nose shape. Both Eqs. (27) and (28) were derived based on linear theory, which is primarily applicable below about $M_\infty = 3.0$. Cole[11] used Newtonian flow theory, in conjunction with the calculus of variations to determine ogive shapes of minimum wave drag for a given length and diameter. He found that the optimum nose shapes could be closely approximated by power law bodies defined by

$$r = r_{\text{ref}}\left(\frac{x}{\ell}\right)^n \tag{29}$$

Here $n = 3/4$ if no centrifugal force effects of the flow around the ogive were accounted for and $n = 2/3$ if centrifugal force effects were accounted for.

Brown attempted to develop shapes of minimum drag when considering both wave and skin friction drag[12] but, as in Refs. 9 to 11, base drag was not considered. However, as seen in Fig. 4.28, base pressure, and hence base drag, can be fairly high for Mach numbers less than 3 to 4. At transonic speeds, it is not unusual for the base drag to account for 40 to 50% of total drag. As a result, the present author derived an approximate shape for minimum total drag.[12] This shape is given in Fig. 8.10. The key elements of this shape are as follows: 1) the ratio of nose length to total length is 0.6. 2) The afterbody consists primarily of a boattail with a base diameter of 70% of the maximum diameter. 3) The nose was slightly blunted with a nose radius that was 20% of the maximum radius. 4) The volume was kept as high as possible while keeping the drag as low as possible. Wind tunnel tests[12] indeed verified that the shape of Fig. 8.10 gave significantly lower drag than an ogive (either Von Karman or 2/3 power law) of the same length and diameter as the total configuration length and diameter of the optimum shape in Fig. 8.10.

Reference 12 developed the optimum projectile shape of Fig. 8.10 primarily from an empirical standpoint. The approximate minimum drag shape Fig.

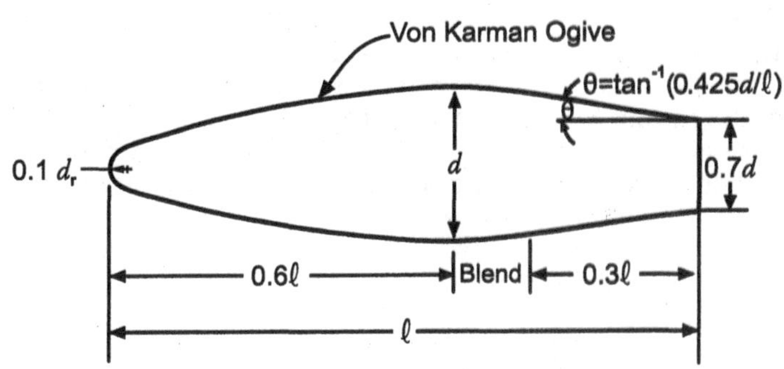

Fig. 8.10 Approximate minimum drag projectile.

8.10 has a 20% blunt Von Karman ogive nose that is 60% of the total length of the body. This is followed by a smooth curve that blends the ogive with the boattail that leads to a base diameter that is 70% of the maximum diameter. This configuration was primarily derived for low supersonic and transonic Mach numbers $0.8 \le M_\infty \le 2.5$. For higher Mach numbers, the bluntness should be reduced and the nose length increased as a percent of the total body length. In no case should the boattail angle exceed about 7 to 8 deg. Later on (see Ref. 13 and 14), an analytical approach was applied to the same problem, where the total drag was defined as

$$C_D = \frac{2\pi}{S_r}\int_0^{\ell} C_p(x)\, r\left[r'(x)\right]^2 \mathrm{d}x + C_{f\infty}\frac{A_W}{A_{\mathrm{ref}}} - C_{P_B}\left(\frac{d_B}{d_r}\right)^2 \tag{30}$$

The first term of Eq. (30) is the wave drag integral; the second term, the skin friction drag; and the third term, the base drag. The pressure coefficient of the wave drag term was estimated using a combination of Newtonian theory and first-order shock expansion theory to simplify the mathematical optimization process. The skin friction and base drag terms were approximated by the methods discussed in Chapter 4. It should be pointed out that the base pressure coefficient of Eq. (30) is that on a conical boattail shape. If it is desired to use the C_{P_B} of a cylindrical afterbody from Fig. 4.28, the $(d_B/d_r)^2$ of Eq. (30) should be $(d_B/d_r)^3$.

Figure 8.11 gives the value of total drag as a function of the forebody length to total length ratio, ℓ_f/ℓ, for several Mach numbers and for a $\ell/d = 5$ configuration, using Eq. (30). Because only first-order methods were used to predict wave drag, the absolute value of drag is low by 10 to 20% in the figure. However, the important point from this figure is the variation in the value of ℓ_f/ℓ where the minimum drag occurs as a function of Mach number. For low Mach numbers, a value of ℓ_f/ℓ of about 0.6 to 0.65 appears to be optimum (lowest drag), closely matching the empirical selection of Fig. 8.10. However, as Mach number increases to 5, the value of ℓ_f/ℓ, where the minimum drag occurs increases to 0.8. This agrees with one's intuition because base drag goes to zero as Mach number increases to infinity and reaches its maximum value around $M_\infty = 1.1$.

The other important point from Fig. 8.11 is the fact that, around the minimum drag point, one can change the value of ℓ_f/ℓ by 5% with only a slight drag penalty. This is illustrated in Fig. 8.12, which plots ℓ_f/ℓ versus Mach number of the minimum drag shape and a drag shape that gives a 1% increase. This figure is quite comforting to the designer because it shows that, if a design is optimized about a given Mach number and for a given ℓ/d, only a slight drag penalty occurs for a slightly nonoptimum shape.

Another point worth illustrating in Fig. 8.10 is the effect on drag of a configuration with a given nose length as a function of nose bluntness. This is illustrated in Fig. 8.13 by plotting the percent increase in forebody drag of

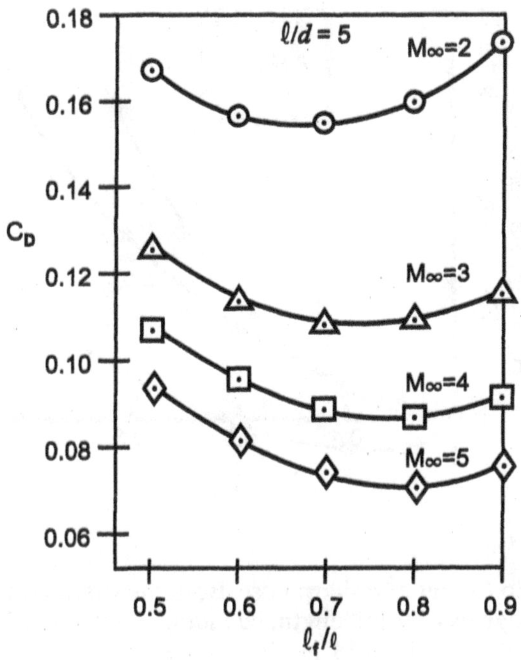

Fig. 8.11 Approximate drag coefficient as function of nose length for ℓ/d = 5.

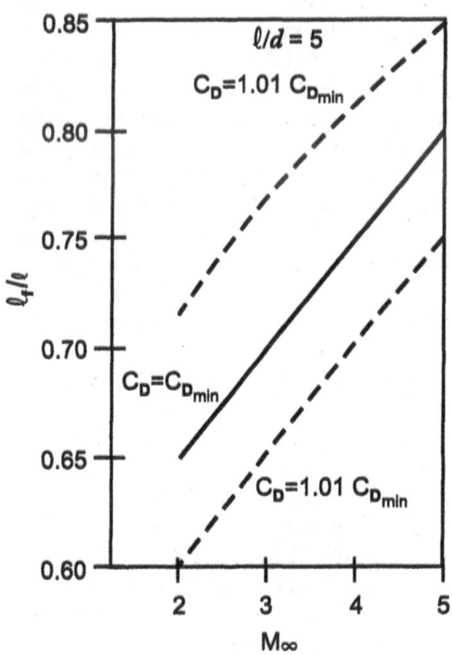

Fig. 8.12 Drag penalty for nonoptimum configurations for ℓ/d = 5.

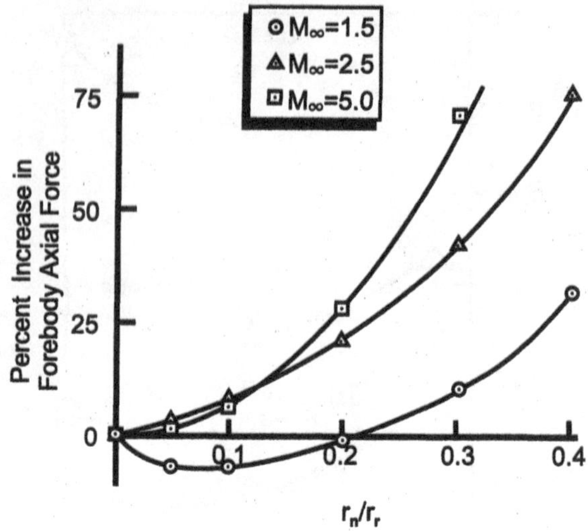

Fig. 8.13 Percent change in forebody axial force coefficient as a function of nose bluntness (ℓ_N = 3.0-caliber tangent ogive).

a nose with some bluntness ratio to that of a sharp nose. Both the sharp and blunt nose shapes are tangent ogives of total length, including bluntness, of 3.0 calibers. However, the results of Fig. 8.13 holds true for other ogive shapes and lengths as well. Only the forebody drag is considered because base drag is constant. These results are plotted as a function of nose bluntness ratio for Mach numbers 1.5, 2.5, and 5.0. The second-order axial force methods of Chapter 4 are used to compute the wave drag and Van Driest II for skin friction drag. Note that, for the M_∞ = 1.5 case, nose bluntness initially decreases the axial force and gives only about a 7.5% increase in axial force for a 30% nose bluntness compared to a sharp ogive. However, for higher Mach numbers, nose bluntness has a much stronger effect on drag. The absolute optimum drag configuration is, therefore, a function of Mach number. However, in most cases, the design engineer is interested in as large a volume obtainable with as little a drag penalty as possible. That is the reason for the 20% bluntness selected for the optimum shape of Fig. 8.10. If one were not concerned about volume and was flying at M_∞ = 2.0 and greater, the percent bluntness should be reduced to 5% or less for minimum drag.

The final point to be made concerning Fig. 8.10 has to do with the afterbody or boattail shape. In viewing Eq. (30), the implication is that, if one chose $d_B = 0$, the base drag would go to zero. However, this is not true due to viscous effects. Experience has shown that the boundary layer will separate at about 7.0 to 8.0 deg boattail angle on a 1.0-caliber boattail. As the boattail length increases, the angle where separation begins to occur also decreases. Thus, to optimize volume while getting the lowest practical value of base drag, short of using boundary-layer suction techniques, a base diameter of about 70% of the maximum diameter is about the lowest diameter for most practical shapes. A diameter smaller than this will give a larger separation region. This larger separation region effectively defines the base diameter for base drag computations.

IV. Multifin Weapon Aerodynamics

A. Introduction and Background

Chapter 1 discussed the requirement for computing aerodynamics of multifin weapons during design tradeoff studies. As a result of this desire to investigate various multi-tail-fin alternatives for providing static stability and aerodynamics on a given design, this section of Chapter 8 will define a method that can be used in conjunction with the planar and cruciform fin methodology of Chapters 3 to 5 to compute aerodynamics of six- and eight-fin configurations.

Typical weapon configuration design and control alternatives, for which aerodynamics are desired, are shown in Fig. 8.14. These configurations define the general requirements to be considered in the analytical development methodology for multifin configurations. In general, one can have a body–tail configuration that is either guided or unguided. If it is unguided, four, six, or eight tail fins can be assumed. On the other hand, if it is tail-controlled, the author is not aware of any tail control alternatives other than for cruciform (four) fins. Hence, this will be the requirement for tail-controlled weapons.

Canard or wing–body–tail configurations have more options for control than body–tail. The control can be from the canards or wings, in which case there will be two or four canards (wings) present and either four, six, or eight tail surfaces. Here, the tail surfaces are used exclusively for stability. For the tail control option, the forward set of lifting surfaces can have two, four, six, or eight fins but, again, the tail controls are assumed to be cruciform. In effect, the above alternative design and control constraints are placed on the aerodynamics methodology from a practical standpoint.

The above set of requirements had to do with practical configurations for which aerodynamics are desired. A second set of requirements in the analytical methodology development has to do with the methodology development approach in Chapters 3 to 5, which is for planar or cruciform fins. This latter set of requirements is driven by how to most cost-effectively integrate multifin computations to a computer code set up for two- or four-fin alternatives. This requirement thus leads one to define factors by which the two- or four-fin aerodynamics can be multiplied to make the minimum amount of changes to an aerodynamics code designed for two or four fins.

The goal is, therefore, to develop factors by which the aerodynamics of two- and four- fin configurations can be multiplied to give acceptable accuracy for aerodynamics of the six- and eight-fin cases shown in Fig. 8.14. While these factors are being derived primarily for integration to a modified version of the AP98,[15] they could be used by other agencies in conjunction with estimates of four-fin aerodynamics to estimate the aerodynamics of six- and eight-fin configurations if desired. The work described in this section is presented in detail in Ref. 16, and the interested reader is referred to that reference for more details.

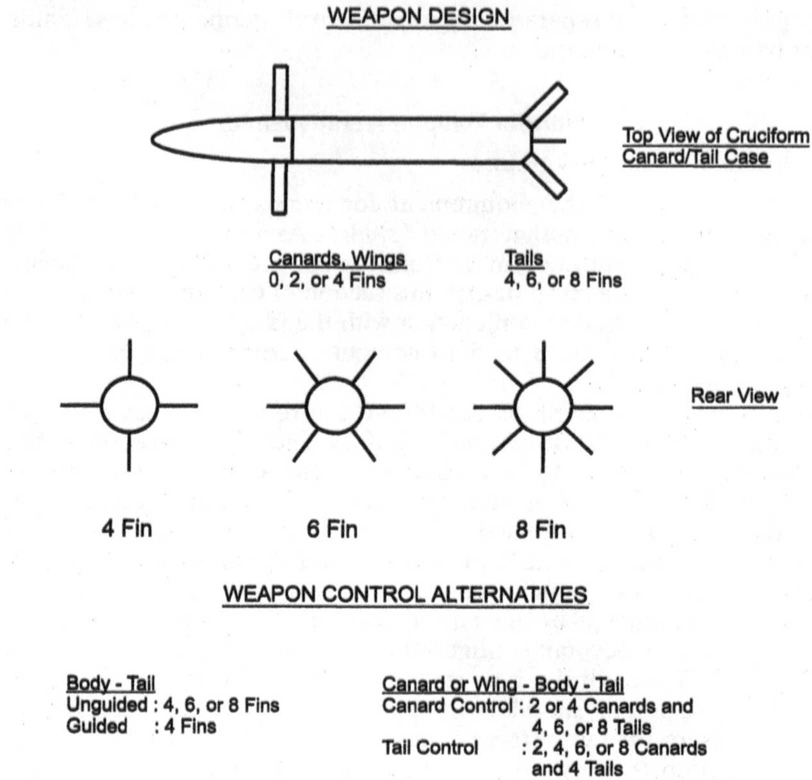

Fig. 8.14 Typical weapon design and control alternatives.

B. Approach and Analysis

The multifin aerodynamics we are interested in include the effect on axial force, normal force, center of pressure, roll damping moment, and pitch damping moment. Magnus moment is presently assumed to be zero for fin-stabilized weapons due to low spin rates (this assumption can be inaccurate if the spin rate is greater than a few cycles per second). Each of these aerodynamics will be considered individually in the following discussion.

Based on past experience with calculation of axial force of two fins versus four fins, we have found reasonable accuracy by taking the axial force coefficient of a single fin and multiplying it by the number of fins of interest. No interference effect between the body and fins (of reasonable thickness), other than the base pressure, has been observed. Hence, this assumption will be made for the axial force of weapons with more than four fins. That is,

$$C_A = C_{A_{\text{BODY}}} + (C_{A_{SF}})(\text{NF}) \tag{31}$$

where NF is the number of fins present.

It will also be assumed that the center of pressure of the total normal force contribution of multifins is the same as that of four fins. The total normal force contribution of the fins includes the normal force of the fins in presence of the body plus the additional normal force on the body as a result of the fins being present.

For roll damping moments, the method of Chapter 3 assumes that the fins go to the centerline of the body, the roll damping of planar fins is computed using linear theory, the number of fins is accounted for by slender body theory, and body interference effects are accounted for by Fig. 8.15. Figure 8.15 is taken from Ref. 17, was later revised by Nielsen,[18] and basically says that for two- or four-fin cases, the wing–body roll damping is nearly independent of r/s for values up to 0.4. After that, the roll damping goes to the body-alone value in a nearly linear fashion as r/s approaches 1.0. Also, slender body theory says that the values of roll damping, $C_{\ell p}$, of eight, six, and four fins compared to that of two fins are 2.3, 2, and 1.62, respectively. These factors are used in conjunction with the roll damping moment of the fin alone and the interference factors of Fig. 8.15 to compute roll damping of multifin configurations. The only assumption made here is that for six or eight fins, the curve of Fig. 8.15 for four fins can be used directly. This assumption is based on the fact that slender body theory shows little difference between two- and four-fin wing–body roll damping as a function of r/s, as seen in Fig. 8.15.

As far as pitch damping moment is concerned, the computational procedure is similar to that of the roll damping. The wings are assumed to extend to the centerline of the body, the pitch damping of planar wings is computed using linear theory, and then the method of Bryson[19] is used to account for the interference effects of the body in the presence of the wing. Because the wings are assumed to extend to the centerline of the body and the number

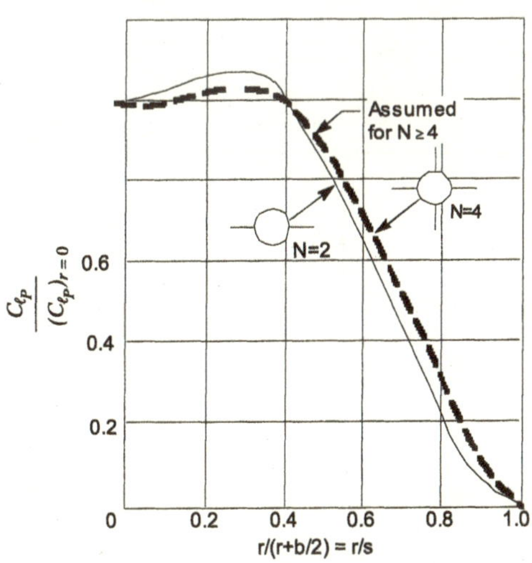

Fig. 8.15 Effect of body radius on damping in roll for fixed span (taken from Refs. 17 and 18).

of wings will be accounted for by the factor of the normal force of the wing alone, this wing–body interference factor will be less than 1. Figure 8.16 gives the slender body theory interference results for pitch damping moment for two, four, six, and eight wings computed by the methods presented in Refs. 18 and 19. As seen in Fig. 8.16, increasing the number of fins from two to eight has very little effect for small values of r/s ($r/s \leq 0.4$), but wing–body interference has an increasing effect for all fins as r/s approaches 0.6 to 0.8.

Slender body theory states that the normal force of six and eight fins is 1.5 and 2.0 times that of four fins. Experience has shown these factors to be high in general, particularly as AOA increases. Because we know all of the nonlinearities for the normal force contributions of the wing, body, and interference terms from Chapter 5, it seems logical to assume that

$$\left[C_{N_{W(B)}}, C_{N_{B(W)}}, C_{N_{T(V)}}\right]_{6,8\text{ fin}} = (F_6, F_8)\left[C_{N_{W(B)}}, C_{N_{B(W)}}, C_{N_{T(V)}}\right]_{4\text{ fin}} \quad (32)$$

where the factors F_6 and F_8 must be determined. They are the factors by which we multiply cruciform fin normal force components for six- and eight-fin configurations, respectively. These factors are the slender body theory factors of 1.5 and 2.0 mentioned previously. However, the slender body theory values will be replaced with values appropriate for a given aspect ratio wing at a given AOA and Mach number. While the factors will be determined explicitly for the wing–body plus the body–wing contributions to normal force, they will also be assumed to apply to the tail downwash $C_{N_{T(V)}}$ and pitch damping moment for the time being. The remainder

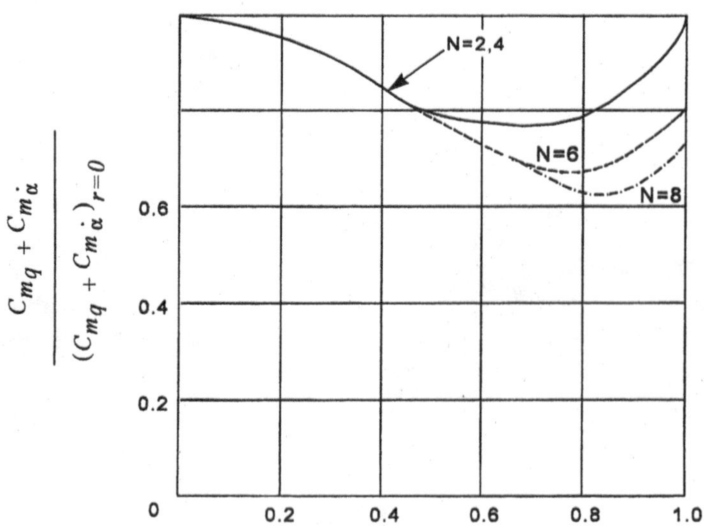

Fig. 8.16 Effect of body radius on damping in pitch for fixed span wings.

of the approach and analysis section will discuss the CFD methods used to calculate the factors F_6 and F_8 of Eq. (32).

C. Computational Fluid Dynamics Predictions for Multifin Aerodynamics

The slender body theory was limited to low AOA for slender wing–body configurations. The results do not account for wing-to-wing shock interactions, wing-to-wing blockage effects that occur at high AOA and Mach number, or wing geometry effects. To address these issues and either confirm slender body theory (which says the multiwing aerodynamics are independent of these effects) or develop an estimate of multifin aerodynamics as a function of the above-mentioned slender body theory assumptions, CFD will be used.

Before discussing the actual CFD codes used to perform the multifin aerodynamic calculations, it is believed appropriate to discuss some of the nonlinear physical phenomena involved to which slender body theory does not account. The first is wing blockage effects. To visualize wing blockage effects, several sketches of the configurations that the flow would see, as AOA is increased from 0 to 90 deg, are shown in Fig. 8.17 for four-, six-, and eight-fin configurations.

Figure 8.17 is for trapezoid fins that have a moderate span and chord. Notice that, at an AOA of 0 deg, no blockage effects occur between fins, only fin-to-fin interference from shock waves. Note that, at an AOA of 90 deg, there appears to be little or no difference among four-, six-, and eight-fin cases. Thus, to assume that slender body theory factors on normal force for the six- (1.5) and eight-fin (2.0) cases go to the four-fin value (1.0) at $\alpha = 90$ deg appears reasonable. Also notice that, for an AOA of 45 deg, there appears to be blockage of the airflow for either the six- or eight-fin cases from adjacent fins, and the body also adversely affects leeward plane fins. The implication of these comments is that wing-to-wing blockage occurs at fairly high AOA for windward plane fins and at moderate AOA for leeward plane fins.

It should be noted that slender body theory factors at low AOA assume that fins in both leeward and windward planes are effective in providing lift. At both the 45- and 90-deg AOA, it is clear from Fig. 8.17 that, even if windward plane fins remain completely or partially effective, leeward plane fins are mostly blocked by the body or fins. Hence, part of the additional lift factor will be reduced because, in the leeward plane, a good portion of the fin is shielded or blocked by the body or windward plane fins.

Another physical phenomenon inherent in multifin aerodynamic computations is wing-to-wing shock interactions. Obviously, this phenomenon is also not accounted for by slender body theory because slender body theory allows for an isentropic flow assumption, which in turn means that very weak shock waves are present. In reality, these shock interactions between wings in Fig. 8.17 are functions of wing geometry, Mach number, and AOA. CFD codes can easily account for this physical phenomenon.

Two CFD codes will be used in the computational process. They are the ZEUS[20] and GASP[21] codes. The ZEUS code is a full Euler Solution,

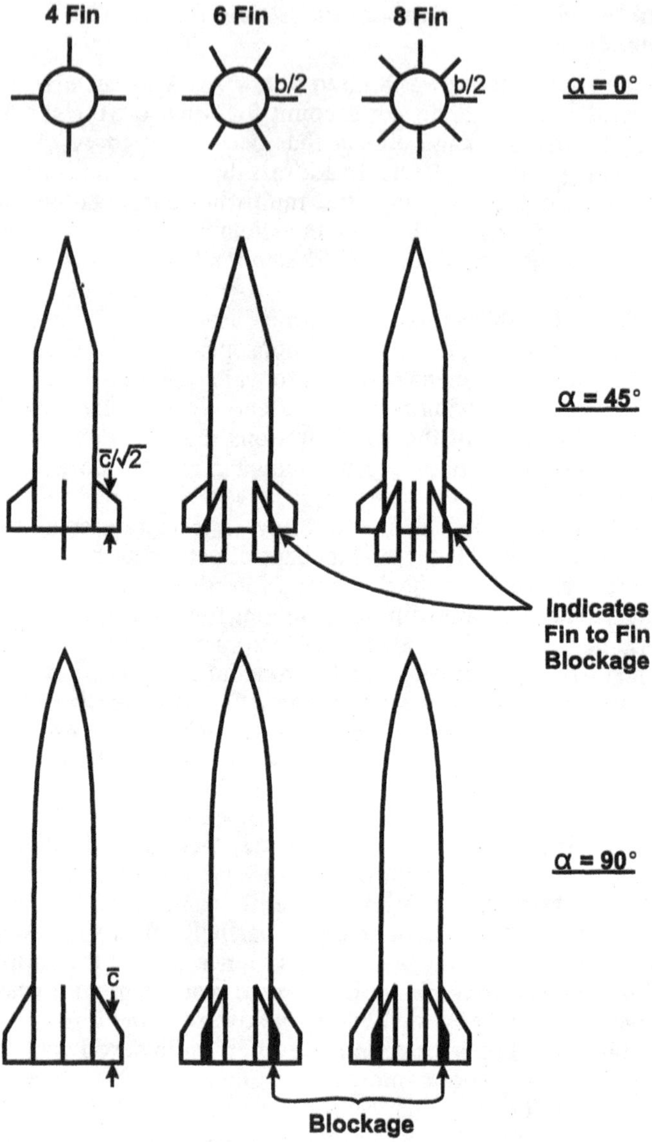

Fig. 8.17 Front view of missile as AOA goes from 0 to 90°, illustrating wing-to-wing blockage effects ($\Lambda_{LE} \neq 0°$).

whereas the GASP code is a full Navier–Stokes Solution with a subsonic Euler Solution option. The ZEUS code uses a marching solution to Euler equations. This means that flow along the axial plane must be supersonic for the code to have hyperbolic flow conditions throughout the computational region. This region encompasses the region from the bow shock to the rear of the body.

The ZEUS code has been recently downloaded to a personal computer with a pre- and postprocessing interface developed.[22] This interface (referred to as ZEUS++) uses much of the logic as used in the AP98 personal computer interface[23] in terms of several options for available body geometries. This greatly simplifies the geometry inputs for many cases and thus decreases the setup time significantly for the ZEUS code. Also, with the higher-speed personal computers now available, computational time for a sharp-nose, wing–body case is quite reasonable for many design computations.

The configuration chosen for the computation of factors F_6 and F_8 of Eq. (32) is the NASA Tri-service model (see Fig. 4.35).[24] The NASA Tri-service model was the basic configuration used because wind tunnel data were available for the four-fin case at a wide range of aspect ratio, Mach number, and AOA. Aspect ratios of 0.25, 0.5, 1.0, 2.0, and 4.0 were considered at Mach numbers of 1.5, 2.0, 3.0, and 4.5. For six- and eight-fin computations, the same geometric configuration was used for each individual fin as in the four-fin case. The hinge line location on the body was held constant for all aspect ratios. The normal force for the body alone was determined first at each Mach number and AOA. This result was subtracted from normal force values computed for the four-, six-, and eight-fin cases at the corresponding freestream conditions. It was assumed that this remainder was the fin normal-force contribution, including all interference effects. The ratio of six- and eight-fin values to those for four fins gave the multiplying factor indicating the effectiveness of the extra fins.

The NASA Tri-service database considered taper ratio as a configuration variable in addition to aspect ratio. However, after performing several computations for factors F_6 and F_8 as a function of taper ratio, it was decided to drop this variable as secondary in importance compared to aspect ratio, Mach number, and AOA. Variations in factors F_6 and F_8 as a function of taper ratio were less than 5% for the cases considered. Hence, a value of 0.5 was used for taper ratio in all the ZEUS++ and GASP Euler calculations for factors F_6 and F_8. Computational times per data point for the ZEUS++ varied from 3 to 15 minutes using a 200-MHz Intel Pentium II computer chip. The higher the AOA, the larger the computational time. Computational times were not obtained for the GASP Euler solutions as they were run on a workstation in conjunction with other codes being run simultaneously. Times on the order of hours versus minutes were typical, however.

As already mentioned, the ZEUS[20] Euler solver is limited by subsonic flow occuring anywhere in the flowfield. To compute normal force factors for multifin cases where subsonic Mach numbers existed, the subsonic Euler solution option of the GASP[21] code was utilized. For these cases, the

ZEUS++ tool was used to generate three-dimensional grids that were then imported to the GASP flowfield software. The computational domains were (51 × 36 × 165), (51 × 36 × 165), (51 × 54 × 165), and (51 × 36 × 165) for aspect ratios of 0.25, 0.5, 1.0, and 2.0, respectively. The first number indicates the number of grid points in the radial direction; the second and third, the number of points in the circumferential and axial directions, respectively. Note that all the cases are symmetric about the pitch plane and, therefore, only 180 deg of the circumferential plane was considered. An example of the results of the normal force computations for factors F_6 and F_8 of both the ZEUS++ and GASP computations are given in Table 8.2 for the aspect ratio 0.5 case. A box is placed around GASP computations so they can be distinguished from those of ZEUS computations. As with ZEUS++ computations, GASP computations were compared to the NASA Tri-service database for the body-alone and four-fin computations before proceeding to six- and eight-fin computations. The results of these comparisons can be found in Table 8.3 for the aspect ratio 0.5 case. Normal force coefficients are shown from the CFD computations and from the wind tunnel database at each point where information was available for both. The percent difference between the two is also given. In the case of the CFD data, a box is placed around the GASP results as before. Results of comparisons were within experimental errors in most cases, so it is believed that six- and eight-fin results of Table 8.2 should prove adequate for the development of a semi-empirical model for multifin aerodynamics. Other aspect ratio data is given in Ref. 16.

An exception to the computations being within experimental error occurred for the Mach 0.6 cases, where the full Euler solution of the GASP code was used. Here, the GASP body-alone solution gave normal force coefficients that were higher than data for moderate AOAs (a = 15 to 30 deg). In analyzing this with the AP98, it was concluded that the crossflow Reynolds number was supercritical, which meant that, instead of a crossflow drag coefficient of 1.2, a value less than that was needed to match experiment. Physically, what is happening when the crossflow drag coefficient decreases rapidly is that the flow around the body remains more attached in the leeward plane as opposed to separating near the maximum diameter of the body in the crossflow plane. The inviscid Euler solution cannot model this without some help. The full Navier–Stokes solution from GASP, given the correct turbulence model, should be able to model this phenomenon. However, time did not permit this approach. As a result, engineering judgment will be used for Euler solutions at low Mach numbers where the crossflow separation model is not accurate. This problem did not appear to occur with the ZEUS++ or the GASP at higher Mach number, where leeward plane pressures are fairly small in comparison to windward plane pressures.

Another problem in GASP Euler solutions occurred for larger aspect ratio fin cases. Here the fins are very small[24] and any errors in the body-alone solution can produce fairly large errors in factors F_6 and F_8. As a result, engineering judgment must be used for large aspect ratios as well.

Results from Table 8.2 for aspect ratio of 0.5 (as well as other aspect ratios not shown) were then plotted as functions of AOA. An example of these

Table 8.2 Euler CFD calculations from ZEUS++[22] and GASP[21] codes

	AR = 0.5				
a	$M_\infty = 0.6$	$M_\infty = 1.5$	$M_\infty = 2.0$	$M_\infty = 3.0$	$M_\infty = 4.5$
Six fins, F_6					
1	1.336	1.235	1.201	1.321	1.485
2		1.232	1.191	1.322	1.487
3		1.215	1.188	1.315	1.486
4		1.211	1.179	1.314	1.484
5			1.180	1.312	1.494
6			1.176	1.308	1.507
7			1.175	1.308	1.522
8			1.174	1.304	1.491
9			1.174	1.300	1.480
10			1.169	1.293	1.479
11			1.171		
12			1.171		
15	1.050	1.087		1.270	1.527
20				1.292	1.533
25				1.294	1.483
30	0.794	0.944	1.070	1.291	1.354
45			0.952	1.002	
60				0.709	
Eight fins, F_8					
1	2.143	1.336	1.273	1.369	1.728
2		1.351	1.274	1.364	1.715
3		1.350	1.269	1.357	1.717
4		1.341	1.262	1.361	1.719
5			1.256	1.368	1.760
6			1.255	1.369	1.773
7			1.252	1.372	1.796
8			1.248	1.371	1.785
9			1.251	1.370	1.793
10			1.252	1.371	1.818
11			1.256		
12			1.258		
15	1.201	1.184		1.396	1.844
20				1.430	1.816
25				1.422	1.741
30	0.750	1.065	1.160	1.418	1.608
45			1.038	1.050	
60				0.943	

Table 8.3 Comparison of CFD results to NASA database for four-fin configuration

M_∞	a	C_N(wind tunnel)	C_N(CFD)	Difference, %
Body alone				
0.6	15	1.00	2.267	126.7
	30	3.09	4.658	50.7
1.5	15	1.39	1.629	17.2
	30	6.11	7.120	16.5
2.0	5	0.32	0.372	16.2
	10	0.82	0.835	1.8
	15	1.86	1.923	3.4
	30	6.47	7.222	11.6
4.5	5	0.43	0.381	−11.4
	10	1.21	1.129	−6.7
	15	2.21	1.997	−9.6
	20	3.28	3.039	−7.3
	25	4.62	4.251	−8.0
	30	6.00	5.565	−7.2
AR = 0.5				
0.6	15	3.62	4.129	14.1
	15	3.67	3.702	0.9
1.5	30	9.57	10.808	12.9
2.0	5	0.86	1.041	21.0
	10	2.00	2.184	9.2
	30	9.62	10.058	4.6
3.0	5	0.81	0.866	6.9
	10	1.95	1.959	0.5
	15	3.35	3.349	0.0
	20	4.81	4.882	1.5
	25	6.54	6.617	1.2
4.5	5	0.65	0.712	9.5
	10	1.62	1.611	−0.6
	15	2.70	2.672	−1.0
	20	4.05	4.017	−0.8
	25	5.68	5.597	−1.5
	30	7.51	7.347	−2.2

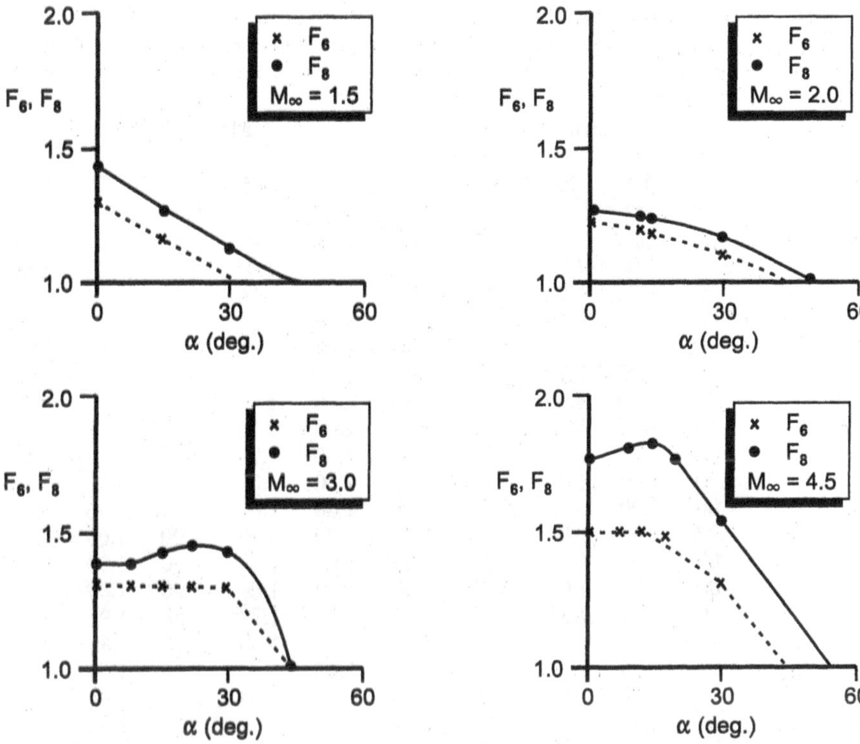

Fig. 8.18 Ratio of normal force of six and eight fins to that of four fins based on CFD (AR = 0.5).

results is shown in Fig. 8.18 for aspect ratio 0.5 (the reader is referred to Ref. 15 for all data on other aspect ratios). It is clear from viewing Fig. 8.18 that, to assume slender body theory factors of 1.5 and 2.0 for the normal force of 6 and 8 fins, respectively, compared to that of four fins will give significant errors, particularly at higher AOAs. These errors are the greatest for the lower aspect ratio cases, as seen in Table 8.4. Then, curves were drawn through the data for factors F_6 and F_8 for use in the AP98. Data from all of these curves are given in Table 8.4. This then is the model that will be used for multifin aerodynamics. Any Navier–Stokes calculations in the future or comparisons to wind tunnel data for configurations outside the database can be used to fine-tune this model.

D. Comparison of New Method for Multifin Aerodynamics to Experiment

The very limited availability of experimental data for normal force on missiles with more than four fins makes validation of the new methodology difficult. One set of ballistic range data for a six-finned projectile, along with

Table 8.4 Approximated values of factors F_6 and F_8 obtained from smoothed values of the ZEUS++ and GASP code computations and engineering judgment

		F_6 Mach number					F_8 Mach number				
AR	α	0.6	1.5	2.0	3.0	4.5	0.6	1.5	2.0	3.0	4.5
0.25	0	1.26	1.37	1.27	1.19	1.22	1.90	1.42	1.4	1.27	1.30
	15	1.00	1.00	1.10	1.19	1.35	1.45	1.03	1.17	1.27	1.46
	30	1.00	1.00	1.00	1.19	1.22	1.00	1.00	1.01	1.27	1.32
	45	1.00	1.00	1.00	1.00	1.00	1.00	1.00	1.00	1.00	1.00
	60	1.00	1.00	1.00	1.00	1.00	1.00	1.00	1.00	1.00	1.00
	75	1.00	1.00	1.00	1.00	1.00	1.00	1.00	1.00	1.00	1.00
	90	1.00	1.00	1.00	1.00	1.00	1.00	1.00	1.00	1.00	1.00
0.50	0	1.35	1.25	1.20	1.30	1.47	2.00	1.36	1.28	1.35	1.72
	15	1.06	1.10	1.15	1.29	1.50	1.50	1.18	1.24	1.40	1.83
	30	1.00	1.00	1.07	1.28	1.36	1.00	1.08	1.16	1.41	1.60
	45	1.00	1.00	1.00	1.00	1.00	1.00	1.00	1.04	1.06	1.20
	60	1.00	1.00	1.00	1.00	1.00	1.00	1.00	1.00	1.00	1.00
	75	1.00	1.00	1.00	1.00	1.00	1.00	1.00	1.00	1.00	1.00
	90	1.00	1.00	1.00	1.00	1.00	1.00	1.00	1.00	1.00	1.00
1.0	0	1.40	1.22	1.35	1.42	1.50	1.92	1.27	1.58	1.96	2.00
	15	1.15	1.13	1.23	1.32	1.50	1.69	1.38	1.38	1.80	2.00
	30	1.07	1.00	1.00	1.21	1.38	1.43	1.28	1.15	1.64	2.00
	45	1.02	1.00	1.00	1.10	1.13	1.20	1.05	1.00	1.48	1.61
	60	1.00	1.00	1.00	1.00	1.00	1.00	1.00	1.00	1.32	1.25
	75	1.00	1.00	1.00	1.00	1.00	1.00	1.00	1.00	1.16	1.00
	90	1.00	1.00	1.00	1.00	1.00	1.00	1.00	1.00	1.00	1.00
2.0	0	1.42	1.50	1.50	1.50	1.50	1.92	1.77	1.97	1.92	1.90
	15	1.31	1.41	1.27	1.39	1.50	1.70	1.95	1.75	1.77	2.00
	30	1.17	1.00	1.03	1.27	1.45	1.47	1.65	1.57	1.62	2.10
	45	1.03	1.00	1.00	1.14	1.23	1.25	1.32	1.27	1.47	1.95
	60	1.00	1.00	1.00	1.00	1.00	1.02	1.00	1.02	1.32	1.62
	75	1.00	1.00	1.00	1.00	1.00	1.00	1.00	1.00	1.17	1.32
	90	1.00	1.00	1.00	1.00	1.00	1.00	1.00	1.00	1.00	1.00

accompanying CFD information, may be found in Ref. 25. The geometric configuration of the model used in these tests is shown in Fig. 8.19. It consists of a cone-cylinder body, 23.14 calibers in total length, with a diameter of 27.05 mm. The cone half-angle is 8 deg and the leading and trailing edges of the fins are blunt. For AP98 runs, Reynolds number was computed based on sea-level conditions and the body diameter. The "wind tunnel model with no boundary-layer trip" option was chosen for the viscous computations.

Comparisons for normal force coefficient slope at zero AOA, axial force coefficient, and center of pressure are shown in Figs. 8.20a, 8.20b, and 8.20c, respectively. For these cases, range data were available from $M_\infty = 3.5$ to 5.3 and CFD computations were made at $M_\infty = 4.41, 5.0$, and 5.88. AP98 results,

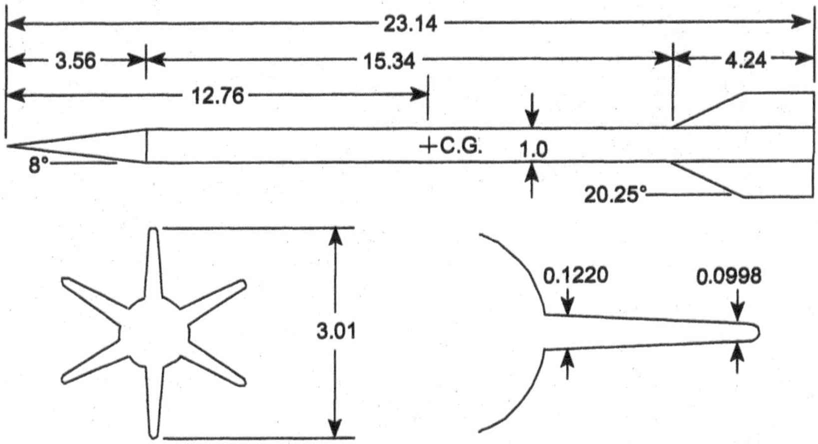

Fig. 8.19 Schematic of M289 projectile configuration (from Ref. 25).

in conjunction with the multifin aerodynamics methods, are shown for M_∞ = 2.0 to 6.0. The large scatter in range data could be the result of AOA motion that is not accounted for in either AP98 or CFD runs. It can be seen that AP98 results agree reasonably well with CFD computations and both fall in the middle of the range data. Figures 8.20d, 8.20e, and 8.20f present comparisons for pitching moment coefficient slope at zero AOA, roll damping moment coefficient, and pitch damping moment coefficient, respectively. Range data are available for the same Mach number range as before, but CFD and AP98 results are shown for M_∞ = 3.0 to 5.5. Once again, AP98 results are in fairly good agreement with CFD computations, except for pitching moment slope, where they tend to be somewhat high. In general, both AP98 and CFD results tend to be high compared to range data.

A second set of experimental range data and CFD computations was available from Ref. 25 for a similar six-finned projectile. In this instance, the cone-cylinder body is 13.94 calibers in length with a diameter of 35.2 mm. The cone half-angle is 8 deg and the leading and trailing edges of the fins are blunt. This configuration is shown in Fig. 8.21. The same computational options were used as in the previous case. Ballistic data were available over a Mach number range from 3.0 to about 4.5. CFD data were given from M_∞ = 3.0 to 5.5 and AP98 computations and new multifin aerodynamics methods were performed over this same Mach number interval.

Comparisons for normal force coefficient and pitching moment coefficient slopes at zero AOA are shown in Figs. 8.22a and 8.22b. AP98 results at lower Mach numbers tend to be somewhat high compared to CFD numbers in both cases, and both tend to lie above range data. Figure 8.22c shows the comparison for axial force coefficient. Good agreement is obtained throughout in this instance. The comparison for pitch damping moment coefficient is

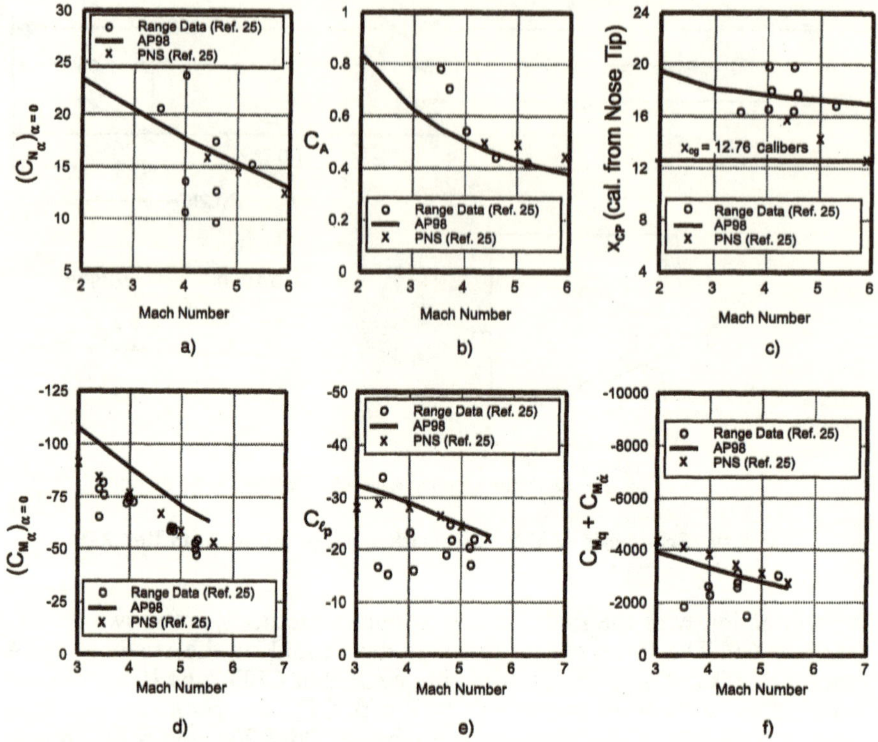

Fig. 8.20 Comparison of new multifin method to PNS and experiment for Fig. 8.19 configuration: a) normal force coefficient, b) axial force coefficient, c) center of pressure, d) pitching moment coefficient slope, e) roll damping moment coefficient, and f) pitch damping moment coefficient.

shown in Fig. 8.22d. Once again, AP98 numbers are somewhat high relative to CFD results and both tend to lie above the majority of the ballistic data.

A third set of data was available from some preliminary wind tunnel tests of Ref. 26. The model used is shown in Fig. 8.23. It consists of a circular body, approximately 12.26 calibers in length, with a 3-caliber von Kármán ogive nose. The body diameter is 2.976 in. Eight small, high aspect ratio pop-out fins are located at the rear of a short boattail section. The model was also tested with four pop-out canards, but this configuration was not considered because the desire was to isolate the effects of the fins. It was necessary to modify the fin geometry to conform to input requirements of AP98. The equivalent fin has a trapezoidal planform with the same area, sweep angle, and aspect ratio as the original. AP98 runs were conducted at indicated Reynolds numbers for each case and the "wind tunnel model with no boundary-layer trip" option was used.

Wind tunnel data were available in this case for the body alone, so it was used to adjust for the effects of crossflow separation and reattachment. This

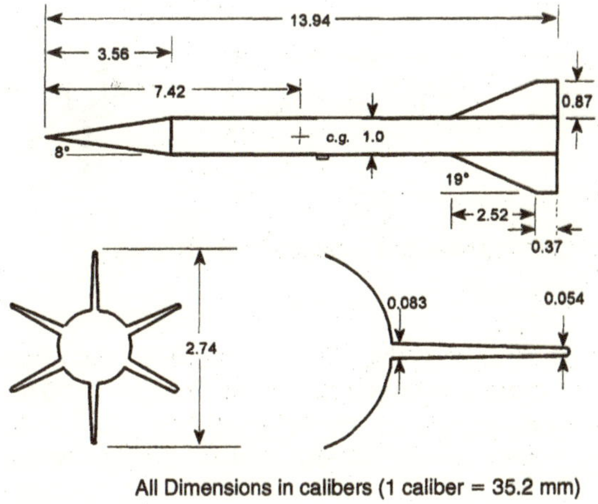

Fig. 8.21 Schematic of M735 projectile configuration (from Ref. 25).

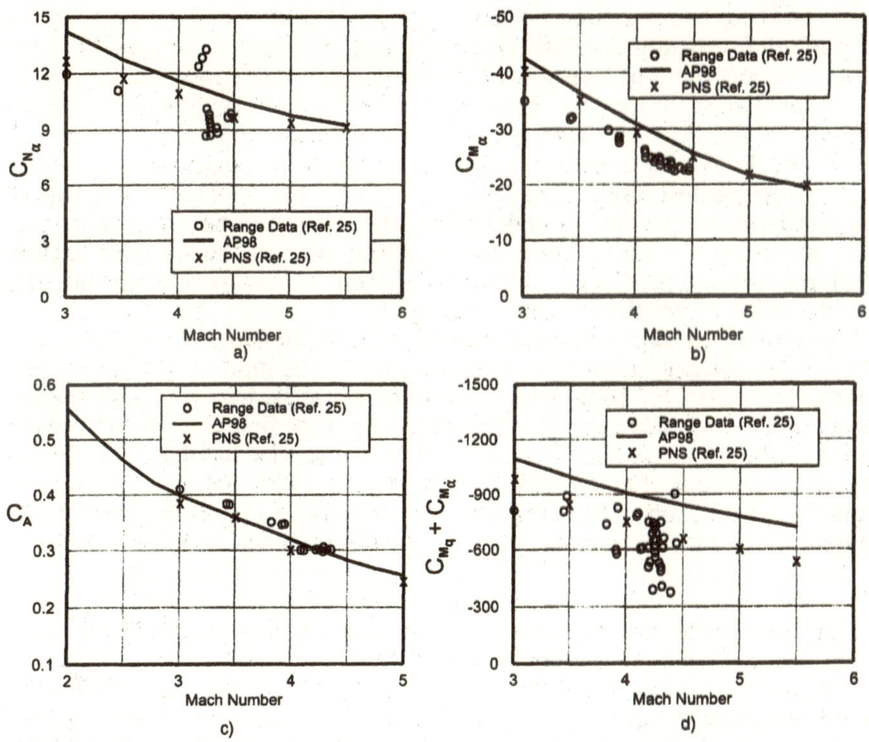

Fig. 8.22 Comparison of new multifin method of PNS and experiment for Fig. 8.21 configuration: a) normal force coefficient slope, b) pitching moment coefficient slope, c) axial force coefficient, and d) pitch damping moment coefficient.

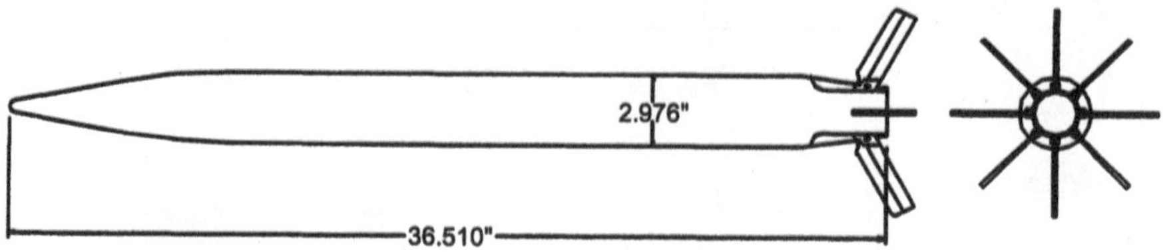

Fig. 8.23 Schematic of eight-fin body–tail concept (from Ref. 26).

adjustment is made in AP98 by changing the critical crossflow Reynolds number and shifting the value of crossflow Mach number at which the "drag bucket" starts. These two parameters are set to obtain a good fit to the experimental body-alone data at each Mach number and are then used for all further computations.

Wind tunnel data were also available for this configuration with only four fins, and normal force comparisons with AP98 for this case are shown in Fig. 8.24. These results are included to provide information on how well AP98 does on these computations because they are used as a basis for the eight-fin model. If, for example, AP98 predictions are low for a given case here, we would expect them to be low for the corresponding eight-fin case.

Comparisons for total normal force for the full eight-fin configuration are shown in Fig. 8.25 for Mach numbers of 0.4, 0.8, 1.05, 1.3, 1.6, and 2.0, respectively. AOAs range up to 15 deg. In general, comparisons are quite good. The greatest disagreement occurs at lower Mach numbers and higher AOAs. Body aerodynamics under these conditions can be very sensitive to the subcritical or supercritical status of the flow in the leeward region, making accurate predictions difficult. Differences may be related primarily to this effect rather than to fin modeling.

In summary, a semi-empirical method to compute aerodynamics of multifin missile configurations has been developed. The method was developed using full Euler CFD codes with wind tunnel databases, slender body theory, and engineering judgment. Euler calculations were first compared to cruciform fin-body calculations from the NASA Tri-service database for normal force calculations of body-alone and fin-body aerodynamics. After this, factors for aerodynamics of configurations with six and eight fins were computed based on the four-fin results. Conclusions from this study were as follows:

1) Agreement between the NASA Tri-service database and CFD computations was quite good except at subsonic Mach numbers
2) It was concluded that the major reason for the discrepancy at subsonic Mach numbers between Euler computations and wind tunnel data was the failure of the Euler solution to adequately predict the correct flow in the leeward plane caused by viscous effects.
3) Full Navier–Stokes solutions, with the appropriate turbulence model, are necessary for adequate solutions of cruciform missile aerodynamics for

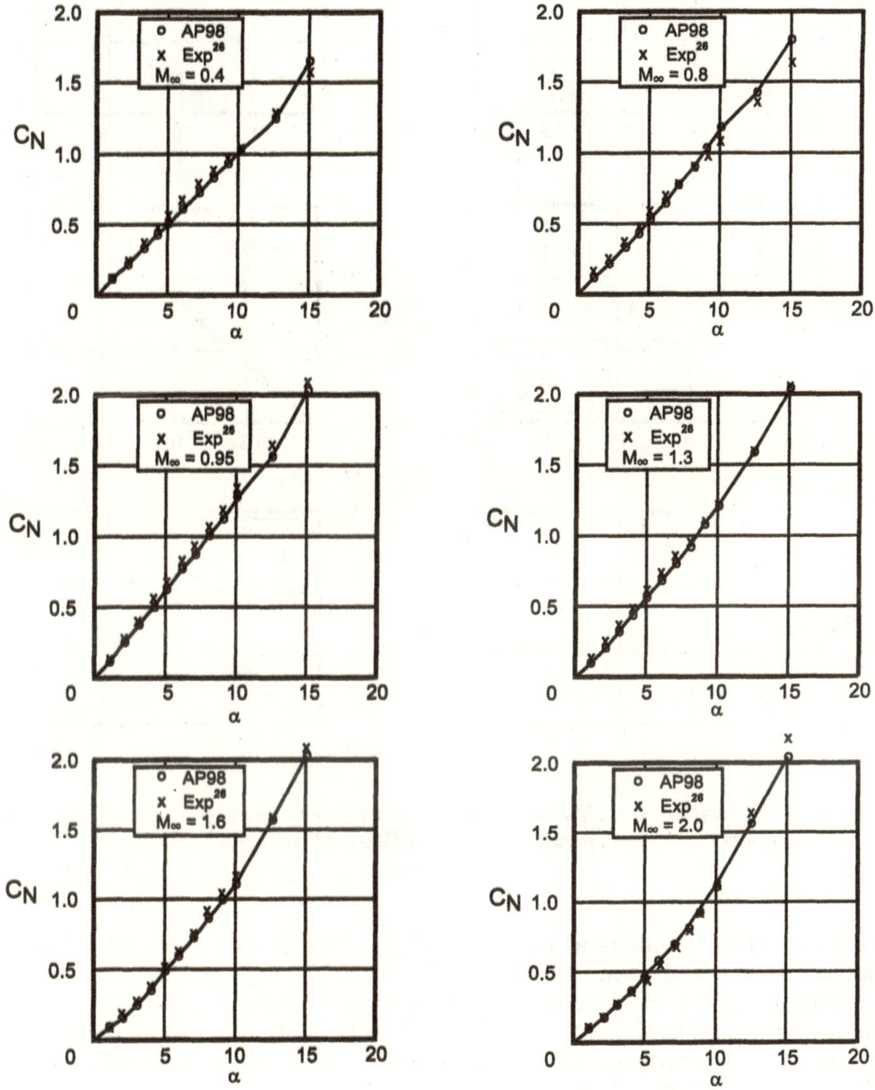

Fig. 8.24 Normal force coefficient comparisons of theory and experiment at various Mach numbers for cruciform body–tail configuration of Fig. 8.23.

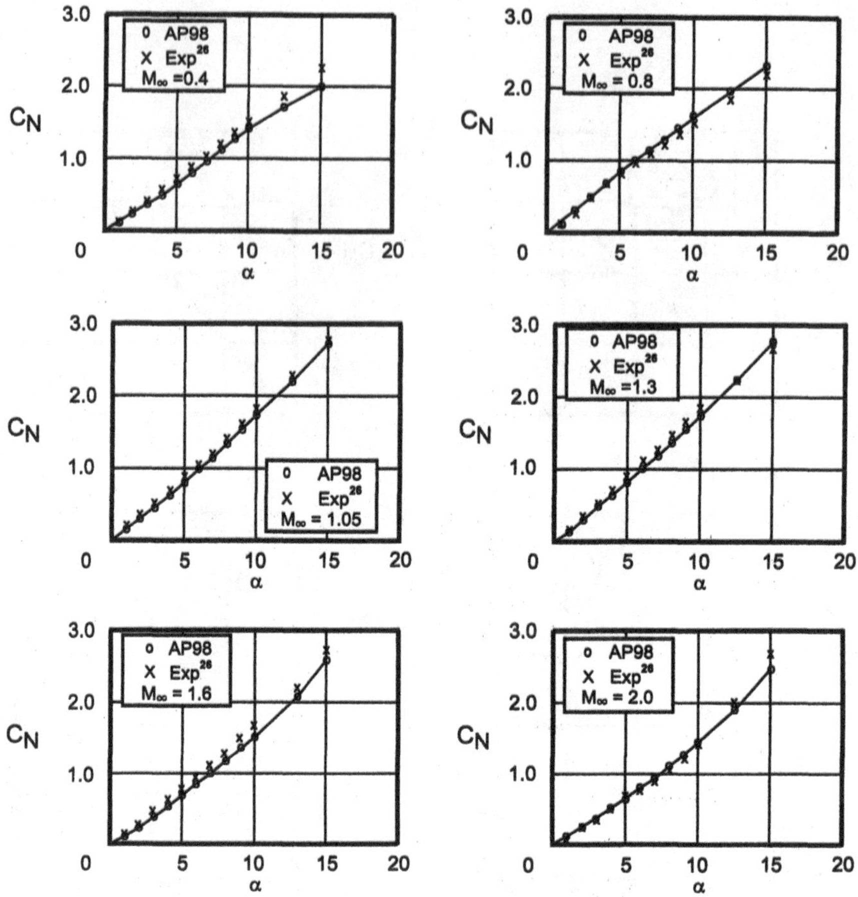

Fig. 8.25 Normal force coefficient comparisons of theory and experiment at various Mach numbers for body–tail configuration of Fig. 8.23 with eight tail fins.

subsonic Mach numbers as AOA increases. Comparisons with experiment could be improved upon at AOAs greater than 30 deg at all Mach numbers.

4) Agreement between the semi-empirical theory and experiment was seen to be quite good for normal force coefficient prediction at low AOA over a range of Mach numbers. Validation at higher AOA and for the other aerodynamic coefficients will require additional data.

5) Comparisons of multifin aerodynamics for axial force, pitching moment, center of pressure, roll, and pitch damping moments between limited experimental data and theory were considered adequate for an engineering method.

6) It is believed that the semi-empirical model developed with full Euler solutions and engineering judgment is adequate. However, it is recom-

mended that full Navier–Stokes computations be performed to improve the accuracy of the semi-empirical method.

V. Weapon Performance

Weapon performance is probably the largest application of aerodynamics by far. Weapon performance information includes range, time of flight, maneuverability, and miss distance. All of these performance criteria inherently include stability and control as well, either implicitly or explicitly. The most simple approximation used in range and time of flight calculations is the so-called particle ballistic or point mass model. A model such as this assumes that the flight vehicle can be represented by a particle in flight with drag but no lift. Typical particle ballistic models are either two-degree-of-freedom (2DOF) or three-degree-of-freedom (3DOF) models. The 2DOF case is a planar model that gives range (x) and altitude (y) as a function of time, whereas the 3DOF case gives range, altitude, and crossrange (z) as well. Some 2DOF models integrate equations of motion in a plane and account separately for drift due to spin, yaw of repose, or crosswinds. In essence, these models are quasi-3DOF models.

Figure 8.26 shows the forces acting on a particle in planar 2DOF flight. Summing the forces in the x and y direction, and using Newton's second law, planar equations of motion for the particle are

$$\ddot{x} + f(M, h) \cos\theta \, (\dot{x}^2 + \dot{z}^2) = T/m \cos\theta \tag{33}$$

$$\ddot{z} + f(M, h) \sin\theta \, (\dot{x}^2 + \dot{z}^2) = T/m \sin\theta - g \tag{34}$$

where $f(M, h) = \dfrac{1}{2} \dfrac{\rho S_{\text{ref}} C_D}{m}$

Equations (33) and (34) represent a set of coupled, nonlinear ordinary differential equations that can be integrated as a function of time (using

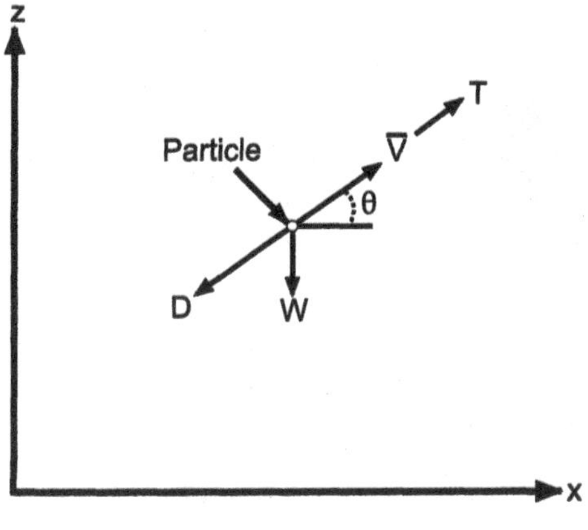

Fig. 8.26 Forces acting on a particle in planar motion (no lift).

something like a fourth-order Runge–Kutta technique) to obtain $\dot{x}$, $\dot{z}$, x, and z. To start the integration, the initial flight path angle and velocities $\dot{x}$ and $\dot{z}$ are needed. Also, if thrust is present, a value of thrust versus time is required as well as a value of mass versus time of the particle. A value of air density is required as a function of altitude. For this simple mathematical model, the only aerodynamic term needed is drag coefficient. The drag coefficient is a strong function of Mach number and a weak function of altitude due to the skin friction drag component. Because drag is such an important part of accurate range and time of flight information, this was the reason Chapters 3 and 4 stressed second-order accuracy in the wave drag calculation. The drag coefficients shown in Eqs. (33) and (34) can include a component due to lift, if the coning angle of the projectile's flight is known or the trim AOA of the weapon's flight is known. However, Eqs. (33) and (34), as written, do not account directly for any change in the trajectory due to any lift force that may be generated by this coning angle.

Also, it should be pointed out that many particle ballistic models do not account for the drag change with altitude and, therefore, use the sea-level drag coefficient. This simplifies inputs significantly, with generally less than a 0.5% error in range and time of flight.

The next use, for the semi-empirical aerodynamic estimates, is for 3DOF performance models. These models assume that the configuration is a rigid body flying in steady equilibrium conditions during flight. By steady equilibrium flight, we mean that the vehicle has trimmed out at some AOA at which lift and weight are balanced and pitching moment is zero. The three degrees of freedom are the same x and z position coordinates of the particle ballistic model, plus a rotational degree of freedom (pitching moment) about the center of gravity of the rigid body. Figure 8.27 depicts the force and moment diagram of a typical body–tail configuration.

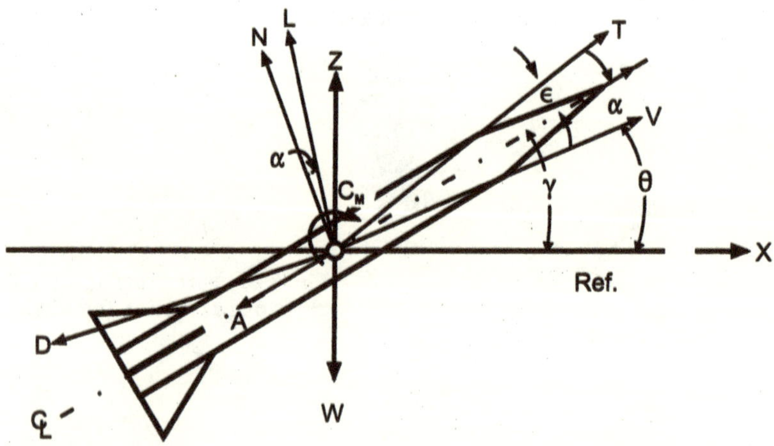

Fig. 8.27 Forces and moments of a body–tail configuration.

In general, the thrust can be vectored or misaligned by some angle ε from the body axis. The angles θ and γ are the angles between the velocity vector and the axis of symmetry, respectively, and some reference axis, the x axis here, such as the inertial axis system of the earth. Once again, applying Newton's second law, we obtain the equations of motion of this vehicle as

$$\ddot{x} + f(M,h)\cos\theta\left(1 + \frac{C_L}{C_D}\tan\theta\right)(\dot{x}^2 + \dot{z}^2) = \frac{T}{m}\cos(\theta + a + \varepsilon) \qquad (35)$$

$$\ddot{z} + f(M,h)\cos\theta\left(\tan\theta - \frac{C_L}{C_D}\right)(\dot{x}^2 + \dot{z}^2) = \frac{T}{m}\sin(\theta + a + \varepsilon) - g \qquad (36)$$

$$\ddot{\gamma} - \frac{1}{2}\frac{\rho_\infty V_\infty^2 S_{\text{ref}}\ell_{\text{ref}}}{I_Z}\left(C_{M_{\dot{\gamma}}}\dot{\gamma} + C_{M_{\dot{a}}}\dot{a} + C_{M_a}a + C_{M_\delta}\delta\right) = 0 \qquad (37)$$

where $\gamma = a + \theta$

Most 3DOF models assume that the rigid body is in equilibrium during its flight path so that, at each integration time step of Eqs. (35) and (36), an AOA is found about which the pitching moment is zero. This is the trim AOA. Referring to Eq. (37), the rotation rate $\ddot{\gamma}$ and the damping derivatives $\dot{\gamma}$ (note the $C_{M_{\dot{\gamma}}}$ is related to C_{M_q} because $q = \dot{\theta}$ and $\dot{\gamma} = \dot{\theta} + \dot{a}$) and $\dot{a}$ are neglected at each interval during the trajectory due to the assumption of being in a state of local equilibrium. For the vehicle in equilibrium, $C_M = 0$ and Eq. (37) reduces to

$$a_{TR} = \frac{C_{M_\delta}\delta}{C_{M_a}} \qquad (38)$$

a_{TR} is the value of a that allows the vehicle to be in a condition of equilibrium. Figure 8.28 illustrates this set of conditions for various values of δ.

The point of discussion of the 3DOF flight trajectory simulation is that, for this case, not only does one need to have drag or axial force information but values of C_L, C_D, a_{TR}, δ_{TR}, and $C_{N_{TR}}$. Aerodynamics are normally computed in terms of axial force, normal force, and pitching moment (see Chapters 3 to 5). C_L and C_D are related to these coefficients through a simple translation. That is,

$$C_L = C_N\cos a - C_A\sin a$$

$$C_D = C_N\sin a + C_A\cos a \qquad (39)$$

The normal process for the flight dynamicist to compute range and time of flight from Eqs. (35) and (36) is to first know the values of thrust versus time, mass of the vehicle versus time, and center of gravity versus time. Also, if the thrust is vectored or misaligned, values of ε versus some control are needed. Finally, an initial launch angle θ is required. The aerodynamicist will provide maps of C_N versus C_M (about some center of gravity location)

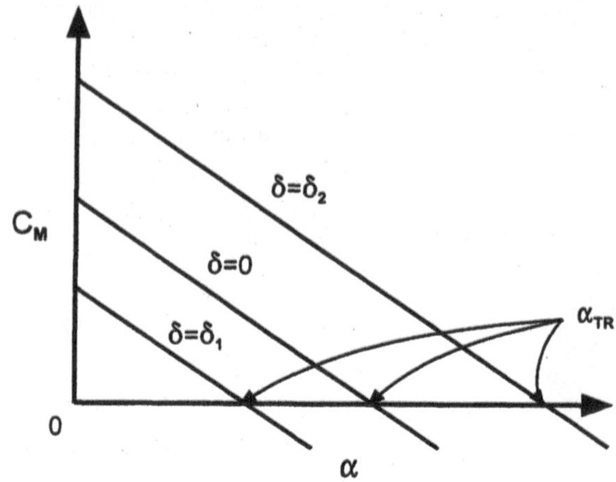

Fig. 8.28 Illustration of trim AOA.

for various α's and δ's at given Mach numbers. Also, values of C_A at sea level plus a ΔC_A due to altitude is provided. Figure 8.29 illustrates an example of a typical process the aerodynamicist is required to go through to provide inputs for a 3DOF trim performance model. Figure 8.29a illustrates a typical example of a body–tail configuration. This case is taken from the NASA-Tri service database of Fig. 4.35 and consists of the 12.33-caliber body with a 3.0-caliber tangent ogive nose with aspect ratio tail of 1.0. Figure 8.29b is a set of pitching moment versus normal force curves for values of α and δ at a freestream Mach number of 3.0. In practice, there would be a series of these curves for various Mach numbers that would cover the anticipated flight envelope, including changes in thrust and center of gravity. Shown on Fig. 8.29b are lines for constant values of $\delta = 0, -20$, and -40 deg. The dashed lines correspond to constant values of α, including $-10, 0, 10, 20, 30$, and 40 deg. The values of δ where $C_M = 0$ are the values that allow the vehicle to be in an equilibrium trim position. That defines the value of $C_{N_{TR}}$. Figures 8.29c, 8.29d, and 8.29e give the values of C_A

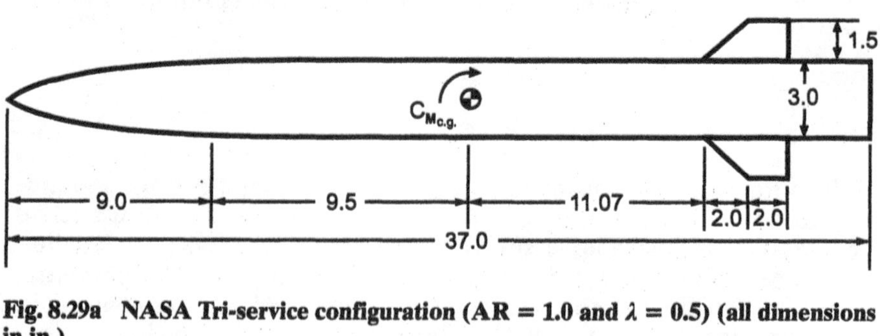

Fig. 8.29a NASA Tri-service configuration (AR = 1.0 and λ = 0.5) (all dimensions in in.).

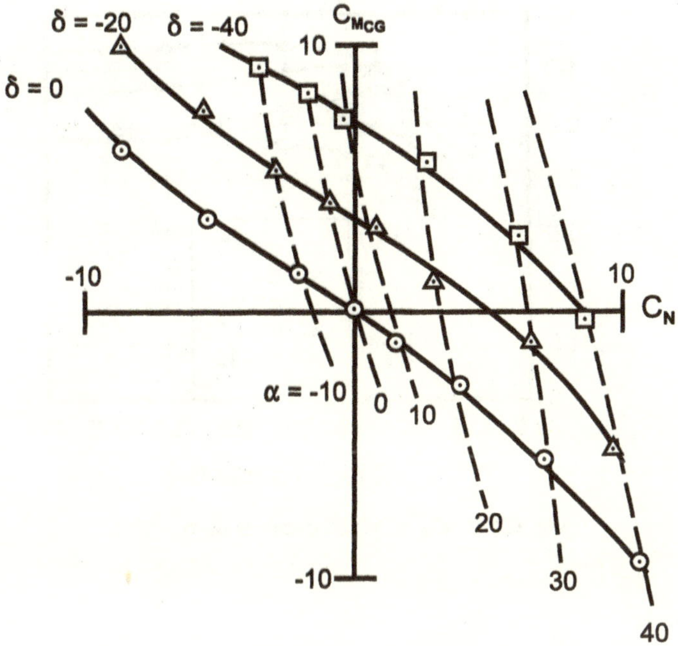

Fig. 8.29b C_M **vs** C_N **for Various** α**'s and** δ**'s at** $M_\infty = 3.0$**.**

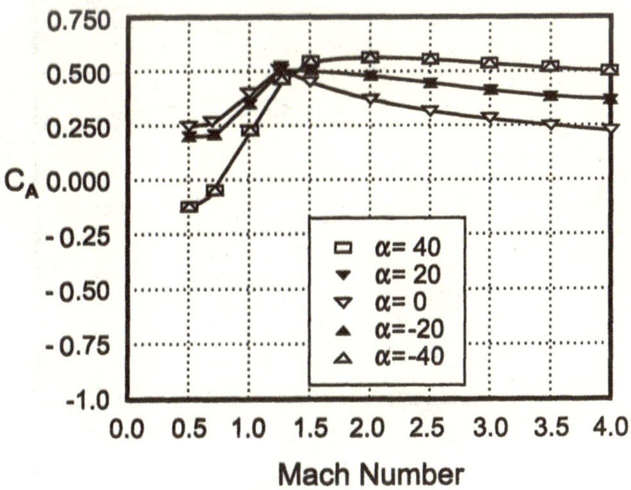

Fig. 8.29c C_A **vs Mach number (**$\alpha = 0°$**).**

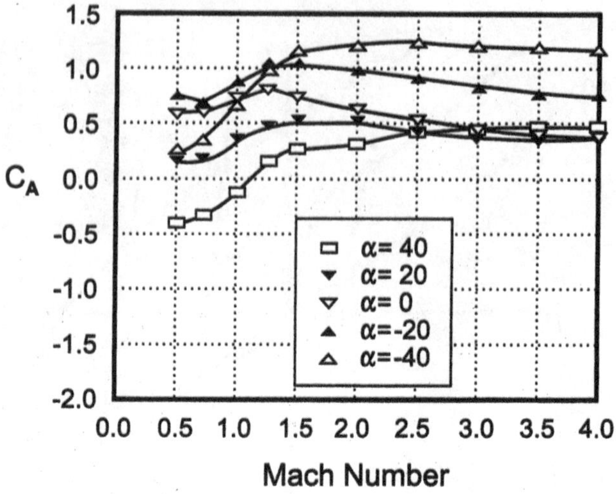

Fig. 8.29d C_A vs Mach number ($\delta = -20°$).

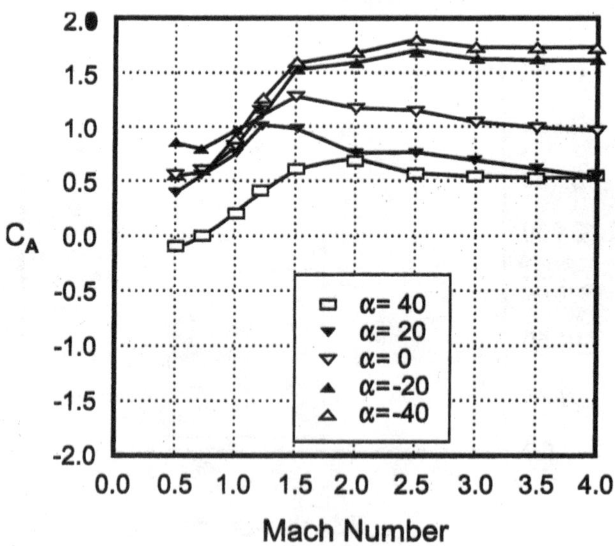

Fig. 8.29e C_A vs Mach number ($\delta = -40°$).

versus Mach number for control deflection of 0, −20, and −40 deg, respectively. Based on the value of a_{TR} and δ_{TR} from Fig. 8.29b, one can interpolate to find the approximate value of C_A at trim. C_L and C_D are obtained from Eq. (38). Equations (35) and (36) can then be integrated to the next time step. In many flight dynamics models, the aerodynamics are changed every few time steps as opposed to every integration step with acceptable accuracy. Axial force change with altitude for this case is approximately linear up to 80,000 ft and increases about 0.02. This is small enough to be neglected if desired.

Another important point to be gained from Fig. 8.29b is the value of $C_{N_{TR}}$ that allows maximum maneuverability. Generally, this will be the value at the upper AOA that the vehicle can fly due to either structural or control constraints. It could also be defined by a maximum allowable control deflection. Assuming the configuration of Fig. 8.29a has the structural integrity to withstand loads at 40-deg AOAs and the controls can be deflected up to 40 deg deflection, then the maximum value of $C_{N_{TR}}$ at $M_\infty = 3.0$ is about 8.6. The maneuverability in g's is defined by

$$\eta = \frac{N_{TR}}{W} = \frac{1/2\,\rho_\infty V_\infty^2 A_{\text{ref}}}{W} + C_{N_{TR}} \tag{40}$$

Note from Eq. (40) that maneuverability varies directly with density. Thus, if the configuration of Fig. 8.29a could pull 50 g's at sea level, it could only pull 3.7 g's at 65,000 altitude. This is an illustration of why it is difficult to defeat a target as altitude increases with only aerodynamic controls.

The other important performance parameter is miss distance. To determine miss distance, the flight dynamicist will have some means of acquiring and tracking the target as well as the kill vehicle's relative position with respect to the target. This information is processed and the controls are deflected to a value sufficient to put the vehicle on an intercept line. Aerodynamics needed for the trajectory simulation are obtained in a process similar to that already discussed.

A final point worthy of note is the fact that, in many cases, one is interested in obtaining maximum range for a given configuration. There are two requirements to this process. First, one needs to have a trajectory optimization code available that can trade off propellant burn rate with trajectory time history. The second part is to utilize a glide angle that allows for maximum lift-to-drag ratio during trim conditions. This latter statement is rather obvious when examining Eq. (35).

VI. Summary of Aerodynamic Prediction Methods

A summary of theoretical and empirical methods used for computing aerodynamics that are discussed in this book (and used in the AP98) are

Component/ Mach number region	Subsonic, $M_\infty < 0.8$	Transonic, $0.8 \le M_\infty \le 1.2$	Low supersonic, $1.2 \le M_\infty \le 1.8$	Moderate/high supersonic, $1.8 \le M_\infty \le 1.8$	Hypersonic, $M_\infty > 6.0$
Nose wave drag	Empirical (4.8.2)	Semiempirical, based on Euler solutions (4.8.1)	Second-order Van Dyke plus MNT[a] (3.4, 4.5)	SOSET plus IMNT[b] (4.3, 4.6)	SOSET plus IMNT Modified for real gases (7.5, 7.6)
Boattail or flare wave drag	—	Wu and Aoyoma (4.8.1)	Second-order Van Dyke (3.4)	SOSET (4.3)	SOSET for Real Gases (7.5, 7.6)
Skin friction drag	Van Driest II (4.7)				
Base drag	Improved empirical method (4.8.5)				
Axial force at α	Improved empirical method (5.7)				
Aeroheating information	—			SOSET plus IMNT for real gases (7.7)	
Inviscid lift and pitching moment	Empirical (4.8.3)	Semiempirical based on Euler solutions (4.8.3)	Tsien first-order crossflow (3.4)	SOSET (4.3, 4.6)	SOSET for real gases (7.6)
Viscous lift and pitching moment	Improved Allen and Perkins crossflow (5.2)				
Nonaxisymmetric body aero ($\phi = 0, 45°$)	Modified Jorgensen (6.0)				
Nonlinear st. loads available ($\phi = 0, 45°$)	No		Yes (8.2)		

[a]MNT = modified Newton theory
[b]IMNT = improved modified Newton theory

Fig. 8.30 AP98 methods for body-alone aerodynamics (numbers in parentheses are the sections in this book in which the methods are discussed).

given in Figs. 8.30 through 8.32. For the reader's convenience, these figures also show the section in the book where each method is discussed. The author would like to again emphasize that these methods are not intended to be all-inclusive of the field of approximate or semi-empirical aerodynamic methods. These methods are the ones that I have found to be the most useful in calculating aerodynamics of many weapon configurations. In many cases, where existing aerodynamic methods proved to be inadequate for rapid, accurate aerodynamic computations, the author developed new methods to meet this need. Figures 8.30 through 8.32 are, therefore, intended to be a useful summary reference for where these methods are discussed in the book.

Component/ Mach number region	Subsonic, $M_\infty < 0.8$	Transonic, $0.8 \le M_\infty \le 1.2$	Low supersonic, $1.2 \le M_\infty \le 1.8$	Moderate/ high supersonic, $1.8 \le M_\infty \le 6.0$	Hypersonic, $M_\infty > 6.0$
Wave drag		Empirical (4.8.1)	Linear theory plus MNT[a] (3.6.1)	Shock expansion plus MNT along strips (4.3)	Shock expansion plus MNT for real gases along strips (4.3)
Skin friction drag	Van Driest II (4.7)				
Trailing edge separation drag	Empirical (4.8.5)				
Body base pressure caused by tail fins	Improved empirical (4.8.5)				
Inviscid lift and pitching moment • Linear	• Lifting surface theory (3.5)	• Empirical (3.6.3)	• 3DTWT[b] (3.6)	• 3DTWT (3.6)	• 3DTWT (3.6)
• Nonlinear	• Empirical (5.3)				
Wing-body, body-wing interference ($\phi = 0, 45°$) • Linear	• Slender body theory or linear theory modified for short afterbodies (3.9.1)				
• Nonlinear	• Empirical (5.4)				
Wing-body, interference due to δ ($\phi = 0, 45°$) • Linear	• Slender body theory (3.9.1)				
• Nonlinear	• Empirical (5.5)				
Wing-tail interference ($\phi = 0, 45°$)	Line vortex theory with modifications for $K_{W(B)}$ term and nonlinearities (3.9.2, 5.6)				
Aeroheating	None present			Shock expansion plus MNT (7.7)	Shock expansion plus MNT real gases (7.7)
Nonaxisymmetric body aero ($\phi = 0, 45°$)	Improved Nelson estimate for AP98 (6.7)				
Nonlinear st. loads available ($\phi = 0, 45°$)	No		Yes (8.2)		

[a]MNT = modified Newton theory
[b]3DTWT = Three-Dimensional Thin Wing Theory

Fig. 8.31 AP98 methods for wing-alone and interference aerodynamics (numbers in parentheses are the sections in this book in which the methods are discussed).

Component/ Mach number region	Subsonic, $M_\infty < 0.8$	Transonic, $0.8 \leq M_\infty \leq 1.2$	Low supersonic, $1.2 \leq M_\infty \leq 1.8$	Moderate/high supersonic, $1.8 \leq M_\infty \leq 6.0$	Hypersonic, $M_\infty > 6.0$
Body alone	Empirical (3.7, 3.8)				
Wing and interference roll damping moment	Lifting surface theory (3.7.1)	Empirical (3.7.3)	Linear thin wing theory (3.7.2)		
Wing magnus moment	Assumed zero				
Wing and interference pitch damping moment	Lifting surface theory (3.8.1)	Empirical (3.8.3)	Linear thin wing theory (3.8.2)		

Fig. 8.32 AP98 methods for dynamic derivatives (numbers in parentheses are the sections in this book in which the methods are discussed).

References

[1]Moore, F. G., McInville, R. M., and Housh, C., "Methods for Distributing Semiempirical, Nonlinear, Aerodynamic Loads on Missile Components," *JRS*, Vol. 34, No. 6, Nov.–Dec 1997, pp. 744–752.

[2]Monta, W. J., "Supersonic Aerodynamic Characteristics of a Sparrow III Type Missile Model with Wing Controls and Comparison with Existing Tail-Control Results," NASA TP 1078, Oct. 1977.

[3]Buning, P. G., et al., "Overflow User's Manual," Ver. 1.6an, Jun. 1995.

[4]McInville, R. M., Moore, F. G., and Housh, C., "Nonlinear Structural Load Distribution Methodology for the Aeroprediction Code," NSWCDD/TR-96/133, Dahlgren, VA, Sept. 1996.

[5]Stallings, R. L., and Lamb, M., "Wing Alone Aerodynamic Characteristics for High Angles of Attack at Supersonic Speeds," NASA TP 1889, July 1981.

[6]Baker, W. B., Jr., "Static Aerodynamic Characteristics of Generalized Slender Bodies With and Without Fins at Mach Numbers from 0.6 to 3.0 and Angles of Attack from 0 to 180°," Arnold Engineering and Development Rept. AEDC-TR-75-124, Vols. 1 and 2, Tullahoma, TN., May 1976.

[7]Nielsen, J. N., Hemsch, M. J., and Smith, C. A., "A Preliminary Method for Calculating the Aerodynamic Characteristics of Cruciform Missiles to High Angles of Attack Including Effects of Roll Angle and Control Deflections," Office Naval Research Rept. ONR-CR215-226-4F, Arlington, VA.

[8]Moore, F. G., and McInville, R. M., "Extension of the NSWCDD Aeroprediction Code to the Roll Position of 45 Degrees," NSWCDD/TR-95/160, Dec. 1995.

[9]Von Karman, Th., "The Problem of Resistance in Compressible Fluids," Galcit Publ. No. 75, 1936.

[10]Haack, W., "Projectile Forms of Minimum Wave Resistance," Douglas Aircraft Company, Rept. No. 288, 1946.

[11]Cole, J. D., "Newtonian Flow Theory for Slender Bodies," U. S. Air Force, Project Rand, R.M. 1633, Feb. 1956.

[12]Moore, F. G., "A Study to Optimize The Aeroballistic Design of Naval Projectiles," NWL TR-2337, Sept. 1969.

[13]Hagar, W. W., DeJarnette, F. R., and Moore, F. G., "Optimal Projectile Shapes for Minimum Total Drag," NSWC/DL TR-3597, May 1977.

[14]Moga, N. J., and Dejarnette, F. R., "Optimal Bodies for Minimum Total Drag at Supersonic Speeds," AIAA Paper 81-0075, 1981.

[15]Moore, F. G., McInville, R. M., and Hymer, T., "The 1998 Version of the NSWC Aeroprediction Code: Part 1—Summary of New Theoretical Methodology," NSWCDD/TR-98/1, April 1998.

[16]Moore, F. G., McInville, R. M., and Robinson, D. I., "A Simplified Method for Predicting Aerodynamics of Multi-Fin Weapons," NSWCDD/TR-99/19, March 1999.

[17]Adams, G., and Dugan, D., "Theoretical Damping in Roll and Rolling Moment Due to Differential Wing Incidence for Slender Cruciform Wings and Wing-Body Combinations," NACA Rept. 1088, 1952.

[18]Nielsen, J., *Missile Aerodynamics*, Nielsen Engineering Research, Mountain View, CA, 1960.

[19]Bryson, A. E., "Stability Derivatives for a Slender Missile with Application to a Wing–Body Vertical Tail Configuration," *Journal of Aeronautical Sciences*, Vol. 20, No. 5, 1953.

[20]Wardlaw, A. B., and Davis, S., "A Second-Order-Gudonov Method for Supersonic Tactical Missiles," NSWC TR 86-506, 1986.

[21]Walters, R. W., Slack, D. C., Cimmella, P., Applebaum, M. P., and Frost, C., "A Users Guide to GASP," Virginia Polytechnic Inst. State Univ., Dept. Aerospace Ocean Engineering, Blacksburg, VA, Nov. 1990.

[22]Robinson, D. F, "ZEUS^{++}—A Graphical User Interface Flowfield Analysis Tool," NSWCDD/TR-98/147 (submitted for publication).

[23]Hymer, T. C., Downs, C., and Moore, F. G., "Users Guide for an Interactive Personal Computer Interface for the 1998 Aeroprediction Code (AP98)," NSWCDD/TR-98/7, June 1998.

[24]NASA Langley Research Center Tri-service missile database, transmitted from NASA/LRC Jerry M. Allen to Naval Surface Warfare Center, Dahlgren Div., Nov. 1991 (Submitted for publication).

[25]Guidos, B. J., "Static Aerodynamics CFD Analysis for 120 MM Hypersonic KE Projectile Design," U.S. Army Research Lab. Rept. ARL-MR-84, Aberdeen Proving Ground, MD, Sept. 1984.

[26]DTRA/SAIC Projectile Development and Test Wind Tunnel Data, Defense Threat Reduction Agency, Alexandria, VA (unpublished).

Chapter 9

Future Direction for Aeroprediction Methodology

THE primary purposes for this book, as stated in the preface, were to

1) Document in a single book all the approximate aerodynamic methods the author has developed or found the most useful during his career. Most of these methods are in current use in the latest version of the aeroprediction code, AP98, which is being used by many agencies around the world. Some of the more recent developments, such as the multifin aerodynamics of Chapter 8, are not a part of the AP98 but will be a part of a future version of the aeroprediction code.
2) Provide engineering methods for the practicing engineer that bridge the gap between the academic textbook (and the many theoretical methods taught) and practical configurations that allow aerodynamic estimates to be made in a reasonable time and for a reasonable cost.
3) Provide a book that can be useful in a missile design course taught in the aerospace curriculum at many universities. Several universities have acquired copies of the aeroprediction code and currently use it in this way.

It is my hope that these objectives have been partially achieved. Only time will tell whether they are fully achieved. Before concluding the book, I believe it is worthwhile to give the author's perspective on the future needs and opportunities in the aeroprediction area. These needs and opportunities will be broken down along the lines of the book. That is, numerical codes and semi-empirical codes. I will cover semi-empirical codes first.

I. Semi-Empirical Code Requirements

Future opportunities in the semi-empirical methods area are placed in two categories. These categories are defined as major technology needs and value-added technology needs. The difference between the definition of "major" and "value-added" has to do with the cost to develop the technology and integrate it to the AP98 or some other aerodynamic prediction code. Generally before any new technology is developed, there needs to be a request from users and sponsor support to fund the technology development for any of these needs.

The two remaining major technology needs are to account for the aerodynamic effects of side jets and an accurate semi-empirical code to predict out-of-the-pitch-plane aerodynamics. Recall in the performance section discussion of Chapter 8 that maneuverability is directly proportional to density of the atmosphere. As weapons go above 100,000 ft altitude, it becomes increasingly difficult to provide adequate maneuverability to successfully defeat high-speed maneuvering targets. As a result, many weapon concepts are considering using side jets for control, in lieu of or to supplement aerodynamic controls. Also, at low altitudes where the acquisition range is limited and response time is short, side jets could also prove useful. To develop a semi-empirical method to address this need requires a level of maturity in the technology similar to the situation where no side jets are present. That is, we need to understand the physical effects of the side jets on aerodynamics, we need a good wind tunnel database (analogous to the NASA Tri-service database) that covers all the geometry and flight parameters of interest, and we need a parallel effort in computational fluid dynamics (CFD) codes to assist and augment semi-empirical modeling and fill in gaps in the wind tunnel database. CFD codes are particularly useful in breaking down the interference of the jets on the body, once they have been adequately validated against good wind tunnel data. As of the writing of this book, the author is not aware of any generic wind tunnel test plans, even though many marketing attempts have been made to address this issue.

The second major technology need for semi-empirical codes is to be able to model out-of-the-pitch-plane aerodynamics with some confidence in the level of accuracy. The major difficulty is in accurately estimating induced roll. This difficulty arises because this term is fairly small and is within the accuracy of many wind tunnel tests not designed explicitly to measure the parameter. Some engineers who perform approximate six-degree-of-freedom (6DOF) performance analysis use pitch plane aerodynamics, assume the yaw plane is the same as the pitch plane, and neglect the induced roll entirely. However, for accurate 6DOF analysis, the author feels comfortable recommending only wind tunnel tests or full CFD analysis at present. An accurate semi-empirical, out-of-the-pitch-plane aerodynamics code presents a major challenge. A less accurate code that attempts to achieve ±25% average accuracy, versus the ±10% average accuracy on C_A and C_N of the pitch-plane aerodynamics, may be achievable, but with difficulty.

In addition to the major technology needs of semi-empirical codes, there are several value-added technology needs which, if accomplished, could make the methods presented in this text even more robust. The first of these is to use CFD to refine many of the methods and assumptions made in the development of these methods as a result of limitations of available wind tunnel data. For example, the largest component wind tunnel database available is the NASA Tri-service database. It was limited to fins with $r/s = 0.5$, in upper Mach number to 4.6, and to only one set of lifting surfaces. More recently, there has been a wind tunnel data set developed[1] that has values of r/s of 0.25 and 0.33 as well as 0.5. This data set could be used to refine assumptions made in the nonlinear angle of attack (AOA) range for r/s.

CFD codes are quite useful, when first validated against an existing database, for examining physical phenomena outside a database. Problems such as internal shock interactions, wing–tail interference, and higher Mach number ($M_\infty > 4.6$) assumptions naturally lend themselves to use of CFD codes to augment development of semi-empirical codes.

Another problem area is an improved power-on base drag predictor that blends together with base bleed methodology. The state of the art in this area at present (to the author's knowledge) is that we can approximate the effects of base bleed (very low exit velocity) on base pressure and, also, when the jet exit Mach number is greater than one, we can approximate the effect on base pressure. What is needed is a method to approximate the effect on base pressure at all jet exit Mach numbers.

The final area where some improvements in semi-empirical methods are needed is in dynamic derivative estimation. At present, no methods are available, to the author's knowledge, that explicitly account for nonlinear aerodynamics in dynamic derivatives, similar to the Chapter 5, which was limited to static aerodynamic estimations.

This list of needs for semi-empirical codes is limited to today's weapons and the author's knowledge of the state of the art in predicting aerodynamics of these weapons. Tomorrow's weapons may look entirely different and, as they change, the requirements for technology development will change. Hence, the list of aerodynamic requirements for semi-empirical codes needs continuous updating and refining.

II. Computational Fluid Dynamics Code Needs

The author will discuss this area from an application perspective, as opposed to a CFD code development perspective.

As exhibited in Section 8.4, Euler codes have progressed to the point of being used in a production mode. The author believes downloading of the ZEUS[2] code to a personal computer, creating a user-friendly pre- and postprocessor[3] analogous to the pre- and postprocessor distributed with the AP98[4] is a major advance in that direction. We, at The Naval Surface Warfare Center, Dahlgren Division in Dahlgren, Virginia, have found that good accuracy for in-plane and out-of-plane aerodynamics, including induced roll, can be obtained with a Euler code at AOAs where flow separation is small and for supersonic Mach numbers. The author believes that this approach is, therefore, the best to get a complete set of aerodynamics for a 6DOF simulation code over a limited set of flight conditions. However, much work remains to make the code robust. Much of this work revolves around getting accurate axial force on a large set of configurations without unduly sacrificing accuracy and computational speed. Combining many semi-empirical techniques in this book with solution of Euler equations is the approach recommended. As personal computers get faster, the times of 3 to 15 minutes per case of Section 8.4 will be reduced by an order of magnitude. Even these times are acceptable in today's environment. One could purchase a high performance personal computer for less than $2500 and let it be used exclusively for 6DOF calculations, while the engineer goes about other tasks on another personal computer.

The author believes that Navier–Stokes numerical solution methods have advanced to the point where only marginal improvements will be made from here for laminar flows. Navier–Stokes codes appear to have adequate modeling for laminar flow, but they appear to still be lacking in terms of transition prediction and turbulence modeling. There are many good Navier–Stokes solvers available. Assuming that the issues of transition and turbulence modeling are adequately addressed, the user still needs two things. First, Navier–Stokes codes need to be made more user friendly and be downloaded to a personal computer with pre- and postprocessing software similar to Refs. 3 and 4. Second, computer speeds need to be at least 1000 times faster than current capabilities. Twenty hours per case on a supercomputer, with the supercomputer being off site, with the hassle involved in uplinking and downloading data from the off-site location, and with the educational and experience level required of the user will continue to limit Navier–Stokes applications to only a few validation cases for the foreseeable future. In other words, Navier–Stokes codes will continue to be used primarily as analysis rather than production tools for the foreseeable future. Looking back 20 years, this was the same situation with Euler codes in the late 1970s and early 1980s. Hence, the author believes that, eventually, Navier–Stokes codes will be used in production but, due primarily to the problems listed, we are several years away from that being a reality today.

References

[1]*Fin-Body Interference Data Set,* Transmitted to Naval Surface Warfare Center, Dahlgren Div. by Jerry M. Allen of NASA/LRC, Aug. 1996 (from submitted to publication).

[2]Wardlaw, A. B., and Davis, S., "A Second-Order-Gudonov Method for Supersonic Tactical Missiles," Naval Surface Warfare Center, Dahlgren Div. Rept. NSWC TR 86-506, 1986.

[3] Robinson, D. F., "ZEUS++ A Graphical User Interface Flowfield Analysis Tool," NSWCDD/TR-98/147 (submitted for publication).

[4]Hymer, T. C., Downs, C., and Moore, F. G., "Users Guide for an Interactive Personal Computer Interface for the 1998 Aeroprediction Code (AP98)," NSWCDD/TR-98/7, June 1998.

Index

www.ingramcontent.com/pod-product-compliance
Lightning Source LLC
Chambersburg PA
CBHW031040100626
45930CB00016B/1
* 9 7 8 1 5 6 3 4 7 3 9 9 9 *